Wild Halophytes: Tools for Understanding Salt Tolerance Mechanisms of Plants and for Adapting Agriculture to Climate Change

Wild Halophytes: Tools for Understanding Salt Tolerance Mechanisms of Plants and for Adapting Agriculture to Climate Change

Editors

Oscar Vicente

Marius-Nicusor Grigore

MDPI • Basel • Beijing • Wuhan • Barcelona • Belgrade • Manchester • Tokyo • Cluj • Tianjin

Editors
Oscar Vicente
Institute for the Conservation and Improvement of Valencian Agrodiversity (COMAV)
Universitat Politècnica de València
Valencia
Spain

Marius-Nicusor Grigore
Faculty of Medicine and Biological Sciences
Stefan cel Mare University of Suceava
Suceava
Romania

Editorial Office
MDPI
St. Alban-Anlage 66
4052 Basel, Switzerland

This is a reprint of articles from the Special Issue published online in the open access journal *Plants* (ISSN 2223-7747) (available at: www.mdpi.com/journal/plants/special_issues/halophytes).

For citation purposes, cite each article independently as indicated on the article page online and as indicated below:

LastName, A.A.; LastName, B.B.; LastName, C.C. Article Title. *Journal Name* **Year**, *Volume Number*, Page Range.

ISBN 978-3-0365-6573-6 (Hbk)
ISBN 978-3-0365-6572-9 (PDF)

Cover image courtesy of Marius-Nicusor Grigore

Contents

About the Editors

Oscar Vicente

Oscar Vicente is Professor of Biochemistry and Molecular Biology at the Department of Biotechnology, Polytechnic University of Valencia, Spain. He obtained his PhD in Sciences from the Autonomous University of Madrid in 1983, spent a 3-year postdoctoral period at the Friedrich-Miescher Institut in Basel, and then as Assistant Professor at the Institute of Microbiology and Genetics, University of Vienna, until his return to Spain in 1996. His research interest focuses on the study of crop and wild plants' abiotic stress tolerance mechanisms, in the context of climate change, using physiological, biochemical and molecular approaches. Oscar Vicente has published over 160 indexed scientific papers. He is vice-president of the European Biotechnology Thematic Network Association (EBTNA) and included in the 'Top-100' Spanish scientists in Plant Science and Agronomy. Oscar Vicente is Doctor Honoris Causa by the University of Agricultural Sciences and Veterinary Medicine, Cluj-Napoca (Romania).

Marius-Nicusor Grigore

Marius-Nicusor Grigore is lecturer at the Faculty of Medicine and Biological Sciences, Stefan cel Mare University, Suceava, Romania. He received his B.Sc. and Ph.D. in Biology from Alexandru Ioan Cuza University, Iași, Romania. He was a postdoctoral fellow in the field of halophytes' ecophysiology at the same university, and an invited researcher at the Polytechnic University of Valencia, Spain. His research interests are the ecology, anatomy and physiology of halophytes and plant responses to abiotic stress. He has published 10 books, 9 book chapters, and over 70 peer-reviewed papers on halophytes. In 2016, he was awarded the Romanian Academy's 'Emil Racoviță' prize for the book "Halophytes: An Integrative Anatomical Study" (Springer, 2014); "Handbook of halophytes. From Molecules to Ecosystems towards Biosaline Agriculture", Springer, was awarded the scientific merit 'Nicolae Giosan' by Academy of Agricultural and Forest Sciences 'Gheorghe Ionescu-Șișești' București, Romania (11/2022).

Preface to "Wild Halophytes: Tools for Understanding Salt Tolerance Mechanisms of Plants and for Adapting Agriculture to Climate Change"

This reprint deals with an amazing group of plants, the halophytes or halophilic plants; the etymological meaning of the name is 'salt-loving plants', in biological contrast to conventional crops, which are glycophytes, glycophilic plants, or 'sweet-loving plants'. This distinction suggests that the former require salt in their lives, whereas the glycophytes should avoid salt and seek fresh water and non-saline environments. The halophytes are, therefore, the subject of this book, considered from different points of view: ecological, physiological, and biochemical. The proposed scope of the reprint covers different research possibilities and uses of halophytes. First, obviously, halophytes are ideal experimental systems for basic studies of salt tolerance mechanisms in plants at the morphological, anatomical, physiological, biochemical, and molecular levels. Apart from being a source of knowledge, halophytes can also provide biotechnological tools—salt-tolerance genes and salt-induced promoters—for the genetic improvement of salt tolerance of conventional crops. Furthermore, some halophytes could represent the basis of sustainable 'saline agriculture', being commercially grown in salinised land and irrigated with brackish water or even seawater, and, therefore, not competing with our conventional crops for these limited resources, fertile land, and good-quality irrigation water.

There are several reasons and motivations for writing and editing this reprint, especially in the current climate change—or rather 'climate emergency'—scenario. Climate change represents the most critical challenge for agricultural production and food security in the foreseeable future. Together with drought, soil salinity causes the most considerable reduction of crop yields worldwide, and climate change contributes to the accelerated loss of irrigated cropland in arid and semiarid regions due to secondary soil salinisation. The most sensible strategy to address this problem should be based on the genetic improvement of crop salt tolerance, which, in turn, requires elucidating the physiological and molecular mechanisms of salt tolerance. As mentioned above, halophytes are the most suitable experimental systems to undertake these studies. In addition, even though all plants use the same set of general, conserved responses to high salinity—such as the control of ion transport, osmolyte biosynthesis or activation of antioxidant systems—the efficiency and relative contribution of those responses to the mechanisms of tolerance vary widely in different species. Therefore, there is no 'single model'for salt tolerance research, as none will provide enough information.

Second, there is a purely scientific interest in halophytes. Halophytes were first scientifically drawn to attention by Goethe (ca. 1790). Still, mentions of this intriguing ecological group of plants date back much earlier, to the era of Theophrastus (ca. 371–ca. 287 BC). It should be expected that this long history has led to a deep and complete knowledge of salt tolerance in halophytes; however, as of today, this is not true. Moreover, despite the accumulated data on halophytes, many aspects remain obscure. Indeed, literature dealing exclusively with halophytes is scarce; a large part of the work on the plant–salinity relationship focuses on salt stress applied to glycophytes rather than halophytes. Indeed, scientists have used almost exclusively cultivated plants (apart from the Arabidopsis thaliana model) for these studies. It is clear that the economic importance of crops, as our primary source of food and feed and suppliers of many other goods, has overwhelmingly surpassed the curiosity of a few botanists. However, interest in halophytes has substantially increased in the past decade, perhaps because many researchers realised that the (primary) mechanisms of salt tolerance must be found in halophytes.

We would expect this reprint to be of interest to all scientists working on different aspects of halophytes'research: ecology and botany; elucidation of abiotic stress tolerance mechanisms biotechnological applications (e.g., in phytoremediation, as a source of metabolites of medical or industrial interest); or their domestication and breeding for saline agriculture. However, the reprint should be attractive also to those interested, more generally, in research on plant abiotic stress tolerance. People involved in environmental and conservation policies; agronomists and plant breeders worried about the adverse effects of climate change on crop yields; or postgraduate students seeking a specific research direction in the field of plant biology, could also gain useful information from this reprint.

Finally, we would like to thank all authors of the articles included in the Special Issue, and the reviewers who helped improve the quality of the accepted papers. With their invaluable contribution, we hope this reprint will open new perspectives in halophytes'research, highlighting new challenges to be addressed in the future.

Oscar Vicente and Marius-Nicusor Grigore
Editors

plants

Editorial

Wild Halophytes: Tools for Understanding Salt Tolerance Mechanisms of Plants and for Adapting Agriculture to Climate Change

Marius-Nicușor Grigore [1,*] and Oscar Vicente [2]

1 Faculty of Medicine and Biological Sciences, "Ștefan cel Mare" University of Suceava, Str. Universității 13, 720229 Suceava, Romania
2 Institute for the Conservation and Improvement of Valencian Agrodiversity (COMAV, UPV), Universitat Politècnica de València, Camino de Vera s/n, 46022 Valencia, Spain
* Correspondence: marius.grigore@usm.ro

Abstract: Halophytes, wild plants adapted to highly saline natural environments, represent extremely useful—and, at present, underutilised—experimental systems with which to investigate the mechanisms of salt tolerance in plants at the anatomical, physiological, biochemical and molecular levels. They can also provide biotechnological tools for the genetic improvement of salt tolerance in our conventional crops, such as salt tolerance genes or salt-induced promoters. Furthermore, halophytes may constitute the basis of sustainable 'saline agriculture' through commercial cultivation after some breeding to improve agronomic traits. All these issues are relevant in the present context of climate emergency, as soil salinity is—together with drought—the most critical environmental factor in reducing crop yield worldwide. In fact, climate change represents the most serious challenge for agricultural production and food security in the near future. Several of the topics mentioned above—mainly referring to basic studies on salt tolerance mechanisms—are addressed in the articles published within this Special Issue.

Keywords: climate emergency; crops' wild relatives; glycophytes; halophytes; phytoremediation; salt stress; salt tolerance mechanisms

Citation: Grigore, M.-N.; Vicente, O. Wild Halophytes: Tools for Understanding Salt Tolerance Mechanisms of Plants and for Adapting Agriculture to Climate Change. *Plants* **2023**, *12*, 221. https://doi.org/10.3390/plants12020221

Received: 16 December 2022
Accepted: 20 December 2022
Published: 4 January 2023

1. Introduction

The increase in crop yields necessary to feed a growing world population, expected to reach 10×10^9 people by 2050, is seriously hampered in the present global warming scenario. Indeed, climate change constitutes the most critical challenge for agricultural production and food security in the next few decades [1–3]. Amongst the multiple environmental stress conditions negatively affecting plant growth and productivity, drought and soil salinity are the major causes of the reduction in crop yields worldwide [4]. Climate change contributes to the increasing loss of cropland because of longer, more frequent and severe drought periods, as well as secondary salinisation, especially in areas cultivated under irrigation schemes in arid and semiarid regions. Some estimations indicate that more than 20% of irrigated farmland is already seriously affected by salinisation, and this figure is expected to increase in the coming years [5,6].

The genetic improvement of crop salt tolerance—by classical breeding or using transgenic or genome editing approaches—appears to be the most sensible strategy with which to address the problem of reduction in yields due to land salinisation (see, for example, [7–9]). This approach, in turn, requires elucidating the physiological, biochemical, and molecular mechanisms underlying salt tolerance. Mostly model species, such as *Arabidopsis thaliana* and a few crops, have been used to investigate these mechanisms, even though these species are rather sensitive to salt stress [10]. It is true that salt tolerance depends on basic, conserved responses to salinity—including the control of ion transport and ion homeostasis, osmolyte biosynthesis for osmotic adjustment, and activation of antioxidant systems—which are triggered in all salt-affected plants, tolerant or not [11–13]. However, the relative efficiency of these responses and the specific mechanisms of tolerance used vary widely in different species. Therefore, it seems evident that salt-tolerant plants are more appropriate for use as experimental systems for dissecting salt tolerance mechanisms, and that there is no single 'model' species that will provide enough information.

In the frame of this Special Issue's topic, it is important to differentiate between a 'glycophyte' and a 'halophyte'. Most plants, including all major crops, are glycophytes, meaning that they are sensitive to relatively low levels of salt in the soil. Halophytes, on the other hand, are wild plants adapted to saline environments that are able to survive and complete their life cycle in habitats with a soil salinity equivalent to at least 200 mM NaCl, although some can withstand salinities even higher than that of seawater [14,15]. Obviously, this is an 'operational' (and useful) definition, but, in fact, plant species show a continuous range of salt tolerance—or sensitivity. Thus, many glycophytes may have different degrees of salt tolerance, whereas there are salt-tolerant plants defined as 'obligatory' or 'facultative' halophytes. In addition, although salt tolerance depends mainly on the genotype, it is affected by many other factors, such as the plant developmental stage or the simultaneous presence of other stressful conditions. Therefore, the distinction between glycophytes and halophytes is not so clear-cut (see [16] for an extended discussion of these issues).

In any case, halophytes are ideal experimental material for fundamental studies of salt tolerance mechanisms in plants at the physiological, biochemical, and molecular levels. They can also provide biotechnological tools for improving the salt tolerance of conventional crops—for example, salt tolerance genes that could enhance this trait when expressed in transgenic plants, or salt-induced promoters used for the expression of those genes. In addition, some wild halophytes could be commercially cultivated, representing the basis of a sustainable 'saline agriculture'. This would require previous domestication and some breeding to improve agronomic characteristics, but the point to be highlighted is that they already possess salt tolerance, the most difficult trait to incorporate via conventional breeding. These 'new' crops could be grown in salinised land and irrigated with brackish or even seawater, not competing with our conventional crops for limited resources: fertile land and good-quality irrigation water. This Special Issue attempts to cover all the aspects mentioned above regarding halophyte basic research and its applications.

2. Special Issue Contents

A suitable strategy by which to elucidate salt tolerance mechanisms is performing comparative analyses of the physiological and biochemical responses to salt stress of genetically related taxa with different degrees of tolerance. The selected genotypes may include glycophytic and halophytic species of the same genus or related genera, or different cultivars, varieties, or accessions of the same species (e.g., [17–19]). This approach is used in several papers included in this Special Issue. Thus, Ghanem et al. [20] analysed the responses to increasing salinity of three obligate halophytes, *Arthrocnemum macrostachyum*, *Sarcocornia fruticosa*, and *Salicornia europaea* (Amaranthaceae), collected from the same natural habitat, at the vegetative phase of development, measuring biomass and several biochemical stress markers—ion, chlorophylls, proline and antioxidant compounds contents, and some antioxidant enzyme activities. Based on the obtained data, the authors suggested that the investigated species adopt two differential strategies: salt tolerance in *S. europaea* appears to be primarily due to the activation of antioxidant enzymes and the biosynthesis of proline, whereas in *A. macrostachyum* and *S. fruticosa*, it is based on the rearrangement of the chlorophylls ratio and the biosynthesis of antioxidant compounds (carotenoids, phenolics, and flavonoids).

Calone et al. [21] subjected plants of the perennial species *Sarcocornia fruticosa* and the annual *Salicornia europaea* and *Salicornia veneta* to 30 days of intense salt stress (700 mM NaCl) and water deficit (complete withholding of irrigation), followed by 15 days of recovery (irrigation with non-saline water). Growth parameters and biochemical stress markers were determined after the stress and recovery treatments. The three species showed high tolerance to salt stress, based on the accumulation of ions (Na^+, Cl^-, Ca^{2+}) in the shoots and the synthesis of organic osmolytes. Interestingly, active ion transport to the shoots and high levels of glycine betaine were also observed in non-stressed control plants and after the recovery period, suggesting the presence of constitutive stress defence mechanisms. The three halophytes were found to be sensitive to water stress, although *S. fruticosa* to a lesser extent than the two annual species; this could be due to adaptation to a drier habitat than that of the *Salicornia* species, but a more gradual stress-induced senescence in perennials may also contribute to the greater drought tolerance of *S. fruticosa*.

The article by Ibraheem et al. [22] refers to another genus of the Amaranthaceae family. The authors compared the physiological and metabolic stress responses of three *Suaeda species* (*S. monoica*, *S. vermiculata*, and *S. schimperi*) in their natural saline environments—contrary to the other papers in this Special Issue, which report studies under controlled greenhouse conditions. Therefore, this work required extensive soil analyses to determine not only the salinity level but also other stress factors affecting the plants in the field, such as deficiency in essential nutrients and the presence of toxic heavy metals. Soil characteristics were then correlated with metabolic parameters in the plants—organic and inorganic nutrients, photosynthetic pigments, amino acid profiles, oxidative stress markers, and antioxidant metabolites, amongst others. The results demonstrated common tolerance mechanisms, such as the use of Na^+ and other inorganic elements as cheap osmotica, as well as species-specific stress responses. In particular, the three *Suaeda* species are promising halophytes for the phytoremediation of heavy metal-contaminated soils, showing some specificity in their capacity to accumulate different heavy metals.

The genus *Plantago* (Plantaginaceae) includes halophytes and glycophytes, as well as drought-tolerant species, and is particularly well suited for investigating plant stress tolerance mechanisms. Ltaeif et al. [23] compared the salt stress responses of two halophytes of the genus, *P. crassifolia* and *P. coronopus*, and two glycophytes, *P. ovata* and *P. afra*. As expected, the biochemical responses were different in the two groups of plants; the halophytes accumulated higher leaf Na^+ and proline contents and showed a lower level of oxidative stress. It was confirmed that *P. coronopus* and *P. crassifolia* are the most tolerant to salt stress, while *P. afra* is the most sensitive of the four species. *Plantago ovata* could not withstand the strongest salt stress treatment (one month in the presence of 800 mM NaCl); nevertheless, it was shown to also be quite resistant to salt stress, apparently through specific responses that differed from those of the halophytes; they include a weaker salt-induced inhibition of photosynthesis, the accumulation of Cl^- to higher concentrations in the leaves, and the activation of K^+ uptake and transport to the leaves under conditions of high external salinity.

Crop wild relatives, generally more resistant to abiotic stress than their cultivated counterparts, constitute an excellent resource for developing new cultivars with enhanced tolerance [24]. Therefore, it is interesting to determine the tolerance mechanisms of wild species of interest. Jekabsone et al. [25] studied the responses of several wild accessions of *Trifolium fragiferum* (Fabaceae) from natural habitats with different salinity levels to controlled salt treatments compared with a commercial cultivar. The authors reported a decrease in plant biomass and changes in partitioning between different organs with increasing salinity, responses that were genotype-specific. In addition, Na^+ and Cl^- accumulation were organ-specific, whereas responses related to mineral nutrition were both genotype- and organ-specific. In several accessions, salinity stimulated reproductive development. The experiments revealed high intraspecies morphological and physiological variability in the responses of the analysed *T. fragiferum* accessions to salinity, meaning that they can be defined as 'ecotypes'.

Ishikawa et al. [26], on the other hand, compared cultivated rice (*Oryza sativa*, salt-sensitive) with a wild relative (*Oryza coarctata*, salt-tolerant) (Poaceae), demonstrating that the two species use different strategies to control Na^+ uptake. At the early stage of the salt stress treatment, wild rice increased its xylem Na^+ loading for a quick and efficient osmotic adjustment but then maintained shoot Na^+ contents at non-toxic levels by activating the high-affinity K^+ transporter HKT1;5 (responsible for xylem Na^+ unloading) and the tonoplast Na^+/H^+ antiporter NHX (for sequestering Na^+ and K^+ into root vacuoles). On the contrary, *O. sativa* initially limited Na^+ uptake and transport to the shoot through the activation of SOS1, the plasma membrane Na^+/H^+ antiporter, in the roots. However, cultivated rice failed to maintain this response in the long term because SOS1-mediated Na^+ exclusion is highly energy-demanding. Therefore, the higher salt tolerance of wild rice seems to rely on efficient Na^+ sequestration in root vacuoles as opposed to Na^+ exclusion.

Similarly, Bigot et al. [27] also compared the salt resistance of a crop, tomato (*Solanum lycopersicum*), and a wild relative, *S. chilense* (Solanaceae), focusing on the reproductive phase, particularly Na localisation in floral organs. Salinity was found to affect reproductive development in the two species, but in different ways. For example, salt stress induced a decrease in the number of inflorescences in both species, but the number of flowers per inflorescence or sepal length was only found to be reduced in cultivated tomatoes. Additionally, the fruit set was not affected by salinity, but fruit size and weight were reduced in *S. lycopersicum*. Growth in the presence of salt decreased the stamen length in *S. chilense*, which was accompanied by a reduction in pollen production and an increase in pollen viability. The work included an extensive analysis of the concentrations and localisation of different ions (Na^+, K^+, Mg^{2+}, and Ca^{2+}) in reproductive structures, which differed in the two studied species. For example, Na^+ was found to be predominantly located in non-reproductive floral organs in *S. lycopersicum* and in the male floral organs of *S. chilense*. The expression of different genes involved in ion transport, analysed by qRT-PCR, also differed in flowers of both species. This study concludes that *S. chilense* was more tolerant to salinity than *S. lycopersicum* during the reproductive phase, which could be associated, at least in part, with the different distribution and transport of ions in their flower organs.

The article by Cárdenas-Pérez et al. [28] also addresses basic salt tolerance mechanisms, although without involving comparative studies of different species but instead a single one, *Salicornia europaea* (Amaranthaceae). Combining morphological, anatomical, and biochemical analyses and advanced statistical methods, this study found that *S. europaea* grows optimally between 200 and 400 mM NaCl and that growth is limited at 0, 800, and 1000 mM NaCl. Almost all analysed traits were found to be dependent on the salinity level but differently affected. The most affected traits included photosynthetic pigments and protein content, plant surface area, peroxidase activity, and anatomical traits related to cell shape. Although this species has been extensively studied, these results significantly expand the present knowledge on the changes in *S. europaea* functional traits in response to salt stress.

Mir et al. [29] investigated the mechanisms of environmental stress tolerance in the threatened halophyte *Limonium angustebracteatum* (Plumbaginaceae), an endemic species of the east and southeast of the Iberian Peninsula of high conservation interest. The study provides new and interesting data on the ultrastructure of salt glands, typical for halophytic members of the family. In addition, several anatomical, physiological, and biochemical responses were assessed in plants subjected to one month of water deficit (complete lack of irrigation) and salt stress (watering with increasing NaCl concentrations, up to 800 mM). The species is highly tolerant to salt stress, but plant growth was found to be significantly inhibited by severe water stress. Apart from salt secretion through salt glands, its salt tolerance is based on the efficient osmotic adjustment by the accumulation of high concentrations of ions (Na^+ and Cl^- as well as K^+ and Ca^{2+}) and the osmolytes proline (Pro) and glycine betaine (GB) in the leaves. The relatively high leaf concentrations of the four ions and GB (but not Pro) in control plants pointed to the presence of constitutive mechanisms of stress tolerance. A large increase in root K^+ concentrations; the active transport of Na^+, Cl^-, and Ca^{2+} to the leaves; and an increase in leaf GB contents were observed in water-stressed plants. Although the responses to water and salt stress differed, K^+ homeostasis was shown to be essential for tolerance to both stress treatments.

Bueno and Cordovilla [30] studied the possible effect of different plant growth regulators—phytohormones such as salicylic acid, gibberellins, cytokinins, and auxins, and the polyamine spermidine—added to a hydroponic culture on plants of the halophyte *Plantago coronopus* subsequently subjected to a salt stress treatment (200 mM NaCl), with a focus on the use of this species in saline agriculture. All these plant regulators improved plant growth in the absence of salt, whereas in salt-treated plants, spermidine application was the most effective pre-treatment, inducing the strongest growth stimulation, osmolyte (sorbitol) accumulation, and an increase in antioxidant metabolites (phenolic compounds and flavonoids). The authors conclude that this treatment, activating defence mechanisms against stress, could contribute to improving salt tolerance in *P. coronopus*.

The paper by Sánchez-Gavilán et al. [31] describes the phytochemical analysis of several Amaranthaceae species of the genera *Sarcocornia* (*S. alpini*, *S. pruinosa*, and *S. perennis*) and *Arthronemum* (*A. macrostachyum*) from different coastal and inland salt marshes of the Iberian Peninsula. Separation by gas chromatography or HPLC, coupled with mass spectrometry, was used to identify bioactive compounds (phenolic acids, flavonoids, lipids) in the plant material. Trans-cinnamic, salicylic, veratric, coumaric, and caffeic acids were present in all analysed species, whereas ferulic acid was only detected in *A. macrostachyum*. The identified flavonoids were cyanidin-3-*O*-arabinoside, luteolin-7-glucoside, dihydroquercetin, and p-coumaroyl-glucoside. Regarding fatty acids, palmitic, linoleic, and oleic acids were detected in *Sarcocornia* as the most abundant, whereas palmitic, linolenic, and stearic acids were the main fatty acids present in *A. macrostachyum*. Apart from the biological function of these secondary metabolites in the mechanisms of stress tolerance, their properties (e.g., as antioxidants) increase the interest in the use of these species for commercial cultivation in the frame of sustainable saline agriculture because of their high nutritional value.

Finally, the article by Carreiras et al. [32] addressed the question of whether heavy metal preconditioning could influence the salinity tolerance of *Spartina patens* (Poaceae), an invasive halophytic grass that represents a severe problem for the biodiversity of Mediterranean salt marshes. The authors compared the responses of plants from two salt marshes of the Tagus estuary (Portugal), one pristine and the other contaminated by heavy metals, to increasing salinity. The analysis of photochemical processes, photosynthetic pigments profiles, antioxidant enzyme activities, and lipid composition in plants of the two populations revealed intraspecific physiological differences, resulting in the better adaptation and tolerance to salt stress of *S. patents* from the contaminated marsh, especially at high salt concentrations. Those differences include, for example, salt-induced increases in the chlorophyll a/b ratio and oleic acid content in plants from the heavy-metal-contaminated area or the stronger generation of ROS, and therefore more intense plant damage, in the population from the pristine marsh. The article also discusses the implications of this variability at the population level for the frequency and distribution of the species in salt marshes in the face of climate change.

3. Conclusions

This Special Issue, covering several topics included in its proposed scope and a wide range of halophytes, comprises 12 published articles. In five of them, the investigated species belong to the Amaranthaceae family—which includes some of the most salt-tolerant taxa known—of the *Arthrocnemum*, *Sarcocornia*, *Salicornia*, and *Suaeda* genera; however, members of the Plantaginaceae, Fabaceae, Poaceae, Solanaceae, and Plumbaginaceae families are also studied in other papers. The selected halophytes include some that have been extensively studied before, such as *Salicornia europaea* or *Plantago coronopus*, as well as a narrow endemic species of high conservation interest (*Limonium angustebracteatum*), or *Spartina patens*, an invasive species that represents a serious risk to the biodiversity of Mediterranean salt marshes. Most of the accepted articles deal with elucidating the mechanisms of salt tolerance, in several cases based on comparative studies of the salt stress responses of related taxa with different degrees of tolerance. Other papers include more applied aspects, such as phytochemical analyses that support the commercial cultivation of some halophytes because of their high contents of healthy antioxidant metabolites or the possibility of using others species in the phytoremediation of heavy metals-contaminated soils. Anatomical, physiological, and biochemical analyses are the most common experimental approaches used in these studies. Altogether, this Special Issue provides a comprehensive and updated overview of the biology of halophytes, contributing to expanding our knowledge of this amazing group of salt-tolerant plants. Other aspects initially included in the scope of the Special Issue, such as studies conducted using molecular biology or 'omics' approaches or the generation of biotechnological tools for the breeding of salt tolerance in conventional crops, have not been directly addressed here, but could be included in a possible 'second edition' of the Special Issue.

Author Contributions: Both authors contributed to the design, writing and editing of this Editorial article. All authors have read and agreed to the published version of the manuscript.

Funding: This research received no external funding.

Acknowledgments: We are indebted to all the authors contributing their work to this Special Issue and to all reviewers whose comments and suggestions helped to improve the quality of the accepted papers.

Conflicts of Interest: The authors declare no conflict of interest.

References

1. FAO; IFAD; UNICEF; WEP; WHO. *The State of Food Security and Nutrition in the World 2018. Building Climate Resilience for Food Security and Nutrition*; Food and Agriculture Organization (FAO) of the United Nations: Rome, Italy, 2018.
2. Raza, A.; Razzaq, A.; Mehmood, S.S.; Zou, X.; Zhang, X.; Lv, Y.; Xu, J. Impact of climate change on crops adaptation and strategies to tackle its outcome: A review. *Plants* **2019**, *8*, 34. [CrossRef] [PubMed]
3. IPCC. Climate Change 2022: Impacts, Adaptation and Vulnerability. In *Contribution of Working Group II to the Sixth Assessment Report of the Intergovernmental Panel on Climate Change*; Pörtner, H.-O., Roberts, D.C., Tignor, M., Poloczanska, E.S., Mintenbeck, K., Alegría, A., Craig, M., Langsdorf, S., Löschke, S., Möller, V., et al., Eds.; Cambridge University Press: Cambridge, UK; New York, NY, USA, 2022. [CrossRef]
4. Boyer, J.S. Plant productivity and environment. *Science* **1982**, *218*, 443–448. [CrossRef] [PubMed]
5. Butcher, K.; Wick, A.F.; DeSutter, T.M.; Chatterjee, A.; Harmon, J. Soil salinity: A threat to global food security. *Agron. J.* **2016**, *108*, 2189–2200. [CrossRef]
6. Mukhopadhyay, R.; Sarkar, B.; Jat, H.S.; Sharma, P.C.; Bolan, N.S. Soil salinity under climate change: Challenges for sustainable agriculture and food security. *J. Environ. Manag.* **2021**, *280*, 111736. [CrossRef] [PubMed]
7. Fita, A.; Rodríguez-Burruezo, A.; Boscaiu, M.; Prohens, J.; Vicente, O. Breeding and domesticating crops adapted to drought and salinity: A new paradigm for increasing food production. *Front. Plant Sci.* **2015**, *6*, 978. [CrossRef]
8. Kotula, L.; García Caparros, P.; Zörb, C.; Colmer, T.D.; Flowers, T.J. Improving crop salt tolerance using transgenic approaches: An update and physiological analysis. *Plant Cell Environ.* **2020**, *43*, 2932–2956. [CrossRef]
9. Qin, H.; Li, Y.; Huang, R. Advances and challenges in the breeding of salt-tolerant rice. *Int. J. Mol. Sci.* **2020**, *21*, 8385. [CrossRef]
10. Zhu, J.-K. Genetic analysis of plant salt tolerance using Arabidopsis. *Plant Physiol.* **2000**, *124*, 941–948. [CrossRef]
11. Munns, R.; Tester, M. Mechanisms of salinity tolerance. *Ann. Rev. Plant Biol.* **2008**, *59*, 651–681. [CrossRef]
12. Volkov, V. Salinity tolerance in plants. Quantitative approach to ion transport starting from halophytes and stepping to genetic and protein engineering for manipulating ion fluxes. *Front. Plant Sci.* **2015**, *6*, 873. [CrossRef]
13. Álvarez, M.E.; Savouré, A.; Szabados, L. Proline metabolism as regulatory hub. *Trends Plant Sci.* **2022**, *27*, 39–55. [CrossRef]
14. Flowers, T.J.; Hajibagheri, M.A.; Clipson, N.J.W. Halophytes. *Q. Rev. Biol.* **1986**, *61*, 313–335.
15. Flowers, T.J.; Colmer, T.D. Salinity tolerance in halophytes. *N. Phytol.* **2008**, *179*, 945–963. [CrossRef]
16. Grigore, M.N. Definition and Classification of Halophytes as an Ecological Group of Plants. In *Handbook of Halophytes*; Grigore, M.N., Ed.; Springer: Cham, Switzerland, 2021. [CrossRef]
17. Volkov, V.; Wang, B.; Dominy, P.J.; Fricke, W.; Amtmann, A. *Thellungiella halophila*, a salt-tolerant relative of *Arabidopsis thaliana*, possesses effective mechanisms to discriminate between potassium and sodium. *Plant Cell Environ.* **2003**, *27*, 1–14. [CrossRef]
18. Moir-Barnetson, L.; Veneklaas, E.J.; Colmer, T.D. Salinity tolerances of three succulent halophytes (*Tecticornia* spp.) differentially distributed along a salinity gradient. *Funct. Plant Biol.* **2016**, *43*, 739–750. [CrossRef]
19. Derbali, W.; Manaa, A.; Goussi, R.; Derbali, I.; Abdelly, C.; Koyro, H.-W. Post-stress restorative response of two quinoa genotypes differing in their salt resistance after salinity release. *Plant Physiol. Biochem.* **2021**, *164*, 222–236. [CrossRef]
20. Ghanem, A.F.M.; Mohamed, E.; Kasem, A.M.M.A.; El-Ghamery, A.A. Differential salt tolerance strategies in three halophytes from the same ecological habitat: Augmentation of antioxidant enzymes and compounds. *Plants* **2021**, *10*, 1100. [CrossRef]

21. Calone, R.; Mircea, D.-M.; González-Orenga, S.; Boscaiu, M.; Lambertini, C.; Barbanti, L.; Vicente, O. Recovery from salinity and drought stress in the perennial *Sarcocornia fruticosa* vs. the annual *Salicornia europaea* and *S. veneta*. *Plants* **2022**, *11*, 1058. [CrossRef]
22. Ibraheem, F.; Al-Zahrani, A.; Mosa, A. Physiological adaptation of three wild halophytic *Suaeda* species: Salt tolerance strategies and metal accumulation capacity. *Plants* **2022**, *11*, 537. [CrossRef]
23. Ltaeif, H.B.; Sakhraoui, A.; González-Orenga, S.; Landa Faz, A.; Boscaiu, M.; Vicente, O.; Rouz, S. Responses to salinity in four *Plantago* species from Tunisia. *Plants* **2021**, *10*, 1392. [CrossRef]
24. Prohens, J.; Gramazio, P.; Plazas, M.; Dempewolf, H.; Kilian, B.; Díez, M.J.; Fita, A.; Herraiz, F.J.; Rodríguez-Burruezo, A.; Soler, S.; et al. Introgressiomics: A new approach for using crop wild relatives in breeding for adaptation to climate change. *Euphytica* **2017**, *213*, 158. [CrossRef]
25. Jekabsone, A.; Andersone-Ozola, U.; Karlsons, A.; Romanovs, M.; Ievinsh, G. Effect of salinity on growth, ion accumulation and mineral nutrition of different accessions of a crop wild relative legume species, *Trifolium fragiferum*. *Plants* **2022**, *11*, 797. [CrossRef] [PubMed]
26. Ishikawa, T.; Shabala, L.; Zhou, M.; Venkataraman, G.; Yu, M.; Sellamuthu, G.; Chen, Z.-H.; Shabala, S. Comparative analysis of root Na^+ relation under salinity between *Oryza sativa* and *Oryza coarctata*. *Plants* **2022**, *11*, 656. [CrossRef] [PubMed]
27. Bigot, S.; Pongrac, P.; Šala, M.; van Elteren, J.T.; Martínez, J.-P.; Lutts, S.; Quinet, M. The halophyte species Solanum chilense Dun. maintains its reproduction despite sodium accumulation in its floral organs. *Plants* **2022**, *11*, 672. [CrossRef]
28. Cárdenas-Pérez, S.; Rajabi Dehnavi, A.; Leszczynski, K.; Lubinska-Mielinska, S.; Ludwiczak, A.; Piernik, A. *Salicornia europaea* L. functional traits indicate its optimum growth. *Plants* **2022**, *11*, 1051. [CrossRef]
29. Mir, R.; Romero, I.; González-Orenga, S.; Ferrer-Gallego, P.P.; Laguna, E.; Boscaiu, M.; Oprică, L.; Grigore, M.-N.; Vicente, O. Constitutive and adaptive traits of environmental stress tolerance in the threatened halophyte *Limonium angustebracteatum* Erben (Plumbaginaceae). *Plants* **2022**, *11*, 1137. [CrossRef]
30. Bueno, M.; Cordovilla, M.d.P. Plant growth regulators application enhance tolerance to salinity and benefit the halophyte *Plantago coronopus* in saline agriculture. *Plants* **2021**, *10*, 1872. [CrossRef]
31. Sánchez-Gavilán, I.; Ramírez Chueca, E.; de la Fuente García, V. Bioactive compounds in *Sarcocornia* and *Arthrocnemum*, two wild halophilic genera from the Iberian Peninsula. *Plants* **2021**, *10*, 2218. [CrossRef]
32. Carreiras, J.; Pérez-Romero, J.A.; Mateos-Naranjo, E.; Redondo-Gómez, S.; Matos, A.R.; Caçador, I.; Duarte, B. Heavy metal pre-conditioning history modulates *Spartina patens* physiological tolerance along a salinity gradient. *Plants* **2021**, *10*, 2072. [CrossRef]

plants

MDPI

Article

Constitutive and Adaptive Traits of Environmental Stress Tolerance in the Threatened Halophyte *Limonium angustebracteatum* Erben (Plumbaginaceae)

Ricardo Mir [1], Ignacio Romero [1], Sara González-Orenga [2], P. Pablo Ferrer-Gallego [3], Emilio Laguna [3], Monica Boscaiu [2], Lăcrămioara Oprică [4], Marius-Nicuşor Grigore [5,*] and Oscar Vicente [1]

[1] Institute for the Conservation and Improvement of Valencian Agrodiversity (COMAV, UPV), Universitat Politècnica de València, Camino de Vera 14, 46022 Valencia, Spain; rimimo@upvnet.upv.es (R.M.); iromloz@posgrado.upv.es (I.R.); ovicente@upvnet.upv.es (O.V.)

[2] Mediterranean Agroforestry Institute (IAM, UPV), Universitat Politècnica de València, Camino de Vera 14, 46022 Valencia, Spain; sagonor@doctor.upv.es (S.G.-O.); mobosnea@eaf.upv.es (M.B.)

[3] Centre for Forestry Research and Experimentation (CIEF), CIEF-Wildlife Service, Generalitat Valenciana, Avda Comarques del País Valencia, 114, 46930 Quart de Poblet, Valencia, Spain; flora.cief@gva.es (P.P.F.-G.); laguna_emi@gva.es (E.L.)

[4] Faculty of Biology, Alexandru Ioan Cuza University of Iasi, Bulevardul Carol I nr. 11, 700506 Iași, Romania; lacramioara.oprica@uaic.ro

[5] Faculty of Medicine and Biological Sciences, "Ștefan cel Mare" University of Suceava, Str. Universității 13, 720229 Suceava, Romania

* Correspondence: marius.grigore@usm.ro

Citation: Mir, R.; Romero, I.; González-Orenga, S.; Ferrer-Gallego, P.P.; Laguna, E.; Boscaiu, M.; Oprică, L.; Grigore, M.-N.; Vicente, O. Constitutive and Adaptive Traits of Environmental Stress Tolerance in the Threatened Halophyte *Limonium angustebracteatum* Erben (Plumbaginaceae). *Plants* **2022**, *11*, 1137. https://doi.org/10.3390/plants11091137

Academic Editor: Fulai Liu

Received: 15 March 2022
Accepted: 20 April 2022
Published: 22 April 2022

Abstract: *Limonium angustebracteatum* is a halophyte endemic to the E and SE Iberian Peninsula with interest in conservation. Salt glands represent an important adaptive trait in recretohalophytes like this and other *Limonium* species, as they allow the excretion of excess salts, reducing the concentration of toxic ions in foliar tissues. This study included the analysis of the salt gland structure, composed of 12 cells, 4 secretory and 8 accessory. Several anatomical, physiological and biochemical responses to stress were also analysed in adult plants subjected to one month of water stress, complete lack of irrigation, and salt stress, by watering with aqueous solutions of 200, 400, 600 and 800 mM NaCl. Plant growth was inhibited by the severe water deficit and, to a lesser extent, by high NaCl concentrations. A variation in the anatomical structure of the leaves was detected under conditions of salt and water stress; plants from the salt stress treatment showed salt glands sunken between epidermal cells, bordered by very large epidermal cells, whereas in those from the water stress treatment, the epidermal cells were heterogeneous in shape and size. In both, the palisade structure of the leaves was altered. Salt excretion is usually accompanied by the accumulation of salts in the foliar tissue. This was also found in *L. angustebracteatum*, in which the concentration of all ions analysed was higher in the leaves than in the roots. The increase of K^+ in the roots of plants subjected to water stress was also remarkable. The multivariate analysis indicated differences in water and salt stress responses, such as the accumulation of Na and Cl, or proline, but K^+ homeostasis played a relevant role in the mechanism of tolerance to both stressful conditions.

Keywords: salt glands; recretohalophytes; endemism; water deficit; salt stress; ion transport; osmolytes accumulation; salt tolerance; drought tolerance; conservation programmes

1. Introduction

The genus *Limonium* Miller (Plumbaginaceae) represents a fascinating halophytic model for understanding the functioning of coastal ecosystems. This genus is one of the most complex taxonomic groups in the Mediterranean flora [1]. Taxonomic research in the Iberian Peninsula revealed the existence of many endemics with narrow distribution ranges [1,2], including recently described new species from southeastern Spain (see, e.g., in the Valencian Community [3–5]).

The Valencian Community is one of the richest European territories in species of the *Limonium* genus [6]. *Limonium angustebracteatum* Erben is a species of sea lavender endemic to E and SE Spain (Alicante, Almeria, Castellón, Murcia and Valencia provinces) growing on argillaceous-sandy soils in littoral salt marsh halophytic communities. This species was described by Erben [7] from material collected in Sagunto, on "Playa del Puig", a locality near Valencia City (Valencia province, Spain).

Limonium angustebracteatum, like other species of this genus, combines morpho-anatomical, biochemical and physiological traits that enable its growth in saline environments. *Limonium* species are known as recretohalophytes, as they have salt glands concentrated mainly in their leaves. This adaptation to saline environments occurs only in a few species of different families, including Plumbaginaceae [8]. Salt glands allow the removal of excess salts and play an essential role in regulating the internal ionic composition of leaves and ensuring osmotic balance, which, together with efficient osmotic adjustments, help prevent the dehydration of leaf cells [9]. Salt glands are reported in 87 species of *Limonium* [10], and they play an essential role in salt tolerance in this genus [11,12]. In addition to eliminating salts through glands, *Limonium* species accumulate toxic ions in their vacuoles, ensuring low-cost osmotic adjustment and avoiding ion toxicity, a common mechanism in dicotyledonous halophytes [13–16]. Under stress conditions, osmotic balance is also ensured by the synthesis and accumulation of osmolytes, or compatible solutes, in the cytoplasm. These are chemically diverse, the most common being proline and other amino acid derivatives, glycine betaine and other quaternary ammonium compounds, soluble sugars and polyols or sugar alcohols [17,18]. In addition to their primary function in osmotic adjustment [19], osmolytes also play many other roles, such as chemical chaperones, signalling molecules, modulators of gene expression or scavengers of "reactive oxygen species" (ROS) [17–20]. Quantifying the levels of ions and osmolytes in plants subjected to increased salt concentrations under controlled greenhouse conditions is of great relevance for understanding salt tolerance mechanisms in halophytes.

It is necessary to distinguish between constitutive stress tolerance mechanisms, present even in the absence of stress, and induced mechanisms, activated in response to stress. Constitutive mechanisms have a genetic basis and are species-specific; they include, for example, succulence or salt glands in some extremophiles [17] and other anatomical characteristics, such as those related to the reduction of water loss under high salinity conditions [21], or root structure [22]. However, halophytes, especially extremophiles, also possess other built-in mechanisms and are metabolically pre-adapted to salinity [23,24]. Studies on the halophyte *Eutrema salsugineum* revealed a phenotypic and metabolic adaptive plasticity not found in the related species *Arabidopsis thaliana* [25]. Many genes induced by salt stress in glycophytes are constitutively expressed at high levels in *E. salsugineum* [26]. The additional activation of induced stress responses at the transcriptional level occurs only at higher salinities, as reported in *E. salsugineum* compared to *Arabidopsis*, in agreement with the big difference in salt tolerance observed between these two species [27]. Metabolic pre-adaptation implies that extremophile species can show, even in the absence of salt, elevated levels of metabolites that are usually salt-induced; in addition, they can also respond to increased levels of salt stress by accumulating additional osmolytes not synthesised at lower salinities, as reported in different taxa, including *Limonium* species [28,29].

The aim of this study was two-fold. Firstly, we undertook a detailed study of the (yet unknown) morphology of the salt glands of *Limonium angustebracteatum* by Cryo-Field Emission Scanning Electron Microscopy (Cryo-FESEM) and toluidine blue-stained leaf sections. Secondly, we aimed to test the species' tolerance to abiotic stress and elucidate its main tolerance mechanisms by analysing the effects of salinity and severe water stress under controlled experimental conditions on the plants' anatomical structures, growth and biochemical parameters. These analyses included the determination of photosynthetic pigments, ions (Na^+, Cl^-, K^+ and Ca^{2+}) and osmolytes (proline, glycine betaine and total soluble sugars) contents and their changes in response to the applied stress treatments.

2. Results

2.1. Effects of Stress Treatments on Plant Growth

One-year-old *Limonium* plants were subjected to different salt and water stress treatments as described in the Material and methods section. After four weeks of treatment, the effects of stress were visually observed in the overall growth of the plants. The strongest growth inhibition was registered in plants subjected to water stress, whereas those from the salt treatments did not show large variations with respect to the control non-stressed plants (Figure 1).

Figure 1. Effect of four weeks of salt treatments with the indicated NaCl concentrations, or water stress (WS), on *Limonium angustebracteatum* plants. The scale bar represents 3 cm.

The number of leaves at the end of the treatments was compared to that at the beginning of the experiments. Except for the control plants, irrigated with non-saline water, where some new leaves were produced (1.8 per plant, on average), the number of leaves decreased in all other treatments, especially in the presence of 800 mM NaCl and under water stress conditions (Figure 2a). The mean leaf area also decreased compared to the control, significantly in plants of the 600 and 800 NaCl and the water stress treatments (Figure 2b). Root fresh weight decreased in all cases with respect to the control, notably in water-stressed plants, where an eight-fold reduction was recorded (4.34 vs. 37.87 g); although significant, the salt-induced reduction in root fresh weight was much less pronounced, down to about 50% of the control, without significant differences between the salt treatments (Figure 2c). Average values of leaf fresh weight were also lower in the stressed plants than in the non-stressed control, but the differences were significant only in the presence of 400 mM NaCl or after the water stress treatment (Figure 2d); in any case, this reduction (30−40% of the control) was not as substantial as for the root fresh weight. Another relevant parameter is the water content of the plants, calculated by the ratio between fresh and dry weight. As expected, water content decreased significantly in the water-stressed plants, especially in the roots (17.1% compared to 75.0% in control plants, Figure 2e) and to a much lower extent in the leaves (64.7% vs. 68.0%, Figure 2f). A significant but lesser reduction was also recorded in the roots of the 400 mM NaCl-treated plants (69.1%, Figure 2e) and in the leaves of the plants subjected to 600 mM NaCl (65.8%, Figure 2f).

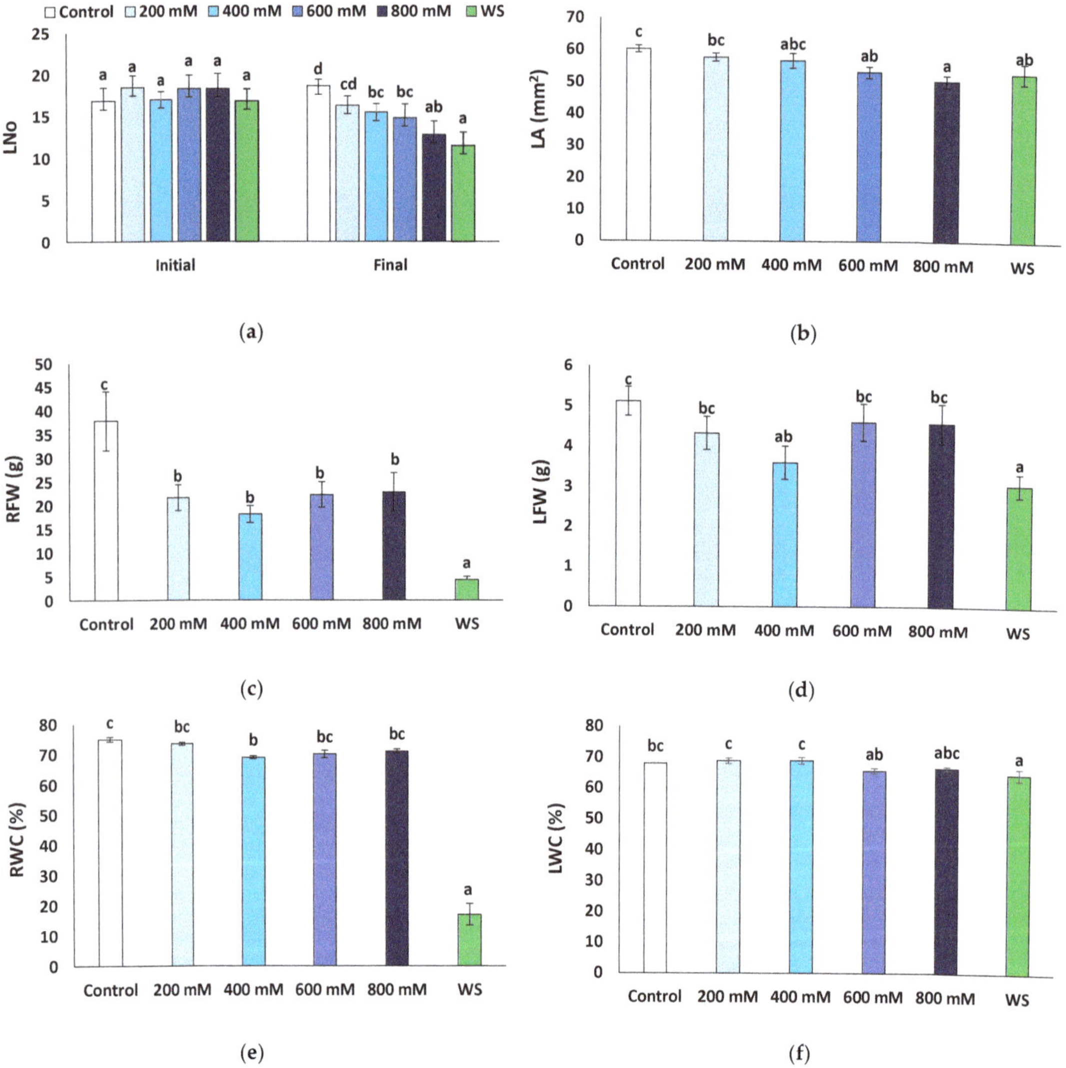

Figure 2. Growth parameters after four weeks of salt treatments with the indicated NaCl concentrations (X-axis), or one month of water stress (WS), of *Limonium angustebracteatum* plants. (**a**) Number of leaves, Lno, (**b**) leaf area, LA, (**c**), root fresh weight, RFW, (**d**), leaf fresh weight, LFW, (**e**) root water content, RWC, and (**f**) leaf water content, LWC. Means ± SE, n = 6. Different lowercase letters above the bars indicate significant differences between treatments, according to Tukey's test ($\alpha = 0.05$).

2.2. Ultrastructure of Salt Glands

Secretory salt glands are described in several genera belonging to the Plumbaginaceae family, including *Limonium* [19]. To better characterise these structures in *Limonium angustebracteatum*, we observed their anatomy using Cryo-FESEM microscopy. Salt glands were located at the epidermis level of leaves, and secreted salt was deposited on the leaf surface (Figure 3a,b). Based on our observations, we suggest that *L. angustebracteatum* salt glands are organised structures formed by 12 specialised cells (Figure 3c,d): four secretory cells containing a secretory pore that form an inner quadrant, an external ring of four evident accessory cells, and likely, just beneath, another quadrant of four accessory cells (Figure 3c,d).

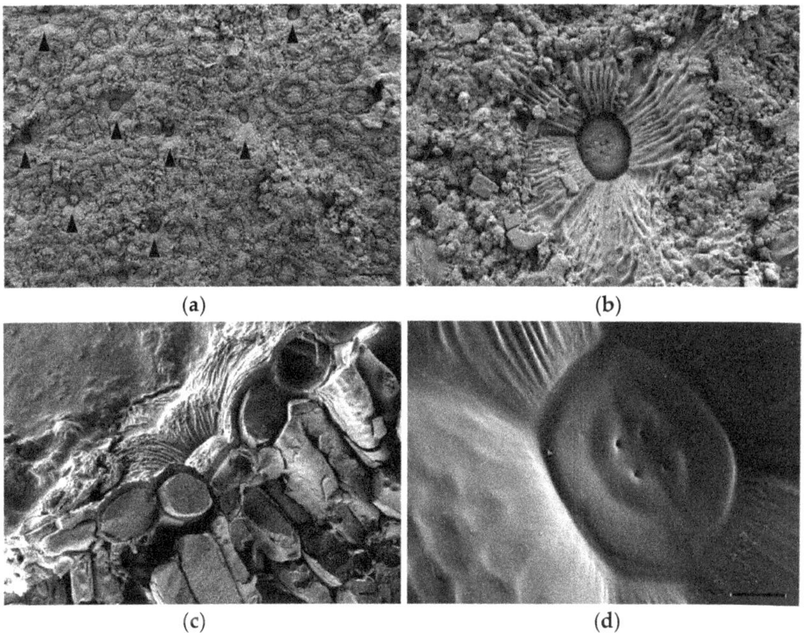

Figure 3. Cryo-FESEM images of salt glands of *Limonium angustebracteaum* grown under control conditions. (**a**) Salt glands (indicated by black arrowheads) located on the leaf epidermis. (**b**) Close look up of a salt gland on the leaf epidermis. Notice salt crystals' excretion. (**c**) Transversal section of a salt gland located on leaf epidermis pits. (**d**) Detailed image of the salt gland core. Only two rings of salt gland cells are easily noticeable on this SEM image: 4 secretory cells in the centre, each of them with a secreting pore; 4 accessory cells surrounding the inner ring of secretory cells; just beneath it, a border of another accessory cell (bottom left) can be observed. Scale bar: 100 μm (**a**), or 10 μm (**b**–**d**).

2.3. *Effects of Stress Treatments on Anatomical Structures*

Toluidine blue-stained leaf sections excised from plants subjected to salt and drought treatments were compared to those of control plants. This experiment allowed us to analyse the overall leaf anatomy, as well as the leaf cell size and morphology. In the control plants, leaf cross-sections were generally thicker than those of salt-treated and drought-stressed plants. Moreover, they were more compact and had a higher degree of tissue organisation (Figure 4a,c,e). The leaf structure of the plants grown under control conditions was bifacial dorsiventral, with palisade tissue towards the upper epidermis and spongy tissue beneath the lower epidermis. In these leaves, palisade tissue consisted of 1–3 layers of long cells, with regular disposition, without air spaces between them, being consequently more compact (Figure 4a). On the contrary, a typical palisade tissue was hardly noticeable in the plants subjected to salt stress, probably due to the lax leaf structure, and, although they presented 1–2 layers of palisade cells, they were short, with a slightly disorganised disposition, and presented air spaces between them (Figure 4c). In plants subjected to the drought treatment, as compared to the control plants, their epidermal leaf cells were heterogeneous in shape and size and ranged from very large and aculeiform-like shaped cells to large cells with no regular shape (Figure 4e). As in the case of the 600 mM NaCl salt treatment, at the analysed cross-section level, the leaf lost its bifacial structure so that the palisade tissue was no longer noticeable; instead, large irregularly shaped cells, with air spaces between them, occupied its position (Figure 4e).

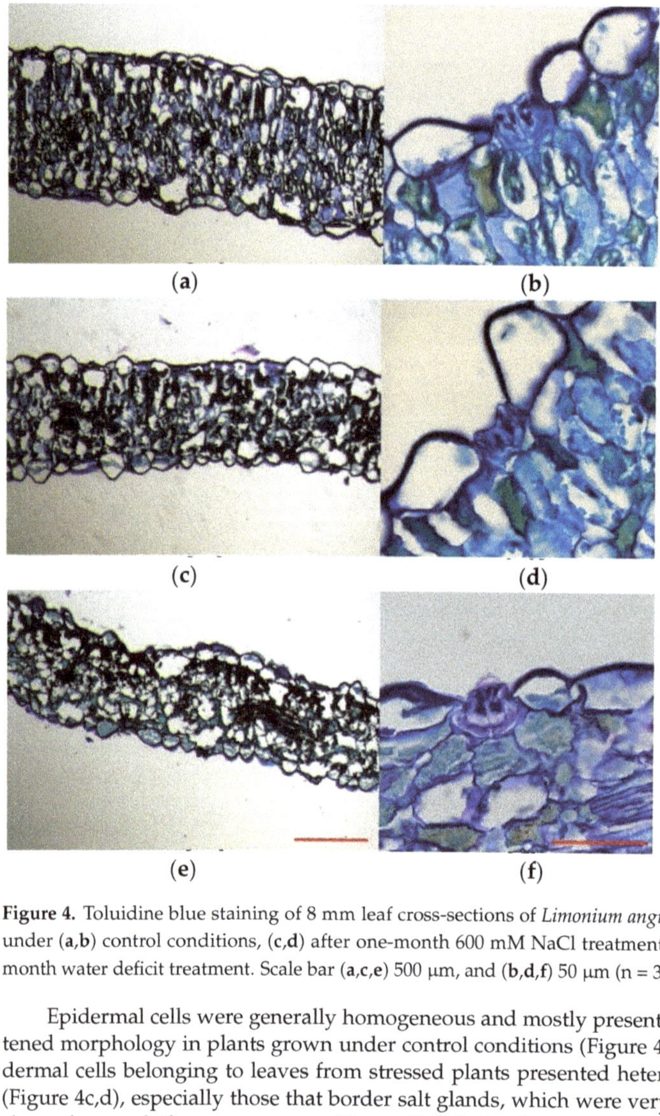

Figure 4. Toluidine blue staining of 8 mm leaf cross-sections of *Limonium angustebracteaum* plants under (**a**,**b**) control conditions, (**c**,**d**) after one-month 600 mM NaCl treatment and (**e**,**f**) after one-month water deficit treatment. Scale bar (**a**,**c**,**e**) 500 μm, and (**b**,**d**,**f**) 50 μm (n = 3).

Epidermal cells were generally homogeneous and mostly presented a thin and flattened morphology in plants grown under control conditions (Figure 4a). However, epidermal cells belonging to leaves from stressed plants presented heterogeneous shapes (Figure 4c,d), especially those that border salt glands, which were very large, and some showed an aculeiform appearance (Figure 4b,d,f). In all cases, stomata were noticeable on both the upper side and the underside of leaves; thus, the lamina was amphistomatic. Stomata appeared to be generally located at the level of epidermal cells and showed a very large substomatic chamber (Figure 4a,c,e). Salt glands were located in the upper epidermis in plants growing under all tested conditions and were sunken in the epidermis (Figure 4b,d,f). This was especially evident for salt-watered plants (Figure 4d). Generally, the epidermal cells surrounding the salt glands were larger in stressed plants than in plants watered with the control solution (Figure 4b,d,f).

2.4. Photosynthetic Pigments

The concentrations of chlorophyll a (Chl a), chlorophyll b (Chl b) and total carotenoids (Car) were measured at the end of the treatments in fresh leaf material (Figure 5). Concentrations of Chl a did not vary significantly in plants from different treatments, but Chl b decreased significantly, down to about 35% of the control in the presence of 400 mM NaCl (0.49 vs. 1.39 mg g^{-1} DW) and to ca. 44% for the water stress treatments (0.61 mg g^{-1} DW). An opposite variation pattern was observed for carotenoids, which showed significantly higher values in the presence of 400 mM NaCl and upon the water stress treatment than in the control.

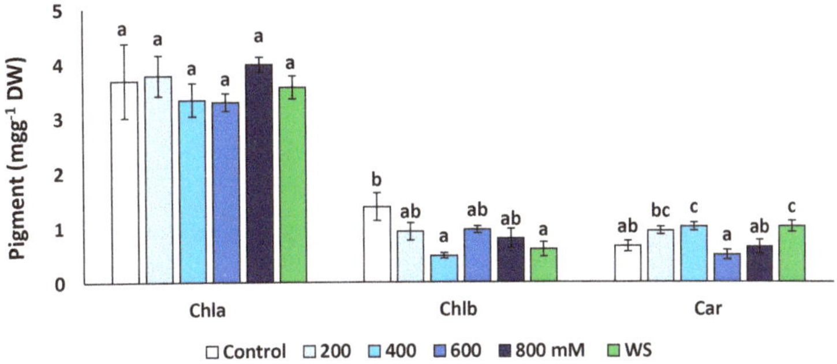

Figure 5. Variation of photosynthetic pigments contents in *Limonium angustebracteatum* leaves after one month of treatment with the indicated NaCl concentrations or one month of water stress (WS). Means ± SE, n = 6. For each pigment, different lowercase letters above the bars indicate significant differences between treatments, according to Tukey's test (α = 0.05). Abbreviations: Chla, chlorophyll a; Chlb, chlorophyll b; Car, carotenoids.

2.5. Effects of Stress Treatments on Ion Contents

Monovalent ions (Na^+, K^+ and Cl^-) and Ca^{2+} concentrations were measured at the end of the treatments in dry root and leaf material. As expected, both Na^+ (Figure 6a) and Cl^- (Figure 6b) increased significantly under salt stress but not in water-stressed plants. A maximum Na^+ concentration was found in the presence of 800 mM NaCl, reaching 7.5-fold higher values than in the non-stressed control in the roots (725 µmol g^{-1} DW, compared to 96 µmol g^{-1} DW) and a 2.5-fold increase (642 vs. 254 µmol g^{-1} DW) in the leaves (Figure 6a). A similar pattern of variation was established for Cl^- (Figure 6b), with maximum measured concentrations of 1408 µmol g^{-1} DW in the roots and 1606 µmol g^{-1} DW in leaves of the 800 mM NaCl-treated plants; these values represent relative increases over the corresponding controls of, approximately, 7-fold and 4.4-fold in roots and leaves, respectively. In general, the Na^+ and Cl^- concentrations in the salt-treated plants were the same in roots and leaves at all external salinities tested. However, interestingly, in non-stressed controls and plants subjected to water stress, the contents of both ions were significantly higher in the leaves than in the roots (Figure 6a,b).

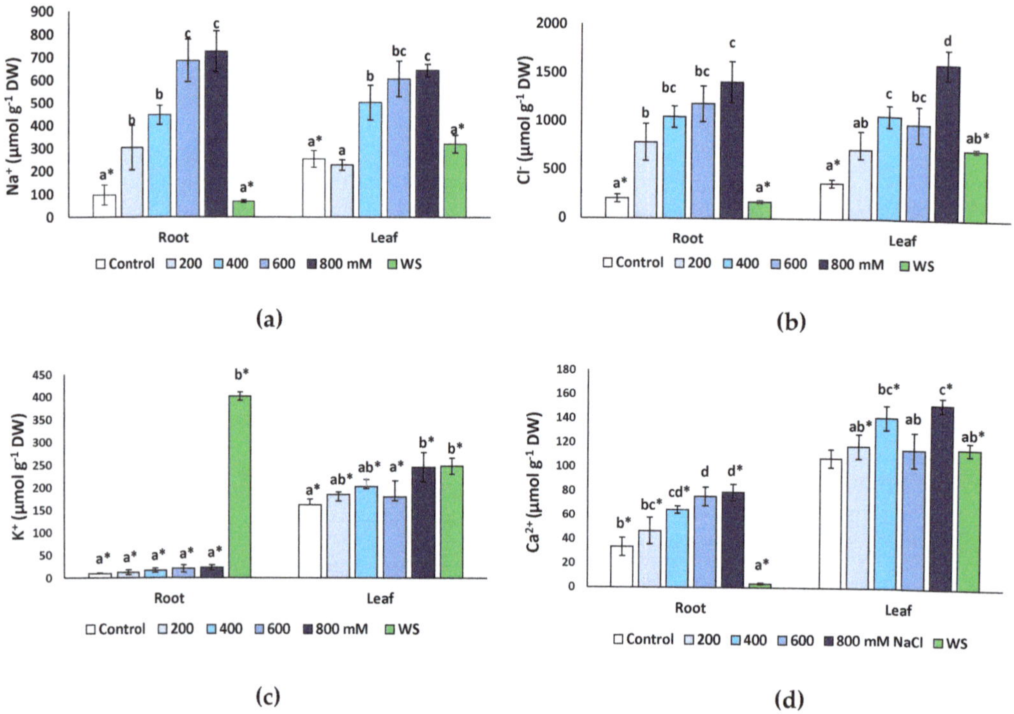

Figure 6. Ion contents in roots and leaves of *Limonium angustebracteatum* plants after one month of treatment with the indicated NaCl concentrations or one month of water stress (WS). (**a**) Na^+, (**b**) Cl^-, (**c**) K^+, (**d**) Ca^{2+}. Means ± SE, n = 6. For each organ, different lowercase letters above the bars indicate significant differences between treatments, according to Tukey's test (α = 0.05). Asterisks indicate significant differences between ion concentration values in roots and leaves for the same treatment.

An unusual pattern of variation was found for K^+ in roots (Figure 6c). Average contents of this cation increased in roots in response to salt stress, up to three-fold higher than in the control in the presence of 800 mM NaCl (ca. 27 vs. 9 μmol g^{-1} DW), but these differences were not statistically significant. However, water deficit stress caused a considerable increase in K^+ concentration to more than 380 μmol g^{-1} DW, or 42-fold over the control values. In the leaves (Figure 6c), a modest (ca. 1.5-fold) but significant increase in K^+ was detected in response to the highest salt concentration tested (800 mM NaCl) and to water stress. Leaf K^+ levels were significantly higher than those in the roots, in control plants and at all NaCl concentrations; only in water-stressed plants, the opposite correlation was observed, with K^+ contents substantially higher in the roots than in the leaves (Figure 6c).

Finally, Ca^{2+} concentrations showed patterns of variation roughly similar, qualitatively, to those of Na^+ and Cl^-, with significant increases over control values, in roots and leaves, in the presence of 400 mM NaCl and higher salt concentrations (Figure 6e). Leaf Ca^{2+} contents did not vary significantly in response to water deficit but decreased sharply, about 10-fold, in the roots of water-stressed plants (Figure 6e). Ca^{2+} concentrations were significantly higher in the leaves than in the roots in the controls, in all salt treatments and, especially, in the water-stressed plants (Figure 6e).

2.6. Effects of Stress Treatments on Osmolyte Accumulation

Proline (Pro), glycine betaine (GB) and total soluble sugars (TSS) represent the most common plant osmolytes, which are synthesised and accumulate in the cells contributing to osmotic adjustment under abiotic stress conditions causing cell dehydration. All three were analysed in fresh leaf material from plants sampled at the end of the experiment after one month of exposure to stress.

Pro contents increased significantly in response to salt stress, at 400 mM and higher NaCl concentrations, from 15.7 to 756.6 μmol g^{-1} DW, in the control and 800 mM NaCl-treated plants, respectively, representing a 48-fold increase. Conversely, leaf Pro contents in plants subjected to the water deficit treatment did not differ significantly from those measured in control plants or at moderate (200 mM NaCl) salinity (Figure 7a).

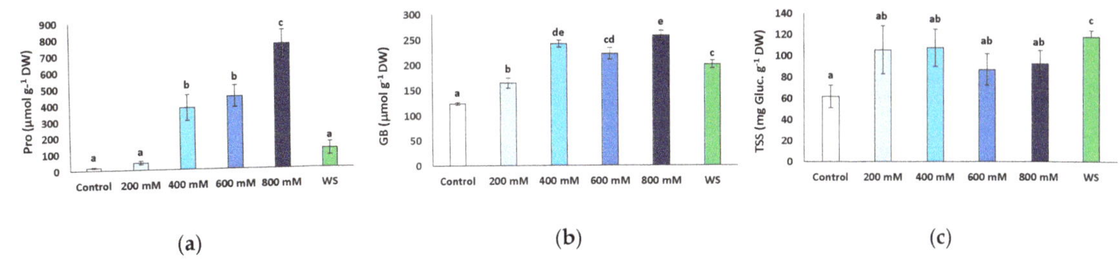

Figure 7. Osmolyte contents in leaves of *Limonium angustebracteatum* plants after one month of treatment with the indicated NaCl concentrations or one month of water stress (WS). (**a**) proline (Pro), (**b**) glycine betaine (GB), (**c**) total soluble sugars (TSS). Means ± SE, n = 6. Different lowercase letters above the bars indicate significant differences between treatments, according to Tukey's test ($\alpha = 0.05$).

GB contents showed a significant increase in response to water deficit and all salt stress treatments with respect to the control. However, the variation was not as marked as that of Pro, reaching no more than twice the control value (Figure 7b). It should be mentioned that this relatively small stress-induced increment in GB levels was primarily due to the presence of the osmolyte at high concentrations in non-stressed plants, ca. 123 μmol g^{-1} DW; that is, almost 10-fold higher than Pro background contents (Figure 7b).

The mean TSS leaf concentrations also increased in the stressed plants relative to the control, but this variation was only significant in the plants subjected to water stress, reaching almost twice the values measured in the control plants (117.7 vs. 61.6 mg eq. gluc g^{-1} DW) (Figure 7c).

2.7. Multivariate Analysis

A principal component analysis (PCA) was performed on all quantitative traits analysed. Four components were detected with eigenvalues greater than one, covering 98.04% of the total variability of the data (Table 1, Figure 8). The first component explained 43.10% of the total variation and correlated (values greater than 0.25, shown in bold font in Table 1) positively with the concentration of the osmolytes glycine betaine (GB) and proline (Pro) and with leaf concentrations of Na^+, K^+, Cl^- and Ca^{2+}, and negatively with the leaf area (LA), leaf number (L No) and chlorophyll b (Chl b). The second component, which explained an additional 36.2% of the variation, was positively correlated (values above 0.25) with growth parameters (leaf and root fresh weight and root water content) and with root Na^+, Cl^- and Ca^{2+} concentrations, and negatively with root K^+ and carotenoids (Caro) levels. The first component separated the scores of control and salt-stressed plants, except for those of the 800 mM NaCl treatment, which were separated at one end of the second component, while those of water stress at the other end of the same axis (Figure 8). Thus, the most separated treatments on the two axes of the biplot were the control, water stress and 800 mM NaCl.

Table 1. Component weights in the PCA performed on *Limonium angustebracteatum* plants subjected to one month of stress treatments. Stronger correlations are shown in bold font. Abbreviations: LNo, number of leaves; LA, leaf area; RFW, root fresh weight; LFW, leaf fresh weight; RWC, root water content; LWC, leaf water content; Chl a, chlorophyll a; Chl b, chlorophyll b; Caro, carotenoids; Na^+_r, sodium in roots; Na^+_l sodium in leaves; Cl^-_r, chlorine in roots; Cl^-_l, chlorine in leaves; K^+_r, potassium in roots; K^+_l, potassium in leaves; Ca^{2+}_r, calcium in roots; Ca^{2+}_l, calcium in leaves; Pro, proline; GB, glycine betaine; TSS, total soluble sugars.

Trait	Component 1	Component 2	Component 3	Component 4
L No	**−0.285**	0.170	0.178	−0.059
LA	**−0.308**	0.006	0.267	−0.034
RFW	−0.212	**0.276**	−0.058	0.116
LFW	−0.156	**0.302**	−0.206	0.124
RWC	−0.090	**0.341**	0.184	0.033
LWC	−0.138	0.147	0.532	0.154
Chl a	0.001	0.053	−0.115	0.789
Chl b	**−0.262**	0.155	−0.304	0.104
Caro	0.017	**−0.268**	0.4445	0.133
Na^+_r	0.222	**0.267**	−0.019	−0.127
Na^+_l	**0.265**	0.194	−0.112	−0.205
Cl^-_r	0.224	**0.259**	0.121	−0.059
Cl^-_l	**0.304**	0.150	0.023	0.151
K^+_r	0.074	**−0.343**	−0.207	0.002
K^+_l	**0.271**	−0.162	−0.088	0.314
Ca^{2+}_r	0.139	**0.328**	0.103	−0.114
Ca^{2+}_l	**0.263**	0.123	0.221	0.281
Pro	**0.279**	0.199	−0.083	0.037
GB	**0.328**	0.062	0.079	−0.104
TSS	0.198	−0.234	0.277	−0.019

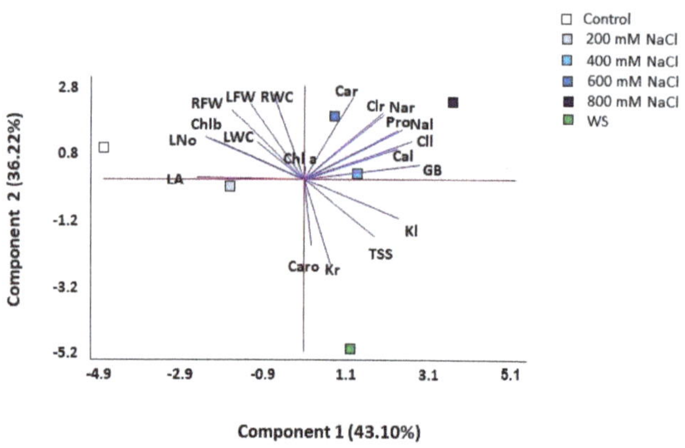

Figure 8. Biplot of the principal component analysis (PCA) conducted with the analysed traits in *Limonium angustebracteatum* plants subjected to one-month stress treatments. Abbreviations: LNo, number of leaves; LA, leaf area; RFW, root fresh weight; LFW, leaf fresh weight; RWC, root water content; LWC, leaf water content; Chla, chlorophyll a; Chl b, chlorophyll b; Caro, carotenoids; Na^+_r, sodium in roots; Na^+_l sodium in leaves; Cl^-_r, chlorine in roots; Cl^-_l, chlorine in leaves; K^+_r, potassium in roots; K^+_l, potassium in leaves; Ca^{2+}_r, calcium in roots; Ca^{2+}_l, calcium in leaves; Pro, proline; GB, glycine betaine; TSS, total soluble sugars.

The hierarchical cluster analysis (HCA) performed together with the heatmap (Figure 9) with traits measured in all plants confirmed the PCA results and revealed a clear separation of the drought-stressed plants in one cluster. In addition, the cluster topology allowed the separation of two other groups, one including plants from the control and the 200 mM NaCl treatments and the other from the 600 and 800 mM NaCl treatments. Plants treated with the intermediate salt concentration (400 mM NaCl) showed a heterogeneous pattern, with individuals falling into different clusters. Plants from the water stress treatment showed high positive correlations with root K^+ contents and negative with root fresh weight, water content and Ca^{2+} concentration. The cluster of plants from the high salt treatments (600 and 800 mM NaCl) had positive correlations with root Na^+ and Cl^- and negative with leaf area, leaf water content and carotenoid levels. Finally, plants from the control and the lowest salt concentration (200 mM NaCl) were positively correlated with growth parameters and negatively with ions and osmolytes contents (Figure 9).

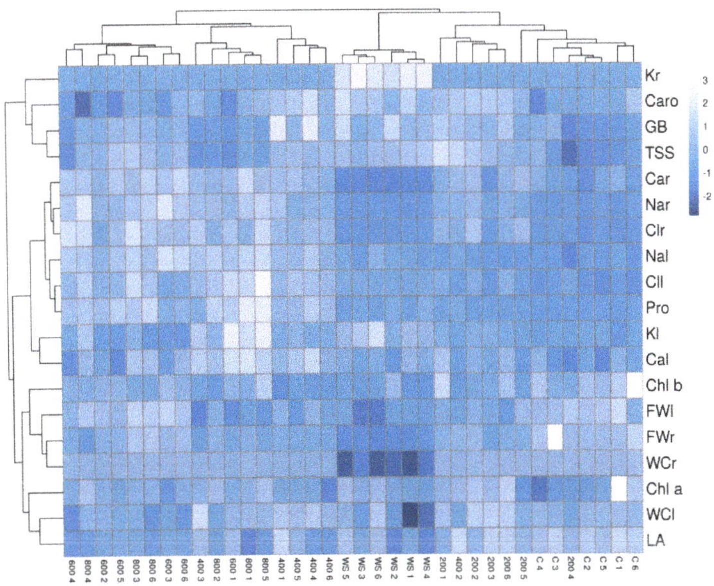

Figure 9. Hierarchical cluster analysis and heatmap of physiological and biochemical parameters in *Limonium angustebracteatum* plants subjected to one-month stress treatments. Abbreviations: LNo, number of leaves; LA, leaf area; RFW, root fresh weight; LFW, leaf fresh weight; RWC, root water content; LWC, leaf water content, Chla, chlorophyll a; Chl b, chlorophyll b; Caro, carotenoids; Na^+_r, sodium in roots; Na^+_l sodium in leaves; Cl^-_r, chlorine in roots; Cl^-_l, chlorine in leaves; K^+_r, potassium in roots; K^+_l, potassium in leaves; Ca^{2+}_r, calcium in roots; Ca^{2+}_l, calcium in leaves; Pro, proline; GB, glycine betaine; TSS, total soluble sugars.

3. Discussion

Limonium angustebracteatum is an endemic species of the Iberian Peninsula with a high conservation interest. Although it is still largely distributed in E and SE Spain, some populations declined significantly during recent decades. In the Valencian region, it is quoted as NT (Near threatened) [30] for the IUCN Red List Categories [31]. In addition, it is a key species for some protected habitats in this region, such as *Limonietalia* salt steppes, becoming particularly dominant in the rare plant association *Artemisio gallicae-Limonietum angustebracteati* Costa and Boira 1981 (see [32,33]).

Like other species of this genus, *L. angustebracteatum* is a recretohalophyte with salt excretion capacity through salt glands located in the aerial parts of the plants, which provide a relevant constitutive mechanism contributing to salt tolerance in these species. Salt glands are epidermal structures specialised in the storage and exclusion of salts present in halophytes of different families; they function by regulating the ionic balance and ensuring a stable osmotic pressure [26,34,35]. The great phylogenetic diversity of recretohalophytes, and their structural differences, suggest that this trait originated independently multiple times during plant evolution [36] and represents a clear example of convergent adaptation [37]. Salt glands were already described in classical plant anatomy studies [38], but recent research sheds light on their evolutionary origin and the physiological, biochemical and molecular mechanisms of salt secretion through salt glands [19,20,34,36,39]. There are many reports on the structure and function of salt glands in the genus *Limonium* [34,40–44] due to its large number of species, many of them endemics, adapted to saline or dry environments.

The ultrastructure of salt glands in Plumbaginaceae is problematic because of the large diversity of structural architectures reported for different genera and species and also the lack of a consistent and standardised language to define different types of component salt gland cells [19,38]. Furthermore, there are technical difficulties in handling anatomical cross-sections and electron microscopy images, which, in fact, can never reveal the entire structure and topography of different types of cells and their connections within a salt gland. For this reason, salt glands structures based on 12, 16 or 20 cells have been proposed, with different relative positions with respect to the epidermis level (summarised in [19]). Based on our analysis of *L. angustebracteatum*, we suggest a salt gland structure composed of 12 cells, 4 secretory cells forming an inner quadrant, an external ring of 4 completely visible accessory cells and just beneath it, another quadrant of 4 accessory cells. A similar 12-cells-based salt gland structure was also reported for other *Limonium* species: *L. aureum* (L.) Hill [45], *L. bellidifolium* (Gouan) Dumort. [46], *L. bocconei* (Lojac.) Litard., *L. pignanttii* Brullo and Di Martino or *L. lojaconoi* Brullo [47].

The number of salt glands is shown to increase with various treatments, such as Ca^{2+} [48], exogenous nitric oxide [49] or melatonin [41]; however, NaCl appears to be the main trigger for salt gland development [34,40,44]. Salt gland density is also correlated with the plants' natural environment; for example, *Limonium* species growing under high salinity conditions have a higher density of glands than those present in less saline habitats [43].

The available data on the anatomical modifications of halophytes under salt and water stress are surprisingly scarce compared to glycophytes. In *L. angustebracteatum*, the large epidermal cells of salt and drought-exposed plants can be regarded in terms of increasing the thickness of, especially, the upper epidermal layer, as it is also reported for plants of the halophyte *Salvadora persica* subjected to high salinity levels [50]. In this species, leaf palisade tissue disappeared in the presence of 750 mM NaCl [50]; the same effect was observed in our experiments with *L. angustebracteatum* in salt-treated plants and, particularly, in plants exposed to drought. This disorganisation of typical palisade tissue might be an adaptation to minimise the photosynthetic energy utilisation under intense stress and could be correlated with the reduction of photosynthetic pigment levels in stressed plants. A palisade tissue with shorter cells was also reported in the halophyte *Juncus acutus* L., in the presence of 400 mM NaCl [51].

Regarding the physiological responses of *L. angustebracteatum*, high salt concentrations and water stress affected plant growth. All plants, including halophytes, respond to severe stress, reducing or even completely stopping their growth, as metabolic precursors and energy resources are used under such conditions to activate defence mechanisms rather than for biomass accumulation [52]. All glycophytes and most halophytes show optimum growth in the absence of salt. Only some extremophiles grow better in the presence of low or moderate salinity; even in these plants, high salt concentrations have inhibitory effects on growth [53]. The growth of halophytes of the genus *Limonium* is generally unaffected by low NaCl concentrations [54] or is, sometimes, even stimulated [55–58]. However, growth inhibition is observed at high salinities, 300 or 400 mM NaCl, even in the most salt-tolerant species [59,60]. Nevertheless, there are also *Limonium* species that are not halophytes but rather susceptible to even low concentrations of NaCl, such as *L. perezii* (Stapf) F.T. Hubb. and *L. sinuatum* (L.) Mill. [61] or *L. dufourii* [62,63]. *Limonium angustebracteatum* had optimal growth in the absence of salt, but no significant inhibition was observed in the presence of 200 M NaCl. The same response was found in other *Limonium* endemics of this geographic area [64,65]. This behaviour enables the species to grow in saline depressions but also in dunes and other elevated, permeable substrates, where soil salinity may be lower due to the leaching of ions.

Of all the tested treatments, the most significant growth inhibition was recorded in the plants subjected to water deficit by complete withholding of irrigation for one month; the effect of water stress was most notably reflected in a drastic reduction in root fresh weight and water content. This response, reported even for *Limonium* species growing in markedly dry areas [66], is explained by the severity of water stress in potted plants, where roots cannot extend in search of more profound and wetter soil layers as they do in their natural habitats [65]. Leaf senescence was accentuated by stress, as revealed by the loss of more than five leaves, on average, in the presence of 800 mM NaCl and the water deficit treatment. The remaining leaves had a smaller surface area, but photosynthetic pigment contents did not decrease substantially; only Chl b showed a significant, although slight, reduction under stress and carotenoid levels even increased, again slightly but significantly, in response to some stress treatments. These results indicate that *L. angustebracteatum* behaves as a typical perennial halophyte, tolerating high salt concentrations far beyond those found in the soil in its natural habitat [63]. Moreover, under controlled greenhouse conditions, it is more susceptible to severe water stress than moderate salinity, even though the plants survived the strong water deficit conditions applied. In their natural habitat, plants of this perennial species also survive the intense summer drought characteristic of the Mediterranean region, where periods of one to more than two months of absolute lack of precipitations are not uncommon.

Like all recretohalophytes, *Limonium* species secrete ions, especially Na^+ and Cl^-, through their glands [39,55,67]. However, these plants also can accumulate these ions in the cells, sequestering them in the vacuoles to avoid their toxic effects in the cytosol; inorganic ions represent a 'cheap' osmoticum used by many halophilic dicots for osmotic adjustment under stress [22,68]. Similar patterns of a concentration-dependent increase of Na^+ and Cl^- in response to the NaCl treatments were found for the two monovalent ions and the two organs, roots and leaves. Interestingly, in non-stressed plants and under water deficit conditions, the concentrations of both ions were significantly higher in the leaves than in the roots; this difference was much more pronounced in the case of water-stressed plants. These data indicate that, even at low external salinity, these ions are actively transported from roots to leaves, where they are used for osmotic balance, as reported in other *Limonium* species [62,64,69,70].

An increase in Na^+ is usually accompanied by a decrease in K^+, as both cations compete for the same binding sites. Moreover, it is well known that excess Na^+ causes the depolarisation of the plasma membrane, inducing the activation of outward rectifier K^+ channels and thus the loss of cellular K^+ [71,72]. K^+ is an essential nutrient for plants, involved in many cellular and metabolic processes, such as cell elongation, stability of membrane integrity, enzyme activation, protein synthesis, photosynthesis, stomatal movement or phloem transport [73]. In stress resistance, the role of K^+ is related to osmotic adjustment [74]. In *Limonium*, the activation of K^+ transport from the roots to the aboveground organs was reported [62], which ensures that its leaf concentration is maintained or decreases only slightly with increasing salinity, thus counteracting, at least partly, Na^+ deleterious effects. The same mechanism was described in other halophytes, such as *Plantago crassifolia* [60] or *Inula crithmoides* [75]. Contrary to the general behaviour, even salt-induced increases of K^+ contents, relative to the control, were reported, for example, in the roots of several Mediterranean *Limonium* species [64] or the leaves of *L. stocksii* under low salinity [70]. Similar responses were observed in *L. angustebracteatum* plants subjected to salt treatments, where leaf K^+ levels increased slightly with increasing salinity and were significantly higher than in roots at each NaCl concentration tested, including the non-stressed controls. This indicates that K^+ uptake and transport to the leaves is an essential mechanism of salt tolerance in this species, contributing to osmotic balance, also in the absence of salt. Indeed, in situ subcellular localisation studies revealed that K^+ accumulated in both the cytoplasm and the nucleus of salt gland cells under saline conditions, which may play an important role in salt secretion [76]. Leaf K^+ contents under water deficit conditions were similar to those measured at the highest salinity, 800 mM NaCl, suggesting the participation of this cation in the responses of the plants to drought. However, an unusual pattern was found for the K^+ concentration in the roots of *L. angustebracteatum* water-stressed plants, which increased more than 40-fold over the control values. The plants seem to use K^+ uptake as a defence mechanism against drought, which strongly affects the root system. K^+ increases in roots (although not so pronounced) and leaves were also reported in other congeners such as *L. girardianum* and *L. narbonense* [66]. Higher K^+ content in leaves than in roots was also reported in plants of this genus sampled in natural environments [65].

Calcium is another essential cation involved in membrane and cell wall stabilisation, the regulation of ion transport and selectivity and enzymatic activities [77]. Under salt stress conditions, the external application of Ca^{2+} reduces the toxic effects of NaCl, presumably by facilitating increased K^+/Na^+ selectivity [77]. Ca^{2+} plays an essential role in stress signalling; cytosolic calcium activates the calcium sensor protein SOS3, which triggers a signalling cascade activating the plasma membrane Na^+/H^+ antiporter, SOS1, leading to Na^+ efflux out of the cytosol to the apoplast and also contributing to Na^+ compartmentalisation in the vacuole through the equivalent tonoplast antiporter, NHX1 [78,79]. Ca^{2+} levels increased in both roots and leaves in several species of this genus [64,70]; in other species, on the contrary, this cation was found to decrease with salinity [64,80]. The strong reduction of Ca^{2+} in the roots of water-stressed plants could be related to the high input of K^+ ions, as the presence of other cations is known to profoundly influence Ca^{2+} uptake, with high levels of K^+ and Mg^{2+} reducing Ca^{2+} uptake [81].

A fundamental mechanism to ensure osmotic balance and compensate for the accumulation of toxic ions in the vacuoles, mostly in dicotyledonous halophytes, is the synthesis of osmolytes [23]. The genus *Limonium* is notable for the variety of compatible solutes used concomitantly, even in the same species [54]. Proline, one of the most common plant osmolytes, is found in all *Limonium* species and can be considered a functionally relevant osmolyte in this genus [54]. Many reports in *Limonium* species indicate an increase in Pro contents under stress, and often higher Pro levels are correlated with higher stress tolerance [63]. However, this cannot be generalised to all *Limonium* taxa since, for example, in *L. latifolium*, the variation in Pro concentrations was considered to be related to damage and successive repair in the mitochondrial step of proline oxidation [82] rather than to its salt tolerance. In *L. angustebracteatum*, a linear increase of Pro contents in parallel to increasing salinity was observed, reaching over 750 μmol g^{-1} DW in the presence of 800 mM NaCl, but not in response to water stress, suggesting that this osmolyte plays a relevant role in the plant adaptation to salinity, but not so much to drought.

Glycine betaine is another common osmolyte in plants, present at high concentrations in the salt-tolerant Amaranthaceae and Poaceae [83] and acting as an osmoregulator under abiotic stress conditions [84]. GB is synthesised primarily from choline, and GB accumulators have particular adaptations in the biogenesis of choline and the methyl group that are not present in other plants [85], allowing accumulations of 4–40 μmol g^{-1} FW in spinach and sugar beet [84], or up to 900 μmol g^{-1} DW in the halophyte *Suaeda fruticosa* [86]. The values recorded in *L. angustebracteatum* (even in control plants) are above those previously reported in *Limonium* species sampled in natural environments [65,87,88]. Since GB concentrations increased significantly under both types of stress, we can conclude that this compatible solute plays a relevant role in the osmotic adjustment of *L. angustebracteatum*.

On the other hand, soluble sugars do not seem to participate in the salt or drought tolerance mechanisms in *L. angustebracteatum*. TSS levels did not vary significantly in plants subjected to the salt treatments, and the increase observed in response to water deficit was too small to have any relevant osmotic effect. Nevertheless, it is always difficult to assess the possible role in stress tolerance mechanisms of changes in sugar concentrations because of their multiple biological functions not directly related to stress responses, as direct products of photosynthesis, metabolic precursors or energy sources [89].

These results suggest that *Limonium angustebracteatum* can behave as a eurioic species, particularly regarding the soil salt concentration. This would imply that the species can easily adapt to environmental changes, even though it has little resistance to prolonged periods of flooding, according to our observations in the field (EL and PFG, pers. obs.). In this respect, *L. angustebracteatum* shows a behaviour similar to that of *L. dufourii* [63] and far from that shown by more water-resistant species such as *L. albuferae* P.P. Ferrer et al. [63]. Furthermore, unlike most *Limonium* species sharing its habitat in the Valencian Community (*L. dufourii*, *L. girardianum* (Guss.) Kuntze, *L. virgatum* (Willd.) Fourr., *L. angustebracteatum* is a much more robust plant, bearing bigger and deeper root systems [11,28], which allows it to colonise dunes or other habitats unable for those species. These physiological and morphological features can explain that *L. angustebracteatum* was more abundant and less threatened than *L. dufourii*, listed as CR (Critically Endangered) under the IUCN classification (see [28,90]). However, even possessing this greater colonising capacity, *L. angustebracteatum* has vanished in some areas where it was abundant in the past, such as the Devesa de l'Albufera in Valencia (EL and PFG, pers. obs.), after abrupt alterations of the local ecosystem due to human activity (see [8,9]). Despite this local population decline, the results presented here allow us to propose *L. angustebracteatum* as a suitable species to be included in future ecological restoration projects because of its higher resilience compared to more delicate, endangered cohabitant taxa (i.e., *L. dufourii*).

4. Materials and Methods

4.1. Plant Material and Stress Treatments

One-year-old *L. angustebracteatum* plants, obtained from the germination of seeds, were used for this study. The plants were grown in a greenhouse with natural illumination, relative humidity of 65% and a 23–30 °C temperature range. Plants were individually placed in 12 cm diameter pots filled with a mixture of commercial peat and vermiculite (3:1) and watered regularly with tap water. Treatments were performed on 36 plants of uniform size, with six replicates per treatment. The following treatments were applied: control (irrigation with tap water, twice per week), water stress (complete withholding of irrigation) and salt stress (irrigation, twice per week, with aqueous solutions of 200, 400, 600 and 800 mM NaCl). After four weeks of treatment, the plants were cut and weighed, and their leaves were scanned to measure the leaf area. Part of the fresh material was frozen and partly dried in an oven at 65 °C for three days until a constant weight was recorded. The water content of the roots and leaves was measured according to the formula:

$$WC\ (\%) = [(FW - DW)/FW] \times 100 \quad (1)$$

4.2. Cryo-FESEM Preparations

Cryo-Field Emission Scanning Electron Microscopy (Cryo-FESEM) was performed at the Electronic Microscopy Service of the Polytechnic University of Valencia (QUORUM TECHNOLOGIES, Model PP3010T, Laughton, UK). Leaf samples were excised from adult plants grown under control conditions and physically fixed at ultra-low temperatures with slush nitrogen at −210 °C (cryogenisation); samples were maintained under low temperatures throughout the whole process in a preparation camera. Some samples were cracked with an inner stick to image transversal sections, and then a sublimation step (−90 °C for 10 min) was performed to eliminate residual liquid water. Finally, the samples were sputtered with platinum particles before imaging.

4.3. Light Microscopy

Sample preparation was carried out at the Microscopy Service of the Institute of Plant Molecular and Cell biology (IBMCP, Polytechnic University of Valencia). First, small sections of the leaves (4 mm^2 approximately) were excised from plants grown under control, salt and drought conditions, collected on FAE (50% ethanol, 3.7% formaldehyde and 5% acetic acid) and subjected to vacuum for 15 min or until the samples were sunk. Then, the

FAE solution was refreshed, and the samples were maintained at 4 °C. Next, the samples were dehydrated and included on paraffin with an automatic tissue processor (TP 1020, Leica, Germany). Briefly, the samples were incubated for 1 h with four increasing ethanol solutions (70, 90, 95, and 100%) and three commercial histoclear solutions. Then, inclusion was performed during two sequential incubations under a vacuum of 1 and 3 h with melted paraffin. Afterwards, samples were individually mounted in paraffin blocks and left to solidify at room temperature. Finally, the included samples were sliced (8 μm) with an RM2025 microtome (Leica Biosystems, Nussloch, Germany) and mounted on polysine-enriched slides.

For toluidine blue (TB) staining, sections on the slides were deparaffinised and rehydrated with sequential 10 min incubations with histoclear, ethanol 100%, 90% and 70%, and finally, with distilled water. Next, sections on the slides were incubated for 2 min with a 0.02% TB solution, rinsed with distillate water and dried. Finally, samples on the slides were mounted with distilled water and images were taken with a microscope Eclipse E1000 (Nikon, Tokyo, Japan).

4.4. Photosynthetic Pigments

Fresh leaves (50 mg) were used for the quantification of chlorophyll a (Chl a), chlorophyll b (Chl b) and carotenoids (Car) by the spectrophotometric method described by Lichtenthaler and Welburn [91]. Extraction was performed with one mL of ice-cold 80% acetone followed by overnight shaking at room temperature under dark conditions. Next, a centrifugation step was performed (13,300× *g*, 10 min at 4 °C). Finally, the absorbance of the supernatants (measured at 470, 646 and 663 nm) was used to estimate the concentrations of the pigments according to the equations previously described [91].

4.5. Quantification of Ions

Sodium (Na^+), potassium (K^+), chloride (Cl^-) and calcium (Ca^{2+}) levels in the roots and leaves were estimated according to Weimberg [92]. Samples (50 mg) of ground dry plant material were suspended in 15 mL of deionised water, heated at 95 °C in a water bath for one hour, followed by cooling on ice and filtration through a 0.45 μm nylon filter. The cations were quantified with a PFP7 flame photometer (Jenway Inc., Burlington, VT, USA) and the anion using a chlorimeter (Sherwood, model 926, Cambridge, UK).

4.6. Quantification of Osmolytes

Proline (Pro) was quantified according to the classical protocol by Bates et al. [93], with some modifications [94]. Pro was extracted from 50 mg of fresh leaves in 3% aqueous sulphosalicylic acid mixed with acid ninhydrin solution and incubated for one h at 95 °C. Next, the mixture was cooled on ice, and then two volumes of toluene were added. The absorbance of the supernatant was read at 520 nm, using toluene as a blank. Samples containing known Pro concentrations were assayed in parallel to obtain a standard curve. Pro concentration was expressed as μmol g^{-1} DW.

Glycine betaine (GB) was determined in 1-mL aqueous extracts prepared from 50 mg dry leaf material, according to published procedures [95,96]. The extract was supplemented with potassium iodide, kept on ice for 90 min and then extracted with 1,2-dichloroethane (pre-cooled at −20 °C). Finally, the absorbance of the sample was measured at 365 nm. GB content was expressed as μmol g^{-1} DW.

Total soluble sugars (TSS) were measured according to the method described by Dubois et al. [97] with some modifications [94]. First, fresh leaf material was ground in liquid N_2 and extracted with 80% (*v*/*v*) methanol and mixed in a rocker shaker for 24 h. Next, samples were centrifuged at 13,300× *g* for 10 min, and supernatants were collected, diluted with water, and supplemented with concentrated sulphuric acid and 5% phenol. After 20 min incubation at room temperature, the absorbance was measured at 490 nm. TSS concentrations were expressed as equivalents of glucose, used as the standard (mg eq. glucose g^{-1} DW).

4.7. Statistical Analysis

Data were analysed using the program SPSS for Windows (SPSS Inc., Chicago, IL, USA). Before the analysis of variance, the Shapiro–Wilk test was used to check for the validity of the normality assumption and Levene's test for the homogeneity of variance. If ANOVA requirements were accomplished, the significance of the differences between treatments was tested with a one-way ANOVA, followed by post hoc comparisons using Tukey's HSD test at a significance level of $p = 0.05$. The mean values of all parameters measured in the plants were used for a principal component analysis (PCA). Hierarchical cluster analysis (HCA) and the corresponding heatmap were performed using the ClustVis 2.0 tool [98]. Rows were centred, and unit variance scaling was applied to rows. Both rows and columns were clustered using correlation distance and average linkage.

5. Conclusions

The results obtained revealed that *L. angustebracteatum* is a recretohalophyte highly resistant to salt stress. In addition to salt secretion through salt glands, its salt tolerance seems to depend on efficient osmotic adjustment by the foliar accumulation of high concentrations of ions (Na^+ and Cl^-, but also K^+ and Ca^{2+}) and the osmolytes proline and glycine betaine. The increase of K^+ contents with increasing salinity also represents an especially remarkable tolerance mechanism as it can partially counteract the Na^+ toxic effects. The contents of all four ions were significantly higher in the leaves than in the roots in non-stressed plants, indicating the presence of constitutive defence mechanisms based on active ion transport to the leaves, even at low external salinity. This constitutive response also included GB (but not Pro) accumulation since high absolute concentrations of the osmolyte were also measured in the leaves of the control plants.

On the contrary, the species is more susceptible to water deficit, but an active transport to the leaves of Na^+, Cl^- and Ca^{2+} and a slight but significant increase in GB (but not Pro) contents were observed in water-stressed plants. These inorganic ions and the organic osmolyte contribute to osmotic balance under water stress conditions.

The large increase (over 40-fold) in K^+ levels in roots of water-stressed plants supports the notion that K^+ homeostasis plays a relevant role in the mechanisms of tolerance to both stressful conditions. Furthermore, Ca^{2+} can also be involved in salt and drought stress responses as an essential signalling molecule besides its osmotic effects.

Author Contributions: Conceptualisation, M.-N.G. and O.V.; methodology, R.M. and I.R.; software, L.O.; validation, P.P.F.-G., E.L. and M.B.; formal analysis, S.G.-O.; investigation, R.M., I.R. and S.G.-O.; resources, R.M. and O.V.; data curation, M.B. and L.O.; writing—original draft preparation, R.M., P.P.F.-G. and M.B.; writing—review and editing, E.L., M.-N.G. and O.V.; visualisation, R.M.; supervision, M.B.; project administration, O.V.; funding acquisition, O.V. All authors have read and agreed to the published version of the manuscript.

Funding: This research received no external funding.

Institutional Review Board Statement: Not applicable.

Informed Consent Statement: Not applicable.

Data Availability Statement: Data are contained within the article.

Acknowledgments: The authors thank Inmaculada Ferrando Pardo for helping in the study and conservation of the seeds in the Centre for Forest Research and Experimentation of the Valencian Region (CIEF). Thanks also to the Devesa-Albufera technical service of the Valencia City Council for all the access facilities for the in situ fieldwork in La Albufera Natural Park. We are indebted to the Electron Microscopy Service of Universitat Politècnica de València and Marisol Gascón (IBMCP Microscopy Service) for their excellent technical support. We also thank José M. Seguí-Simarro for using the optical microscope Nikon Eclipse E1000. RM holds a CDEIGENT (2018/023) fellowship from Generalitat Valenciana. IR's participation results from a mobility stay funded by the Erasmus+—KA1 Erasmus Mundus Joint Master Degrees Programme of the European Commission under the PLANT HEALTH Project.

Conflicts of Interest: The authors declare no conflict of interest.

References

1. Erben, M. Limonium Mill. In *Flora Iberica*; Castroviejo, S., Aedo, C., Cirujano, S., Laínz, M., Montserrat, P., Morales, R., Muñoz Garmendia, F., Navarro, C., Paiva, J., Soriano, C., Eds.; Real Jardín Botánico, C.S.I.C.: Madrid, Spain, 1993; Volume 3, pp. 2–143.
2. Sáez, L.; Rosselló, J.A. Is *Limonium cavanillesii* Erben (*Plumbaginaceae*) really an extant species? *An. Jard. Bot. Madr.* **1999**, *57*, 47–55. [CrossRef]
3. Crespo, M.B. A new coastal species of *Limonium* (*Plumbaginaceae*) from Southeastern Spain. *Folia Geobot.* **2009**, *44*, 177–190. [CrossRef]
4. Ferrer-Gallego, P.P.; Navarro, A.; Pérez, P.; Roselló, R.; Rosselló, J.A.; Rosato, M.; Laguna, E. A new polyploid species of *Limonium* (*Plumbaginaceae*) from the Western Mediterranean basin. *Phytotaxa* **2015**, *234*, 263–270. [CrossRef]
5. Ferrer-Gallego, P.P.; Roselló, R.; Rosato, M.; Rosselló, J.A.; Laguna, E. *Limonium albuferae* (*Plumbaginaceae*), a new polyploidy species from the Eastern Iberian Peninsula. *Phytotaxa* **2016**, *252*, 114–122. [CrossRef]
6. Laguna, E.; Fos, S.; Ferrando-Pardo, I.; Ferrer-Gallego, P.P. Endangered halophytes and their conservation lessons from Eastern Spain. In *Handbook of Halophytes: From Molecules to Ecosystems towards Biosaline Agriculture*; Grigore, M.N., Ed.; Springer Nature Switzerland AG: Cham, Switzerland, 2020; pp. 661–723.
7. Erben, M. Die Gattung *Limonium* im Südwestmediterranien Raum. *Mitt. Bot. Staatssamml. München* **1978**, *14*, 361–631.
8. Yuan, F.; Leng, B.; Wang, B. Progress in studying salt secretion from the salt glands in recretohalophytes: How do plants secrete salt? *Front. Plant Sci.* **2016**, *7*, 977. [CrossRef] [PubMed]
9. Shabala, S.; Bose, Y.; Hedrich, R. Salt bladders: Do they matter? *Trends Plant Sci.* **2014**, *19*, 687–691. [CrossRef]
10. Caperta, A.D.; Róis, A.S.; Teixeira, G.; Garcia-Caparros, P.; Flowers, T.J. Secretory structures in plants: Lessons from the Plumbaginaceae on their origin, evolution and roles in stress tolerance. *Plant Cell Environ.* **2020**, *43*, 2912–2931. [CrossRef]
11. Yuan, F.; Lyum, M.J.; Leng, B.Y.; Zhu, X.G.; Wang, B.S. The transcriptome of NaCl-treated *Limonium bicolor* leaves reveals the genes controlling salt secretion of salt gland. *Plant Mol. Biol.* **2016**, *91*, 241–256. [CrossRef]
12. Zhang, M.; Chen, Z.; Yuan, F.; Wang, B.; Chen, M. Integrative transcriptome and proteome analyses provide deep insights into the molecular mechanism of salt tolerance in *Limonium bicolor*. *Plant Mol. Biol.* **2022**, *108*, 127–143. [CrossRef]
13. Wyn Jones, R.; Storey, R.; Leigh, R.A.; Ahmad, N.; Pollard, A. A hypothesis on cytoplasmic osmoregulation. In *Regulation of Cell Membrane Activities in Plants*; Marre, E., Ciferri, O., Eds.; Elsevier: Amsterdam, The Netherlands, 1977; pp. 121–136.
14. Yeo, A.R. Salinity resistance: Physiologies and prices. *Physiol. Plant.* **1983**, *58*, 214–222. [CrossRef]
15. Glenn, E.P.; Brown, J.J.; Blumwald, E. Salt tolerance and crop potential of halophytes. *Critic. Rev. Plant Sci.* **1999**, *18*, 227–255. [CrossRef]
16. Flowers, T.J.; Munns, R.; Colmer, T.D. Sodium chloride toxicity and the cellular basis of salt tolerance in halophytes. *Ann. Bot.* **2014**, *115*, 419–431. [CrossRef] [PubMed]
17. Flowers, T.J.; Colmer, T.D. Salinity tolerance in halophytes. *New. Phytol.* **2008**, *179*, 945–963. [CrossRef] [PubMed]
18. Slama, I.; Abdelly, C.; Bouchereau, A.; Flowers, T.; Savouré, A. Diversity, distribution and roles of osmoprotective compounds accumulated in halophytes under abiotic stress. *Ann. Bot.* **2015**, *115*, 433–447. [CrossRef]
19. Hasewaga, P.; Bressan, R.A.; Zhu, J.K.; Bohnert, J. Plant cellular and molecular responses to high salinity. *Annu. Rev. Plant Physiol. Plant Mol. Biol.* **2000**, *51*, 463–499. [CrossRef]
20. Alvarez, M.; Savoure, A.; Szabados, L. Proline metabolism as regulatory hub. *Trends Plant Sci.* **2022**, *27*, 39–55. [CrossRef]
21. Xu, X.; Feng, J.; Lü, S.; Lohrey, G.T.; An, H.; Zhou, Y.; Jenks, M.A. Leaf cuticular lipids on the Shandong and Yukon ecotypes of saltwater cress, *Eutrema salsugineum*, and their response to water deficiency and impact on cuticle permeability. *Physiol. Plant.* **2014**, *151*, 446–458. [CrossRef] [PubMed]
22. Inan, G.; Zhang, Q.; Li, P.; Wang, Z.; Cao, Z.; Zhang, H.; Zhang, C.; Quist, T.M.; Goodwin, S.M.; Zhu, J.; et al. Salt cress: A halophyte and cryophyte *Arabidopsis* relative model system and its applicability to molecular genetic analyses of growth and development of extremophiles. *Plant Physiol.* **2004**, *135*, 1718–1737. [CrossRef]
23. Sanchez, D.H.; Siahpoosh, M.R.; Roessner, U.; Udvardi, M.K.; Kopka, J. Plant metabolomics reveals conserved and divergent metabolic responses to salinity. *Physiol. Plant.* **2008**, *132*, 209–219. [CrossRef]
24. Rahman, M.M.; Mostofa, M.G.; Keya, S.S.; Siddiqui, M.N.; Ansary, M.M.U.; Das, A.K.; Rahman, M.A.; Tran, L.S.-P. Adaptive mechanisms of halophytes and their potential in improving salinity tolerance in plants. *Int. J. Mol. Sci.* **2021**, *22*, 10733. [CrossRef] [PubMed]
25. Guevara, D.R.; Champigny, M.J.; Tattersall, A.; Dedrick, J.; Wong, C.E.; Li, Y.; Labbe, A.; Ping, C.L.; Wang, Y.; Nuin, P.; et al. Transcriptomic and metabolomic analysis of Yukon *Thellungiella* plants grown in cabinets and their natural habitat show phenotypic plasticity. *BMC Plant Biol.* **2012**, *12*, 175. [CrossRef] [PubMed]
26. Kazachkova, Y.; Eshel, G.; Pantha, P.; Cheeseman, J.M.; Dassanayake, M.; Barak, S. Halophytism: What have we learnt from *Arabidopsis thaliana* relative model systems? *Plant Physiol.* **2018**, *178*, 972–988. [CrossRef] [PubMed]
27. Amtmann, A. Learning from evolution: *Thellungiella* generates new knowledge on essential and critical components of abiotic stress tolerance in plants. *Mol. Plant* **2009**, *2*, 3–12. [CrossRef] [PubMed]
28. Sévin, D.C.; Stählin, J.N.; Pollak, G.R.; Kuehne, A.; Sauer, U. Global metabolic responses to salt stress in fifteen species. *PLoS ONE* **2016**, *11*, e0148888. [CrossRef] [PubMed]

29. Liu, X.; Grieve, C. Accumulation of chiro-inositol and other non-structural carbohydrates in *Limonium* species in response to saline irrigation waters. *J. Am. Soc. Hortic. Sci.* **2009**, *134*, 8. [CrossRef]
30. Laguna, E. (Ed.) *Flora Endémica, Rara o Amenazada de la Comunidad Valenciana;* Generalitat Valenciana: Valencia, Spain, 1998.
31. IUCN. *The IUCN Red List Categories and Criteria, Version 3.1*, 2nd ed.; IUCN: Gland and Cambridge, UK, 2012.
32. Laguna, E. (Ed.) *Priority Habitats of the Valencian Community;* Generalitat Valenciana: Valencia, Spain, 2003.
33. Fabregat, C.; Ranz, J. *Manual de Identificación de los Hábitats Protegidos de la Comunitat Valenciana (Decreto 70/2009);* Generalitat Valenciana: Valencia, Spain, 2015.
34. Feng, Z.; Sun, Q.; Deng, Y.; Sun, S.; Zhang, J.; Wang, B. Study on pathway and characteristics of ion secretion of salt glands of *Limonium bicolor*. *Acta Physiol. Plant.* **2014**, *36*, 2729–2741. [CrossRef]
35. Santos, J.; Al-Azzawi, M.; Aronson, J.; Flowers, T.J. eHALOPH a database of salt-tolerant plants: Helping put halophytes to work. *Plant Cell Physiol.* **2016**, *57*, e10. [CrossRef]
36. Flowers, T.J.; Galal, H.K.; Bromham, L. Evolution of halophytes: Multiple origins of salt tolerance in land plants. *Funct. Plant Biol.* **2010**, *37*, 604–612. [CrossRef]
37. Dassanayake, M.; Larkin, J.C. Making plants break a sweat: The structure, function, and evolution of plant salt glands. *Front. Plant Sci.* **2017**, *8*, 406. [CrossRef]
38. Grigore, M.N.; Toma, C. Morphological and Anatomical Adaptations of Halophytes: A Review. In *Handbook of Halophytes: From Molecules to Ecosystems towards Biosaline Agriculture;* Grigore, M.N., Ed.; Springer Nature Switzerland AG: Cham, Switzerland, 2021. [CrossRef]
39. Lu, C.; Yuan, F.; Guo, J.; Han, G.; Wang, C.; Chen, M.; Wang, B. Current understanding of role of vesicular transport in salt secretion by salt glands in recretohalophytes. *Int. J. Mol. Sci.* **2021**, *22*, 2203. [CrossRef]
40. Leng, B.Y.; Yuan, F.; Dong, X.X.; Wang, J.; Wang, B.S. Distribution pattern and salt excretion rate of salt glands in two recretohalophyte species of *Limonium* (*Plumbaginaceae*). *S. Afr. J. Bot.* **2018**, *115*, 74–80. [CrossRef]
41. Li, J.; Yuan, F.; Liu, Y.; Zhao, Y.; Wang, B.; Chen, M. Exogenous melatonin enhances salt secretion from salt glands by upregulating the expression of ion transporter and vesicle transport genes in *Limonium bicolor*. *BMC Plant Biol.* **2020**, *20*, 1–11. [CrossRef] [PubMed]
42. Gao, Y.; Zhao, B.; Jiao, X.; Chen, M.; Wang, B.; Yuan, F. Coupled development of salt glands, stomata, and pavement cells in *Limonium bicolor*. *Front. Plant Sci.* **2021**, *12*, 745422. [CrossRef] [PubMed]
43. Mi, P.; Yuan, F.; Guo, J.; Han, G.; Wang, B. Salt glands play a pivotal role in the salt resistance of four recretohalophyte *Limonium* Mill. species. *Plant Biol.* **2021**, *23*, 1063–1073. [CrossRef]
44. Xu, X.; Zhou, Y.; Mi, P.; Wang, B.; Yuan, F. Salt-tolerance screening in *Limonium sinuatum* varieties with different flower colors. *Sci. Rep.* **2021**, *11*, 14562. [CrossRef]
45. Ni, X.-L.; Tan, L.-L.; Shen, X.-D. Developmental and anatomical studies of the salt gland in *Limonium aureum*. *Acta Bot. Boreal. Occid. Sin.* **2012**, *32*, 1587–1591.
46. de Fraine, E. The morphology and anatomy of the genus *Statice*, as represented at Blake ney point. I. *Statice binervosa*, G.E Smith and *Statice bellidifolia* D.C. (= *S. reticulata*). *Ann. Bot.* **1916**, *30*, 239–282. [CrossRef]
47. Colombo, P. Morpho-anatomical and taxonomical remarks on *Limonium* (*Plumbaginaceae*) in Sicily. *Flora Medit.* **2002**, *12*, 389–412.
48. Ding, F.; Chen, M.; Sui, N.; Wang, B.S. Ca^{2+} significantly enhanced development and salt-secretion rate of salt glands of *Limonium bicolor* under NaCl treatment. *S. Afr. J. Bot.* **2010**, *76*, 95–101. [CrossRef]
49. Ding, F. Effects of salinity and nitric oxide donor sodium nitroprusside (SNP) on development and salt secretion of salt glands of *Limonium bicolor*. *Acta Physiol. Plant.* **2013**, *35*, 741–747. [CrossRef]
50. Parida, A.K.; Veerabathini, S.K.; Kumari, A.; Agarwal, P.K. Physiological, anatomical and metabolic implications of salt tolerance in the halophyte *Salvadora persica* under hydroponic culture condition. *Front. Plant Sci.* **2016**, *22*, 351. [CrossRef] [PubMed]
51. Al Hassan, M.; Gohari, G.; Boscaiu, M.; Vicente, O.; Grigore, M.N. Anatomical modifications in two *Juncus* species under salt stress conditions. *Not. Bot. Horti Agrobo.* **2015**, *43*, 501–506. [CrossRef]
52. Munns, R.; Tester, M. Mechanisms of salinity tolerance. *Annu. Rev. Plant Biol.* **2008**, *59*, 651–681. [CrossRef] [PubMed]
53. Flowers, T.J.; Hajibagheri, M.A.; Clipson, N.J.W. Halophytes. *Q. Rev. Biol.* **1986**, *61*, 313–335. [CrossRef]
54. González-Orenga, S.; Grigore, M.-N.; Boscaiu, M.; Vicente, O. Constitutive and induced salt tolerance mechanisms and potential uses of *Limonium* Mill. species. *Agronomy* **2021**, *11*, 413. [CrossRef]
55. Morales, M.A.; Olmos, E.; Torrecillas, A.; Sánchez-Blanco, M.J.; Alarcon, J.J. Differences in water relations, leaf ion accumulation and excretion rates between cultivated and wild species of *Limonium* sp. grown in conditions of saline stress. *Flora* **2001**, *196*, 345–352. [CrossRef]
56. Ding, F.; Song, J.; Ruan, Y.; Wang, B.S. Comparison of the effects of NaCl and KCl at the roots on seedling growth, cell death and the size, frequency and secretion rate of salt glands in leaves of *Limonium sinense*. *Acta Physiol. Plant.* **2009**, *31*, 343–350. [CrossRef]
57. Xianzhao, L.; Chunzhi, W.; Qing, S. Screening for salt tolerance in eight halophyte species from Yellow River Delta at the two initial growth stages. *ISRN Agron.* **2013**, *2013*, 592820. [CrossRef]
58. Souid, A.; Gabriele, M.; Longo, V.; Pucci, L.; Bellani, L.; Smaoui, A.; Abdelly, C.; Hamed, K. Salt tolerance of the halophyte *Limonium delicatulum* is more associated with antioxidant enzyme activities than phenolic compounds. *Funct. Plant Biol.* **2016**, *43*, 607–619. [CrossRef]

59. Hameed, A.; Gulzar, S.; Aziz, I.; Hussain, T.; Gul, B.; Khan, M.A. Effects of salinity and ascorbic acid on growth, water status and antioxidant system in a perennial halophyte. *AoB Plants* **2015**, *7*, plv004. [CrossRef] [PubMed]
60. Al Hassan, M.; Pacurar, A.; López-Gresa, M.P.; Donat-Torres, M.P.; Llinares, J.V.; Boscaiu, M.; Vicente, O. Effects of salt stress on three ecologically distinct *Plantago* species. *PLoS ONE* **2016**, *11*, e0160236. [CrossRef] [PubMed]
61. Grieve, C.M.; Poss, J.A.; Grattam, S.R.; Sheuse, P.J.; Lieth, J.H.; Zeng, L. Productivity and mineral nutrition of *Limonium* species irrigated with saline wastewaters. *Hort. Sci.* **2005**, *40*, 654–658. [CrossRef]
62. González-Orenga, S.; Ferrer-Gallego, P.P.; Laguna, E.; López-Gresa, M.P.; Donat-Torres, M.P.; Verdeguer, M.; Vicente, O.; Boscaiu, M. Insights on salt tolerance of two endemic *Limonium* species from Spain. *Metabolites* **2019**, *9*, 294. [CrossRef] [PubMed]
63. González-Orenga, S.; Donat-Torres, M.; Llinares, J.V.; Navarro, A.; Collado, F.; Ferrer-Gallego, P.; Laguna, E.; Vicente, O.; Boscaiu, M. Multidisciplinary studies supporting conservation programmes of two rare, endangered *Limonium* species from Spain. *Plant Soil* **2021**, *466*, 505–524. [CrossRef]
64. Al Hassan, M.; Estrelles, E.; Soriano, P.; López-Gresa, M.P.; Bellés, J.M.; Boscaiu, M.; Vicente, O. Unraveling salt tolerance mechanisms in halophytes: A comparative study on four Mediterranean *Limonium* species with different geographic distribution patterns. *Front. Plant Sci.* **2017**, *17*, 1438. [CrossRef]
65. González-Orenga, S.; Llinares, J.V.; Al Hassan, M.; Fita, A.; Collado, F.; Lisón, P.; Vicente, O.; Boscaiu, M. Physiological and morphological characterisation of *Limonium* species in their natural habitats: Insights into their abiotic stress responses. *Plant Soil* **2020**, *449*, 267–284. [CrossRef]
66. González-Orenga, S.; Al Hassan, M.; Llinares, J.V.; Lisón, P.; López-Gresa, M.P.; Verdeguer, M.; Vicente, O.; Boscaiu, M. Qualitative and quantitative differences in osmolytes accumulation and antioxidant activities in response to water deficit in four Mediterranean *Limonium* species. *Plants* **2019**, *8*, 506. [CrossRef]
67. Tabot, P.T.; Adams, J.B. Salt secretion, proline accumulation and increased branching confer tolerance to drought and salinity in the endemic halophyte *Limonium linifolium*. *S. Afr. J. Bot.* **2014**, *94*, 64–73. [CrossRef]
68. Flowers, T.J.; Yeo, A.R. Ion relation of salt tolerance. In *Solute Transport in Plant Cells and Tissues*; Baker, D.A., Hall, J.L., Eds.; Longman Scientific and Technical: Harlow, UK, 1988; pp. 39–413.
69. Alarcón, J.J.; Morales, M.A.; Torrecillas, A.; Sánchez-Blanco, M.J. Growth, water relations and accumulation of organic and inorganic solute in the halophytes *Limonium latifolium* cv. Avignon and its interspecific hybrid *Limonium caspia* × *Limonium latifolium* cv. Bettlaard during salt stress. *J. Plant Physiol.* **1999**, *154*, 795–801. [CrossRef]
70. Zia, S.; Egan, T.P.; Khan, M.A. Growth and selective ion transport of *Limonium stocksii* Plumbaginacea under saline conditions. *Pak. J. Bot.* **2008**, *40*, 697–709.
71. Flowers, T.J.; Troke, P.; Yeo, A.R. Mechanism of salt tolerance in halophytes. *Annu. Rev. Plant Physiol.* **1977**, *28*, 89–121. [CrossRef]
72. Greenway, H.; Munns, R. Mechanisms of salt tolerance in non-halophytes. *Annu. Rev. Plant Physiol.* **1980**, *31*, 149–190. [CrossRef]
73. Marschner, P. *Marschner's Mineral Nutrition of Higher Plants*, 3rd ed.; Academic Press: London, UK, 2012; pp. 178–189.
74. Wang, M.; Zheng, Q.; Shen, Q.; Guo, S. The critical role of potassium in plant stress response. *Int. J. Mol. Sci.* **2013**, *14*, 7390. [CrossRef]
75. Al Hassan, M.; Chaura, J.; López-Gresa, M.P.; Borsai, O.; Daniso, E.; Donat-Torres, M.P.; Mayoral, O.; Vicente, O.; Boscaiu, M. Native-invasive plants vs. halophytes in Mediterranean salt marshes: Stress tolerance mechanisms in two related species. *Front. Plant Sci.* **2016**, *7*, 473. [CrossRef]
76. Feng, Z.T.; Deng, Y.Q.; Zhang, S.C.; Liang, X.; Yuan, F.; Hao, J.L.; Zhang, J.C.; Sun, S.F.; Wang, B.S. K(+) accumulation in the cytoplasm and nucleus of the salt gland cells of *Limonium bicolor* accompanies increased rates of salt secretion under NaCl treatment using NanoSIMS. *Plant Sci.* **2015**, *238*, 286–296. [CrossRef]
77. Parida, A.K.; Das, A.B. Salt tolerance and salinity effects on plants: A review. *Ecotoxicol. Environ. Saf.* **2005**, *60*, 324–349. [CrossRef]
78. Manishankar, P.; Wang, N.; Köster, P.; Alatar, A.A.; Kudla, J. Calcium signaling during salt stress and in the regulation of ion homeostasis. *J. Exp. Bot.* **2018**, *24*, 4215–4226. [CrossRef]
79. Seifikalhor, M.; Aliniaeifard, S.; Shomali, A.; Azad, N.; Hassani, B.; Lastochkina, O.; Li, T. Calcium signaling and salt tolerance are diversely entwined in plants. *Plant Signal. Behav.* **2019**, *14*, 1665455. [CrossRef]
80. Carter, C.T.; Grieve, C.M.; Poss, J.A. Salinity effects on emergence, survival, and ion accumulation of *Limonium perezii*. *J. Plant Nutr.* **2005**, *28*, 1243–1257. [CrossRef]
81. Aliniaeifard, S.; Shomali, A.; Seifikalhor, M.; Lastochkina, O. Calcium signaling in plants under drought. In *Salt and Drought Stress Tolerance in Plants. Signaling and Communication in Plants*; Hasanuzzaman, M., Tanveer, M., Eds.; Springer: Cham, Switzerland, 2020. [CrossRef]
82. Gagneul, D.; Aïnouche, A.; Duhazé, C.; Lugan, R.; Larher, F.R.; Bouchereau, A. A reassessment of the function of the so-called compatible solutes in the halophytic Plumbaginaceae *Limonium latifolium*. *Plant Physiol.* **2007**, *144*, 1598–1611. [CrossRef] [PubMed]
83. Rhodes, D.; Hanson, A.D. Quaternary ammonium and tertiary sulfonium compounds in higher plants. *Annu. Rev. Plant Physiol. Plant Mol. Biol.* **1993**, *44*, 357–384. [CrossRef]
84. Giri, J. Glycinebetaine and abiotic stress tolerance in plants. *Plant Signal. Behav.* **2011**, *6*, 1746–1751. [CrossRef] [PubMed]
85. Rhodes, D.; Nadolska-Orczyk, A.; Rich, P. Salinity, osmolytes and compatible solutes. In *Salinity: Environment-Plants-Molecules*; Läuchli, A., Lüttge, U., Eds.; Springer: Dordrecht, The Netherlands, 2002.

86. Khan, M.A.; Ungar, I.A.; Showalter, A.M. The effect of salinity on the growth, water status, and ion content of a leaf succulent perennial halophyte, *Suaeda fruticosa* (L.) Forssk. *J. Arid Environ.* **2000**, *45*, 73–84. [CrossRef]
87. Tipirdamaz, R.; Gagneul, D.; Duhazé, C.; Aïnouche, A.; Monnier, C.; Özkum, D.; Larher, F. Clustering of halophytes from an inland salt marsh in Turkey according to their ability to accumulate sodium and nitrogenous osmolytes. *Environ. Exp. Bot.* **2006**, *57*, 139–153. [CrossRef]
88. Furtana, G.B.; Duman, H.; Tipirdamaz, R. Seasonal changes of inorganic and organic osmolyte content in three endemic *Limonium* species of Lake Tuz (Turkey). *Turk. J. Bot.* **2013**, *37*, 455–463. [CrossRef]
89. Gil, R.; Boscaiu, M.; Lull, C.; Bautista, I.; Lid, N.A.; Vicente, O. Are soluble carbohydrates ecologically relevant for salt tolerance in halophytes? *Funct. Plant Biol.* **2013**, *40*, 805–818. [CrossRef]
90. Aguilella, A.; Fos, S.; Laguna, E. (Eds.) *Catálogo Valenciano de Especies de Flora Amenazada*; Generalitat Valenciana: Valencia, Spain, 2010.
91. Lichtenthaler, H.K.; Wellburn, A.R. Determinations of total carotenoids and chlorophylls a and b of leaf extracts in different solvents. *Biochem. Soc. Trans.* **1983**, *11*, 591–592. [CrossRef]
92. Weimberg, R. Solute adjustments in leaves of two species of wheat at two different stages of growth in response to salinity. *Physiol. Plant.* **1987**, *70*, 381–388. [CrossRef]
93. Bates, L.S.; Waldren, R.P.; Teare, I.D. Rapid determination of free proline for water-stress studies. *Plant Soil* **1973**, *39*, 205–207. [CrossRef]
94. González-Orenga, S.; Leandro, M.E.D.A.; Tortajada, L.; Grigore, M.N.; Llorens, A.A.; Ferrer-Gallego, P.P.; Laguna, E.; Boscaiu, M.; Vicente, O. Comparative studies on the stress responses of two *Bupleurum* (*Apiaceae*) species in support of conservation programmes. *Environ. Exp. Bot.* **2021**, *191*, 104616. [CrossRef]
95. Grieve, C.M.; Grattan, S.R. Rapid assay for determination of water soluble quaternary ammonium compounds. *Plant Soil* **1983**, *70*, 303–307. [CrossRef]
96. Nawaz, K.; Ashraf, M. Exogenous application of glycine betaine modulates activities of antioxidants in maize plants subjected to salt stress. *J. Agron. Crop Sci.* **2010**, *196*, 28–37. [CrossRef]
97. Dubois, M.; Gilles, K.A.; Hamilton, J.K.; Reberd, P.A.; Smith, F. Colorimetric method for determination of sugars and related substances. *Anal. Chem.* **1956**, *28*, 350–356. [CrossRef]
98. Metsalu, T.; Vilo, J. ClustVis: A web tool for visualizing clustering of multivariate data using Principal Component Analysis and heatmap. *Nucleic Acids Res.* **2015**, *43*, W566–W570. [CrossRef] [PubMed]

plants

MDPI

Article

Recovery from Salinity and Drought Stress in the Perennial *Sarcocornia fruticosa* vs. the Annual *Salicornia europaea* and *S. veneta*

Roberta Calone [1,2], Diana-Maria Mircea [2,3], Sara González-Orenga [2], Monica Boscaiu [4], Carla Lambertini [5], Lorenzo Barbanti [1,*] and Oscar Vicente [2]

1 Department of Agricultural and Food Sciences, Alma Mater Studiorum, University of Bologna, Viale Fanin 44, 40127 Bologna, Italy; roberta.calone3@unibo.it

2 Institute for Conservation and Improvement of Valencian Agrodiversity (COMAV), Universitat Politècnica de València, Camino de Vera 14, 46022 Valencia, Spain; dmircea@doctor.upv.es (D.-M.M.); sagonor@doctor.upv.es (S.G.-O.); ovicente@upvnet.upv.es (O.V.)

3 Department of Horticulture and Landscape, University of Agricultural Sciences and Veterinary Medicine of Cluj-Napoca, 3-5 Manastur St., 400372 Cluj-Napoca, Romania

4 Mediterranean Agroforestry Institute (IAM), Universitat Politècnica de València, Camino de Vera 14, 46022 Valencia, Spain; mobosnea@eaf.upv.es

5 Dipartimento di Bioscienze, Università di Milano, Via Celoria 26, 20133 Milano, Italy; carla.lambertini@unimi.it

* Correspondence: lorenzo.barbanti@unibo.it; Tel.: +39-051-20-9-6643

Abstract: Current agricultural problems, such as the decline of freshwater and fertile land, foster saline agriculture development. *Salicornia* and *Sarcocornia* species, with a long history of human consumption, are ideal models for developing halophyte crops. A greenhouse experiment was set up to compare the response of the perennial *Sarcocornia fruticosa* and the two annual *Salicornia europaea* and *S. veneta* to 30 days of salt stress (watering with 700 mM NaCl) and water deficit (complete withholding of irrigation) separate treatments, followed by 15 days of recovery. The three species showed high tolerance to salt stress, based on the accumulation of ions (Na^+, Cl^-, Ca^{2+}) in the shoots and the synthesis of organic osmolytes. These defence mechanisms were partly constitutive, as active ion transport to the shoots and high levels of glycine betaine were also observed in non-stressed plants. The three halophytes were sensitive to water stress, albeit *S. fruticosa* to a lesser extent. In fact, *S. fruticosa* showed a lower reduction in shoot fresh weight than *S. europaea* or *S. veneta*, no degradation of photosynthetic pigments, a significant increase in glycine betaine contents, and full recovery after the water stress treatment. The observed differences could be due to a better adaptation of *S. fruticosa* to a drier natural habitat, as compared to the two *Salicornia* species. However, a more gradual stress-induced senescence in the perennial *S. fruticosa* may contribute to greater drought tolerance in this species.

Keywords: *Sarcocornia fruticosa*; *Salicornia europaea*; *Salicornia veneta*; halophytes; salt stress; drought stress; stress recovery; osmolytes; ion transport; oxidative stress markers

Citation: Calone, R.; Mircea, D.-M.; González-Orenga, S.; Boscaiu, M.; Lambertini, C.; Barbanti, L.; Vicente, O. Recovery from Salinity and Drought Stress in the Perennial *Sarcocornia fruticosa* vs. the Annual *Salicornia europaea* and *S. veneta*. *Plants* **2022**, *11*, 1058. https://doi.org/10.3390/plants11081058

Academic Editors: Josefa M. Alamillo and Anna Aksmann

Received: 14 February 2022
Accepted: 9 April 2022
Published: 13 April 2022

Publisher's Note: MDPI stays neutral with regard to jurisdictional claims in published maps and institutional affiliations.

1. Introduction

In response to the current increase in world population, agriculture is called to address two major but opposite needs: increasing food production while decreasing its negative environmental impacts. Boosting food security through sustainable agricultural practices represents a priority objective for the 2030 Agenda for Sustainable Development [1], a goal that, to date, is even more urgent, considering that, in 2020, the number of undernourished people worldwide has increased by 83–132 million due to the COVID-19 pandemic [2]. However, the growing competition for land and water caused by the dramatic expansion of cities [3], in conjunction with the increasingly recurrent phenomena of soil erosion, water scarcity, and loss of agrobiodiversity, are posing serious obstacles to achieving this objective.

The Mediterranean basin is amongst the areas most threatened by salinisation in the world due to climate change [4]. According to the Intergovernmental Panel on Climate Change, in the Mediterranean region, temperatures will rise by 2–4 °C, and rainfall will decrease between 4% and 30% by 2050 [5], whereas sea level is expected to increase by approximately 35 cm by 2100 [6]. The projected climate changes will also exacerbate the salt accumulation processes driven by seawater intrusion in the coastal shallow aquifers, which in turn will constrain soil fertility and crop productivity.

In 2009, the World Bank introduced the concept of climate-smart agriculture (CSA), referring to an integrated approach to address the complex nexus of climate change, food security, and sustainable development [7]. Today, the FAO Strategic Framework 2022–2031 considers the transition to CSA imperative to improve agricultural resilience and productivity and lower its climate footprint and costs [8]. The CSA approach is implemented through three priority lines of action: firstly, boosting sustainable agricultural production to support increased incomes and food security; secondly, increasing agroecosystems' adaptive capacity; and thirdly, reducing greenhouse gas emissions while increasing carbon sequestration [9].

The CSA applications are context-specific, depending on the local socio-political, financial, and environmental context, and encourage the integration of new technologies and practices such as precision farming tools, decision support systems for land and water management, conservative and organic crop practices, integrated pest and disease management, and the introduction of drought-, salt-, and flood-tolerant crops [10]. In this last regard, the Mediterranean region represents a precious hotspot of biodiversity, with a remarkable richness in cultivated and native wild plants that have adapted to various unfavourable conditions such as prolonged drought, salinity, and flooding.

Halophytes are extremophile plants that can tolerate harsh conditions and salinity levels toxic to most plants. Within the CSA framework, the study of halophytes' stress tolerance mechanisms is an outlooking strategy for improving crop resilience to environmental stress. Besides providing valuable scientific models, these plants can be cultivated for the direct production of food, fodder, biomass and medicinal compounds, as well as for soil phytoremediation, carbon sequestration, and landscaping purposes, including the recovery of marginal saline soils and water [11,12]. About 1100 halophyte species occur in the Mediterranean Basin, when considered in its broadest meaning, i.e., from the Aral Sea to the Atlantic Ocean [13]. Taxonomical, biological, and ecological diversity is high here, and there are traditional and new potential uses of these plants.

The subfamily *Salicornioideae* includes around 100 species of succulent halophytes, the *Sarcocornia/Salicornia* lineage being one of the most important in terms of species diversity [14]. This lineage consists of hygro-halophytes diversified during the Middle Miocene [15] and was confirmed by transcribed spacer (ITS) and atpB–rbcL spacer sequences as monophyletic, being clearly separated from other taxa [15]. Molecular phylogenetic studies based on external transcribed spacer (ETS) sequence revealed that this lineage comprises three primary clades: *Salicornia*, American-Eurasian *Sarcocornia*, and South African-Australian *Sarcocornia* [16]. The genus *Sarcocornia* A.J. Scott was separated from *Salicornia* L. and *Arthrocnemum* Moq. on the basis of morphological characters [17]. The *Salicornia* and *Sarcocornia* genera are morphologically similar and can be distinguished only by inflorescence characters and their life form, the former including only annuals and the latter only perennials. *Salicornia* is clearly a monophyletic genus, as revealed by ETS sequence data [16], whereas *Sarcocornia* remains unresolved as possibly paraphyletic [14]. Annual *Salicornia* species evolved from the perennial *Sarcocornia* during Miocene, and their high self-fertility allowed their rapid expansion, colonising coastal and inland remote habitats [14,16].

Three species of the *Sarcocornia/Salicornia* lineage were selected for this study. *Salicornia europaea* L. belongs to a diploid clade including genotypes that show a wide geographical distribution. *S. veneta* Pign. et Lausi is a member of the well-supported monophyletic group of *Salicornia dolichostachya* Moss with very little genetic variation among its taxa [16,18]. The species is endemic to NE Italy in the area of the Lagoon of Venice and West Slovenia and is classified as vulnerable according to the UICN criteria [19]. The third species under study, *Sarcocornia fruticosa* (L.) A.J. Scott with Mediterranean distribution, belongs to the Eurasian clade of *Sarcocornia* [14]. The three species are morphologically similar, with succulent and articulate stems, reduced leaves, and inflorescences of minute reduced flowers. Their young, fleshy tips are edible and commercialised with the name of "samphire", "sea asparagus", "pickleweed", or "poor man's asparagus" [20]. Thanks to the crunchy texture and salty taste, their succulent shoots are highly appreciated in gourmet cuisine [21–23]. Moreover, they are a good source of fibre, antioxidants, and anti-inflammatory metabolites, such as vitamin C and polyphenolic compounds, making them an ideal nutraceutical supplement [23,24]. These species are also appreciated as oil-seed crops. Indeed, oil extracted from their seeds is rich in polyunsaturated fatty acids, particularly oleic and linoleic acid, having valuable health properties [25]. Furthermore, these species can produce high amounts of biomass rich in lignocellulosic materials suitable for bioethanol production [26]. The high biomass production, combined with the high phytoextraction capacity, also makes these species very attractive for the phytoremediation of saline and heavy metal-contaminated soils [27]. Finally, several studies have demonstrated their suitability for the regreening of marginal areas to increase carbon sequestration and relieve soil erosion [28,29].

Without salt glands or salt bladders, the strategy of glassworts to tolerate the ionic and osmotic components of salt stress relies largely on the massive accumulation and vacuolar compartmentalisation of Na^+ and Cl^- [30–33], which allow them to maintain the osmotic potential necessary to drive water uptake into cells while preventing ion-related cytotoxic effects. Moreover, they have evolved the ability to increase succulence in shoots diluting the accumulated ions [34], synthesise compatible solutes for osmotic adjustment, especially glycine betaine [34–37], produce ROS-scavenging enzymes and compounds [38,39], maintain high K-Na selectivity [33], and effectively regulate ammonium detoxification processes under stress conditions [40]. Furthermore, glassworts have the ability to transit from green to reddish colouration through the accumulation of red-violet pigments and betacyanins, which allow them to cope with excessive light energy in the photosystems when the plants experience osmotic stress and photosynthesis declines by dissipating excess excitation energy into heat [41].

In their natural habitats, halophytes are subjected to wide seasonal oscillations in precipitations and temperature, and therefore in soil moisture and salinity, which result in periods of high and low stress intensity that alternate during the year [42]. Significantly stressful conditions at the field level, however, are often only transient and rarely cause plant death as more favourable conditions usually return, although they often result in reduced crop yield [43]. However, basic studies on stress tolerance in halophytes have generally focused on their responses to different applied stress treatments, and very little is known on the equally important mechanisms of stress recovery, which are essential for ensuring sustainable crop production under intermittent stress events.

The focus of the present study was to analyse differences between the three aforementioned Salicornioideae species in their responses to stress and stress recovery treatments, which could be due to differences in the plants' life cycle or native environments. For this, we determined growth parameters in plants of the investigated species after applying controlled salt and water deficit treatments in a greenhouse, followed by irrigation with non-saline water. To obtain insights into their stress tolerance mechanisms, growth responses were correlated with changes in the levels of specific biochemical stress markers, such as photosynthetic pigments, different mono and divalent ions and organic osmolytes, oxidative stress markers, and antioxidant compounds.

2. Results

2.1. Substrate Electric Conductivity and Moisture

During the stress period, the substrate electric conductivity (EC) increased significantly in the pots subjected to salt stress, reaching over 15 dS m^{-1} for all three halophytes, with a maximum of 21 dS m^{-1} in *S. fruticosa*, whereas the water stress treatment did not cause any change in the control EC values (Figure 1A). After 15 days of watering the pots with non-saline water ('recovery' treatment), the substrate EC in salt-treated pots decreased to control values (for *S. europaea* and *S. veneta*) or even slightly (but significantly) below the control for *S. fruticosa*. However, substrate salinity in the pots previously subjected to the withholding of irrigation remained similar to the controls after recovery (Figure 1A).

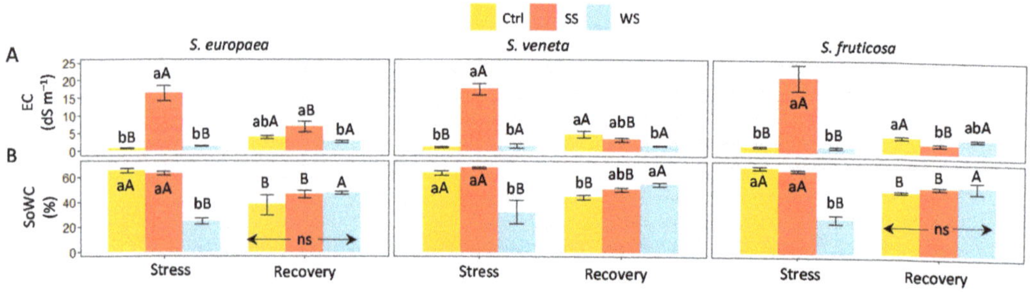

Figure 1. Effect of 30 days of stress treatments (Stress), followed by watering with non-saline water for 15 days (Recovery) on (**A**) Substrate electrical conductivity (soil EC) and (**B**) water content (soil WC). Ctrl, control; SS, salt stress (watering with 700 mM NaCl); WS, water stress (complete withholding of irrigation). For each species and sampling (Stress or Recovery), different lowercase letters over the bars indicate significant differences between treatments (Ctrl, SS, and WS) at $p \leq 0.05$. Different uppercase letters indicate significant differences between the two sampling times (Stress and Recovery) for each species and treatment, at $p \leq 0.05$. Vertical bars indicate standard error (n = 4). ns: non-significant.

Contrary to the EC data, the substrate water content, with control values of about 65% for all three halophytes, was not affected by the salt treatment; however, soil moisture decreased significantly under water deficit conditions, down to between 25 and 30%, depending on the species (Figure 1B). After recovery from water stress, substrate moisture increased to reach values equal or even higher (in *S. veneta*) than the controls, whereas recovery from salt stress did not alter the soil water content when compared to the corresponding controls (Figure 1B).

2.2. Plant Growth

Plant height and the number of branches were measured in all plants at the beginning (T0) and every 15 days during the experiments; that is, after 15 and 30 days of water or salt stress and at the end of the 'recovery' treatment (Table 1). Both parameters increased significantly during the stress treatments in control and stressed plants. The salt treatment did not cause significant growth inhibition in any of the three species. In contrast, compared to the control, water deficit induced a significant plant height reduction in the two *Salicornia* species and also a reduction (down to 57% of the control) in the number of branches in *S. europaea*. However, this inhibitory effect was only observed after 30 days of withholding irrigation, not at day 15 of the treatment (Table 1). These data indicate a strong tolerance of the three species to salinity, even at very high salt concentrations (700 mM NaCl), and a slightly higher drought sensitivity of the two *Salicornia* species compared to *Sarcocornia fruticosa*.

After 15 days of recovery, the plant height and the number of branches of *S. europaea* and *S. veneta* plants were statistically homogeneous in all treatments (control and water and salt stress); the same result was observed for plant height in *S. fruticosa*. The number of branches increased during recovery in the latter species but to a lesser extent in the previously stressed plants, which did not reach the control values (Table 1).

Table 1. Plant height (cm) and number of branches in the three halophytes (SE, *S. europaea*; SV, *S. veneta*; SF, *S. fruticosa*) measured at the beginning (T0) and after 15 (T15), 30 (T30), or 45 (T45) days of starting the stress treatments. Ctrl, control; SS, salt stress (watering with 700 mM NaCl); WS, water stress (complete withholding of irrigation). The values shown are means ± SE (n = 4). For each species, different lowercase letters in a column indicate significant differences between the three treatments within the same sampling time, whereas different uppercase letters in each row indicate significant differences between sampling times for the same treatment, at $p \leq 0.05$.

		Plant Height (PH) (cm)				Number of Branches (No. B)			
		T0	Stress (T15)	Stress (T30)	Recovery (T45)	T0	Stress (T15)	Stress (T30)	Recovery (T45)
SE	Ctrl	5.8 ± 0.3 aC	8.7 ± 0.5 aB	12.9 ± 1.0 aA	12.3 ± 1.0 aA	5.7 ± 0.5 aC	10.2 ± 0.7 aB	18.1 ± 1.8 aA	23.8 ± 3.3 aA
	SS	5.7 ± 0.3 aC	8.5 ± 0.4 aB	11.4 ± 0.5 aA	10.5 ± 0.6 aA	6.3 ± 0.5 aC	11.0 ± 0.7 aB	18.2 ± 1.2 aA	20.8 ± 1.5 aA
	WS	5.3 ± 0.3 aC	7.6 ± 0.4 aB	7.0 ± 0.6 bA	10.4 ± 0.9 aA	6.0 ± 0.5 aB	12.0 ± 2.1 aAB	10.3 ± 1.5 bA	16.9 ± 2.4 aA
SV	Ctrl	9.8 ± 0.4 aC	14.7 ± 0.7 aB	21.6 ± 2.0 aA	22.6 ± 1.8 aA	2.1 ± 0.3 aC	6.9 ± 0.6 aB	11.8 ± 1.6 aA	13.0 ± 1.7 aA
	SS	10.2 ± 0.4 aC	15.8 ± 0.5 aB	20.6 ± 0.8 abA	19.9 ± 1.5 aA	1.5 ± 0.3 aC	8.0 ± 0.6 aB	10.1 ± 0.9 aA	15.0 ± 3.4 aA
	WS	9.6 ± 0.5 aC	15.1 ± 0.5 aB	16.1 ± 0.6 bA	19.6 ± 1.5 aA	1.5 ± 0.3 aC	7.3 ± 0.5 aB	9.5 ± 0.8 aB	9.3 ± 2.4 aA
SF	Ctrl	5.6 ± 0.3 aC	8.9 ± 0.5 aB	13.1 ± 1.3 aA	14.6 ± 1.3 aA	0.4 ± 0.2 aD	8.1 ± 1.0 aC	18.9 ± 3.0 aB	28.5 ± 1.5 aA
	SS	5.1 ± 0.4 aC	9.2 ± 0.4 aB	11.6 ± 0.7 aA	13.1 ± 0.6 aA	0.4 ± 0.2 aC	8.8 ± 1.0 aB	18.4 ± 2.3 aA	21.7 ± 1.7 bA
	WS	5.0 ± 0.3 aC	8.3 ± 0.5 aB	10.0 ± 0.9 aA	12.6 ± 0.7 aA	0.5 ± 0.2 aD	8.7 ± 1.0 aC	14.0 ± 2.4 aB	20.8 ± 2.5 bA

After the stress and recovery periods, plants were harvested to determine shoot fresh weight (FW) and water content percentage (WC) as the most reliable parameters to assess the treatment effects on plant growth. Salt stress did not affect the shoot FW or WC of the *Salicornia* species significantly, whereas *S. fruticosa* plants appeared to be slightly more affected, with a more accentuated (but still non-significant) reduction in the mean FW and a slight (but significant) reduction in WC (Figure 2A,B). On the other hand, water stress strongly reduced shoot FW in the three species (Figure 2A), partly due to plant dehydration, as it was accompanied by a small but significant WC decrease compared to the control plants (Figure 2B).

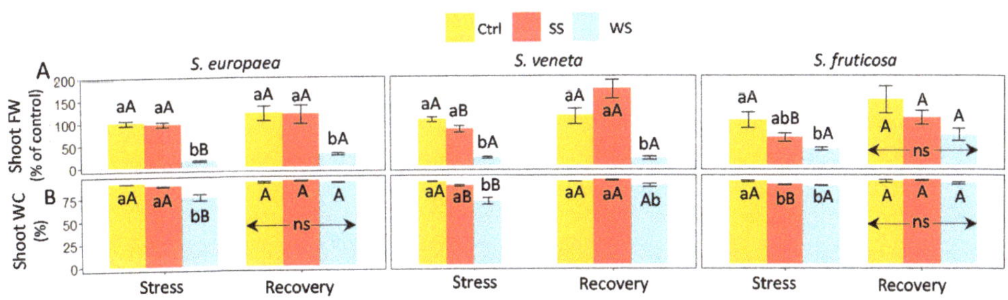

Figure 2. Effect of 30 days of stress treatments (Stress), followed by watering with non-saline water for 15 days (Recovery) on (**A**) shoot fresh weight (FW) and (**B**) shoot water content (SWC) in the three halophytes. Ctrl, control; SS, salt stress (watering with 700 mM NaCl); WS, water stress (complete withholding of irrigation). For each species and sampling (Stress or Recovery), different lowercase letters over the bars indicate significant differences between treatments (Ctrl, SS, and WS), whereas different uppercase letters indicate significant differences between the two samplings (Stress and Recovery) for each species and treatment, at $p \leq 0.05$. Vertical bars indicate standard error (n = 4). Values in (**A**) are shown as percentages of shoot FW of control plants (Ctrl, Stress), taken as 100%; the corresponding absolute values for *S. europaea*, *S. veneta*, and *S.fruticosa* were 13.3, 10.1, and 5.6 g plant^{-1}, respectively. ns: non-significant.

After recovery, the salt-stressed plants of the three halophytes maintained a shoot FW and WC similar to their corresponding controls. However, watering with non-saline water had distinct effects on plants previously subjected to water deficit, depending on the species. Thus, *S. europaea* plants showed a significant increase in FW upon recovery, but with values still well below those of the control plants and the complete rehydration of the shoots; in contrast, no significant effects were observed in *S. veneta*. Only in *S. fruticosa* did shoot FW not show any statistically significant differences from the control after recovery, although the mean value was lower (Figure 2). Therefore, confirming the measurements of other growth parameters, *S. fruticosa* appears to be more tolerant to drought than the *Salicornia* species and also shows better recovery from the water deficit treatment.

2.3. Photosynthetic Pigments

Mean values of photosynthetic pigment contents showed a decreasing trend in response to the salt treatment in plants of the two annual *Salicornia* species (Figure 3); however, the differences with the non-stressed plants were only significant for chlorophyll a (Chl. a) in *S. europaea* (Figure 3A) and carotenoids (Caro) in *S. veneta* (Figure 3C), whereas no variations in chlorophyll b (Chl. b), the second most abundant chlorophyll in oxygenic photosynthetic organisms, were recorded. After irrigation with non-saline water, no significant differences with the controls were found for any pigment. In contrast, water deficit caused a significant reduction in the levels of the three pigments in both annual species; in all cases, mean pigment contents increased after the recovery treatment, reaching values not significantly different from the controls. On the other hand, in the perennial *S. fruticosa*, neither salt nor water stress induced any significant variation in pigment concentrations, and the recovery treatment had no effect, except for a slight yet significant increase in Caro levels in salt-treated plants. However, it should be mentioned that the pigment levels in the *S. fruticosa* control plants were lower than those determined in *S. europaea* and *S. veneta* (Figure 3). These responses agree with the observed stress-induced changes in growth parameters, confirming the high salt tolerance of the three species, the relatively higher drought tolerance of *S. fruticosa* compared to the annual species, and the effectiveness of the recovery treatment.

2.4. Ion Accumulation

Root and shoot Na^+ and Cl^- concentrations increased significantly in response to the salt stress treatment in the three halophytes, as expected, whereas water deficit did not have any effect on the ions levels. The recovery treatment reduced the contents of both ions in roots of salt-stressed plants down to control levels, except for Na^+ in *S. veneta*, which showed a still significant but less accentuated decrease. In contrast, no differences were observed in shoot Na^+ or Cl^- contents before and after recovery, except *for S. europaea*, in which Cl^- content increased slightly but significantly in the control. Under all tested conditions, the concentrations of both ions were substantially higher in shoots than in roots (Figure 4A,B).

Variations of K^+ concentrations showed different patterns, depending on the species and the treatments (Figure 4C). First, control levels in the roots of non-stressed plants differed substantially between species, being the highest in *S. veneta*—about 1.7-fold higher than in *S. europaea* and three-fold higher than in *S. fruticosa*, approximately. Shoot K^+ contents were similar to those in roots in *S. europaea*, whereas they were higher in shoots than in roots in *S. veneta* and *S. fruticosa*. The stress treatments did not cause changes in the root K^+ concentration, except for the significant decrease observed in salt-stressed *S. veneta* plants. At the shoot level, mean K^+ concentrations decreased upon salt treatment, although the difference with the control was non-significant in *S. europaea*. Under water stress, K^+ contents increased, decreased, and remained the same as in the controls in *S. europaea*, *S. veneta*, and *S. fruticosa*, respectively. After recovery, K^+ concentrations were generally lower than control values in the roots and shoots of salt-stressed plants and not significantly different from the controls in plants previously subjected to water stress, although some exceptions to this general behaviour were observed in *S. europaea* (Figure 4C).

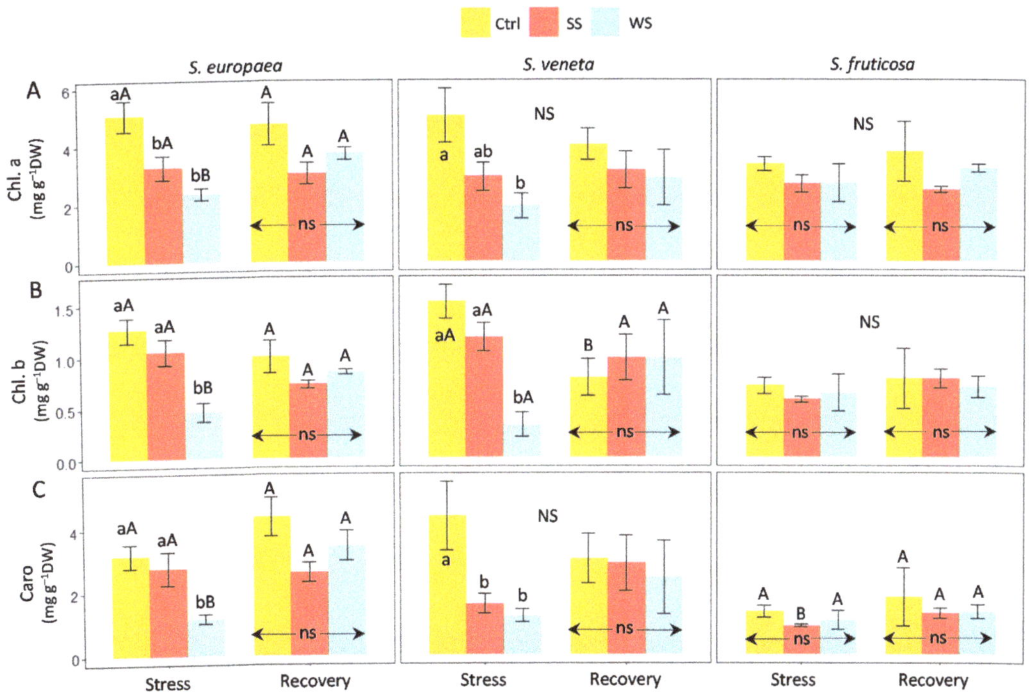

Figure 3. Effect of 30 days of stress treatments (Stress), followed by watering with non-saline water for 15 days (Recovery) on (**A**) chlorophyll a (Chl. a), (**B**) chlorophyll b (Chl. b), and (**C**) carotenoids (Caro) in the three halophytes. Ctrl, control; SS, salt stress (watering with 700 mM NaCl); WS, water stress (complete withholding of irrigation). For each species and sampling (Stress or Recovery), different lowercase letters over the bars indicate significant differences between treatments (Ctrl, SS, and WS), at $p \leq 0.05$; ns: non-significant. Different uppercase letters indicate significant differences between the two samplings (Stress and Recovery) for each species and treatment, at $p \leq 0.05$; NS: non-significant. Vertical bars indicate standard error (n = 4).

The patterns of Ca^{2+} variation in the roots of the three species were similar to those observed for Na^+ and Cl^-, that is, a significant increase in response to salt stress and no effect of water stress except for an increase in *S. europaea* (Figure 4D). Shoot Ca^{2+} concentration significantly increased in the salt-treated plants of *S. veneta* and *S. fruticosa*, but not of *S. europaea*, with no effect of water stress. After the recovery period, root Ca^{2+} concentration in the salt-stressed plants decreased but remained significanlty higher than in control plants, and was statistically comparable with the water-stressed plants. In shoots, the Ca^{2+} concentration did not vary after recovery, except for an increase in the salt-treated plants of *S. veneta* (Figure 4D).

2.5. Osmolytes, Oxidative Stress Markers and Antioxidants

Common osmolytes, glycine betaine (GB), proline (PRO), and total soluble sugars (TSS) were determined and showed distinct accumulation patterns in the shoots of the selected species (Figure 5). Neither salt stress nor water deficit caused any significant change in GB contents in *S. europaea*; they augmented three-fold over control values in salt-stressed *S. veneta* and about 2.5-fold in *S. fruticosa* plants subjected to water stress.

After the recovery period, the GB level increased significantly in non-stressed *S. europaea* and *S. veneta* plants and decreased in those of *S. fruticosa* that underwent the water deficit treatment. Nevertheless, no significant differences between treatments were found in the shoot GB contents of any of the three halophytes after recovery (Figure 5A).

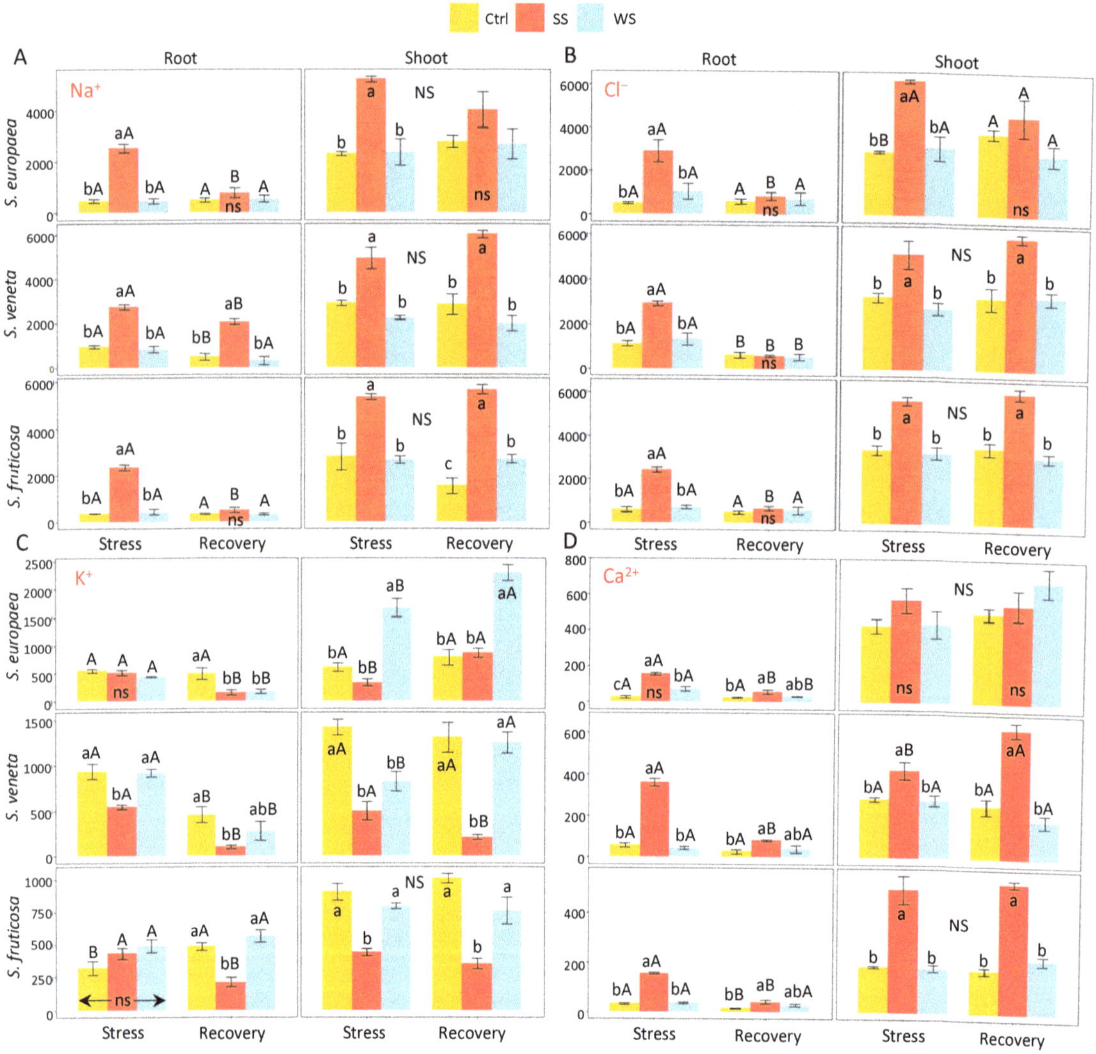

Figure 4. Effect of 30 days of stress treatments (Stress) followed by watering with non-saline water for 15 days (Recovery) on the root and shoot concentration (in µmol g^{-1} DW) of ions: (**A**) sodium (Na^+), (**B**) chloride (Cl^-), (**C**) potassium (K^+), and (**D**) calcium (Ca^{2+}) in the three halophytes. Ctrl, control; SS, salt stress (watering with 700 mM NaCl); WS, water stress (complete withholding of irrigation). For each species and sampling (Stress or Recovery), different lowercase letters over the bars indicate significant differences between treatments (Ctrl, SS, and WS), at $p \leq 0.05$; ns: non-significant. Different uppercase letters indicate significant differences between the two samplings (Stress and Recovery) for each species and treatment, at $p \leq 0.05$; NS: non-significant. Vertical bars indicate standard error (n = 4).

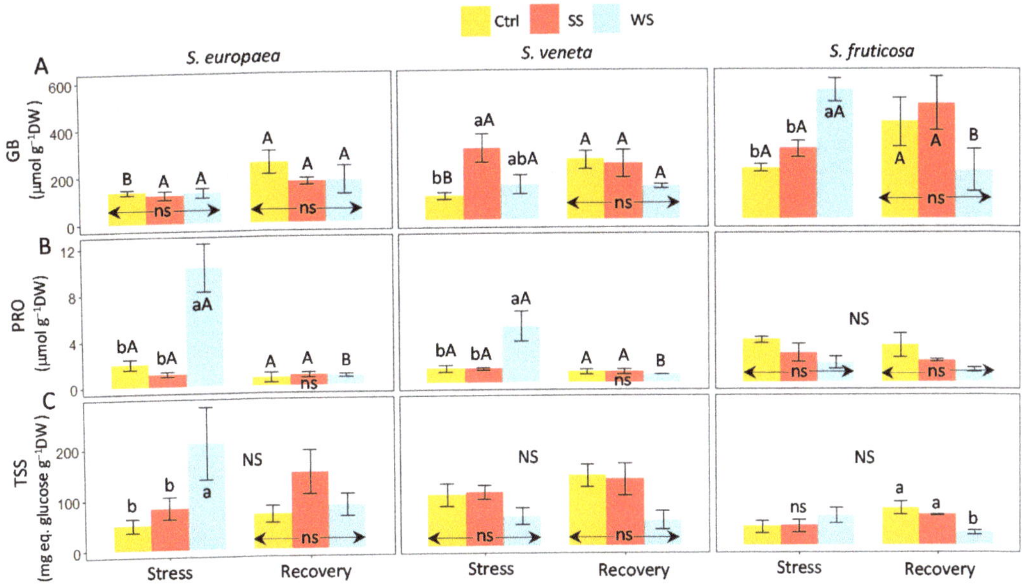

Figure 5. Effect of 30 days of stress treatments (Stress) followed by watering with non-saline water for 15 days (Recovery) on shoot concentration of (**A**) glycine betaine (GB), (**B**) proline (PRO), and (**C**) Total Soluble Sugars (TSS) in the three halophytes. Ctrl, control; SS, salt stress (watering with 700 mM NaCl); WS, water stress (complete withholding of irrigation). For each species and sampling (Stress or Recovery), different lowercase letters over the bars indicate significant differences between treatments (Ctrl, SS, and WS), at $p \leq 0.05$; ns: non-significant. Different uppercase letters indicate significant differences between the two samplings (Stress and Recovery) for each species and treatment, at $p \leq 0.05$; NS: non-significant. Vertical bars indicate standard error (n = 4).

PRO contents did not vary in any species in response to salt stress but increased in the water-stressed plants of *S. europaea* (about five-fold over the control) and, to a lesser extent, *S.veneta* (ca. four-fold). In these two *Salicornia* species, PRO levels decreased to control values after the recovery period, so that, in all cases, the differences between treatments became non-significant. In *S. fruticose*, no variation in PRO contents was observed, for any of the samples, after the stress treatments and after recovery (Figure 5B). Under all experimental conditions, PRO concentrations in molar terms were much lower than those of GB in the three species. GB contents ranged between 100 and more than 500 µmol g^{-1} DW, whereas the maximum measured PRO level (in water-stressed *S. europaea* plants) was only ca. 10 µmol g^{-1} DW (Figure 5A,B).

Only the water-stressed *S. europaea* plants showed a significant increase in shoot TSS levels; all other differences between control and stressed plants in the stress and recovery treatments, or between the two samplings, were non-significant (Figure 5C).

To assess the possible generation of secondary oxidative stress in the plants subjected to salt or water stress treatments, the contents of two reliable biochemical markers, malondialdehyde (MDA) and hydrogen peroxide (H_2O_2), were determined in the shoots of all plants (Figure 6). No increase in MDA or H_2O_2 levels was detected in any of the samples from the stressed plants in relation to the non-stressed controls. MDA contents even decreased in some cases, namely under salt stress in *S. europaea* and under water stress in *S. veneta*. In contrast, no differences in H_2O_2 content between stressed and control plants were detected in the three species. A significant increase in MDA concentration was observed after the recovery period in the salt-stressed plants of *S. europaea* and *S. fruticosa* and in the water-stressed plants of *S. europaea* and *S. veneta*. On the other hand, H_2O_2 levels increased after recovery in the salt-treated plants of *S. veneta* and *S. fruticosa* (Figure 6).

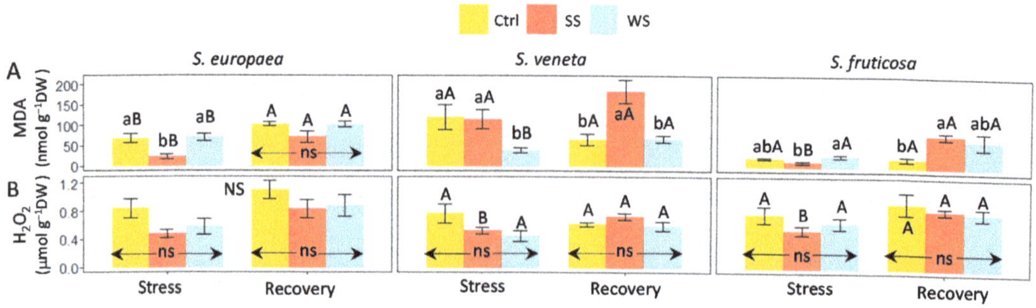

Figure 6. Effect of 30 days of stress treatments (Stress) followed by watering with non-saline water for 15 days (Recovery) on shoot concentration of (**A**) Malondialdehyde (MDA) and (**B**) hydrogen peroxide (H_2O_2) in the three halophytes. Ctrl, control; SS, salt stress (watering with 700 mM NaCl); WS, water stress (complete withholding of irrigation). For each species and sampling (Stress or Recovery), different lowercase letters over the bars indicate significant differences between treatments (Ctrl, SS, and WS), at $p \leq 0.05$; ns: non-significant. Different uppercase letters indicate significant differences between the two samplings (Stress and Recovery) for each species and treatment, at $p \leq 0.05$; NS: non-significant. Vertical bars indicate standard error ($n = 4$).

In agreement with the lack of a detectable generation of oxidative stress under high salinity and water deficit conditions, the activation of the synthesis of common antioxidant compounds, such as phenolic compounds (TPC) and, particularly, the subgroup of flavonoids (TF), was also not observed. Indeed, differences in TPC and TF contents between treatments during the stress and recovery periods were generally non-significant, except for the TF reduction in response to salt in *S. fruticosa*. Moreover, no differences were detected between samplings for each treatment (Figure 7).

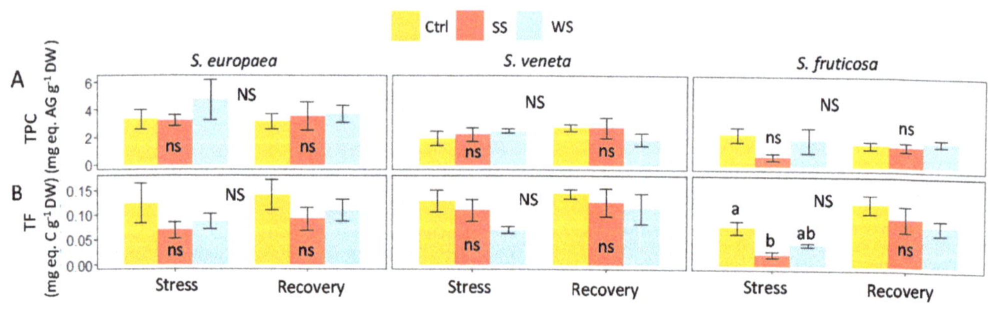

Figure 7. Effect of 30 days of stress treatments (Stress) followed by watering with non-saline water for 15 days (Recovery) on shoot concentration of (**A**) total phenolic compounds (TPC) and (**B**) total flavonoids (TF) in the three halophytes. Ctrl, control; SS, salt stress (watering with 700 mM NaCl); WS, water stress (complete withholding of irrigation). For each species and sampling (Stress or Recovery), different lowercase letters over the bars indicate significant differences between treatments (Ctrl, SS, and WS), at $p \leq 0.05$; ns: non-significant. Different uppercase letters indicate significant differences between the two samplings (Stress and Recovery) for each species and treatment, at $p \leq 0.05$. NS: non-significant. Vertical bars indicate standard error ($n = 4$).

2.6. Physiological Traits Relationships and Results of the Multivariate Analysis

In the three surveyed species, some common trait patterns could be observed (Figure 8). The pigments, namely Chl. a, Chl. B, and Caro, were positively correlated with each other in all three species, indicating their covariation. The potassium shoot concentration, K(s), instead, always resulted in being negatively correlated with Na(r), Na(s), and Cl(s). Plant FW was consistently positively correlated with the shoot water content (SWC), which was positively associated with Chl. a and Caro contents in the two annual plants. Furthermore, SWC in the two *Salicornia* species was negatively correlated with PRO, as the content of this osmolyte mostly increased under water stress, when the plant SWC was the lowest.

The near absence of significant correlations between TPC and other growth-related traits confirmed that salinity and water deficit, under our experimental conditions, did not generate a substantial degree of oxidative stress in the plants.

Two principal component analyses (PCAs) were performed to further evaluate the relationships among traits after the stress (PCAstress) and recovery (PCArecovery) treatments and to quantify the strength and direction of correlations between the original traits and the extrapolated principal components (PCs). The first three PCs (eigenvalues are reported in Table S1 of the Supplementary Materials) explained 62% and 53% of the total variance in PCAstress and PCArecovery, respectively, and were used for PCA interpretation. The correlation circles and the biplots of the first two components, PC1 and PC2, and the variables measured after the 30 days of stress (PCAstress) and the 15 days of recovery (PCArecovery) are reported in Figure 9.

In PCAstress, PC1 accounted for the differences between the salt stress treatment, whose barycentre was located on the positive side of PC1, and the water stress treatment, whose barycentre was located on the negative side of PC1 (Figure 9B). PC1 was positively correlated with Na(r) (0.87), Na(s) (0.85), Cl(s) (0.82), Cl(r) (0.79), Ca(r) (0.75), and FW (0.62), and negatively correlated with K(s) (−0.63) and PRO (−0.56) (Figure 9A), meaning that the accumulation of Na, Cl, and Ca is the primary mechanism helping to sustain plant growth under salt stress, whereas PRO production and K(s) accumulation are the main mechanisms adopted under water stress.

PC2 showed the relationship between Na^+ and Cl^- accumulation, pigment production, and oxidative stress. PC2, indeed, presented the strongest positive correlations with Caro (0.84), Chl. a (0.83), Chl. b (0.75), and the highest negative correlations with Na(s) (−0.39) and Cl(s) (−0.39) (Figure 9A), meaning that the accumulation of these ions interfered with the production of pigments. Interestingly, the barycentres of the two annual species were located on the positive side of the PC2 axes, whereas the barycentre of *S. fruticosa* was located on the negative side (Figure 9B), indicating that pigment production was less affected by ion accumulation in this latter species.

PC3, finally, summarised the relationship between the plant species and the osmolytes. This third component was positively correlated with TSS (0.78), PRO (0.54), and TPC (0.42), and negatively correlated with GB (−0.51) (Table S2 of Supplementary Materials). *S. europaea* and *S. veneta* barycentres were placed on the positive side of the PC3 axis, whereas *S. fruticosa* was in the negative one (Table S3 of Supplementary Materials). This may suggest that the annual species rely on the production of sugars, proline, and phenolic compounds for osmotic adjustment under stress conditions, whereas the perennial species depends more on glycine betaine accumulation for its stress tolerance.

The PCArecovery outlined some evident changes: as in the PCAstress, the PC1 accounted for the different effects of the stress treatments, with the salt stress barycentre placed on the positive side of the PC1 axis and the water stress and control barycentres clustered on the negative side (Figure 9D), suggesting that, after recovery, water-stressed plants behaved similarly to control plants. PC1 was correlated positively with Na(r) (0.81), whose concentration decreased after recovery, especially in salt-treated plants, and negatively with K(r) (−0.55) (Figure 9C), whose concentration decreased after recovery, especially in the annual water-stressed plants.

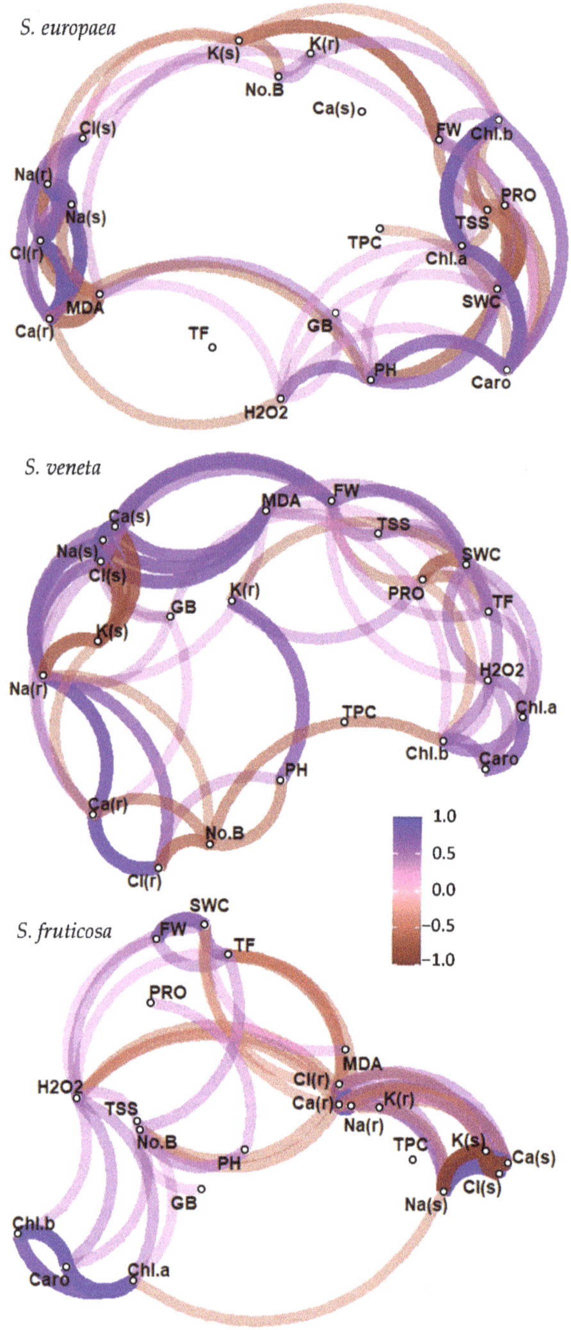

Figure 8. Correlation network diagram showing significant correlations ($p < 0.05$) between the 22 measured traits within each halophyte species, based on the calculation of the Pearson correlation coefficients. Each measured trait represents a node, and highly correlated traits are clustered together. Each path represents a correlation between the two variables it joins. A blue path represents a positive

correlation and a red path represents a negative correlation. Only significant correlations are represented. The width and transparency of the line represent the strength of the correlation (wider and less transparent = stronger correlation). Abbreviations: fresh weight (FW), shoot water content (SWC), plant height (PH), number of branches (No.B), chlorophyll a (Chl. a), chlorophyll b (Chl. b), carotenoids (Caro), root sodium concentration (Na(r)), shoot sodium concentration (Na(s)), root chloride concentration (Cl(r)), shoot chloride concentration (Cl(s)), root potassium concentration (K(r)), shoot potassium concentration (K(s)), root calcium concentration (Ca(r)), shoot calcium concentration (Ca(s)), glycine betaine (GB), proline (PRO), total soluble sugars (TSS), malondialdehyde (MDA), hydrogen peroxide (H_2O_2), total phenolic compounds (TPC), total flavonoids (TF).

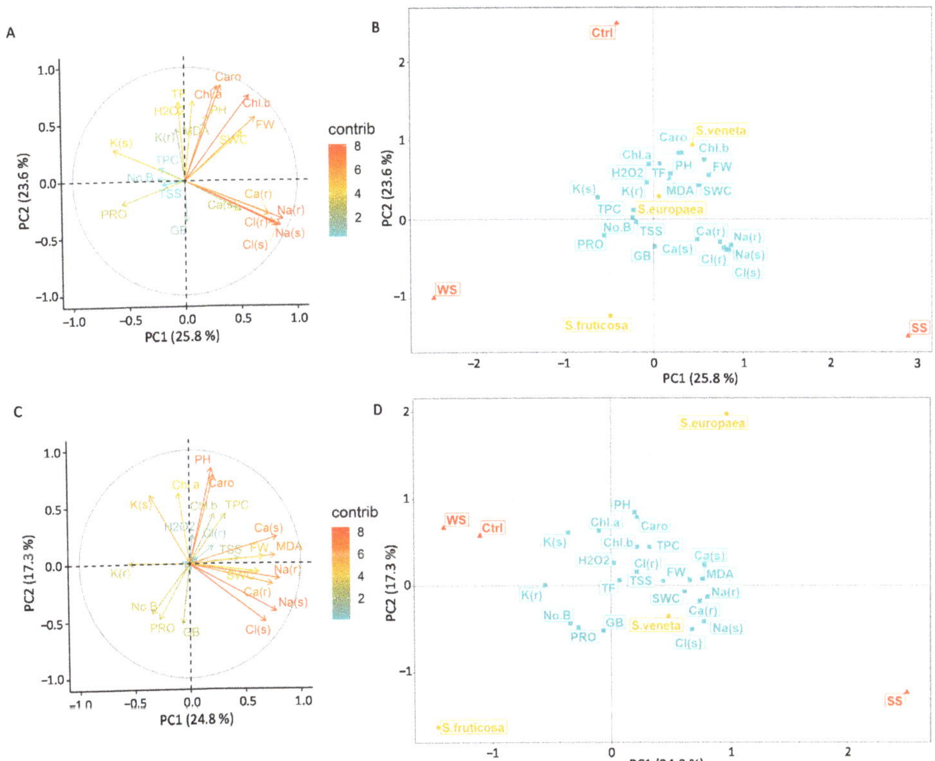

Figure 9. PCA correlation circles of the 22 measured parameters: (**A**) after 30 days of stress treatments (PCAstress) and (**C**) after 15 days of watering with non-saline water (PCArecovery). The increasing arrow lengths and shades of colour from light blue to red indicate the increasing contribution of variables to the definition of the first two principal components. PCA biplot of variables (**B**) after 30 days of stress treatments (Stress) and (**D**) after 15 days of watering with non-saline water (Recovery). Yellow circles show the barycentres of the three halophyte species (*S. europaea*, *S. veneta*, *S. fruticosa*), orange triangles show the barycentres of the three experimental treatments (Ctrl, control; SS, salt stress (watering with 700 mM NaCl water solution); WS, water stress (complete withholding of irrigation)), and the light blue squares show the quantitative variables, i.e., the measured traits (fresh weight (FW), shoot water content (SWC), plant height (PH), number of branches (No.B), chlorophyll a (Chl. a), chlorophyll b (Chl. b), carotenoids (Caro), root sodium concentration (Na(r)), shoot sodium concentration (Na(s)), root chloride concentration (Cl(r)), shoot chloride concentration (Cl(s)), root potassium concentration (K(r)), shoot potassium concentration (K(s)), root calcium concentration (Ca(r)), shoot calcium concentration (Ca(s)), glycine betaine (GB), proline (PRO), total soluble sugars (TSS), malondialdehyde (MDA), hydrogen peroxide (H_2O_2), total phenolic compounds (TPC), total flavonoids (TF).

The PC2 highlighted the differences between the annual *S. europaea* and the perennial *S: fruticosa*, with *S. veneta* showing an intermediate behaviour between the two other species. The barycentre of *S. europaea* was placed on the positive side of the PC2 axis (Figure 9D), which was positively correlated with PH (0.85), Caro (0.78), Chl. a (0.63), and Chl. b (0.44) (Figure 9C), whereas the barycentre of *S. fruticosa* was on the negative side. This placement reflects the fact that the recovery of these traits was more pronounced in *S. europaea* than in *S. fruticosa*, since these traits were compromised more seriously in the annual than in the perennial species under water stress.

Finally, the third component differentiated the control treatment, standing on the positive PC3 side (Table S3 of Supplementary Materials), from the water stress treatment, standing on the negative PC3 side. PC3 was positively correlated with Chl. a (0.64) and Chl. b (0.57), as control plants showed the highest pigment content even at the recovery stage and was negatively correlated to K(s) (−0.35) (Table S2 of Supplementary Materials), which increased in water-stressed plants after the recovery, especially in the two annual halophytes.

3. Discussion

Cultivating drought- and salt-tolerant crops can build resilience to climate change and enhance farm productivity and livelihoods in drought- and salt-prone areas. Generally, salinity and drought regimes are not stable but fluctuate seasonally and geographically, depending on the climate and hydrological conditions of each specific environment. Thus, the extent to which a species can cope with these fluctuations is an important trait that can be selected for saline agriculture.

Salicornia europaea, *S. veneta*, and *Sarcocornia fruticosa* are three halophytic species already traded in the market as leafy vegetables and oil-seed crops, thanks to their high content of nutritional compounds with valuable health-related properties. The natural saline habitats of these species are especially sensitive to climate change effects, which will include more frequent, more intense, and longer drought periods and higher soil salinity levels, albeit with wide seasonal variations [44].

From a general overview of our results, all three species were shown to be remarkably tolerant to salinity but sensitive to water deficit, albeit to a lesser extent in *S. fruticosa*, which showed higher resistance to dehydration and greater ability to recover after drought exposition. Our findings are supported by the ecology and the evolutionary trends within this lineage of species. In the Mediterranean, the two genera grow in close sympatry but are separated ecologically [16]. *Salicornia* dominates inland or coastal lagoons which may remain flooded for longer periods after winter rains. By acquiring an annual life cycle, *Salicornia* species were able to adapt to more unstable habitats and to expand to colder northern areas [16]. European *Sarcorcornia* are frost sensitive and grow only in winter-mild Atlantic coasts or drier Mediterranean areas [14].

The surveyed *S. fruticosa* seeds were collected from a semiarid zone (La Albufera Natural Park, Valencia, Spain), with a mean annual temperature, precipitation, and evapotranspiration of 17.5 °C, 488 mm, and 1199 mm, respectively. On the other hand, the *S. europaea* and *S. veneta* seeds were sampled from a more humid area (Piallassa della Baiona, Ravenna, Italy), having mean annual temperature, precipitation, and evapotranspiration of 14.6 °C, 576 mm, and 828 mm, respectively. This difference in environmental conditions may be the primary reason for developing a more robust drought tolerance in *S. fruticosa*. However, the slower metabolism of perennial plants could represent an advantageous adaptive strategy for survival under stress conditions since it allows for the saving of water and resource consumption while enhancing the synthesis of protective compounds [45]. This may have contributed to the better performance of the perennial *S. fruticosa* under water stress with respect to the annual *S. europaea* and *S. veneta*.

Photosynthetic pigment contents in *S. fruticosa* were not affected by salinity or drought stress, whereas a reduction in pigment contents was recorded in *S. europaea* and *S. veneta*, being generally modest under salt stress but severe in response to water deficit. Here again, these differences could be a consequence of the better adaptation of *S. fruticosa* to semiarid conditions or dependent on its life cycle type. When exposed to stress, annual plants hasten the transition from the vegetative to the reproductive stage, activating a process of stress-induced senescence that shifts nutrient allocation to developing seeds [46,47]. The stress-induced senescence is regulated differently and occurs more gradually in the perennial plants, since they can also propagate vegetatively. When they experience stress, perennial plants prioritise biomass accumulation in roots, whose contribution to stress avoidance is fundamental, protect photosynthetic tissues to sustain C assimilation and boost the source strength, and enhance the conservation of meristematic tissues, which are essential for recovering after the stress period [48,49]. This basic distinction may also explain the different variations in pigment contents under stressful conditions between the perennial *S. fruticosa* and the two annual *S. europaea* and *S. veneta*. In any case, the two annual species were able to restore their pigment pools during the recovery phase.

Similar ion accumulation patterns were observed in all three species, with a consistent increase in Na^+ and Cl^- concentrations at the root and shoot level in response to high salinity. This response is in line with the finding that halophytes can take up and efficiently compartmentalise the ions naturally present in the growth media to conserve the water potential gradient and maintain water uptake [50]. The salt-treated plants retained their high content of Na^+ and Cl^- in the shoots notwithstanding the recovery treatment, since the transport of these ions, to be used as inorganic osmolytes, is energetically cheaper than the *de novo* synthesis of organic osmolytes [51]. It should also be pointed out that Na^+ and Cl^- content in shoots were very high, and much higher than in roots, in the absence of salt; that is, in the control and water stress treatments. This result indicates the active transport of these ions to the aboveground organs, even at low external salinity, so that Na^+ and Cl^- can contribute to cellular osmotic balance also in non-stressed and water-stressed plants.

Salinity, however, caused a decrease in K^+ translocation to the shoots, likely related to the antagonism between K^+ and Na^+ ions, which are physicochemically similar [52]. This is evident in the PCAstress correlation circle, where the Na and Cl arrows are opposite to the K(s) arrow, implying that an increase in the former ions caused a decrease in the latter ion. The significant increase in K^+ shoot allocation under water stress suggested that this ion is a key osmoticum used to maintain water status in *Salicornia* and *Sarcocornia* spp. under water stress conditions. Indeed, water-stressed plants held a high K^+ shoot content even after recovery.

The significant increase in Ca^{2+} concentration under high salinity conditions in both below- and aboveground organs supports the notion that Ca, being involved in a diverse array of sensor proteins, plays a central role in orchestrating the whole-plant response to salt stress [53,54]. Indeed, Ca^{2+} content was positively correlated with Na^+ and Cl^- contents in the PCAstress correlation circle. The ability to preserve Ca uptake and retention under salinity seems to be a common feature of halophytes, since it was also reported in other salt-tolerant species such as *Sarcobatus vermiculatus, Climacoptera turcomanica, Salicornia persica, Halimocnemis pilifera, Petrosimonia glauca*, and *Atriplex verrucifera* [55].

To sum up, the effects of recovery on ion contents were relevant on roots, which are the organs more directly and dynamically in contact with the external environment, whereas ion remobilisation within shoots was not substantially affected by the recovery treatment.

Besides accumulating inorganic ions, glassworts species synthesise several organic osmolytes under osmotic stress, which contribute to cellular osmotic adjustment, free radical scavenging, and the activation of specific signalling pathways.

In both the stressed and non-stressed plants of the two genera, *Salicornia* and *Sarcocornia*, relatively high absolute values of GB were quantified, suggesting that GB accumulation is a constitutive defence mechanism against osmotic stress. Responses of these plants to abiotic stress probably rely more on changes in GB subcellular compartmentalisation, i.e., GB redistribution from the vacuole to the cytoplasm, rather than its de novo synthesis. There is indeed evidence for stress-induced changes in the intracellular localisation of compatible solutes in halophytes, for example, in *Limonium latifolium* [56]; however, data on these putative mechanisms are still scarce. Still, GB concentration can increase in response to stress, as observed under salinity in *S. veneta* and, mostly, in water-stressed *S. fruticosa* plants, suggesting that the higher drought tolerance of this latter species is partly due to a relatively higher GB accumulation.

Proline (PRO) is probably the most common compatible solute in plant species [57]. Nevertheless, no significant change in PRO concentration was detected in our experiments, except for the increase under water stress in *S. europaea* and *S. veneta*. However, the measured absolute PRO concentrations were too low to have any relevant osmotic effect when compared to GB or ion contents in the shoots. Still, PRO could have contributed to enhanced stress tolerance through its additional ability to scavenge ROS, directly stabilise proteins and other cellular structures, and provide cellular redox potential [58].

Comparing these outcomes, it appears that GB is the major organic osmolyte contributing to drought tolerance in *S. fruticosa*, whereas PRO plays a relatively more relevant role *in S. europaea* and *S. veneta*. Indeed, after recovery from water stress, a drop in GB concentration was observed in *S. fruticosa*, and PRO levels decreased significantly in *S. europaea* and *S. veneta*. These results are in agreement with the findings reported by Gil et al. [41], who measured high (>400 µmol g^{-1} DW) GB and very low (1–2 µmol g^{-1} DW) PRO concentrations in *S. fruticosa* under field conditions in the aforementioned semiarid La Albufera Natural Park, and with the results of Parida and Jha [59], who found PRO to be the main organic osmolyte accumulated in response to drought stress in *Salicornia brachiata*.

This supports the assumption that typical GB-accumulating species generally contain low PRO levels and vice versa [60], as already observed in many species, including both halophytes and glycophytes. For example, in the halophyte *Spartina alternifolia*, in the presence of 600 mM NaCl, GB contents were 10-fold higher than those of PRO (ca. 150 vs. 15 µmol g^{-1} FW, respectively) [61]. The differences were much more pronounced in another halophyte, *Halocnemum strobilaceum*, showing GB values > 200-fold greater than those of PRO (700 vs. 3 µmol g^{-1} DW) under 690 mM NaCl [62]. A similar pattern, although with much lower absolute values, was found in the glycophyte *Spinacia oleracea* in the presence of 170 mM NaCl, showing GB concentrations (3.25 µmol g^{-1} FW) about four-fold higher than those of PRO (0.78 µmol g^{-1} FW) [63]. Conversely, PRO appears to contribute relatively more to osmotic balance under drought conditions (200–400 µmol g^{-1} DW) than GB (40–60 µmol g^{-1} DW) in the genus *Capsicum* [64]. The halophyte *Juncus maritimus* also accumulated PRO rather than GB in response to salt stress (400 mM NaCl): ca. 130 vs. 25 µmol g^{-1} DW, respectively [65]. Similarly, a preferential accumulation of PRO over GB was observed in the halophyte *Limonium santapolense* under drought stress (ca. 120 vs. 23 µmol g^{-1} DW, respectively) [66].

The accumulation of the total soluble sugars (TSS) may enhance drought tolerance in *S. europaea*, since TSS levels increased in response to the water stress treatment; however, their contribution to *S. veneta* and *S. fruticosa* stress resistance was negligible. This result is in contrast to previous studies that have reported TSS accumulation as the primary mechanism for osmotic adjustment in *S. fruticosa* [20] and *Salicornia persica* [67]. However, as discussed by Gil et al. [68], sugar accumulation should be interpreted with caution. In fact, unlike other osmolytes occurring in plants at very low levels, unless stressful conditions stimulate their biosynthesis, soluble sugars are components of primary metabolism that play different functional roles unrelated to stress responses. This may be the reason why no significant changes in TSS contents were observed after stress recovery in any of the three studied species.

The fact that the stress treatments did not increase the levels of oxidative stress markers, i.e., MDA and H_2O_2, revealed that no oxidative stress was generated by salt or water stress in any of the three species. In some cases—salt stress in *S. europaea* and water stress in *S. veneta*—the contents of the oxidative stress markers, i.e., MDA and H_2O_2, even decreased with respect to the non-stressed controls. This response may be due to the increased activity of peroxidase, which is generally stored in the peroxisome and vacuoles, and plays an active role in reducing oxidative stress decreasing lipid peroxidation [69].

Consequently, we did not detect a significant accumulation of non-enzymatic, antioxidant compounds, i.e., total phenolic (TPC) or flavonoid (TF) compounds. This is reflected in the PCAstress correlation circle, in which the short and faded MDA, H_2O_2, TPC, and TF arrows denote a weak contribution of these traits to the variability of the whole dataset.

Taken together, these results suggest that the stress responses based on ion transport control and osmolyte accumulation were efficient enough to avoid or even reduce oxidative stress under our experimental conditions. However, we must note that the absence of oxidative stress may also result, at least in part, from efficient enzymatic ROS-detoxifying machinery, based on the activity of antioxidant enzymes such as superoxide dismutase, catalase, ascorbate peroxidase, glutathione peroxidase, and peroxiredoxin [70], among others, which were not specifically addressed in this study.

4. Materials and Methods

4.1. Plant Material

Seeds of *Salicornia europaea* and *Salicornia veneta* were collected from Pialassa della Baiona, a coastal lagoon located within the Po Delta Regional Park in Italy. Seeds of *Sarcocornia fruticosa* were collected from 'La Albufera' Natural Park, located near the city of Valencia, Eastern Spain. Mean annual values of climatic parameters from 2006 to 2021 in the two sampling areas are reported in Table 2. The experiments were carried out in the laboratories and greenhouses of the Institute for the Conservation and Improvement of Valencian Agrodiversity (COMAV), Polytechnic University of Valencia, Spain.

Table 2. Historical weather data (from 2006 to 2021) of the areas of 'La Albufera' Natural Park (Spain) and Piallassa della Baiona (Italy), provided, respectively, by the Spanish Agroclimatic Information System for Irrigation (SIAR) and the Italian Arpae-Simc meteorological network [71,72]. T: temperature; RH: relative humidity; Eto: evapotranspiration. Eto data of Piallassa della Baiona were calculated applying the Thornthwaite method [73].

Year	'La Albufera' Natural Park				Piallassa Della Baiona			
	Mean T (°C)	Mean RH (%)	Rainfall (mm)	ET_0 (mm)	Mean T (°C)	Mean RH (%)	Rainfall (mm)	ET_0 (mm)
2006	17.53	69.13	464.40	1189.38	14.40	77.64	337.65	814.71
2007	16.81	68.13	894.40	1164.50	14.20	73.18	490.00	809.25
2008	16.88	68.35	674.40	1194.10	14.20	73.63	491.13	804.14
2009	17.34	68.60	446.20	1215.26	14.19	72.79	555.86	816.07
2010	16.78	68.31	565.00	1206.22	13.23	74.09	450.00	776.35
2011	17.57	70.32	472.00	1166.73	14.76	71.36	346.60	846.35
2012	17.31	67.58	503.61	1208.25	14.71	69.98	563.60	864.97
2013	17.55	63.26	263.80	1245.42	14.49	72.86	870.20	822.93
2014	18.32	65.32	224.40	1278.22	15.60	73.91	740.00	833.27
2015	17.76	70.02	401.26	1169.08	15.20	77.18	616.80	860.61
2016	17.85	68.66	259.57	1218.41	14.71	80.86	829.40	825.33
2017	17.59	68.51	307.26	1238.82	14.84	76.69	641.80	851.52
2018	17.60	68.06	684.02	1225.71	15.32	78.53	613.60	870.93
2019	17.79	66.59	427.00	1243.83	15.03	81.94	780.80	839.65
2020	18.09	72.95	731.94	1186.44	14.70	76.76	556.40	808.83
2021	17.50	75.40	494.72	1039.10	14.45	75.75	335.00	809.89
Mean	*17.52*	*68.70*	*488.37*	*1199.34*	*14.63*	*75.45*	*576.18*	*828.42*

Seeds were sown manually in plastic trays filled with commercial peat, placed into a growth chamber with a 16/8-h light/dark cycle, day/night temperatures of 25/22 °C, and 70–80% relative humidity and watered thrice per week with tap water. Forty days after sowing, seedlings of each species of uniform size and shape were transplanted into plastic pots (12 cm diameter) filled with 500 g of a mix of commercial peat (26% organic carbon, pH_{H_2O} = 7.0, and EC = 0.6 dS m^{-1}) and perlite (80:20 v/v). Three seedlings were transplanted to each pot. The pots were transferred into the controlled environment of a greenhouse, placed over benches, and irrigated manually with tap water thrice per week. During the experimental period in the greenhouse, temperatures ranged between 21.3 ± 1.6 and 28.6 ± 1.8 °C and RH between 67.5 ± 9.9 and 92.6 ± 2.9%.

4.2. Experimental Design and Stress Treatments

Four weeks after transplanting, when the plantlets were fully established, the pots with individuals of each species were randomly divided into three groups and subjected to the following treatments: control (Ctrl, irrigation with tap water thrice per week), salt stress (SS, irrigation with a 700 mM NaCl aqueous solution, thrice per week), and water stress (WS, complete withholding of irrigation). Pots were placed in trays and were watered from the bottom, i.e., filling the trays, considering a volume of 0.13 L pot^{-1}. After one month of treatment, the stressed plants were allowed to recover during the following fifteen days through intensive pot washings with tap water in the salt stress treatment and through the restoring of the soil moisture level up to 80% in the drought-stress treatment. In this phase, pots were watered from the top (0.13 L pot^{-1} for Ctrl and 0.50 L pot^{-1} for SS and WS) and, only in the SS treatments, the drainage water was always discarded to remove the leached salt. The amount of water (L pot^{-1}) distributed per each treatment during the Stress and Recovery phases are shown in Table 3.

Table 3. Amount of water distributed per pot during the stress period (Stress) and the recovery period (Recovery) in the three treatments (Ctrl, control; SS, irrigation with 700 mM NaCl; WS, complete withholding of irrigation).

	Stress (L pot^{-1})	Recovery (L pot^{-1})	Total (L pot^{-1})
Ctrl	1.75	1	2.75
SS	1.75	4	5.75
WS	0	2	2

The three factors, plant species (PS, 3 levels), stress treatments (ST, 3 levels), and harvesting time (HT, 2 levels), were cross-combined, resulting in 18 treatments. Four completely randomised replicates were set up, totalling 72 pots. This number of replicates is quite commonly adopted in pot experiments on this topic [29,74–76].

The plants were harvested twice, the first half after the thirty days of stress treatments (T30) and the second half after the fifteen days of recovery (T45). Morphological parameters were determined on all individual plants (n = 12 per species and treatment). Samples of the aboveground biomass, i.e., of the leafless succulent green stems, were used for biochemical analysis; in this case, the shoots of the three plants grown in each pot were pooled (n = 4 per species and treatment, but each sample was a pool of three independent plants).

4.3. Plant Growth

The three surveyed species are characterised by strongly reduced leaves, which are embedded to form articulated, photosynthetically active succulent stems appearing to be composed of jointed segments (Figure 10). The number of branches (excluding the main branch) and plant height were determined at the beginning of the treatments (T0), after fifteen (T15) or thirty (T30) days of the stress treatments and after 15 days of recovery; that is, 45 days from the beginning of the experiment (T45). At both harvests, 'Stress' and 'Recovery', the aboveground biomass of each plant was separated from the root and weighed (fresh weight, FW). Roots were cleaned with a brush and weighed. Portions of the shoots and the root material were oven-dried at 65 °C until a constant weight was reached (ca. 72 h) and were then weighed again (dry weight, DW) to determine the water content percentage according to the following formula:

$$WC\ (\%) = \frac{FW - DW}{FW} \times 100 \quad (1)$$

Figure 10. Picture of the three halophytes species after thirty days of stress treatments: control; water stress (complete withholding of irrigation); salt stress (watering with 700 mM NaCl).

Fresh shoot material was flash-frozen in liquid N_2 and stored at −75 °C, and dry material was stored at room temperature in tightly closed paper envelopes. Pot substrate was collected at each harvest time to determine moisture and electrical conductivity (EC) in the laboratory. Substrate moisture was calculated gravimetrically, as described above for the plant samples (Equation (1)). For EC measurements, a 1:5 suspension of the dry substrate and deionised water was prepared and mixed for one hour at 600 rpm and 21 °C before being filtered. The EC was measured with a Crison 522 conductivity meter and expressed in dS m^{-1}.

4.4. Photosynthetic Pigments

The concentrations (mg g^{-1} DW) of chlorophyll a (Chl. a), chlorophyll b (Chl. b), and carotenoids (Caro) in the plant tissues were measured spectrophotometrically, according to a previously described method [77]. Fresh ground shoot material (ca. 0.05 g) was extracted with 1 mL of ice-cold 80% acetone. The samples were mixed during 12 h in a shaker in the dark and then centrifuged at 13,300× *g* for 10 min at 4 °C. The supernatant absorbance was measured at 470, 646, and 663 nm, and the pigment concentrations were calculated, applying the equations described by Lichtenthaler and Wellburn [77].

4.5. Ion Quantification

The concentrations of Na^+, Cl^-, K^+, and Ca^{2+} were calculated separately for roots and shoots following the procedure described by Weimberg [78]. Two mL of Milli-Q water were added to ca 0.1 g of dry plant material, vortexed, and then mixed for 24 h in a shaker. The samples were then incubated in a water bath for 30 min at 95 °C, cooled on ice, and filtered through a 0.45 μm nylon filter. The cations were quantified with a PFP7 flame photometer (Jenway Inc., Burlington, VT, USA), whereas the anions were measured using a chlorimeter (Sherwood, model 926, Cambridge, UK).

4.6. Quantification of Osmolytes

The concentration of glycine betaine (GB) was determined as described by Grieve and Grattan [79], with some modifications [80]. Fresh shoot material (0.15 g) was shaken for 24 h at 4 °C with 1.5 mL Mili Q water and then centrifuged at 13,300× *g* for 10 min. The supernatant was mixed (1:1) with a 2N H_2SO_4 solution and stored in ice for 1 h. Then, 125 μL of the sample were supplemented with 50 μL of ice-cold KI-I_2 solution, which induces glycine betaine precipitation in the form of golden crystals. All the following steps were completed in the dark. The samples were maintained at 4 °C for 16 h and then centrifuged at 13,300× *g* for 45 min at 0 °C. The supernatant was carefully removed, and the glycine betaine crystals were dissolved into 1.4 mL of cold 1,2-dichloroethane; the tubes were kept for 2.5 h under dark and cold conditions, and, finally, their absorbance was recorded at 365 nm. Glycine betaine concentration was calculated against a GB standard calibration curve and expressed as μmol g^{-1} DW.

Proline (PRO) was quantified following the protocol of Bates et al. [81]. Fresh aboveground material (ca. 0.05 g) was extracted in 3% (*w*/*v*) aqueous sulpho-salicylic acid and subsequently supplemented with acid ninhydrin, incubated in a water bath for 1 h at 95 °C, cooled on ice, and then extracted with two volumes of toluene. The absorbance of the organic phase was read with a spectrophotometer at 520 nm, using toluene as a blank. A standard curve was obtained by running parallel assays with known PRO amounts. PRO concentration was expressed as μmol g^{-1} DW.

Total soluble sugars (TSS) were measured from ca. 0.05 g of ground fresh material extracted with 2 mL 80% (*v*/*v*) methanol, according to the method described by Dubois et al. [82]. After mixing in a shaker for 24 h, the samples were centrifuged at 13,300× *g* for 10 min; the supernatants, appropriately diluted with water, were mixed with 95% sulphuric acid and 5% phenol. After 20 min incubation at room temperature, the absorbance was measured at 490 nm. TSS concentration was expressed as equivalents of glucose, used as the standard (mg eq. glucose g^{-1} DW).

4.7. Determination of Oxidative Stress Markers and Antioxidant Compounds

Malondialdehyde (MDA), total phenolic compounds (TPC), and total flavonoids (TF) were quantified in the same methanol extracts prepared for TSS measurements.

The method defined by Hodges et al. [83] was used for MDA quantification, with some modifications [84]. Extracts were mixed with 0.5% thiobarbituric acid (TBA) prepared in 20% trichloroacetic acid (TCA)—or with 20% TCA without TBA for the controls—and then incubated at 95 °C for 20 min, cooled on ice, and centrifuged at 13,300× *g* for 10 min at 4 °C. The supernatant absorbance was measured at 532 nm. The non-specific absorbance at 600 and 440 nm was subtracted, and the MDA concentration was computed, applying the equations proposed by Taulavuori et al. [84]. MDA contents were expressed as nmol g^{-1} DW.

Hydrogen peroxide content in plants was quantified as previously described [85]. Fresh plant material (0.05 g) was extracted with a 0.1% (*w*/*v*) trichloroacetic acid (TCA) solution. After centrifugation, the supernatant was mixed with one volume of 10 mM potassium phosphate buffer (pH 7.0) and two volumes of 1 M potassium iodide. The absorbance of the samples was determined at 390 nm. Reaction mixtures containing known concentrations of H_2O_2 were assayed in parallel to obtain a standard curve, and H_2O_2 concentrations were expressed as μmol g^{-1} DW.

TPC were measured by reaction with the Folin–Ciocalteu reagent, following the method previously [86]. The methanol extracts were mixed with Na_2CO_3, incubated at room temperature in the dark for 90 min, and the absorbance was read at 765 nm. Gallic acid (GA) was used as standard, and the measured TPC concentrations were expressed as GA equivalents (mg eq. GA g^{-1} DW).

TF were quantified by a previously described protocol [87], namely by sample incubation with $NaNO_2$, followed by a reaction with $AlCl_3$. After the reaction, the sample absorbance was determined at 510 nm, and TF contents were expressed as equivalents of the catechin standard (mg eq. C g^{-1} DW).

4.8. Statistical Analysis

The data of the measured traits within each plant species (PS) were subjected to two separated one-way ANOVAs for the respective stress treatments (ST) and harvesting times (HT). The Tukey's honestly significant difference (HSD) post hoc test at $p < 0.05$ was applied to indicate significant differences among levels in significant ANOVA sources. A two-way ANOVA was then performed to assess the interaction between stress treatment (ST) and harvesting time (HT). The two-way ANOVA results are reported in Table S4 of Supplementary Materials.

We investigated the relationships between the 22 traits measured within each halophyte species by computing the Pearson correlation coefficients (r) and then testing their significance with $\alpha = 0.05$. For each species, the correlation matrix is shown as a network diagram where each entity of the dataset represents a node, and highly correlated variables are clustered together. Each path represents a correlation between the two variables it joins. A blue path represents a positive correlation, and a red path represents a negative correlation. Only significant correlations ($p < 0.05$) are represented. The width and transparency of the line represent the strength of the correlation (wider and less transparent = stronger correlation).

Two principal component analyses were carried out on the data collected at the first (PCAstress) and second harvest time (PCArecovery) to summarise the performances outlined by the three genotypes under the Stress and Recovery periods with a multivariate approach.

The principal components (PCs) were obtained from centred and scaled quantitative variables through the diagonalisation of the correlation matrix and extraction of the associated eigenvectors and eigenvalues. All 22 measured traits were set as active quantitative variables, whereas the three halophyte species (*S. europaea*, *S. veneta*, and *S. fruticosa*) and the three treatments (Ctrl, SS, WS) were used as supplementary categorical variables, i.e., variables that were not used in the computation of PCs. The Pearson correlation coefficients were determined between the PCs and each quantitative variable (the 22 measured traits). The associated *p*-values were calculated to classify the variables according to their relevance (Table S2 of Supplementary Materials).

All the statistical analyses were performed with the R 6.3.6 statistical software, using Car [88] and Emmeans [89] packages for the analysis of variance and post hoc test, and the FactoMineR package for principal component analysis [90]. Charts were created with the ggplot2 [91] and corrr [92] R packages.

5. Conclusions

The three investigated halophytes, the annual *S. europaea* and *S. veneta* and the perennial *S. fruticosa*, are highly tolerant to salinity but sensitive to water stress, although the latter species to a lesser extent. Salt tolerance seems to depend mainly on the salt-induced accumulation of ions (Na^+, Cl^- and Ca^{2+}) and the shoot biosynthesis of organic osmolytes, both contributing to osmotic adjustment under stress. Active transport of these ions to the aerial part of the plants and high concentrations of glycine betaine have also been detected in the control, non-stressed plants, indicating that these defence mechanisms against stress are at least partially constitutive.

The higher drought tolerance of *S. fruticosa*, compared to its annual counterparts, was reflected in a relatively lower reduction in shoot fresh weight and the absence of a decrease in photosynthetic pigment content under water deficit conditions and was attributed to the relatively higher accumulation of glycine betaine. *Sarcocornia fruticosa* also showed total recovery capacity after the water stress treatment, whereas the fresh weight of the water-stressed plants of *S. europaea* and *S. veneta* remained at values significantly lower than the controls after the recovery period.

Neither salinity nor drought stress generated oxidative stress. Consequently, the presence of stress response mechanisms based on the activation of antioxidant systems was not expected; indeed, no significant increase in the levels of antioxidant compounds was detected in any of the three halophytes. However, further studies should be carried out to assess the possible contribution of enzymatic antioxidant activities to the whole antioxidant network of these species.

The higher drought tolerance observed in *S. fruticosa* with respect to the two *Salicornia* species could be based on differences in the environmental conditions of the plants' natural habitats, as it is drier for *S. fruticosa*. However, a more gradual process of stress-induced senescence in the perennial *S. fruticosa* compared to the annual *S. europaea* and *S. veneta*, might have allowed water-stressed plants to preserve their pool of photosynthetic pigments and recover to control fresh weight after rewatering. Further studies will be required to confirm this hypothesis, including, for instance, the assessment of the responses to water deficit of annual and perennial plants growing in the same natural habitat.

Supplementary Materials: The following are available online at https://www.mdpi.com/article/10.3390/plants11081058/s1. Table S1. Eigen analysis of PCAstress and PCArecovery correlation matrix; Table S2. Correlation coefficients between the first three PCs (PC1, PC2, PC3) and the quantitative variables traits (fresh weight (FW), shoot water content (SWC), plant height (PH), number of branches (No.B), chlorophyll a (Chl. a), chlorophyll b (Chl. b), carotenoids (Caro), root sodium concentration (Na(r)), shoot sodium concentration (Na(s)), root chloride concentration (Cl(r)), shoot chloride concentration (Cl(s)), root potassium concentration (K(r)), shoot potassium concentration (K(s)), root calcium concentration (Ca(r)), shoot calcium concentration (Ca(s)), glycine betaine (GB), proline (PRO), total soluble sugars (TSS), malondialdehyde (MDA), hydrogen peroxide (H_2O_2), total phenolic compounds (TPC), total flavonoids (TF). The PCs were computed using 22 input data. Significance codes: ^{ns}, $^{(+)}$, *, **, and *** mean, respectively, not significant and significant at $p \leq 0.1$, $p \leq 0.05$, $p \leq 0.01$ and $p \leq 0.001$; Table S3. Coordinates of the barycentres of the supplementary categorical variables in PCAstress and PCArecovery biplots, respectively; Table S4. Two-way analysis of variance (ANOVA) of stress treatments (ST), harvesting time (HT), and their interactions (STxHT) for the three halophyte species, for the 22 measured traits (fresh weight (FW), shoot water content (SWC), plant height (PH), number of branches (No.B), chlorophyll a (Chl. a), chlorophyll b (Chl. b), carotenoids (Caro), root sodium concentration (Na(r)), shoot sodium concentration (Na(s)), root chloride concentration (Cl(r)), shoot chloride concentration (Cl(s)), root potassium concentration (K(r)), shoot potassium concentration (K(s)), root calcium concentration (Ca(r)), shoot calcium concentration (Ca(s)), glycine betaine (GB), proline (PRO), total soluble sugars (TSS), malondialdehyde (MDA), hydrogen peroxide (H_2O_2), total phenolic compounds (TPC), total flavonoids (TF). Significance codes: ^{ns}, $^{(+)}$, *, **, and *** mean, respectively, not significant and significant at $p \leq 0.1$, $p \leq 0.05$, $p \leq 0.01$, and $p \leq 0.001$.

Author Contributions: Conceptualization, M.B., O.V. and R.C.; methodology, M.B. and S.G.-O.; software, R.C.; validation, O.V. and L.B.; formal analysis, R.C.; investigation, R.C., D.-M.M. and S.G.-O.; resources, O.V.; data curation, R.C. and S.G.-O.; writing—original draft preparation, R.C.; writing—review and editing, C.L., O.V. and L.B.; visualization, R.C. and O.V.; supervision, O.V.; project administration, O.V.; funding acquisition, O.V. and M.B. All authors have read and agreed to the published version of the manuscript.

Funding: This research received no external funding.

Institutional Review Board Statement: Not applicable.

Informed Consent Statement: Not applicable.

Data Availability Statement: Data is contained within the article or supplementary material.

Conflicts of Interest: The authors declare no conflict of interest.

References

1. United Nations. *Transforming Our World: The 2030 Agenda for Sustainable Development*; Springer: New York, NY, USA, 2015.
2. Bongaarts, J.; FAO; IFAD; UNICEF; WFP; WHO. The State of Food Security and Nutrition in the World 2020. In *Tranforming Food Systems for Affordable Healty Diets*; FAO: Rome, Italy, 2020; ISBN 9789251329016.
3. United Nations, Department of Economic and Social Affairs, Population Division. *World Urbanization Prospects: The 2018 Revision*; United Nations: New York, NY, USA, 2019; ISBN 9789211483192.
4. Cuttelod, A.; García, N.; Malak, D.A.; Templeand, H.; Vineet, K. The Mediterranean: A Biodiversity Hotspot under Threat. In *The 2008 Review of The IUCN Red List of Threatened Species*; Vié, J.-C., Hilton-Taylor, C., Stuart, S.N., Eds.; IUCN: Gland, Switzerland, 2019.
5. Stocker, T.F.; Qin, D.; Plattner, G.-K.; Tignor, M.; Allen, S.K.; Boschung, J.; Nauels, A.; Xia, Y.; Bex, V.; Midgley, P.M. IPCC, 2013: Climate Change 2013: The Physical Science Basis. In *Contribution of Working Group I to the Fifth Assessment Report of the Intergovernmental Panel on Climate Change*; Cambridge University Press: Cambridge, UK; New York, NY, USA, 2013.
6. Rosa, R.; Marques, A.; Nunes, M.L. Impact of climate change in Mediterranean aquaculture. *Rev. Aquac.* **2012**, *4*, 163–177. [CrossRef]
7. World Bank Group. *Climate-Smart Agriculture Indicators*; World Bank: Washington, DC, USA, 2016.
8. Food and Agriculture Organization of the United Nations. *Strategic Framework 2022-31*; FAO: Rome, Italy, 2021.
9. Hussain, S.; Amin, A.; Mubeen, M.; Khaliq, T.; Shahid, M.; Hammad, H.M.; Sultana, S.R.; Awais, M.; Murtaza, B.; Amjad, M.; et al. Climate Smart Agriculture (CSA) Technologies. In *Building Climate Resilience in Agriculture: Theory, Practice and Future Perspective*; Jatoi, W.N., Mubeen, M., Ahmad, A., Cheema, M.A., Lin, Z., Hashmi, M.Z., Eds.; Springer International Publishing: Cham, Switzerland, 2022; pp. 319–338. ISBN 978-3-030-79408-8.
10. Campbell, B.M.; Thornton, P.; Zougmoré, R.; van Asten, P.; Lipper, L. Sustainable intensification: What is its role in climate smart agriculture? *Curr. Opin. Environ. Sustain.* **2014**, *8*, 39–43. [CrossRef]
11. Rozema, J.; Schat, H. Salt tolerance of halophytes, research questions reviewed in the perspective of saline agriculture. *Environ. Exp. Bot.* **2013**, *92*, 83–95. [CrossRef]
12. Panta, S.; Flowers, T.; Lane, P.; Doyle, R.; Haros, G.; Shabala, S. Halophyte agriculture: Success stories. *Environ. Exp. Bot.* **2014**, *107*, 71–83. [CrossRef]
13. Le Houérou, H.N. forage halophytes and salt-tolerant fodder crops in the Mediterranean Basin. In *Halophytes as a Resource for Livestock and for Rehabilitation of Degraded Lands. Tasks for Vegetation Science*; Squires, V.R., Ayoub, A.T., Eds.; Springer: Dordrecht, The Netherlands, 1994; pp. 123–137.
14. Steffen, S.; Ball, P.; Mucina, L.; Kadereit, G. Phylogeny, biogeography and ecological diversification of *Sarcocornia* (Salicornioideae, Amaranthaceae). *Ann. Bot.* **2015**, *115*, 353–368. [CrossRef] [PubMed]
15. Kadereit, G.; Mucina, L.; Freitag, H. Phylogeny of Salicornioideae (Chenopodiaceae): Diversification, biogeography, and evolutionary trends in leaf and flower morphology. *Taxon* **2006**, *55*, 617–642. [CrossRef]
16. Kadereit, G.; Ball, P.; Beer, S.; Mucina, L.; Sokoloff, D.; Teege, P.; Yaprak, A.E.; Freitag, H. A taxonomic nightmare comes true: Phylogeny and biogeography of glassworts (*Salicornia* L., Chenopodiaceae). *Taxon* **2007**, *57*, 1143–1170. [CrossRef]
17. Scott, A.J. Reinstatement and revision of Salicorniaceae J. Agardh (Caryophyllales). *Bot. J. Linn. Soc.* **1977**, *75*, 357–374. [CrossRef]
18. Papini, A.; Trippanera, G.B.; Maggini, F.; Filigheddu, R.; Biondi, E. New insights in *Salicornia* L. and allied genera (Chenopodiaceae) inferred from nrDNA sequence data. *Plant Biosyst.* **2004**, *138*, 215–223. [CrossRef]
19. Foggi, B.; Daoud-Bouattour, A.; Gammar-Ghrabi, Z.; Limam-Ben Saad, S.; Muller, S.D. The IUCN Red List of Threatened Species. *Salicornia veneta*. Available online: http://dx.doi.org/10.2305/IUCN.UK.2011-1.RLTS.T164320A5824288.en (accessed on 26 March 2022).
20. Ventura, Y.; Sagi, M. Halophyte crop cultivation: The case for *Salicornia* and *Sarcocornia*. *Environ. Exp. Bot.* **2013**, *92*, 144–153. [CrossRef]

21. Ventura, Y.; Wuddineh, W.; Myrzabayeva, M.; Alikulov, Z.; Khozin-Goldberg, I.; Shpigel, M.; Samocha, T.; Sagi, M. Effect of seawater concentration on the productivity and nutritional value of annual *Salicornia* and perennial *Sarcocornia* halophytes as leafy vegetable crops. *Sci. Hortic.* **2011**, *128*, 189–196. [CrossRef]
22. Antunes, M.D.; Gago, C.; Guerreiro, A.; Sousa, A.R.; Julião, M.; Miguel, M.G.; Faleiro, M.L.; Panagopoulos, T. Nutritional characterization and storage ability of *Salicornia ramosissima* and *Sarcocornia perennis* for fresh vegetable salads. *Horticulturae* **2021**, *7*, 6. [CrossRef]
23. Custódio, L.; Rodrigues, M.J.; Pereira, C.G.; Castañeda-Loaiza, V.; Fernandes, E.; Standing, D.; Neori, A.; Shpigel, M.; Sagi, M. A Review on *Sarcocornia* Species: Ethnopharmacology, Nutritional Properties, Phytochemistry, Biological Activities and Propagation. *Foods* **2021**, *10*, 2778. [CrossRef] [PubMed]
24. Cárdenas-Pérez, S.; Piernik, A.; Chanona-Pérez, J.J.; Grigore, M.N.; Perea-Flores, M.J. An overview of the emerging trends of the *Salicornia* L. genus as a sustainable crop. *Environ. Exp. Bot.* **2021**, *191*, 104606. [CrossRef]
25. Loconsole, D.; Cristiano, G.; De Lucia, B. Glassworts: From wild salt marsh species to sustainable edible crops. *Agriculture* **2019**, *9*, 14. [CrossRef]
26. Ali, M.; Mustafa, A.; Abideen, Z.; Gul, B. Bioenergy Production from Halophytes Crops for Sustainable Development. In *Energy and Environmental Security in Developing Countries*; Springer International Publishing: Heidelberg, Germany, 2021; pp. 571–586. ISBN 9783030636548.
27. Caparrós, P.G.; Ozturk, M.; Gul, A.; Batool, T.S.; Pirasteh-Anosheh, H.; Unal, B.T.; Altay, V.; Toderich, K.N. Halophytes have potential as heavy metal phytoremediators: A comprehensive review. *Environ. Exp. Bot.* **2022**, *193*, 104666. [CrossRef]
28. Gispert, M.; Kuliush, T.; Dyachenko, L.; Kharytonov, M.; Emran, M.; Verdaguer, D.; Llorens, L.; Carrasco-Barea, L. Appraising soil carbon storage potential under perennial and annual Chenopodiaceae in salt marsh of NE Spain. *Estuar. Coast. Shelf Sci.* **2021**, *252*, 107240. [CrossRef]
29. Carrasco Barea, L. Soil and Vegetation Carbon Dynamics (Stocks and Fluxes) and Litter Decomposition in a Mediterranean Non-Tidal Salt Marsh. Doctorals Thesis, Departament de Ciències Ambientals, Universitat de Girona, Girona, Spain, 2020; p. 158.
30. Ahmed, H.A.I.; Shabala, L.; Shabala, S. Understanding the mechanistic basis of adaptation of perennial *Sarcocornia quinqueflora* species to soil salinity. *Physiol. Plant.* **2021**, *172*, 1997–2010. [CrossRef]
31. Parks, G.E.; Dietrich, M.A.; Schumaker, K.S. Increased vacuolar Na^+/H^+ exchange activity in *Salicornia bigelovii* torr. in response to NaCl. *J. Exp. Bot.* **2002**, *53*, 1055–1065. [CrossRef]
32. Katschnig, D.; Jaarsma, R.; Almeida, P.; Rozema, J.; Schat, H. Differences in proton pumping and Na/H exchange at the leaf cell tonoplast between a halophyte and a glycophyte. *AoB Plants* **2014**, *6*, plu023. [CrossRef]
33. Flowers, T.J. Physiology of halophytes. *Plant Soil* **1985**, *89*, 41–56. [CrossRef]
34. García-Caparrós, P.; Llanderal, A.; Pestana, M.; Correia, P.; Lao, M. Nutritional and physiological responses of the dicotyledonous halophyte *Sarcocornia fruticosa* to salinity. *Aust. J. Bot.* **2017**, *65*, 573–581. [CrossRef]
35. Pessarakli, M. (Ed.) *Handbook of Plant and Crop Physiology*, 2nd ed.; Marcel Dekker Inc.: New York, NY, USA, 2002; ISBN 0824705467.
36. Gorham, J.; Hughes, L.L.; Wyn Jones, R.G. Chemical composition of salt-marsh plants from Ynys Môn (Anglesey): The concept of physiotypes. *Plant. Cell Environ.* **1980**, *3*, 309–318. [CrossRef]
37. Moghaieb, R.; Saneoka, H.; Fujita, K. Effect of salinity on osmotic adjustment, glycinebetaine accumulation and the betaine aldehyde dehydrogenase gene expression in two halophytic plants, *Salicornia europaea* and *Suaeda maritima*. *Plant Sci.* **2004**, *166*, 1345–1349. [CrossRef]
38. Lee, S.J.; Jeong, E.-M.; Ki, A.Y.; Oh, K.-S.; Kwon, J.; Jeong, J.-H.; Chung, N.-J. Oxidative defense metabolites induced by salinity stress in roots of *Salicornia herbacea*. *J. Plant Physiol.* **2016**, *206*, 133–142. [CrossRef]
39. Abd El-Maboud, M.M.; Elsharkawy, E.R. Ecophysiological responses of the genus *Sarcocornia* A. J. Scott growing at the Mediterranean Sea Coast, Egypt. *Pakistan J. Bot.* **2021**, *53*, 517–523. [CrossRef]
40. Ma, J.; Cirillo, V.; Zhang, D.; Maggio, A.; Wang, L.; Xiao, X.; Yao, Y. Regulation of Ammonium Cellular Levels is An Important Adaptive Trait for the Euhalophytic Behavior of *Salicornia europaea*. *Plants* **2020**, *9*, 257. [CrossRef]
41. Duarte, B.; Caçador, I.; Matos, A.R. Lipid landscape remodelling in *Sarcocornia fruticosa* green and red physiotypes. *Plant Physiol. Biochem.* **2020**, *157*, 128–137. [CrossRef]
42. Gil, R.; Bautista, I.; Boscaiu, M.; Lidón, A.; Wankhade, S.; Sánchez, H.; Llinares, J.; Vicente, O. Responses of five Mediterranean halophytes to seasonal changes in environmental conditions. *AoB Plants* **2014**, *6*, plu049. [CrossRef]
43. Morari, F.; Meggio, F.; Lunardon, A.; Scudiero, E.; Forestan, C.; Farinati, S.; Varotto, S. Time course of biochemical, physiological, and molecular responses to field-mimicked conditions of drought, salinity, and recovery in two maize lines. *Front. Plant Sci.* **2015**, *6*, 314. [CrossRef]
44. Intergovernmental Panel on Climate Change. *Global Warming of 1.5 °C: An IPCC Special Report on the Impacts of Global Warming of 1.5 °C above Pre-Industrial Levels and Related Global Greenhouse Gas Emission Pathways, in the Context of Strengthening the Global Response to the Threat of Climate Change, Sustainable Development, and Efforts to Eradicate Poverty*; Intergovernmental Panel on Climate Change: Geneva, Switzerland, 2018.
45. Keep, T.; Sampoux, J.-P.; Barre, P.; Blanco-Pastor, J.; Dehmer, K.; Durand, J.; Hegarty, M.; Ledauphin, T.; Muylle, H.; Roldán-Ruiz, I.; et al. to grow or survive: Which are the strategies of a perennial grass to face severe seasonal stress? *Funct. Ecol.* **2021**, *35*, 1145–1158. [CrossRef]

46. Albacete, A.; Martinez Andujar, C.; Pérez-Alfocea, F. Hormonal and metabolic regulation of source–sink relations under salinity and drought: From plant survival to crop yield stability. *Biotechnol. Adv.* **2014**, *32*, 12–30. [CrossRef] [PubMed]
47. Schippers, J.H.M.; Schmidt, R.; Wagstaff, C.; Jing, H.-C. Living to Die and Dying to Live: The Survival Strategy behind Leaf Senescence. *Plant Physiol.* **2015**, *169*, 914–930. [CrossRef] [PubMed]
48. Zwicke, M.; Picon-Cochard, C.; Morvan-Bertrand, A.; Homme, M.-P.; Volaire, F. What functional strategies drive drought survival and recovery of perennial species from upland grassland? *Ann. Bot.* **2015**, *116*, 1001–1015. [CrossRef] [PubMed]
49. Griffiths, C.; Gaff, D.; Neale, A. Drying without senescence in resurrection plants. *Front. Plant Sci.* **2014**, *5*, 36. [CrossRef]
50. Flowers, T.J.; Munns, R.; Colmer, T.D. Sodium chloride toxicity and the cellular basis of salt tolerance in halophytes. *Ann. Bot.* **2015**, *115*, 419–431. [CrossRef]
51. Raven, J.A. Tansley Review No. 2: Regulation of pH and Generation of Osmolarity in Vascular Plants: A Cost-Benefit Analysis in Relation to Efficiency of Use of Energy, Nitrogen and Water. *New Phytol.* **1985**, *101*, 25–77. [CrossRef]
52. Hu, Y.; Schmidhalter, U. Drought and salinity: A comparison of their effects on mineral nutrition of plants. *J. Plant Nutr. Soil Sci.* **2005**, *168*, 541–549. [CrossRef]
53. Julkowska, M.M.; Testerink, C. Tuning plant signaling and growth to survive salt. *Trends Plant Sci.* **2015**, *20*, 586–594. [CrossRef]
54. Manishankar, P.; Wang, N.; Köster, P.; Alatar, A.A.; Kudla, J. Calcium signaling during salt stress and in the regulation of ion homeostasis. *J. Exp. Bot.* **2018**, *69*, 4215–4226. [CrossRef]
55. Rathore, A.P.; Chaudhary, D.R.; Jha, B. Biomass production, nutrient cycling, and carbon fixation by *Salicornia brachiata* Roxb.: A promising halophyte for coastal saline soil rehabilitation. *Int. J. Phytoremediation* **2016**, *18*, 801–811. [CrossRef] [PubMed]
56. Gagneul, D.; Aïnouche, A.; Duhazé, C.; Lugan, R.; Larher, F.R.; Bouchereau, A. A Reassessment of the Function of the So-Called Compatible Solutes in the Halophytic Plumbaginaceae *Limonium latifolium*. *Plant Physiol.* **2007**, *144*, 1598–1611. [CrossRef] [PubMed]
57. Szabados, L.; Savouré, A. Proline: A multifunctional amino acid. *Trends Plant Sci.* **2010**, *15*, 89–97. [CrossRef] [PubMed]
58. Ben Rejeb, K.; Abdelly, C.; Savouré, A. How reactive oxygen species and proline face stress together. *Plant Physiol. Biochem.* **2014**, *80*, 278–284. [CrossRef]
59. Parida, A.K.; Jha, B. Physiological and Biochemical Responses Reveal the Drought tolerance Efficacy of the Halophyte *Salicornia brachiata*. *J. Plant Growth Regul.* **2013**, *32*, 342–352. [CrossRef]
60. Tipirdamaz, R.; Gagneul, D.; Duhazé, C.; Aïnouche, A.; Monnier, C.; Özkum, D.; Larher, F. Clustering of halophytes from an inland salt marsh in Turkey according to their ability to accumulate sodium and nitrogenous osmolytes. *Environ. Exp. Bot.* **2006**, *57*, 139–153. [CrossRef]
61. Cavalieri, A.J. Proline and glycinebetaine accumulation by *Spartina alterniflora* Loisel. in response to NaCl and nitrogen in a controlled environment. *Oecologia* **1983**, *57*, 20–24. [CrossRef]
62. Pujol, J.A.; Calvo, J.F.; Ramírez-Díaz, L. Seed germination, growth, and osmotic adjustment in response to NaCl in a rare succulent halophyte from Southeastern Spain. *Wetlands* **2001**, *21*, 256–264. [CrossRef]
63. Di Martino, C.; Delfine, S.; Pizzuto, R.; Loreto, F.; Fuggi, A. Free amino acids and glycine betaine in leaf osmoregulation of spinach responding to increasing salt stress. *New Phytol.* **2003**, *158*, 455–463. [CrossRef]
64. Escalante-Magaña, C.; Aguilar-Caamal, L.F.; Echevarría-Machado, I.; Medina-Lara, F.; Cach, L.S.; Martínez-Estévez, M. Contribution of glycine betaine and proline to water deficit tolerance in pepper plants. *HortScience* **2019**, *54*, 1044–1054. [CrossRef]
65. Al Hassan, M.; López-Gresa, M.D.P.; Boscaiu, M.; Vicente, O. Stress tolerance mechanisms in *Juncus*: Responses to salinity and drought in three *Juncus* species adapted to different natural environments. *Funct. Plant Biol.* **2016**, *43*, 949–960. [CrossRef] [PubMed]
66. González-Orenga, S.; Al Hassan, M.; Llinares, J.V.; Lisón, P.; López-Gresa, M.P.; Verdeguer, M.; Vicente, O.; Boscaiu, M. Qualitative and Quantitative Differences in Osmolytes Accumulation and Antioxidant Activities in Response to Water Deficit in Four Mediterranean *Limonium* Species. *Plants* **2019**, *8*, 506. [CrossRef] [PubMed]
67. Ahmad, S.T.; Sima, N.A.K.K.; Mirzaei, H.H. Effects of sodium chloride on physiological aspects of *Salicornia persica* growth. *J. Plant Nutr.* **2013**, *36*, 401–414. [CrossRef]
68. Gil, R.; Boscaiu, M.; Lull, C.; Bautista, I.; Lid, A.N.; Vicente, Ó. Are soluble carbohydrates ecologically relevant for salt tolerance in halophytes? *Funct. Plant Biol.* **2013**, *40*, 805–818. [CrossRef]
69. Ghanem, A.F.M.; Mohamed, E.; Kasem, A.M.M.A.; El-Ghamery, A.A. Differential Salt tolerance Strategies in Three Halophytes from the Same Ecological Habitat: Augmentation of Antioxidant Enzymes and Compounds. *Plants* **2021**, *10*, 1100. [CrossRef]
70. Choudhury, F.K.; Rivero, R.M.; Blumwald, E.; Mittler, R. Reactive oxygen species, abiotic stress and stress combination. *Plant J.* **2017**, *90*, 856–867. [CrossRef]
71. Arpae-Simc DEXT3R—Servizio IdroMeteoClima. Available online: https://simc.arpae.it/dext3r/ (accessed on 7 February 2022).
72. Ministerio de Agricultura y Pesca, Alimentación y Medio Ambiente. The Agroclimatic Information System for Irrigation (SiAR). Available online: www.siar.es (accessed on 7 February 2022).
73. Thornthwaite, C.W. An Approach toward a Rational Classification of Climate. *Geogr. Rev.* **1948**, *38*, 55. [CrossRef]
74. Ksouri, R.; Megdiche, W.; Debez, A.; Falleh, H.; Grignon, C.; Abdelly, C. Salinity effects on polyphenol content and antioxidant activities in leaves of the halophyte *Cakile maritima*. *Plant Physiol. Biochem.* **2007**, *45*, 244–249. [CrossRef]
75. Ozturk, O.F.; Shukla, M.K.; Stringam, B.; Picchioni, G.A.; Gard, C. Irrigation with brackish water changes evapotranspiration, growth and ion uptake of halophytes. *Agric. Water Manag.* **2018**, *195*, 142–153. [CrossRef]

76. Heidari-Sharifabad, H.; Mirzaie-Nodoushan, H. Salinity-induced growth and some metabolic changes in three *Salsola* species. *J. Arid Environ.* **2006**, *67*, 715–720. [CrossRef]
77. Lichtenthaler, H.K.; Wellburn, A.R. Determinations of total Carotenoids and Chlorophylls A and B of Leaf Extracts in Different Solvents. *Biochem. Soc. Trans.* **1983**, *11*, 591–592. [CrossRef]
78. Weimberg, R. Solute Adjustments in Leaves of Two Species of Wheat at Two Different Stages of Growth in Response to Salinity. *Physiol. Plant* **1987**, *70*, 381–388. [CrossRef]
79. Grieve, C.M.; Grattan, S.R. Rapid Assay for Determination of Water Soluble Quaternary Ammonium Compounds. *Plant Soil* **1983**, *70*, 303–307. [CrossRef]
80. Valadez-Bustos, M.G.; Aguado-Santacruz, G.A.; Tiessen-Favier, A.; Robledo-Paz, A.; Muñoz-Orozco, A.; Rascón-Cruz, Q.; Santacruz-Varela, A. A Reliable Method for Spectrophotometric Determination of Glycine Betaine in Cell Suspension and other Systems. *Anal. Biochem.* **2016**, *498*, 47–52. [CrossRef] [PubMed]
81. Bates, L.S.; Waldren, R.P.T. Rapid Determination of Free Proline for Water—Stress Studies. *Plant Soil* **1973**, *39*, 205–207. [CrossRef]
82. Dubois, M.; Gilles, K.A.; Hamilton, J.K.; Rebers, P.A.; Smith, F. Colorimetric Method for Determination of Sugars and Related Substances. *Anal. Chem.* **1956**, *28*, 350–356. [CrossRef]
83. Hodges, D.M.; Delong, J.M.; Forney, C.F.; Prange, R.K. Improving The Thiobarbituric Acid-Reactive-Substances Assay for Estimating Lipid Peroxidation in Plant Tissues Containing Anthocyanin and other Interfering Compounds. *Planta* **1999**, *207*, 604–611. [CrossRef]
84. Taulavuori, E.; Hellström, E.; Taulavuori, K.; Laine, K. Comparison of Two Methods Used to Analyse Lipid Peroxidation From *Vaccinium myrtillus* (L.) During Snow Removal, Reacclimation and Cold Acclimation. *J. Exp. Bot.* **2002**, *52*, 2375–2380. [CrossRef]
85. Loreto, F.; Velikova, V. Isoprene Produced By Leaves Protects the Photosynthetic Apparatus against Ozone Damage, Quenches Ozone Products, and Reduces Lipid Peroxidation of Cellular Membranes. *Plant Physiol.* **2001**, *127*, 1781–1787. [CrossRef]
86. Blainski, A.; Lopes, G.; Mello, J. Application and Analysis of the Folin Ciocalteu Method for the Determination of the total Phenolic Content from *Limonium brasiliense* L. *Molecules* **2013**, *18*, 6852–6865. [CrossRef]
87. Zhishen, J.; Mengcheng, T.; Jianming, W. The Determination of Flavonoid Contents in Mulberry and Their Scavenging Effects on Superoxide Radicals. *Food Chem.* **1999**, *64*, 555–559. [CrossRef]
88. Fox, J.; Weisberg, S. *An R Companion to Applied Regression*, 3rd ed.; Sage Publications, Inc.: Thousand Oaks, CA, USA, 2019; ISBN 978-1412975148.
89. Lenth, R.V.; Buerkner, P.; Herve, M.; Love, J.; Riebl, H.; Singmann, H.; Emmeans: Estimated Marginal Means, Aka Least-squares Means. R Pack. Available online: https://cran.r-project.org/web/packages/emmeans/emmeans.pdf (accessed on 12 December 2021).
90. Lê, S.; Josse, J.; Husson, F. Factominer: An R Package for Multivariate Analysis. *J. Stat. Softw.* **2008**, *25*, 1–18. [CrossRef]
91. Wickham, H. *Ggplot2: Elegant Graphics for Data Analysis*; Springer: New York, NY, USA, 2009; ISBN 978-0-387-98140-6.
92. Kuhn, M.; Jackson, S.; Cimentada, J. Corrr: Correlations in R. *R Package Version 0.4*. Available online: https://cran.r-project.org/web/packages/corrr/corrr.pdf (accessed on 12 December 2021).

plants

Article

Salicornia europaea L. Functional Traits Indicate Its Optimum Growth

Stefany Cárdenas-Pérez [1], Ahmad Rajabi Dehnavi [1,2], Karol Leszczyński [1], Sandra Lubińska-Mielińska [1], Agnieszka Ludwiczak [1] and Agnieszka Piernik [1,*]

1 Department of Geobotany and Landscape Planning, Faculty of Biological and Veterinary Sciences, Nicolaus Copernicus University in Torun, Lwowska 1, 87-100 Torun, Poland; cardenasperez@umk.pl (S.C.-P.); ahmad.rajabi.dehnavi@gmail.com (A.R.D.); karol.leszczynski96@gmail.com (K.L.); sanlub@doktorant.umk.pl (S.L.-M.); agnieszka.lud@umk.pl (A.L.)

2 Department of Agronomy and Plant Breeding, College of Agriculture, Isfahan University of Technology, Isfahan 84156-83111, Iran

* Correspondence: piernik@umk.pl; Tel.: +48-56-611-2537

Abstract: *Salicornia europaea* L. grows in areas periodically flooded by salty or brackish water. It has potential economic value, because it can be used as food, forage, or biofuel, and has potential in pharmaceuticals and cosmetics. Increasing interest in *S. europaea* is due to its extreme salt tolerance and well growth in marginal saline soils. However, the variation in its functional traits in response to environmental conditions is still poorly studied. There are still questions regarding the optimal level of salinity for different traits. Therefore, we worked to address the question if *S. europaea* traits from different scales are controlled by salinity level. Based on performed pot experiment, we found that almost all traits are salinity dependent but affected in different ways. We demonstrated that morphological, biomass, and anatomical properties indicate optimum growth between 200 and 400 mM NaCl and growth limitations at 0, 800, and 1000 mM NaCl. Moreover, we found the most affected traits which include photosynthetic pigments and protein content, plant surface area, peroxidase activity, and anatomic traits related to cell shape. Our results significantly expanded the knowledge about *S. europaea* functional traits variation in response to salinity, which can be important for discovering regulating processes and for possible future agricultural applications.

Keywords: halophytes; salinity; morphology; anatomy; catalase; peroxidase; hydrogen peroxide; chlorophyll content

Citation: Cárdenas-Pérez, S.; Rajabi Dehnavi, A.; Leszczyński, K.; Lubińska-Mielińska, S.; Ludwiczak, A.; Piernik, A. *Salicornia europaea* L. Functional Traits Indicate Its Optimum Growth. *Plants* **2022**, *11*, 1051. https://doi.org/10.3390/plants11081051

Academic Editors: Oscar Vicente, Marius-Nicusor Grigore and Roberto Barbato

Received: 18 February 2022
Accepted: 8 April 2022
Published: 12 April 2022

1. Introduction

Salicornia europaea L. belongs to the *Amaranthaceae* family (formerly *Chenopodiaceae*), subfam. *Salicornioideae*. The genus *Salicornia* is widespread in temperate and subtropical regions of the Northern Hemisphere but absent in South America and Australia [1]. Presence in habitats with changing seasonal and even daily dynamics has led to high physiological plasticity in plants [2]. This results in phenotypic variability and problems with establishing acceptable systematics [3]. Despite the high phenotypic variability, several common features can be distinguished [4,5]. *S. europaea* has an erect, highly branched stem. Secondary shoots are formed on the primary cylindrical shoots. The plant has strongly reduced leaves, and the assimilation area is located in the shoots. It is green most of its life, but at the end of life cycle, the stems turn red due to chlorophyll destruction which reduces photosynthesis and finally affects nutrient loss, biomass, and hydric balance [6]. Branches have spikes consisting of three flowers, one main and two laterals. Seeds are small, dark, ellipsoidal, and characterized by heteromorphism, i.e., color, shape, and size variability [7].

S. europaea grows in areas periodically flooded by salty or brackish water [8]. In Central Europe this species has been recognized as *Salicornia ramosissima* J. Woods (=*S. herbacea* L.) [9]. It has potential economic value because it is edible, either raw or cooked [10]. In addition, it can be used as a forage for animals or a biofuel and has potential in pharmaceuticals and cosmetics [11,12]. It also has a high fatty acid content in its seeds, which increases its nutritional value [13]. Moreover, *S. europaea* ash can be used to produce glass and soap [14]. The healing properties should also be mentioned because *S. europaea* is rich in tungsten acids, quercetin, and isorhamnetin, which have anti-inflammatory and antioxidant properties. It also contains polysaccharides that play a role in the treatment of constipation, obesity, diabetes, and cancer [14,15].

This species belongs to extreme halophytes [16], which is a group of species strongly adapted to saline environments. They evolved some specific mechanisms to cope with saline environment like reduction of the Na^+ levels, compartmentalization, and excretion of sodium ions [17,18]. The first *S. europaea* mechanism to overcome high Na^+ concentrations is the water storage in the parenchyma, which dilutes the accumulated salts and contributes to maintaining cellular turgor. This allows the plant to cope efficiently with high salinity [16]. However, reduction of growth of this species have been reported at high salinities [19,20]. The adaptation to different levels of salinity can affect water storage intensity and therefore some anatomical properties, e.g., the size and shape of stem-cortex cells [21]. Moreover, in the salinity gradient plants induce some physiological responses related to osmotic adjustment as proline accumulation, and increasing antioxidant enzyme activities, e.g., catalase (CAT) and peroxidase (POD) [19,22]. Salinity can also affect photosynthetic activity by changes in chlorophyll and carotenoids content [23,24] and different proteins content, playing important roles for plant salt tolerance ability [25].

Increasing interest in *S. europaea* is due to its extreme salt tolerance over 1000 mM NaCl [8]. Results of recent studies at the International Center for Biosaline Agriculture (ICBA) in the United Arab Emirates (UAE) show that some varieties of *S. europaea* cultivated with good agronomic practices grow well in marginal soils and can be economically viable [8]. As an obligatory halophyte, it is believed by definition to be able to complete its life cycle in a salt concentration of around 200 mM NaCl (ca. 20 dSm^{-1}) or more under conditions like those that might be encountered in the natural environment [17]. Although, halophytes are species that can live and reproduce successfully under salt stress; it is still not so clear if they need salt for development. Regarding *S. europaea*, Snow and Vince [26] reported better growth of this species outside their home zone in salt marsh habitats and its presence at high salinity because of low competitive ability with other species. It was partly confirmed by Piernik [27] who found, under field conditions, the good growth of this species at lower salinity than in its home vegetation zone. Mucolo et al. [2] found that the very high final germination in distilled water (control) suggests that these taxa do not necessarily have a physiological requirement for salt to germinate.

There are also still few studies reporting or focusing on *S. europaea*'s optimum growth assessment. Lv et al. [28] reported *S. europaea* optimal growth and photosynthetic rates at 200–400 mM NaCl. Araus et al. [8] reported that the best irrigation regime in terms of biomass and seed yield involved brackish water of 25 dSm^{-1}. Similar results were obtained by Singh et al. [29], i.e., a notable amount of biomass for *S. ramosissima* using artificial seawater containing 257 mM NaCl. Except for biomass and seeds yield, there is still a lack of knowledge of *S. europaea*'s morphological and anatomical trait adaptations to different salinity levels and their optimum growth [4,5,16,21,30].

Plant functional traits are defined as any morphological, anatomical, physiological, and phenological plant characteristics affecting overall plant fitness through their influence on survival, growth, and reproduction [31]. They determine how primary producers respond, among others, to environmental factors, affect trophic levels, influence ecosystem processes and services, and provide a link from species richness to ecosystem functional diversity [32]. Plant functional trait data, in the form of species-level trait measurements, are increasingly accessible from large databases [33,34]. However, the variation in functional traits in response to environmental conditions is still poorly understood [35,36]. Variability of functional traits is important because it can play a role in adaptive and non-adaptive processes under changing environments [37–39]. In case of *S. europaea*, few studies report physiological traits responses [19,20,22,40]. The biochemical parameters, although useful, are black boxes in terms of anatomical and structural changes, which are not directly visualized. This is currently a gap in the literature for this species. To our knowledge, there are no comprehensive studies considering traits in the context of the adaptation and optimum growth of *S. europaea* in the salinity gradient. The level of salinity is still under question, which can be considered as optimal for *S. europaea* trait development. Moreover, it is still unknown which *S. europaea* functional traits are the most affected by salinity.

Therefore, to fill this gap in the knowledge the overarching question, we worked to address whether *S. europaea* traits are from different scales controlled by salinity level. To answer this question, we performed complex research on the morphological, anatomical, and physiological traits at different salinity levels. For morphological and anatomical assessments, we applied a novel image analysis method [5]. To present a complex example of the plant trait functional linkage, we applied similarity analysis between different salt treatments [41]. We also selected the most affected by salinity functional traits by the means of discriminant analysis [42]. We hypothesized that: (a) salinity affects plant morphological, anatomical, and physiological responses in different ways, and (b) plant trait responses can indicate optimum growth in the salinity gradient. Understanding complex mechanisms of salt stress adaptation of *S. europaea*, an extreme halophyte species, is possible only based on traits from different functional levels. The determination of key functional traits in salinity adaptations and *S. europaea* optimum growth is also important because of the possible future agricultural application perspectives.

2. Results

2.1. Growth Responses to Different Salinity Levels

To investigate the morphological trait responses to salt stress, we measured plant height, number of branches, plant surface area, and shoot diameter. We found differences in the morphological features dependent on the salt concentrations (Figures 1 and 2). Plants which grew under extreme salinity (1000 mM NaCl) were smaller than those grown in other salt treatments. They had the smallest plant surface area (ca. 105 cm^2) and were significantly shorter (ca. 4.6 cm) and thinner (diameter ca. 0.248 cm) than the others (Figure 2). Moreover, they had the lowest number of branches (ca. 9). There were not significant differences in plant height and shoot diameter between 0, 200, 400, and 800 mM NaCl treatments (Figure 2a,d). The plant surface area was higher in 400 (283 cm^2) and 800 mM NaCl (ca. 289 cm^2) (Figure 2c). The highest number of branches (ca. 26) was noted at 800 mM, but there were no significant differences between 800, 400, and 200 mM treatments (Figure 2b). Therefore, the relationship between this trait and salinity is not so clear.

For the effect of salinity on biomass accumulation, we measured shoot, root, and total fresh and dry weight. The highest values for fresh and dry weight (SFW 38.5 g, RFW 12.1 g, TFW 50.6 g, SDW 5.4 g, RDW 4.1 g, and TDW 9.5 g) were obtained in 400 mM NaCl, although for all these traits there were no statistically significant differences between 400 and 200 mM NaCl (Figure 3). The lowest values for fresh and dry weights (SFW 2.03 g, RFW 0.44 g, TFW 2.47 g, SDW 0.26 g, RDW 0.08 g, and TDW 0.34 g) were obtained in 1000 mM NaCl. In general, for measurements of fresh and dry weights, there were not significant differences between 0, 800, and 1000 mM NaCl (Figure 3). These results confirm morphological traits assessments, i.e., under 1000 mM NaCl, *S. europaea* cannot grow well. Moreover, growth limitations were detected at 0 and 800 mM NaCl.

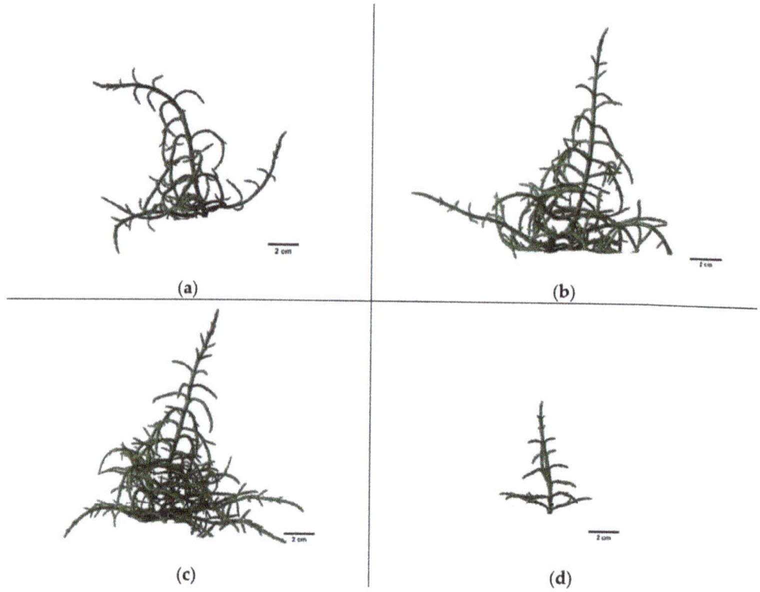

Figure 1. Images of *Salicornia europaea* L. grown in different NaCl concentrations: (**a**) 0 mM; (**b**) 400 mM; (**c**) 800 mM; (**d**) 1000 mM.

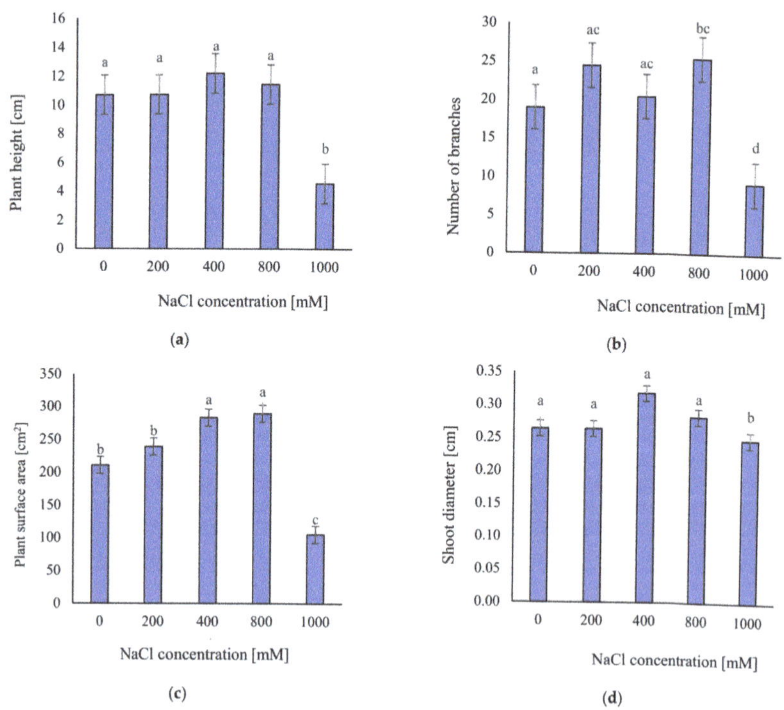

Figure 2. The average parameters of: (**a**) height; (**b**) number of branches; (**c**); surface area; (**d**) shoot diameter of the analyzed plants; ±SD (standard deviation) in the tested samples. ANOVA $p < 0.001$, significant differences based on post hoc Tukey's test are marked with different letters.

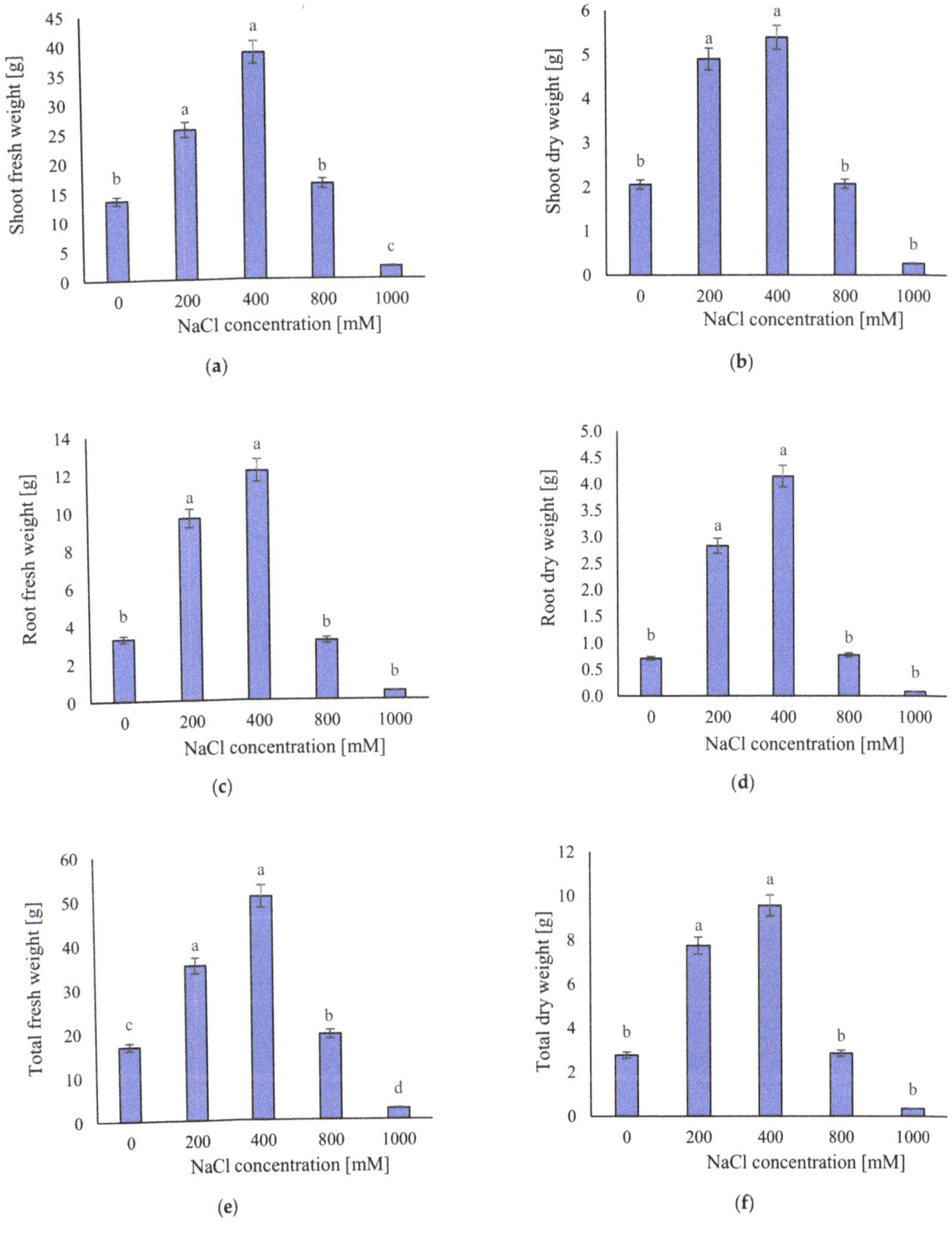

Figure 3. The average parameters of: (**a**) shoot fresh weight; (**b**) shoot dry weight; (**c**) root fresh weight; (**d**) root dry weight; (**e**) total fresh weight; and (**f**) total dry weight in the tested plants; ±SD in the tested samples. ANOVA $p < 0.001$, significant differences based on post hoc Tukey's test are marked with different letters.

2.2. Anatomical Responses to Different Salinity Levels

We based the effect of salinity on anatomical traits on measurements of area, perimeter, diameter, roundness, and aspect ratio of stem-cortex cells (Figure 4). Our results showed that in the control and extremely saline condition (0 and 1000 mM NaCl), the plant cells area, cell perimeter, and cell diameter were lower than in moderate and high saline treatments (200, 400, and 800 mM NaCl) (Figure 5a–c). The plant cells area in the non-saline condition and 1000 mM NaCl was ca. 8394 and 9027 μm^2 respectively, the plant cell perimeter was ca. 345 and 346 μm, and the cell diameter was ca. 130 and 126 μm. There were not statistically significant differences between 200–800 mM NaCl treatments and cells area ranged between 13.360 and 15.967 μm^2, cell perimeter between 431 and 479 μm, and cell diameter between 157 and 177 μm (Figure 5a–c). Moreover, we observed that cells of plants grown under non-saline conditions were significantly less spherical (cell roundness of ca. 0.70), and according to the high aspect ratio (1.53) the most elongated (Figure 5d,e).

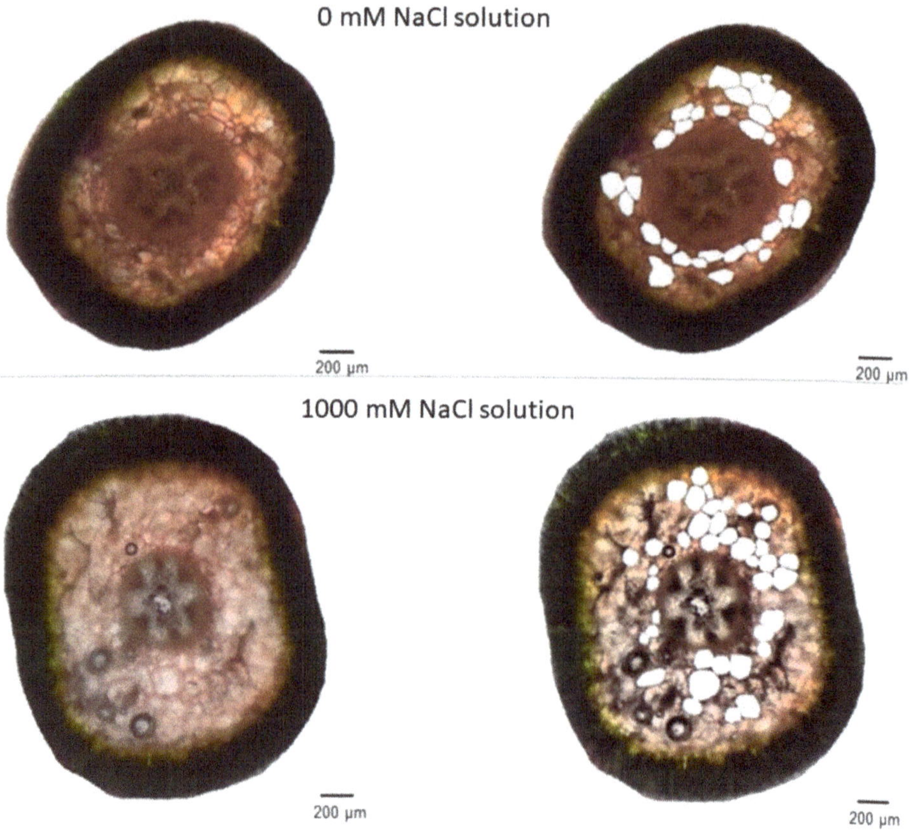

Figure 4. Stem cross-section of *Salicornia europaea* L. from plants treated with 0 and 1000 mM NaCl solutions. At the right side, example cortex cells for measurements by ImageJ software are marked as empty.

These results suggested that under non-saline and extreme saline conditions, stem-cortex cells are smaller probably because of growth stress. We did not find statistically significant differences in anatomical traits between plants grown in 200–800 mM NaCl.

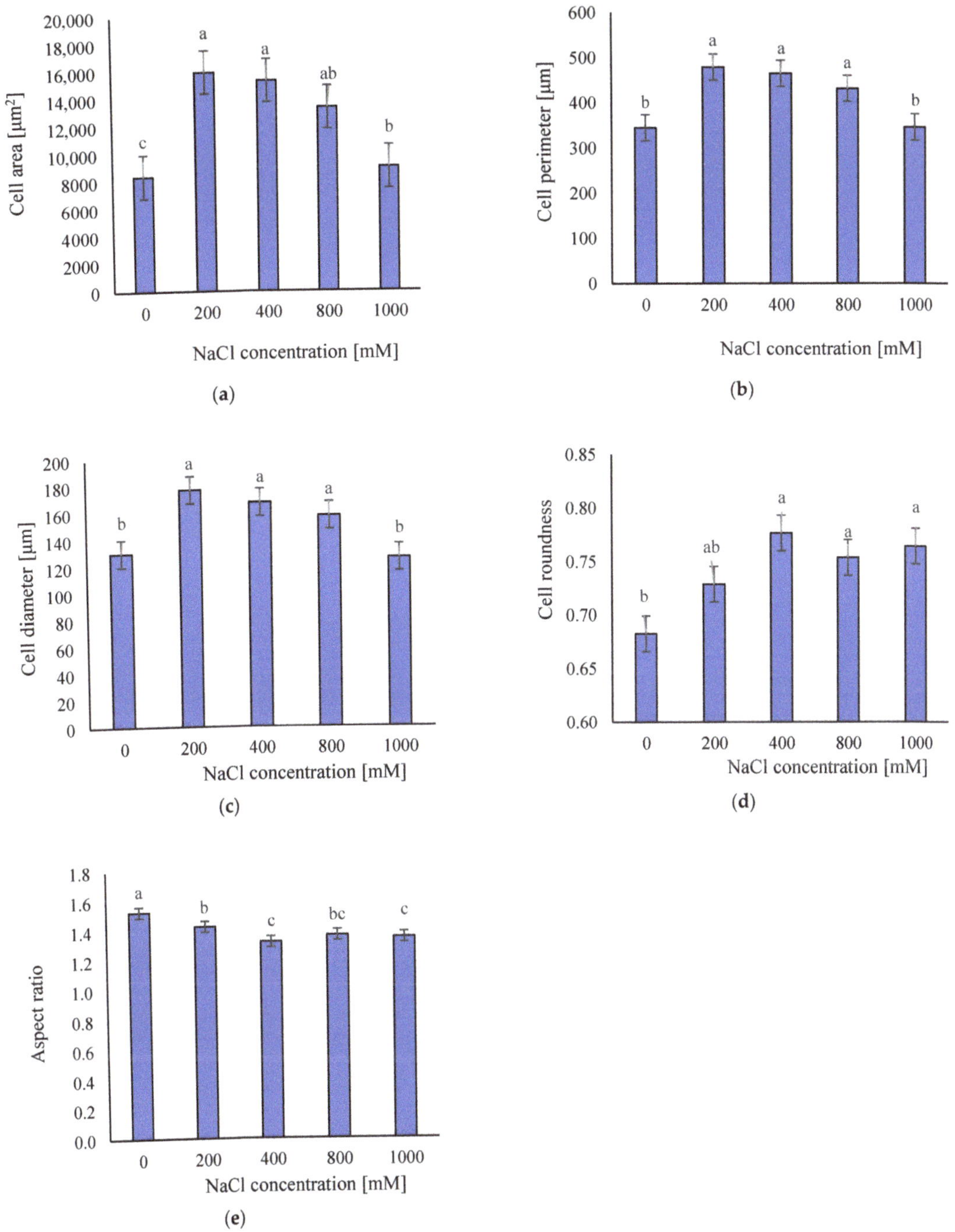

Figure 5. The average parameters of: (**a**) area; (**b**) perimeter; (**c**) diameter; (**d**) roundness; (**e**) aspect ratio of the *Salicornia europaea* L. cells; ±SD in the tested samples. ANOVA $p < 0.001$, significant differences based on post hoc Tukey's test are marked with different letters.

2.3. Biochemical Response to Different Salinity Levels

2.3.1. Photosynthetic Pigments and Soluble Protein Content

Photosynthetic pigments are traits that can affect photosynthetic performance, plant growth, and development. We measured concentrations of chlorophyll a and b and carotenoids. Results demonstrated decreasing photosynthetic pigments content with increasing NaCl concentrations (Figure 6a–c). The highest content was found in control condition (chlorophyll a: 0.715 mg/g FW; chlorophyll b: 0.427 mg/g FW and carotenoids 0.254 mg/g FW). The lowest content was found in 1000 mM NaCl (chlorophyll a: 0.241 mg/g FW; chlorophyll b: 0.151 mg/g FW and carotenoids 0.092 mg/g FW).

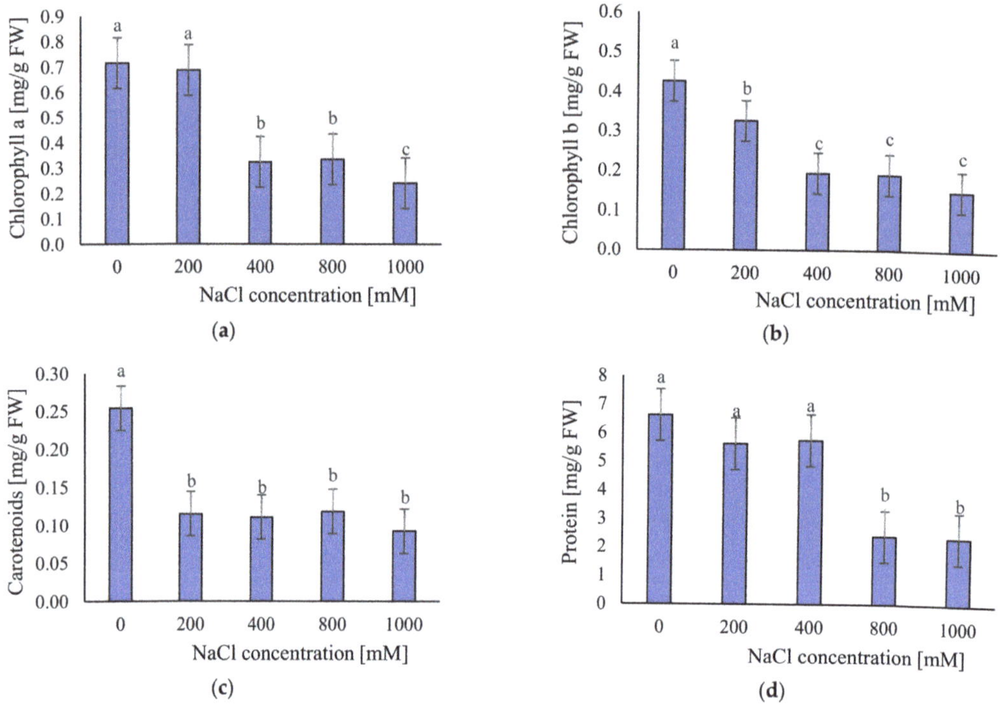

Figure 6. The average concentration of: (**a**) chlorophyll a; (**b**) chlorophyll b; (**c**) carotenoid; and (**d**) soluble protein in the tested plants; ±SD in the tested samples. ANOVA $p < 0.001$, significant differences based on post hoc Tukey's test are marked with different letters.

Soluble protein content was more stable than pigments because, up to 400 mM NaCl, there were not significant changes in their content (ca. 6.0 mg/g FW) (Figure 6d). However, a significant decrease was observed in 800 (ca. 2.39 mg/g FW) and 1000 mM NaCl (ca. 2.35 mg/g FW) (Figure 6d).

2.3.2. Hydrogen Peroxide (H_2O_2) Content

Hydrogen peroxide is a strong ROS (reactive oxygen species) which is overproduced in plant cells under stressful conditions. To investigate how different salinity levels can affect ROS production we measured H_2O_2 content. We observed that under 0 (ca. 4.4 nmol/g FW) to 800 mM NaCl (ca. 14.2 nmol/g FW) the level of H_2O_2 did not differ significantly (Figure 7a). However, we observed a statistically significant increase of the H_2O_2 content in extreme salinity—1000 mM NaCl (ca. 26.2 nmol/g FW). It seems that *S. europaea* has a high tolerance to increasing salinity levels and can prevent H_2O_2 overproduction.

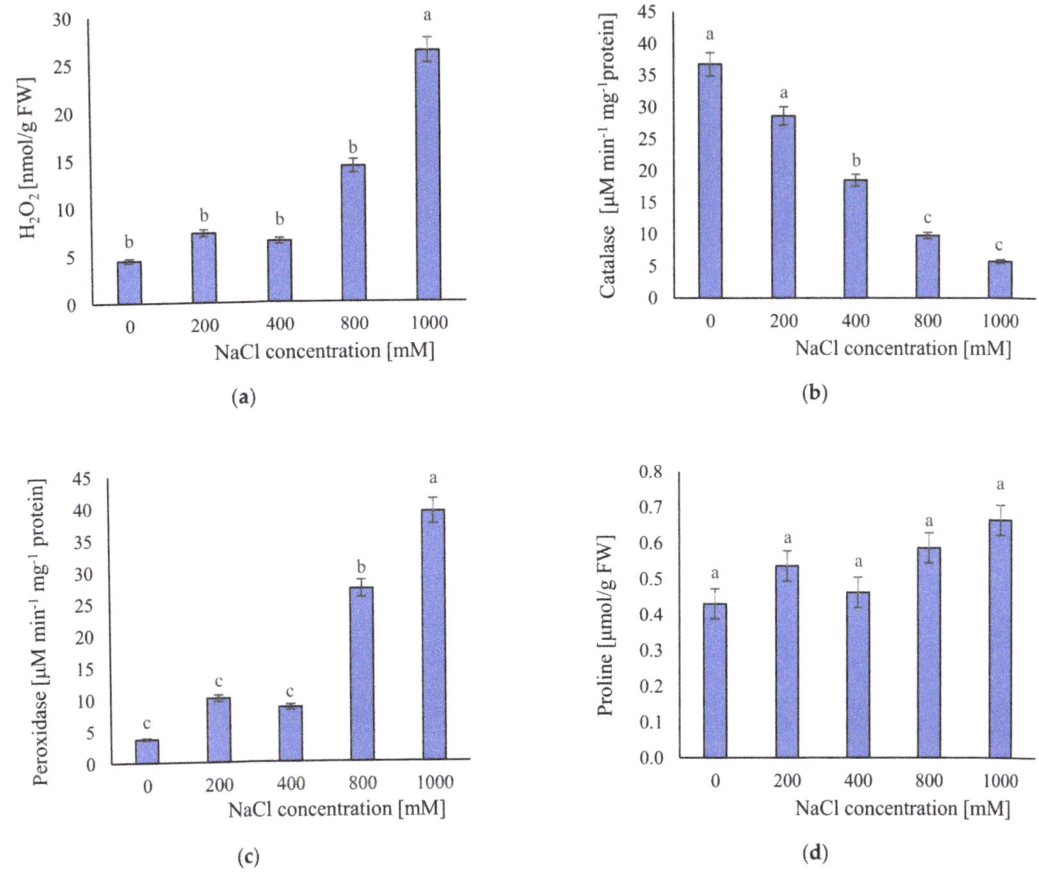

Figure 7. The average parameters of: (**a**) H_2O_2 content; (**b**) catalase activity; (**c**) peroxidase activity; and (**d**) proline content in the tested plants; ±SD in the tested samples. ANOVA $p < 0.001$, significant differences based on post hoc Tukey's test are marked with different letters.

2.3.3. Antioxidant Enzymes Activities and Proline Content

Antioxidant enzymes activities and osmolyte accumulation are two of the most important defense mechanisms against salinity. Therefore, the evaluation of these traits can help in a better understanding of *S. europaea* adaptive response. Thus, we measured activities of two antioxidant enzymes (CAT and POD) and the concentration of proline as an osmolyte accumulation marker. We found that with increasing NaCl concentration, CAT activity decreases. The highest activity of catalase was recorded in plants growing without salinity (36.7 µM min^{-1} mg^{-1} protein), and the lowest in 1000 mM NaCl (5.7 µM min^{-1} mg^{-1} protein) (Figure 7b). These results suggested that catalase in *S. europaea* is sensitive to salt concentration and in low salinity has high activity and efficiency. In contrast, POD activity (Figure 7c) increased with increasing salinity. The lowest POD activity was in plants treated with 0 mM NaCl (3.7 µM min^{-1} mg^{-1} protein), and the highest activity in 1000 mM NaCl (39.2 µM min^{-1} mg^{-1} protein). In the case of proline, we found any statistically significant differences among all salinity levels (Figure 7d). The proline content in plant tissues ranged from ca. 0.4 µmol/g FW in non-saline conditions to 0.7 µmol/g FW in 1000 mM NaCl.

The results suggested that *S. europaea* without NaCl in the medium had a stable proline concentration in tissue which helped it to cope with osmotic stress induced by soil salinity.

3. Discussion

Halophytes are plants that have evolved to survive under high salinity conditions, and many of them are thought to require salt exposure for better growth [43]. These plants can regulate their energy metabolism under saline conditions [43,44]. Thus, a better understanding of halophyte functional traits adaptations is important not only for understanding the natural salt marsh ecosystem function [45,46], but also for reconsidering salt tolerance in glycophyte plants to better manage salinity problems in the future agriculture [47]. As it was already mentioned, up to our knowledge, there is a lack of research involving salt stress responses of complex traits from different functioning levels. Addressing our research to answer the question if *S. europaea* traits from different scales are controlled by salinity level, we found that almost all of them were salinity dependent. However, we confirmed our first assumption, i.e., that traits are affected in different ways.

Analysis of growth responses expressed by morphological properties and biomass accumulation demonstrated that the extreme saline condition of 1000 mM negatively affected these traits. Moreover, biomass accumulation was also reduced at 0 and 800 mM compared with 200 and 400 mM NaCl. This is in line with research that already reported stimulation of *S. europaea* and other halophytes biomass accumulation at moderate salinity [48,49]. Research by Lv et al. [28] demonstrated limitation of plant height, stem diameter, and plant biomass in lower salinity than recorded by us, i.e., of ca. 700 mM NaCl. On the other hand, in line with our findings, they also proved *S. europaea* growth traits limitations in the substrate without salinity. Similar results were obtained by Piernik [27], who reported lower *S. europaea* height and biomass at a very low salinity of ca. 2 dSm^{-1} (ca. 20 mM NaCl) in field experiment. However, in this treatment, opposite to our findings, higher number of branches were noted. Additionally, the results of field research by Silva et al. [50] demonstrated a higher number of *S. europaea* branches at low salinity. It is worth emphasizing that morphological parameters under field conditions can be affected not only by salinity but also by plant density and competition between individuals [51]. Obtained results may suggest that plant biomass is more directly salt stress-dependent than morphological traits.

In the present study, an analysis of the anatomical traits demonstrated that the *S. europaea* stem-cortex cells under non-saline conditions and in 1000 mM NaCl were significantly smaller than in other salt treatments. However, we observed that under control conditions, plant cells were more elongated, while in higher salinities, they were rounder. We did not find statistically significant differences in most anatomical traits between plants grown in 200–800 mM NaCl. Our results are in line with other studies on halophytes that show the increasing cells size in water-absorbing tissues under moderately saline conditions [16]. The water storage in the parenchyma is the first *S. europaea* mechanism to overcome high Na^+ concentrations. This dilutes the accumulated salts and contributes to maintaining cellular turgor, allowing the plant to cope efficiently with high salinity [12,16]. In addition, it was reported that salinity induces vacuolization in many halophytes and may be responsible for the plant adaptation to salt stress [52]. For example, Akcin et al. [52] found that under stressful conditions the juice of halophytes increases. This response is associated with an increase in cell volume due to extra water being stored in the vacuoles for the plant to survive. Moreover, it was reported that *S. europaea* cells can remain turgid and continue proper cell function by ion compartmentalization in cell vacuoles. In this sense, the cytoplasm and organelles of the cell are protected from salt [12]. That is why our results also proved that *S. europaea*, by enlarging its cells under moderately saline conditions, can adapt to high salt stress. On the other hand, the reduction of cell size at extreme salinities can be explained by reduced ability to osmoregulate because of saturation of the system with Na^+ and Cl^- or deficiency of Ca^{2+} and K^+, which are involved in almost all reactions related to plant development [2].

Our results showed that salinity also affects biochemical traits of *S. europaea*. First of all it negatively affects photosynthetic pigments. We observed that the highest concentration of photosynthetic pigments was under non-saline conditions, and the lowest under the highest salinity level (1000 mM NaCl). These results suggested that salinity significantly affects traits which are responsible for photosynthesis performance in plants. Reduction in photosynthetic pigments due to salt stress has been well documented in numerous papers [5,53,54]. This happens by the inactivation of enzymes involved in the synthesis of photosynthetic pigments [55,56]. Furthermore, this reduction can be due to ROS generation and increasing of chlorophyllase enzyme activity [56,57]. However, we found that photosynthetic pigment concentrations were relatively stable at levels above 400 mM NaCl, indicating that the plant can protect, to a certain extent, these traits under extreme saline conditions. High content of photosynthetic pigments in treatments without salinity was not related to high biomass accumulation, which suggests mechanisms of growth limitations other than those directly related to photosynthetic ability. We also calculated Chlorophyll a/b ratio to test if the rearrangement of chlorophyll contents can explain observed by the pattern in *S. europaea* growth performance, as it was reported, for example, in *Arthrocnemum macrostachyum* and *Sarcocornia fruticosa* by Ghanem et al. [58] (Supplementary Figure S1). Based on the obtained results, we can conclude that such a rearrangement was not a strategy of the investigated species.

We also observed that salinity negatively affects soluble protein content of *S. europaea*. This response may be due to the toxic effect of NaCl on protein synthesis, or the proteolysis of proteins caused by proteases induced by salt stress [59]. It is well documented that high concentrations of NaCl destroy the hydration layer of protein, causing its aggregation and denaturation [60]. However, soluble protein compounds were more stable than pigments and maintained a similar level up to 400 mM NaCl.

Hydrogen peroxide, as already stated, is one of the reactive oxygen species responsible for oxidative stress. On the other hand, H_2O_2 is widely generated in many biological systems and mediates various physiological and biochemical processes in plants [61]. Our results showed that the H_2O_2 content was the lowest in the control condition, and by increasing salt levels up to 800 mM NaCl, no significant increase was observed. Only at 1000 mM NaCl was its level significantly higher. The opposite pattern, i.e., increasing level of H_2O_2 in plant tissues together with increasing salinity, has been observed for glycophytic species [54,62,63]. It seems that halophytic *S. europaea* can prevent H_2O_2 overproduction. Nevertheless, in extreme saline conditions, impairment of H_2O_2 production and scavenging can affect membrane structural integrity and peroxidation of lipids and limits plants growth and development [19].

In present study, we also monitored traits related to antioxidant enzymes activities and proline content changes. We found that the highest activity of catalase was under non-saline conditions and decreased together with increasing salt levels. Lower activity of CAT in high salt concentrations indicated CAT as a less effective scavenger of H_2O_2. Furthermore, the CAT has a poor affinity for H_2O_2 and exhibits photo-inactivation and subsequent degradation [64,65]. A completely opposite activity had POD with the lowest values at 0 mM, and the highest at 1000 mM NaCl. In addition, up to 400 mM NaCl POD activity was relatively stable. Increase of the activity of POD at higher salinity levels indicated that this enzyme plays a key role in defense mechanisms of *S. europaea* by scavenging ROS from cells.

Although it is reported that accumulation of proline in *S. europaea* cells is an important adaptive response to salinity [18,43,66], in this study we did not prove a statistically significant increase in proline content by increasing salt levels. Some halophytes produce proline analogues, e.g., glycine betaine under salt stress to survive due to their ability to protect the protein turnover machinery, stabilize proteins, and prevent enzymes from denaturation [66,67]. It seems that *S. europaea* normally has stable proline levels in tissues, which helps it to cope with osmotic stress. However, as proved by Pellegrini et al. [68], proline could be involved in stress tolerance in the *Salicornia* genus regardless of the intensity and duration of the stress. For glycophytic species, a frequently increasing level of proline is observed together with increasing salinity [22,54].

We also referred biochemical traits to dry weight instead of fresh weight to test if differences observed between samples in terms of FW could be partly due to differential loss of water under stress (Supplementary Figures S1 and S2). The tendency in response to salinity level was maintained, and therefore we can conclude that the differences between samples are due to the different levels of stress to which they were subjected.

To conclude on salinity impact, based on all investigated traits and to test our second assumption about indication of plant optimum growth, we performed non-metric multidimensional scaling (NMDS) with Bray-Curtis dissimilarity measure between treatments. Within the analysis we asked for original variables related to the resulting NMDS space as supplementary ones to make the interpretation of the results more clear (Figure 8). Supplementary response variables do not affect the definition of the ordination NMDS axes determined by similarity between samples and can be added to an existing ordination by projection, i.e., by regressing its data on the existing ordination axes [69]. NMDS results demonstrated that some traits can indicate the best *S. europaea* growth expressed by morphological, biomass, and anatomical properties between 200 and 400 mM NaCl (Figure 8). The role of morphological traits adaptations were not stressed up to now. Plants grown in 200 and 400 mM NaCl were the most similar to each other. The first ordination axis explained 98% of the variability between these treatments located at the left side of the diagram and 0 and 1000 mM NaCl located at the right side (Figure 8). The second ordination axis related to the differences between plant traits in 0 and 1000 mM NaCl explained only 1.7% of traits variability, related mostly to oxidative stress. This findings demonstrated by morphological and biomass traits are in line with the field studies of Piernik [27] and the laboratory observations of Lv et al. [28], Cárdenas-Pérez et al. [5], Muscolo et al. [2], and Rozema and Schatb [70] who reported *S. europaea*'s optimum growth at moderate salinity and growth limitation under non-saline conditions. Based on field research and soil sampling on inland salt marshes in Central Europe, Piernik [71] reported the optimum growth of *S. europaea* at 38.1 dSm^{-1}, which corresponds to ca. 380 mM NaCl and is in the range of our findings. The growth stimulation at low and moderate salinity in halophytes such as *S. europaea* may be attributed to the improvement in shoot osmotic status because of increased ion uptake [72]. However, the very high salinity imposed a reduction of the growth, which is probably associated with the reduced turgor and the high energy cost of massive salt secretion and osmoregulation [20,48,73].

Based on discriminant analysis with a forward selection procedure and Monte Carlo permutation test, we selected the most affected traits by salinity. There were chlorophyll a, carotenoids, and protein content explaining ca. 23–25% of the variability between all treatments (Table 1), with the highest values related to low saline conditions. A similar amount of variability, i.e., ca. 25% was explained by plant surface area related to higher salinities. Photosynthetic area, represented in *S. europaea* by plant surface area, determines light interception and is an important parameter in determining plant productivity [74]. Of lower importance but statistically significant in the separation between treatments were POD activity related to the highest salinities, explaining ca. 0.8% of variability, and aspect ratio related to non-saline treatment, explaining 0.7%, and the number of branches and cell perimeter related to the moderate salinities, explaining respectively 0.6% and 0.2% of variability between treatments (Table 1). However, as was already mentioned and discussed, the number of branches can be strongly modified under field conditions not only by salinity but also by competition between individuals for resources [51]. We focused here on conditional effects, which exclude the effect of the most correlated traits. Conditional effects summarize the partial effect of each predictor, representing the variation (and its significance) explained by a predictor after accounting for the effect of predictors already selected [69]. The predictors were chosen in the order of their decreasing explained variation. We skipped simple effects, which summarize the independent effects of all explanatory variables, because they reflected significance assessed by already presented ANOVA.

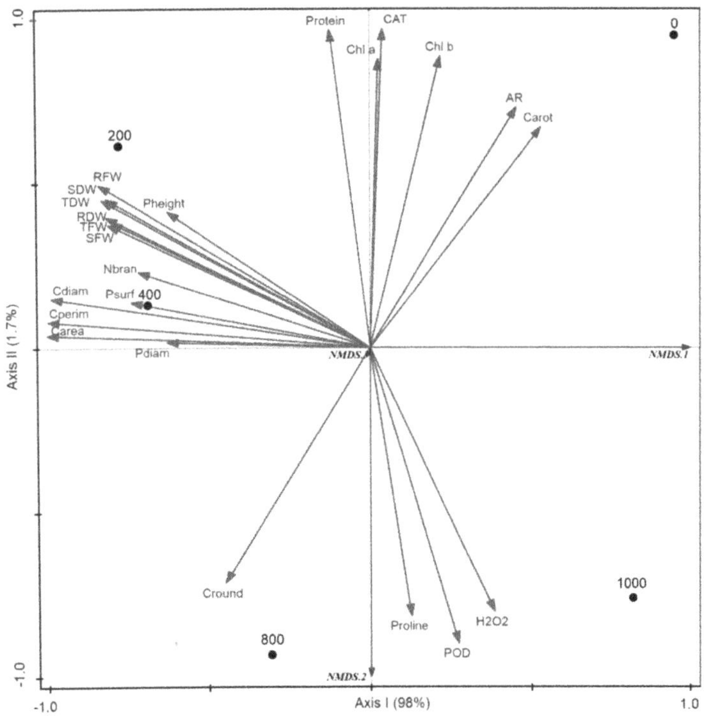

Figure 8. Non-metric multidimentional scaling (NMDS) results—comparison between treatments based on Bray-Curtis dissimilarity measure. Plant functional traits projected as supplementary. Functional traits: Chl a—chlorophyll a, Carot—carotenoids, Psurf—plant surface area, Protein—soluble protein content, POD—peroxidase activity, Nbran—number of branches, AR—aspect ratio, Cperim—cell perimeter, Proline—proline conten, CAT—catalase activity, H_2O_2—hydrogen peroxide, Cdiam—cell diameter, Sdiam—shoot diameter, Carea—cell area, SFW—shoot fresh weight, TDW—total dry weight, Pheight—plant height, Chl b—chlorophyll b, RFW—root fresh weight, SDW—shoot dry weight, Cround—cell roundness. Treatments are marked by dots: 0–1000 mM NaCl. Stress = 0.

Our results demonstrated that though photosynthetic traits are the most affected by salinity, it does not guarantee high productivity of *S. europaea* in non-saline environments and does not affect productivity to the certain salinity level. This can explain why many halophytes are known as light-required species [75]—because they cope with reduced photosynthetic pigments under saline condition. Fortunately, a relatively higher *S. europaea* area under moderate salinity is advantageous for increasing photosynthetic capacity through capturing light at the expense of increased construction costs and produces large boundary layers responsible for reducing transpiration rates and thus heat exchange and carbon dioxide diffusion to the surrounding air [76]. Important findings in our research are also addressed to the anatomic traits most affected by salinity and related to cell shape. As proved by aspect ratio with increasing salinity, cells become rounder, optimizing their surface area and, as was evidenced by cell perimeter, they become enlarged. However, all regulatory mechanisms behind cell adaptations are still not recognized [21]. Our findings can be the key starting points for their future identification.

Table 1. Results of discriminant analysis (CVA) with forward selection and Monte Carlo permutation test demonstrating relative importance and statistical significance of plant traits in the separation of salt treatments 0, 200, 400, 800, and 1000 mM NaCl. Multi-correlated traits have been automatically excluded.

Conditional Effects			
Variable	**V %**	**Pseudo-F**	***p***
Chl a	24.8	7.2	0.002
Carot	24.7	10.3	0.002
Psurf	24.5	18.8	0.002
Protein	23.1	147	0.002
POD	0.8	6.5	0.014
Nbran	0.6	6.4	0.012
AR	0.7	11.3	0.004
Cperim	0.2	3.9	0.03
Proline	<0.1	1	0.372
CAT	<0.1	1.4	0.286
H_2O_2	<0.1	0.8	0.516
Cdiam	<0.1	1.3	0.304
Sdiam	<0.1	0.6	0.598
Carea	<0.1	1	0.396
SFW	<0.1	0.9	0.452
TDW	<0.1	1.2	0.334
Pheight	<0.1	0.6	0.624
Chl b	<0.1	2.2	0.144
RFW	<0.1	0.7	0.5
SDW	<0.1	0.8	0.466
Cround	<0.1	0.2	0.852

Functional traits: Chl a—chlorophyll a, Carot—carotenoids, Psurf—plant surface area, Protein—soluble protein content, POD—peroxidase activity, Nbran—number of branches, AR—aspect ratio, Cperim—cell perimeter, Proline—proline conten, CAT—catalase activity, H_2O_2—hydrogen peroxide, Cdiam—cell diameter, Sdiam—shoot diameter, Carea—cell area, SFW—shoot fresh weight, TDW—total dry weight, Pheight—plant height, Chl b—chlorophyll b, RFW—root fresh weight, SDW—shoot dry weight, Cround—cell roundness. V—variability.

Finally, we would like to highlight that *S. europaea* is a species covering different ecotypes/subspecies, which can be specifically adapted to the local environments [5,12,21]. That is why we reported, in the introduction, *Salicornia ramosissima* J. Woods (= *S. herbacea* L.) from Central Europe [9] as a synonym/reference for the *S. europaea* population investigated by us.

4. Materials and Methods

4.1. Laboratory Plant Material and Plantlets Preparation

S. europaea seeds were collected from the nature reserve of halophytes "Ciechocinek" located in north-central Poland (52°53′ N, 18°47′ E; Central Europe) under permission WOP.6400.12.2020.JC. Matured plants spikes were harvested at the beginning of November, and the seeds were shaken from the spikes. After threshing, 250 seeds that were healthy and uniform in appearance were selected and planted in Petri dishes with Whatman no. 2 filter paper. The Petri dishes were watered with 5 mL distilled water and were placed in a growth chamber with 22 °C and 16 h light period for 9 days. After full germination, uniform plantlets were selected and planted into the pots that contained a mixture of vermiculite and sand (1:1). In total, 60 pots were prepared and divided into five groups (12 pots per each salt concentrations: 0 mM—distilled water, 200 mM, 400 mM, 800 mM, and 1000 mM NaCl). Before planting, each group of 12 pots was located on individual tray lacking drainage and were saturated to their full capacity with relevant NaCl solution (ca. 0.5 L per treatment, pot size: height 5.3 cm, diameter 5.5 cm). All pots were placed in a growth chamber with 22 °C and a 16 h light period and each group was irrigated through pouring distillate water once per week. To prevent nutrient deficiencies, plants were watered with Hoagland's solution once per two weeks [77]. In our previous research we focused on germination and early growth assessments of *S. europaea* [2]. Therefore, this time we focused on plantlets, which were grown for 60 days to the end of the experiment period.

4.2. Growth Analysis

Growth analysis covered morphological and biomass assessments. Two months after starting salt treatments, the plants were subjected to morphological analysis. For this purpose, plant images were taken by a Canon camera with resolution of 5 to 25 MP for morphological analysis. Images were taken from four sides of each sample (see Figure 1) in four replications. We used ImageJ version 1.47 program (National Institutes of Health, Bethesda, MD, USA) for image analysis to measure plant morphological parameters: height, number of branches, surface area, and shoot diameter. The results were calculated according to the number of pixels covering the plant converted to the appropriate metric units.

At the end of the experiment, shoot fresh weight (SFW), root fresh weight (RFW), and total fresh weight (TFW) were measured. After the samples were oven dried for 72 h at 72 °C, shoot dry weight (SDW), root dry weight (RDW), and total dry weight (TDW) were determined.

4.3. Anatomical Analysis

The anatomical analysis was performed based on the cross-sections of plant shoots (middle primary branch—fleshy segment shoot) taken from plants growing in five NaCl concentrations. Slices of fresh tissue were obtained by cutting them with a sharp bi-shave blade and slices of approximately 0.5 mm were used for analysis. The samples were stained with alkaline fuchsin and malachite green for microscopic observations. *S. europaea* cross-sections photos for each NaCl concentration were taken using light microscopy. Photos were used to observe the cortex parenchyma cells, because water storage in the parenchyma is the first *S. europaea* mechanism to overcome high Na^+ concentrations [28]. We used ImageJ version 1.47 (National Institutes of Health, Bethesda, MD, USA) for image analysis. For each sample, more than 200 cells were analyzed to determine the following parameters: (1) cell area (A), which was calculated through the number of pixels inside the borderline; (2) cell perimeter, which was calculated due to the prescribed limits; (3) cell diameter, which was defined by the distance between two points; (4) aspect ratio (AR), which was defined as the quotient between the minimum and maximum diameter, determining the uniformity of the cell; and (5) cell roundness (R), which determines the circularness of a cell using Equation (1). In this equation, for perfectly round cells R = 1 [78].

$$R = (4A)/(\pi(MD)^2), \quad (1)$$

where: A is cell area; MD is cell diameter.

4.4. Biochemical Analyzes

To evaluate the effect of different salinity levels on the biochemical traits of *S. europaea* plants, we measured traits related to photosynthetic activity, osmotic adjustment, oxidative stress, and enzymatic activities.

4.4.1. Photosynthetic Pigments Content

Chlorophylls (Ch a and Ch b) and carotenoids were extracted from fresh plant stems (100 mg) using 80% acetone for 6 h in darkness, and then centrifuged at 10,000 rpm, 10 min. Supernatants were quantified spectrophotometrically. The absorbances at 646, 663, and 470 nm wavelength were measured. The total content of chlorophyll a and b [79] and carotenoids [80], when 80% of acetone was used as dissolvent, were calculated according to Equations (2)–(4) and reported as mg-per-gram fresh weight.

$$\text{Chlorophyll (a)} = [(12.21 \times \text{Abs663}) - (2.81 \times \text{Abs646}) \times \text{mLAcetone}]/\text{mgSteam} \quad (2)$$

$$\text{Chlorophyll (b)} = [(20.13 \times \text{Abs646}) - (2.81 \times \text{Abs663}) \times \text{mLAcetone}]/\text{mgSteam} \quad (3)$$

$$\text{Carotenoids} = (([(1000 \times \text{Abs470} - 3.27(\text{Chla}) - 104(\text{Chlb}))/227] \times \text{mLAcetone}))/\text{mgSteam} \quad (4)$$

4.4.2. Soluble Protein Content

Plant material (0.5 g) was placed in a mortar chilled with liquid nitrogen, then 7 mL of 50 mM phosphate buffer was added and homogenized. After fine grinding, the sample was centrifuged at 20,000× *g* for 25 min at 4 °C. The supernatant was collected for determination of protein content at 595 nm using the Bradford method [81]. The amount of protein was determined using a bovine serum albumin as a standard.

4.4.3. Hydrogen Peroxide (H_2O_2) Content

Hydrogen peroxide was measured according to the methods described by Velikova et al. [82]. Three samples were prepared for each NaCl concentration. Stem tissues (0.5 g) were homogenized in an ice bath with 5 mL of 0.1% TCA (trichloroacetic acid). Then, the homogenate was centrifuged at 12,000× *g* for 15 min at 4 °C. The supernatant (0.5 mL) was transferred to a new tube and 0.5 mL of 10 mM potassium phosphate (pH 7) and 2 mL of 1 M KI were added. The solution was incubated in the dark for one hour and then the absorbance was read at 390 nm wavelength and the hydrogen peroxide concentration was given based on the standard curve from 0 to 40 nM and equation y = 0.0188x + 0.046, R^2 = 0.987 in nM-per-gram fresh weight.

4.4.4. Peroxidase (POD) Activity

The peroxidase activity was measured according to the method of Chance and Maehly [83]. For the measurement, a 3-mL reaction mixture was prepared with the following composition: 50 mM potassium phosphate buffer (pH 7), 20 mM guaiacol, 40 mM H_2O_2, and 0.1 mL of enzyme extract obtained from plant stems. The reaction was initiated by adding the enzyme extract. The increase in absorbance of the reaction solution was measured at 470 nm wavelength. The reading was recorded every 20 s. One unit of peroxidase activity was defined as an absorbance change of 0.01 units per minute. The enzyme activity was expressed on a protein basis [84].

4.4.5. Catalase (CAT) Activity

Catalase was determined by measuring residual H_2O_2 with a titanium reagent [85]. The reaction mixture was 3 mL in total and consisted of: 1 mL of 6 mM H_2O_2 and 1.9 mL of 0.1 M phosphate buffer (pH 7). To start the reaction, 0.1 mL of diluted enzyme extract was added to the test tube. The reaction was stopped after 5 min by adding 4 mL of titanium reagent, which formed a colored complex with residual H_2O_2. Mixtures without enzyme were used as control. The colored complex was centrifuged at 10,000× *g* for 10 min. The absorbance was measured at 415 nm. The residual H_2O_2 content was calculated from the standard curve [86].

4.4.6. Proline Content

The proline content of the leaves was determined according to the method of Bates et al. [87]. Three samples were prepared for each NaCl concentration; 0.5 g of each sample were pulverized on ice and homogenized in a mortar with 5 mL of 3% aqueous sulfosalicylic acid. The homogenate was centrifuged at 18,000× *g*, 10 min at 1 °C, and the supernatant was collected. About 1 mL of the supernatant was added into a test tube and 2 mL of glacial acetic acid, 2 mL of ninhydrin reagent, and 1 mL of sulfosalicylic acid 3% were added to the tube. The reaction mixture was boiled in a 100 °C water bath for one hour. Then, the test tubes were put on ice and 4 mL of toluene were added to them. The samples were transferred to a separating funnel and after thorough mixing, the toluene containing the chromophore was separated. The absorbances were measured at 520 nm wavelength. The amount of proline was determined using the standard curve in the concentration range of 0 to 40 μg mL^{-1} and equation y = 0.0467x − 0.0734, R^2 = 0.963.

4.5. Statistical Analyzes

To compare all treatments, one-way analysis of variance (ANOVA) was performed followed by post hoc analysis using Tukey's test. The PAST 4.0 program [88] was used for statistical analyses. To discuss *S. europaea*'s optimum growth based on the whole set of the traits, we applied non-metric multidimensional scaling (NMDS) with Bray-Curtis dissimilarity measure to demonstrate similarities between plants coming from different salinity treatments. To select the most affected traits by salinity treatments, we applied discriminant analysis with a forward-selection procedure and Monte Carlo permutation test. Only conditional effects were taken into account to exclude the effect of the most correlated traits [42]. For these analyses, the Canoco 5.0 program was applied [69].

5. Conclusions

Addressing our research to answer the question if *S. europaea* traits from different scales are controlled by salinity level, we found that almost all of them were salinity dependent. However, we proved that functional traits were affected by salinity in the different ways and demonstrated significant differences at different salinity levels. Moreover, we did not find a statistically significant relationship between proline levels and increasing salinity, which was not expected based on reported findings regarding glycophyte species. Based on analysis of all investigated traits, we demonstrated that morphological, biomass, and anatomical properties indicated optimum growth between 200 and 400 mM NaCl and growth limitations at 0, 800, 1000 mM NaCl. Moreover, we can conclude that the most affected traits by salinity include photosynthetic pigments, protein content, plant surface area, peroxidase activity, and anatomic traits related to cell shape. Our results significantly expand the knowledge about *S. europaea* functional traits variation in response to salinity, which can be important for discovering regulating processes and for possible future agricultural applications.

Supplementary Materials: The following supporting information can be downloaded at: https://www.mdpi.com/article/10.3390/plants11081051/s1, Figure S1: The average concentration of: (a) chlorophyll a; (b) chlorophyll b; (c) chlorophyll a/b ratio, (d) carotenoid and (e) soluble proteins in the tested plants; Figure S2: The average parameters of: (a) H_2O_2 content; (b) proline content in the tested plants.

Author Contributions: Conceptualization, A.P. and S.C.-P.; data curation, S.C.-P., A.R.D. and A.P.; formal analysis, S.C.-P., K.L., A.R.D. and A.P.; investigation, K.L. and S.C.-P.; methodology, S.C.-P. and A.P.; software, S.C.-P. and A.P.; supervision, A.P.; validation, A.R.D., A.P. and S.L.-M.; visualization, A.R.D., A.P. and S.C.-P.; writing—original draft, A.R.D., S.L.-M. and A.P.; writing—review and editing, A.L. and S.C.-P. All authors have read and agreed to the published version of the manuscript.

Funding: This research received no external funding.

Institutional Review Board Statement: Not applicable.

Informed Consent Statement: Not applicable.

Data Availability Statement: The data presented in this study are available on request from the corresponding authors. The data are not publicly available due to ongoing research.

Acknowledgments: This study was supported by funds for the Nicolaus Copernicus University in Toruń, Poland.

Conflicts of Interest: The authors declare no conflict of interest.

References

1. Santos, J.; Al-Azzawi, M.; Aronson, J.; Flowers, T.J. eHALOPH a database of salt-tolerant plants: Helping put halophytes to work. *Plant Cell Physiol.* **2016**, *57*, e10. [CrossRef] [PubMed]
2. Muscolo, A.; Panuccio, M.R.; Piernik, A. Ecology, Distribution and Ecophysiology of *Salicornia Europaea* L. In *Sabkha Ecosystems*; Khan, M.A., Böer, B., Öztürk, M., Al Abdessalaam, T.Z., Clüsener-Godt, M., Gul, B., Eds.; Springer: Dordrecht, The Netherlands, 2014; Volume 47, pp. 233–240.
3. Kadereit, G.; Ball, P.; Beer, S.; Mucina, L.; Sokoloff, D.; Teege, P.; Yaprak, A.E.; Freitag, H. A taxonomic nightmare comes true: Phylogeny and biogeography of glassworts (*Salicornia* L., *Chenopodiaceae*). *Taxon* **2007**, *56*, 1143–1170. [CrossRef]
4. Cárdenas-Pérez, S.; Chanona-Pérez, J.J.; de Jesús Perea-Flores, M.; Calderon, H.; Piernik, A.; López-Soto, K.D.; González, C.B.G. Microstructure of *Salicornia bigelovii* stems under photonic and electron microscopy. *Microsc. Microanal.* **2020**, *26*, 360–362. [CrossRef]
5. Cárdenas-Pérez, S.; Piernik, A.; Ludwiczak, A.; Duszyn, M.; Szmidt-Jaworska, A.; Chanona-Pérez, J. Image and fractal analysis as a tool for evaluating salinity growth response between two *Salicornia europaea* populations. *BMC Plant Biol.* **2020**, *20*, 467. [CrossRef]
6. Negrão, S.; Schmöckel, S.; Tester, M. Evaluating physiological responses of plants to salinity stress. *Ann. Bot.* **2017**, *119*, 1–11. [CrossRef]
7. Loconsole, D.; Cristiano, G.; De Lucia, B. Glassworts: From wild salt marsh species to sustainable edible crops. *Agriculture* **2019**, *9*, 14. [CrossRef]
8. Araus, J.L.; Rezzouk, F.Z.; Thushar, S.; Shahid, M.; Elouafi, I.A.; Bort, J.; Serret, M.D. Effect of irrigation salinity and ecotype on the growth, physiological indicators and seed yield and quality of *Salicornia europaea*. *Plant. Sci.* **2021**, *304*, 110819. [CrossRef]
9. Krüger, A.; Hellwig, F.; Oberprieler, C. Genetic diversity in natural and anthropogenic inland populations of salt-tolerant plants: Random amplified polymorphic DNA analyses of *Aster tripolium* L. (*Compositae*) and *Salicornia ramosissima* Woods (*Chenopodiaceae*). *Mol. Ecol.* **2002**, *11*, 1647–1655. [CrossRef]
10. Antunes, M.D.; Gago, C.; Guerreiro, A.; Sousa, A.R.; Julião, M.; Miguel, M.G.; Faleiro, M.L.; Panagopoulos, T. Nutritional characterization and storage ability of *Salicornia ramosissima* and *Sarcocornia perennis* for fresh vegetable salads. *Horticulturae* **2021**, *7*, 6. [CrossRef]
11. Rhee, M.H.; Park, H.-J.; Cho, J.Y. *Salicornia herbacea*: Botanical, chemical and pharmacological review of halophyte marsh plant. *J. Med. Plants Res.* **2009**, *3*, 548–555. [CrossRef]
12. Cárdenas-Pérez, S.; Piernik, A.; Chanona-Pérez, J.J.; Grigore, M.N.; Perea-Flores, M.J. An overview of the emerging trends of the *Salicornia* L. genus as a sustainable crop. *Environ. Exp. Bot.* **2021**, *191*, 104606. [CrossRef]
13. Liu, X.G.; Xia, Y.G.; Wang, F.; Sun, M.; Jin, Z.J.; Wang, G.T. Analysis of fatty acid composition of *Salicornia europaea* L. seed oil. *Food Sci.* **2005**, *2*, 42.
14. Gunning, D. *Cultivating Salicornia Europaea (Marsh Samphire)*; Irish Sea Fisheries Board: Dublin, Ireland, 2016; Volume 4, pp. 1–95.
15. Patel, S. Salicornia: Evaluating the halophytic extremophile as a food and a pharmaceutical candidate. *3 Biotech* **2016**, *6*, 104. [CrossRef] [PubMed]
16. Grigore, M.-N.; Toma, C. Morphological and anatomical adaptations of halophytes: A review. In *Handbook of Halophytes: From Molecules to Ecosystems towards Biosaline Agriculture*; Springer: Cham, Switzerland, 2021; pp. 1079–1221. [CrossRef]
17. Flowers, T.J.; Colmer, T.D. Salinity tolerance in halophytes. *New Phytol.* **2008**, *179*, 945–963. [CrossRef] [PubMed]
18. Flowers, T.J.; Hajibagher, M.A.; Yeo, A.R. Ion accumulation in the cell walls of rice plants growing under saline conditions: Evidence for the Oertli hypothesis. *Plant Cell Environ.* **1991**, *14*, 319–325. [CrossRef]
19. Aghaleh, M.; Niknam, V.; Ebrahimzadeh, H.; Razavi, K. Effect of salt stress on physiological and antioxidative responses in two species of Salicornia (*S. persica* and *S. europaea*). *Acta Physiol. Plant.* **2011**, *33*, 1261–1270. [CrossRef]
20. Mohammadi, H.; Kardan, J. Morphological and physiological responses of some halophytes to salinity stress. *Ann. Univ. Mariae Curie Sklodowska Sect. C Biol.* **2016**, *70*, 31. [CrossRef]
21. Cárdenas-Pérez, S.; Niedojadło, K.; Mierek-Adamska, A.; Dąbrowska, G.B.; Piernik, A. Maternal salinity influences anatomical parameters, pectin content, biochemical and genetic modifications of two *Salicornia europaea* populations under salt stress. *Sci. Rep.* **2022**, *12*, 2968. [CrossRef]
22. Mohamed, M.; El-Maboud, A. Seasonal Physiological Response of *Salicornia Europaea* L. *Res. Rev. J. Bot. Sci.* **2021**, *10*, 118–124.
23. Akcin, A.; Yalcin, E. Effect of salinity stress on chlorophyll, carotenoid content, and proline in *Salicornia prostrata* Pall. and *Suaeda prostrata* Pall. subsp. *prostrata* (Amaranthaceae). *Braz. J. Bot.* **2016**, *39*, 101–106. [CrossRef]
24. Rodríguez-Álvarez, M.; Ledea-Rodríguez, J.L.; Murillo-Amador, B.; Mazón-Suástegui, J.M. Morphometry and chlorophyll content of *Salicornia bigelovii* (*Torr*) under sea water treatments and *Natrum muriaticum* as a saline stress mitigator. *Trop. Subtrop. Agroecosystems* **2022**, *25*, 5.
25. Wang, X.; Fan, P.; Song, H.; Chen, X.; Li, X.; Li, Y. Comparative proteomic analysis of differentially expressed proteins in shoots of *Salicornia europaea* under different salinity. *J. Proteome Res.* **2009**, *8*, 3331–3345. [CrossRef] [PubMed]
26. Snow, A.A.; Vince, S.W. Plant zonation in Alaskan salt-marsh. II. An Experimental study of the role of edaphic conditions. *J. Ecol.* **1984**, *72*, 669–684. [CrossRef]
27. Piernik, A. Growth of three meadow species along a salinity gradient in an inland saline habitat: Transplant experiment. *Pol. J. Ecol.* **2006**, *54*, 117–126.

28. Lv, S.; Jiang, P.; Chen, X.; Fan, P.; Wang, X.; Li, Y. Multiple compartmentalization of sodium conferred salt tolerance in *Salicornia europaea*. *Plant Physiol. Biochem.* **2012**, *51*, 47–52. [CrossRef]
29. Singh, D.; Buhmann, A.K.; Flowers, T.J.; Seal, C.E.; Papenbrock, J. Salicornia as a crop plant in temperate regions: Selection of genetically characterized ecotypes and optimization of their cultivation conditions. *AoB Plants* **2014**, *6*, 6. [CrossRef]
30. Rao, G.M.N.; Murty, P.P. Morphological and anatomical features of *Salicornia brachiata Roxb*. *J. Biol. Chem. Res.* **2013**, *30*, 887–891.
31. Violle, C.; Navas, M.L.; Vile, D.; Kazakou, E.; Fortunel, C.; Hummel, I.; Garnier, E. Let the concept of trait be functional! *Oikos* **2007**, *116*, 882–892. [CrossRef]
32. Chapin, F.S., III; Zavaleta, E.S.; Eviner, V.T.; Naylor, R.L.; Vitousek, P.M.; Reynolds, H.L.; Hooper, D.U.; Lavorel, S.; Sala, O.E.; Hobbie, S.E.; et al. Consequences of changing biodiversity. *Nature* **2000**, *405*, 234–242. [CrossRef]
33. Enquist, B.J.; Condit, R.; Peet, R.K.; Schildhauer, M.; Thiers, B.M. Cyberinfrastructure for an integrated botanical information network to investigate the ecological impacts of global climate change on plant biodiversity. *PeerJ Prepr.* **2016**, *4*, e2615v2.
34. Kattge, J.; Bönisch, G.; Dìaz, S.; Lavorel, S.; Prentice, I.C.; Leadley, P.; Tautenhahn, S.; Werner, G.D.A.; Aakala, T.; Abedi, M.; et al. TRY plant trait database—Enhanced coverage and open access. *Glob. Chang. Biol.* **2020**, *26*, 119–188. [CrossRef] [PubMed]
35. Zirbel, C.R.; Bassett, T.; Grmanand, E.; Brudvig, L.A. Plant functional traits and environmental conditions shape community assembly and ecosystem functioning during restoration. *J. Appl. Ecol.* **2017**, *54*, 1070–1079. [CrossRef]
36. Bu, W.; Huang, J.; Xu, H.; Zang, R.; Ding, Y.; Li, Y.; Lin, M.; Wang, J.; Zhang, C. Plant Functional Traits Are the Mediators in Regulating Effects of Abiotic Site Conditions on Aboveground Carbon Stock-Evidence from a 30 ha Tropical Forest. *Plot. Front. Plant Sci.* **2019**, *9*, 1958. [CrossRef] [PubMed]
37. Helsen, K.; Acharya, K.P.; Brunet, J.; Cousins, S.A.O.; Decocq, G.; Kolb, A.; Lemke, I.H.; Lenoir, J.; Plue, J.; Verheyen, K.; et al. Biotic and abiotic drivers of intraspecific trait variation within plant populations of three herbaceous plant species along a latitudinal gradient. *BMC Ecol.* **2017**, *17*, 38. [CrossRef]
38. Pérez-Harguindeguy, N.; Díaz, S.; Garnier, E.; Lavorel, S.; Poorter, H.; Jaureguiberry, P.; Cornelisson, J.H.C. New handbook for standardised measurement of plant functional traits worldwide. *Aust. J. Bot.* **2013**, *61*, 167–234. [CrossRef]
39. Raffard, A.; Lecerf, A.; Cote, J.; Buoro, M.; Lassus, R.; Cucherousset, J. The functional syndrome: Linking individual trait variability to ecosystem functioning. *Proc. R. Soc. B Biol. Sci.* **2017**, *284*, 20171893. [CrossRef]
40. Fan, P.; Nie, L.; Jiang, P.; Feng, J.; Lv, S.; Chen, X.; Bao, H.; Guo, J.; Tai, F.; Wang, J.; et al. Transcriptome Analysis of *Salicornia europaea* under Saline Conditions Revealed the Adaptive Primary Metabolic Pathways as Early Events to Facilitate Salt Adaptation. *PLoS ONE* **2013**, *8*, e80595. [CrossRef]
41. Kaleem, M.; Hameed, M. Functional traits for salinity tolerance in differently adapted populations of *Fimbristylis complanata* (Retz.). *Int. J. Phytoremediation* **2021**, *23*, 1319–1332. [CrossRef]
42. Šmilauer, P.; Lepš, J. *Multivariate Analysis of Ecological Data using CANOCO 5*, 2nd ed.; Cambridge University Press: Cambridge, UK, 2014. [CrossRef]
43. Kumar, A.; Mann, A.; Lata, C.; Kumar, N.; Sharma, P. Salinity-induced physiological and molecular responses of halophytes. In *Research Developments in Saline Agriculture*; Springer: Singapore, 2019; pp. 331–356.
44. Winicov, I.; Bastola, D.R. Salt tolerance in crop plants: New approaches through tissue culture and gene regulation. *Acta Physiol. Plant.* **1997**, *19*, 435–449. [CrossRef]
45. Ulrich, W.; Kubota, Y.; Piernik, A.; Gotelli, N.J. Functional traits and environmental characteristics drive the degree of competitive intransitivity in European saltmarsh plant communities. *J. Ecol.* **2018**, *106*, 865–876. [CrossRef]
46. Ulrich, W.; Hulisz, P.; Mantilla-Contreras, J.; Elvisto, T.; Piernik, A. Compensatory effects stabilize the functioning of Baltic brackish and salt marsh plant communities. *Estuar. Coast. Shelf Sci.* **2019**, *231*, 106480. [CrossRef]
47. Shabala, S. Learning from halophytes: Physiological basis and strategies to improve abiotic stress tolerance in crops. *Ann. Bot.* **2013**, *112*, 1209–1221. [CrossRef] [PubMed]
48. Aghaleh, M.; Niknam, V.; Ebrahimzadeh, H.; Razavi, K. Salt stress effects on growth, pigments, proteins and lipid peroxidation in *Salicornia persica* and *S. europaea*. *Biol. Plant.* **2009**, *53*, 243–248. [CrossRef]
49. Orlovsky, N.; Japakova, U.; Zhang, H.; Volis, S. Effect of salinity on seed germination, growth and ion content in dimorphic seeds of *Salicornia europaea* L. (*Chenopodiaceae*). *Plant Divers* **2016**, *38*, 183–189. [CrossRef] [PubMed]
50. Silva, H.; Caldeira, G.; Freitas, H. *Salicornia ramosissima* population dynamics and tolerance of salinity. *Ecol. Res.* **2007**, *22*, 125–134. [CrossRef]
51. Craine, J.M.; Dybzinski, R. Mechanisms of plant competition for nutrients, water and light. *Funct. Ecol.* **2013**, *27*, 833–840. [CrossRef]
52. Akcin, T.A.; Akcin, A.; Yalcın, E. Anatomical changes induced by salinity stress in *Salicornia freitagii* (*Amaranthaceae*). *Braz. J. Bot.* **2017**, *40*, 1013–1018. [CrossRef]
53. Gong, D.H.; Wang, G.Z.; Si, W.T.; Zhou, Y.; Liu, Z.; Jia, J. Effects of salt stress on photosynthetic pigments and activity of ribulose-1, 5-bisphosphate carboxylase/oxygenase in *Kalidium foliatum*. *Russ. J. Plant Physiol.* **2018**, *65*, 98–103. [CrossRef]
54. Rajabi Dehnavi, A.; Zahedi, M.; Ludwiczak, A.; Piernik, A. Foliar Application of Salicylic Acid Improves Salt Tolerance of Sorghum (*Sorghum bicolor* (L.) Moench). *Plants* **2022**, *11*, 368. [CrossRef]
55. Ashraf, M. The effect of NaCl on water relations, chlorophyll, and protein and proline contents of two cultivars of blackgram (*Vigna mungo* L.). *Plant Soil* **1989**, *119*, 205–210. [CrossRef]

56. Arif, Y.; Singh, P.; Siddiqui, H.; Bajguz, A.; Hayat, S. Salinity induced physiological and biochemical changes in plants: An omic approach towards salt stress tolerance. *Plant Physiol. Biochem.* **2020**, *156*, 64–77. [CrossRef] [PubMed]
57. Arulbalachandran, D.; Ganesh, K.S.; Subramani, A. Changes in metabolites and antioxidant enzyme activity of three *Vigna* species induced by NaCl stress. *Am. Eurasian J. Agron.* **2009**, *2*, 109–116.
58. Ghanem, A.F.M.; Mohamed, E.; Kasem, A.M.M.A.; El-Ghamery, A.A. Differential Salt Tolerance Strategies in Three Halophytes from the Same Ecological Habitat: Augmentation of Antioxidant Enzymes and Compounds. *Plants* **2021**, *10*, 1100. [CrossRef] [PubMed]
59. Parida, A.K.; Das, A.B.; Mittra, B.; Mohanty, P. Salt-stress induced alterations in protein profile and protease activity in the *mangrove Bruguiera parviflora*. *Z. Für Nat. C* **2004**, *59*, 408–414. [CrossRef] [PubMed]
60. Fogarty, A.C.; Duboué-Dijon, E.; Sterpone, F.; Hynes, J.T.; Laage, D. Biomolecular hydration dynamics: A jump model perspective. *Chem. Soc. Rev.* **2013**, *42*, 5672–5683. [CrossRef] [PubMed]
61. Li, J.-T.; Qiu, Z.-B.; Zhang, X.-W.; Wang, L.-S. Exogenous hydrogen peroxide can enhance tolerance of wheat seedlings to salt stress. *Acta Physiol. Plant.* **2011**, *33*, 835–842. [CrossRef]
62. Liu, L.; Huang, L.; Lin, X.; Sun, C. Hydrogen peroxide alleviates salinity-induced damage through enhancing proline accumulation in wheat seedlings. *Plant Cell Rep.* **2020**, *39*, 567–575. [CrossRef]
63. Hernandez, M.; Fernandez-Garcia, N.; Diaz-Vivancos, P.; Olmos, E. A different role for hydrogen peroxide and the antioxidative system under short and long salt stress in *Brassica oleracea* roots. *J. Exp. Bot.* **2010**, *61*, 521–535. [CrossRef]
64. Shang, W.; Feierabend, J. Dependence of catalase photoinactivation in rye leaves on light intensity and quality and characterization of a chloroplast-mediated inactivation in red light. *Photosynth. Res.* **1999**, *59*, 201–213. [CrossRef]
65. Amor, N.B.; Jiménez, A.; Megdiche, W.; Lundqvist, M.; Sevilla, F.; Abdelly, C. Response of antioxidant systems to NaCl stress in the halophyte *Cakile maritima*. *Physiol. Plant.* **2006**, *126*, 446–457. [CrossRef]
66. Moghaieb, R.E.A.; Saneoka, H.; Fujita, K. Effect of salinity on osmotic adjustment, glycinebetaine accumulation and the betaine aldehyde dehydrogenase gene expression in two halophytic plants, *Salicornia europaea* and *Suaeda maritima*. *Plant Sci.* **2004**, *166*, 1345–1349. [CrossRef]
67. Shoukat, E.; Ahmed, M.Z.; Abideen, Z.; Azeem, M.; Ibrahim, M.; Gul, B.; Khan, M.A. Short and long term salinity induced differences in growth and tissue specific ion regulation of *Phragmites karka*. *Flora* **2020**, *263*, 151550. [CrossRef]
68. Pellegrini, E.; Forlani, G.; Boscutti, F.; Casolo, V. Evidence of non-structural carbohydrates-mediated response to flooding and salinity in *Limonium narbonense* and *Salicornia fruticosa*. *Aquat. Bot.* **2020**, *166*, 103265. [CrossRef]
69. Ter Braak, C.J.F.; Smilauer, P. *Canoco Reference Manual and User's Guide: Software for Ordination, Version 5.0*; Microcomputer Power: Ithaca, NY, USA, 2012.
70. Rozema, J.; Schat, H. Salt tolerance of halophytes, research questions reviewed in the perspective of saline agriculture. *Environ. Exp. Bot.* **2013**, *92*, 83–95. [CrossRef]
71. Piernik, A. *Ecological Pattern of Inland Salt Marsh Vegetation in Central Europe*; Wydawnictwo Naukowe Uniwersytetu Mikołaja Kopernika: Toruń, Poland, 2012.
72. Naidoo, G.; Naidoo, Y. Salt tolerance in *Sporobolus virginicus*: The importance of ion relations and salt secretion. *Flora* **1998**, *193*, 337–344. [CrossRef]
73. Kong, Y.; Zheng, Y. Potential of Producing *Salicornia bigelovii* Hydroponically as a Vegetable at Moderate NaCl Salinity. *HortScience Horts* **2014**, *49*, 1154–1157. [CrossRef]
74. Koester, R.P.; Skoneczka, J.A.; Cary, T.R.; Diers, B.W.; Ainsworth, E.A. Historical gains in soybean (*Glycine max* Merr.) seed yield are driven by linear increases in light interception, energy conversion, and partitioning efficiencies. *J. Exp. Bot.* **2014**, *65*, 3311–3321. [CrossRef]
75. Ellenberg, H.; Weber, H.E.; Düll, R.; Wirth, V.; Werner, W.; Paulißen, D. Indicator values of plants in Central Europe. *Scr. Geobot.* **1992**, *18*, 1–248.
76. Díaz, S.; Kattge, J.; Cornelissen, J.H.C.; Wright, I.J.; Lavorel, S.; Dray, S.; Reu, B.; Kleyer, M.; Wirth, C.; Prentice, I.C.; et al. The global spectrum of plant form and function. *Nature* **2016**, *529*, 167–171. [CrossRef]
77. Hoagland, D.R.; Arnon, D.I. *The Water-Culture Method for Growing Plants without Soil. Circular California Agricultural Experiment Station*; University of California: Berkeley, CA, USA, 1950; Volume 347.
78. Cárdenas-Pérez, S.; Chanona-Pérez, J.; Méndez-Méndez, J.; Calderón-Domínguez, G.; López-Santiago, R.; Arzate-Vázquez, I. Nanoindentation study on apple tissue and isolated cells by atomic force microscopy, image and fractal analysis. *Innov. Food Sci. Emerg. Technol.* **2016**, *34*, 234–242. [CrossRef]
79. Arnon, D.I. Copper enzymes in isolated chloroplasts. Polyphenoloxidase in *Beta vulgaris*. *Plant Physiol.* **1949**, *24*, 1. [CrossRef] [PubMed]
80. Lichtenthaler, H.K.; Wellburn, A.R. Determinations of total carotenoids and chlorophylls a and b of leaf extracts in different solvents. *Biochem. Soc. Trans.* **1983**, *11*, 591–592. [CrossRef]
81. Bradford, M.M. A rapid and sensitive method for the quantitation of microgram quantities of protein utilizing the principle of protein-dye binding. *Anal. Biochem.* **1976**, *72*, 248–254. [CrossRef]
82. Velikova, V.; Yordanov, I.; Edreva, A. Oxidative stress and some antioxidant systems in acid rain-treated bean plants: Protective role of exogenous polyamines. *Plant Sci.* **2000**, *151*, 59–66. [CrossRef]
83. Chance, B.; Maehly, A. The assay of catalases and peroxidases. *Methods Biochem. Anal.* **1955**, *1*, 357–424. [CrossRef]

84. Rached-Kanouni, M.; Alatou, D. Change in activity of antioxidative enzymes in leaves of *Acacia retinodes*, *Biota orientalis* and *Casuarina equisetifolia* under heat stress condition. *Eur. Sci. J.* **2013**, *9*, 402–410.
85. Teranishi, Y.; Tanaka, A.; Osumi, M.; Fukui, S. Catalase activities of hydrocarbon-utilizing *Candida* yeasts. *Agric. Biol. Chem.* **1974**, *38*, 1213–1220. [CrossRef]
86. Sairam, R.K.; Rao, K.V.; Srivastava, G. Differential response of wheat genotypes to long term salinity stress in relation to oxidative stress, antioxidant activity and osmolyte concentration. *Plant Sci.* **2002**, *163*, 1037–1046. [CrossRef]
87. Bates, L.S.; Waldren, R.P.; Teare, I. Rapid determination of free proline for water-stress studies. *Plant Soil* **1973**, *39*, 205–207. [CrossRef]
88. Hammer, Ø.; Harper, D.A.; Ryan, P.D. PAST: Paleontological statistics software package for education and data analysis. *Palaeontol. Electron.* **2001**, *4*, 9.

plants

MDPI

Article

Effect of Salinity on Growth, Ion Accumulation and Mineral Nutrition of Different Accessions of a Crop Wild Relative Legume Species, *Trifolium fragiferum*

Astra Jēkabsone [1], Una Andersone-Ozola [1], Andis Karlsons [2], Māris Romanovs [1] and Gederts Ievinsh [1,*]

1 Department of Plant Physiology, Faculty of Biology, University of Latvia, 1 Jelgavas Str., LV-1004 Rīga, Latvia; astra.jekabsone@lu.lv (A.J.); una.andersone-ozola@lu.lv (U.A.-O.); maris.romanovs@lu.lv (M.R.)
2 Institute of Biology, University of Latvia, 4 Ojāra Vācieša Str., LV-1004 Rīga, Latvia; andis.karlsons@lu.lv
* Correspondence: gederts.ievins@lu.lv

Abstract: Crop wild relatives represent a valuable resource for the breeding of new crop varieties suitable for sustainable productivity in conditions of climate change. The aim of the present study was to assess salt tolerance of several wild accessions of *T. fragiferum* from habitats with different salinity levels in controlled conditions. Decrease of plant biomass and changes in partitioning between different organs was a characteristic response of plants with increasing substrate salinity, but these responses were genotype-specific. In several accessions, salinity stimulated reproductive development. The major differences in salinity responses between various *T. fragiferum* genotypes were at the level of dry biomass accumulation as well as water accumulation in plant tissues, resulting in relatively more similar effect on fresh mass. Na^+ and Cl^- accumulation capacity were organ-specific, with leaf petioles accumulating more, followed by leaf blades and stolons. Responses of mineral nutrition clearly were both genotype- and organ-specific, but several elements showed a relatively general pattern, such as increase in Zn concentration in all plant parts, and decrease in Ca and Mg concentration. Alterations in mineralome possibly reflect a reprogramming of the metabolism to adapt to changes in growth, morphology and ion accumulation resulting from effect of NaCl. High intraspecies morphological and physiological variability in responses of *T. fragiferum* accessions to salinity allow to describe them as ecotypes.

Keywords: forage legumes; growth; ions; mineral nutrition; salinity tolerance; strawberry clover

Citation: Jēkabsone, A.; Andersone-Ozola, U.; Karlsons, A.; Romanovs, M.; Ievinsh, G. Effect of Salinity on Growth, Ion Accumulation and Mineral Nutrition of Different Accessions of a Crop Wild Relative Legume Species, *Trifolium fragiferum*. *Plants* **2022**, *11*, 797. https://doi.org/10.3390/plants11060797

Academic Editors: Oscar Vicente and Marius Nicusor Grigore

Received: 19 February 2022
Accepted: 15 March 2022
Published: 17 March 2022

1. Introduction

Only relatively recently a concept of crop wild relative (CWR) plant species has been established [1,2] and it has been verified that CWRs represent extremely valuable potential resource for breeding new crop varieties [3]. In a light of global climate changes, with predicted increase in severity of deviations in environmental constraints, cultivated plants need to possess higher adaptive plasticity towards a range of suboptimal abiotic factors, allowing them to maintain productivity in highly heterogeneous conditions [4]. In this respect, CWRs can be used as a source of resilience-associated characteristics due to generally higher abiotic stress tolerance [3,5].

Soil salinization represents one of the most urgent problems in agriculture [6], and its negative effects on crop productivity are anticipated to become more severe on a background of global climate changes [7]. Because of their symbiosis with nitrogen-fixing rhizobacteria, salt-tolerant forage legume species are especially important for saline marginal lands with characteristically low response to nitrogen fertilizers [8]. Several species from genus *Trifolium* are commonly used in permanent temperate grasslands, and *Trifolium pratensis* and *Trifolium repens* are considered as especially important CWRs in Europe [9]. *Trifolium fragiferum* is a perennial stoloniferous clover species native to Europe, Mediterranean region, Middle East and West Asia [10]. Due to the relative rarity of *T. fragiferum* in

Northern Europe, *T. fragiferum* is legally protected in several countries, including Latvia [11]. While not used commercially in Europe, *T. fragiferum* has been cultivated as forage legume crop in temperate grasslands of Australia and USA [12,13]. The first successful cultivar of *T. fragiferum*, 'Palestine', had been developed in Australia from a material collected near the Dead Sea and used commercially since 1938 [13].

Resilience of *T. fragiferum* has been associated both with clonal type of growth as well as high abiotic stress tolerance of the species. Monopodially branching creeping shoots (stolons) have an ability to fom roots at the nodes [14]. Together with moderate tolerance to soil salinity and alkalinity, *T. fragiferum* also has good flooding tolerance [15], an ability to withstand continuous grazing [16] and repeated trampling [17]. Potential suitability of different wild accessions of *T. fragiferum* to saline conditions is of special interest, as it was established that a wide genetic diversity exists within a species in respect to degree of salt tolerance [18]. As in the Northern Europe *T. fragiferum* is exclusively associated with an endangered habitat 'Baltic coastal meadow' [19], experimental assessment of populations around the Baltic Sea seems to be extremely promising in order to find highly salt-tolerant physiological types of the species useful for further breeding purposes.

Aspects of plant mineral nutrition have been often related with their salinity tolerance, as mineral imbalance resulting due to salinity treatment can be considered as one of indications of general metabolic disorder [20]. More specifically, Na^+ accumulation in plant tissues due to increased substrate NaCl can affect their K^+ status, and consequently, result in disruption of cellular functions. The strategy of active salt exclusion from photosynthetic tissues is a possible adaptive mechanism for salt-tolerant glycophytes and monocotyledonous halophytes [21]. For other halophytes, vacuolar sequestration of Na^+ and Cl^- and maintenance of stable cytosolic K^+, as an avoidance mechanism, together with accumulation of nonionic osmolytes, leads to stabilization of osmotic homeostasis [22,23], concomitantly with readjusting of cellular mineral balance according to the needs of salinity-altered metabolism [24]. Therefore, assessment of salt-induced changes in mineral element concentration in plant tissues can provide information on adaptive cellular responses possibly related to differences in the degree of salinity tolerance.

Evaluation of local diversity of CWRs is an important constituent in a system of sustainable use of biological resources [25]. A number of geographically-isolated micropopulations of *T. fragiferum* associated with natural water reservoirs have been identified in Latvia recently [26]. Tolerance of several of these accessions of *T. fragiferum* against soil waterlogging and flooding, trampling as well as cutting have been evaluated [17]. All accessions appeared to be relatively tolerant to these factors, but accession-specific differences found suggested existence of different physiological types. The aim of the present study was to assess the salinity tolerance of several wild Latvian accessions of *T. fragiferum* from habitats with different salinity levels in comparison to commercial cultivar 'Palestine' as well as *T. fragiferum* accession from a relatively highly saline meadow in Bornholm, Denmark. It was hypothesized that the accessions from habitats with higher soil salinity would be more salinity tolerant in controlled conditions.

2. Results

Morphological differences were observed between control plants of different *T. fagiferum* accessions during cultivation and at the end of the experiment. Thus, plants of accession TF9 had the lowest shoot biomass (Figure 1) but the highest number of stolons and leaves (Table 1). Plants of TF8 (cv. 'Palestine') had the highest shoot biomass in control conditions (Figure 1) and the lowest number of stolons (Table 1). In addition, the longest total length of stolons was evident for plants of accession TF7, but the shortest was seen for accession TF4 (Table 1). Biomass of roots for control plants showed less variance between different accessions, significantly lower values of fresh and dry mass was evident only for TF7 (Figure 2).

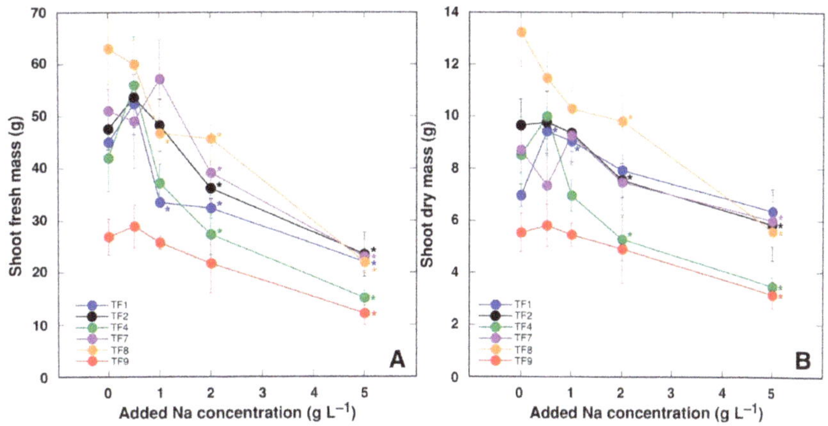

Figure 1. Effect of increasing substrate salinity on shoot fresh mass (**A**) and shoot dry mass (**B**) of *Trifolium fragiferum* plants of different accessions. Asterisks of respective color indicate statistically significant differences from control ($p < 0.05$).

Table 1. Effect of increasing substrate salinity on morphological parameters of *Trifolium fragiferum* plants of different accessions.

Treatment	TF1	TF2	TF4	TF7	TF8	TF9
			Number of stolons (n)			
0	19.8 ± 3.0 a	25.0 ± 2.7 a	18.8 ± 2.1 a	26.8 ± 3.6 a	18.0 ± 2.4 a	30.8 ± 5.8 a
0.5	20.9 ± 4.3 a	21.8 ± 3.3 a	17.6 ± 2.6 a	22.6 ± 4.2 a	16.8 ± 1.0 a	28.6 ± 3.1 a
1	18.8 ± 3.7 a	25.8 ± 3.0 a	18.2 ± 1.9 a	37.6 ± 8.0 a	16.8 ± 1.9 a	24.2 ± 2.3 a
2	12.7 ± 5.5 a	16.2 ± 2.0 b	16.2 ± 1.9 a	18.0 ± 2.6 b	18.2 ± 3.0 a	18.4 ± 6.0 b
5	12.1 ± 1.4 b	14.8 ± 3.6 b	10.2 ± 1.3 b	14.0 ± 2.0 b	13.2 ± 1.3 b	14.2 ± 3.8 b
			Total stolon length (m)			
0	7.38 ± 0.91 a	7.43 ± 1.00 a	3.92 ± 0.70 a	10.28 ± 1.03 a	5.18 ± 1.16 a	6.35 ± 1.27 a
0.5	6.67 ± 0.81 a	7.10 ± 1.10 a	4.05 ± 0.57 a	8.23 ± 2.14 a	4.74 ± 0.36 a	4.97 ± 0.68 a
1	4.56 ± 0.67 b	6.29 ± 0.83 a	2.87 ± 0.41 a	10.54 ± 2.04 a	3.72 ± 0.64 ab	4.11 ± 0.32 ab
2	3.63 ± 0.93 bc	3.94 ± 0.32 b	1.88 ± 0.30 b	4.88 ± 0.59 b	2.65 ± 0.49 b	2.70 ± 0.99 bc
5	2.12 ± 0.20 c	2.21 ± 0.41 c	0.82 ± 0.09 c	2.43 ± 0.34 c	1.35 ± 0.27 c	1.54 ± 0.35 c
			Number of leaves (n)			
0	272 ± 33 a	321 ± 24 a	175 ± 21 a	348 ± 47 a	274 ± 45 a	397 ± 92 a
0.5	366 ± 43 a	346 ± 49 a	204 ± 33 a	314 ± 40 a	241 ± 27 a	464 ± 48 a
1	229 ± 46 a	338 ± 33 a	175 ± 19 a	443 ± 58 a	247 ± 33 a	374 ± 44 a
2	289 ± 78 a	247 ± 16 b	151 ± 14 a	294 ± 25 a	291 ± 40 a	322 ± 61 a
5	168 ± 23 b	206 ± 34 b	82 ± 10 b	153 ± 20 b	133 ± 16 b	214 ± 52 b

Different letters for each parameter within a column indicate statistically significant differences ($p < 0.05$) between treatments for the respective accession.

When treated with low level of NaCl, several accessions showed a tendency for increased mass of shoots (Figure 1), but only for TF1 dry mass of shoots significantly increased at 0.5 and 1 g L^{-1} (Figure 1B). Plants of accessions TF1 and TF8 exhibited significant decrease of shoot fresh mass already at 1 g L^{-1} Na, but all accessions except TF9 had significant decrease of shoot fresh mass at 2 g L^{-1} Na (Figure 1A). In respect to shoot dry mass, significant decrease for TF2, TF4 and TF8 was evident already at 2 g L^{-1} Na^+, but all accessions except TF1 exhibited significant biomass reduction at 5 g L^{-1} (Figure 1B). Both fresh and dry mass of roots was significantly stimulated at low Na^+ concentration only for TF1 (Figure 2). While fresh mass of roots significantly decreased for all accessions at 5 g L^{-1} Na (Figure 2A), root dry mass of TF1 and TF7 did not decrease at this concentration of Na^+ (Figure 2B). There was no stimulative effect of low Na^+ on number of stolons, total length of stolons, and number of leaves for any of accessions of *T. fragiferum* (Table 1). These parameters were significantly reduced by 5 g L^{-1} Na^+ treatment for all accessions, or even at lower concentrations for several accessions.

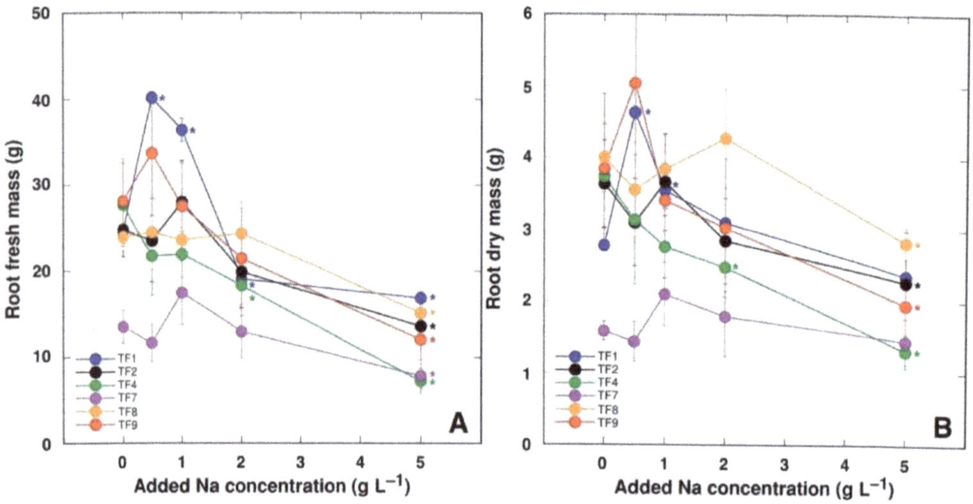

Figure 2. Effect of increasing substrate salinity on root fresh mass (**A**) and root dry mass (**B**) of *Trifolium fragiferum* plants of different accessions. Asterisks of respective color indicate statistically significant differences from control ($p < 0.05$).

It appeared that increasing NaCl concentration in substrate reduced initial differences in fresh and dry mass of shoots and roots between accessions, and this phenomenon was clearly evident by the results of multivariate analysis (Figure 3A). The largest differences between the genotypes in respect to biomass accumulation in roots and shoots were between TF7 and TF9 (Figure 3B), but the largest similarity between TF2 and TF7, and TF1 and TF9 (Figure 4). Analysis of changes in biomass partition also confirmed genotype specificity of salinity effects (Figure 5). Thus, increasing salinity stimulated generative reproduction, and this effect increased in an order TF1 < TF8 < TF2 < TF7 < TF9 < TF4, but partition to roots was enhanced in TF7 and TF8.

Analysis of summed relative effect of salinity revealed that the major differences in salinity responses between various *T. fragiferum* genotypes were at the level of dry biomass accumulation (Figure 6C) as well as water accumulation in plant tissues (Figure 6D), resulting in relatively more similar effect on fresh mass (Figure 6A). Moreover, effect on morphological indices (number of stolons and leaves, as well as stolon length) was rather consistent between different genotypes (Figure 6B).

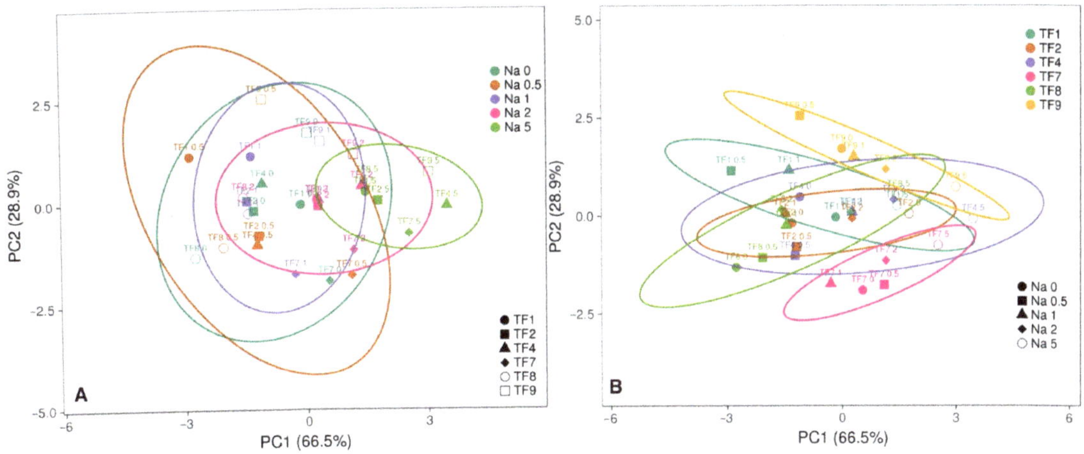

Figure 3. Principal component analysis on effect of increasing substrate salinity on shoot and root fresh mass and dry mass of *Trifolium fragiferum* plants of different accessions. (**A**), grouping according salinity levels; (**B**), grouping according accessions. Prediction ellipses are such that with probability 0.95, a new observation from the same group will fall inside the ellipse. Unit variance scaling was applied to rows; singular value decomposition with imputation was used to calculate principal components. X and Y axes show principal component one and principal component two that explain 66.5% and 28.9% of the total variance, respectively.

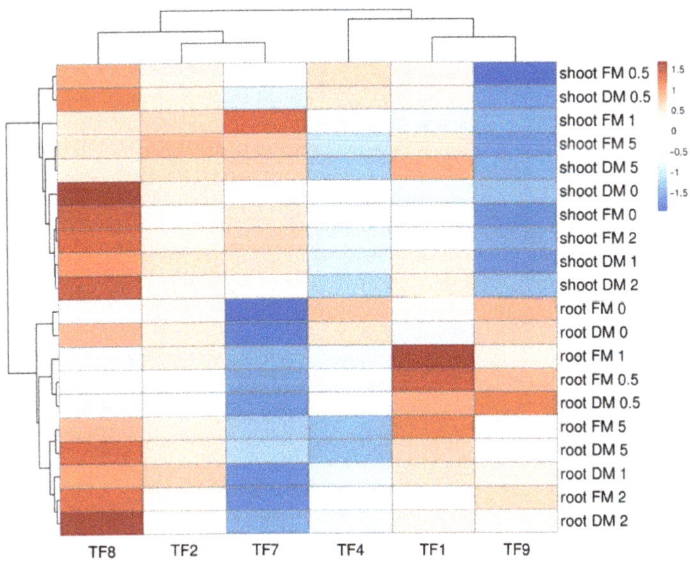

Figure 4. Generated heat map and cluster analysis on effect of increasing substrate salinity on shoot and root fresh mass and dry mass of *Trifolium fragiferum* plants of different accessions. Hierarchical clusters were generated by average linkage method with correlation distance. Color scale shows relative intensity of normalized parameter values. FM, fresh mass; DM, dry mass.

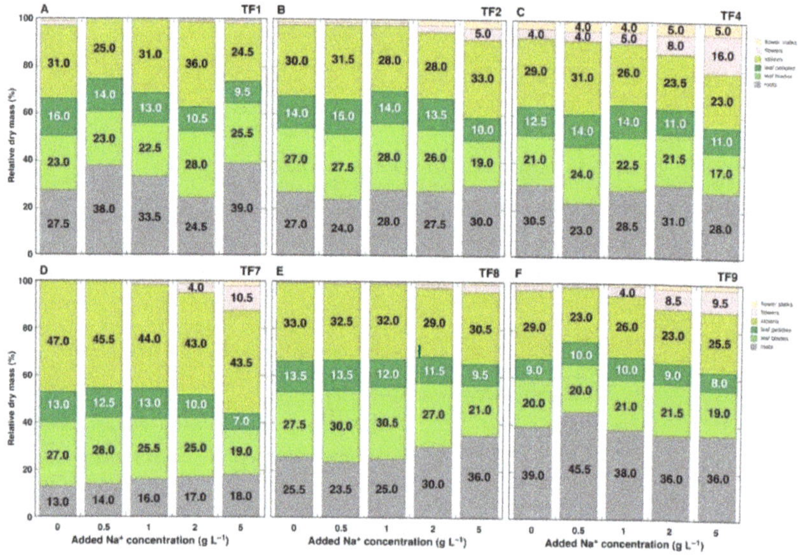

Figure 5. Changes in biomass partitioning in *Trifolium fragiferum* plants of different accessions due to increasing substrate salinity. (**A**), accession TF1; (**B**), accession TF2; (**C**), accession TF4; (**D**), accession TF7; (**E**), accession TF8; (**F**), accession TF9.

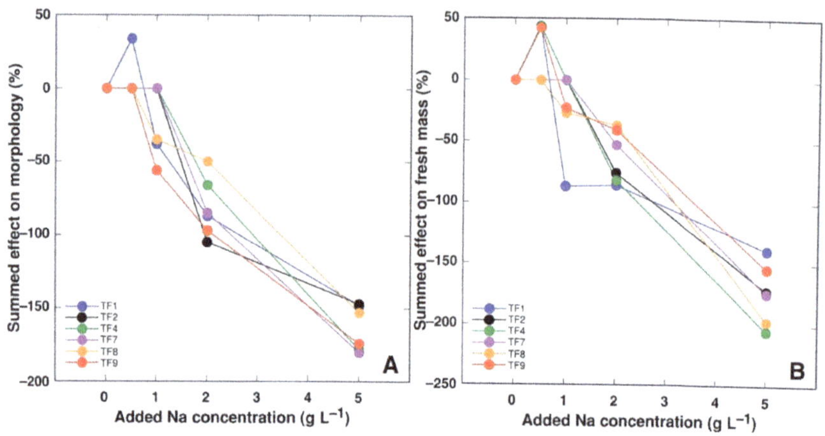

Figure 6. *Cont.*

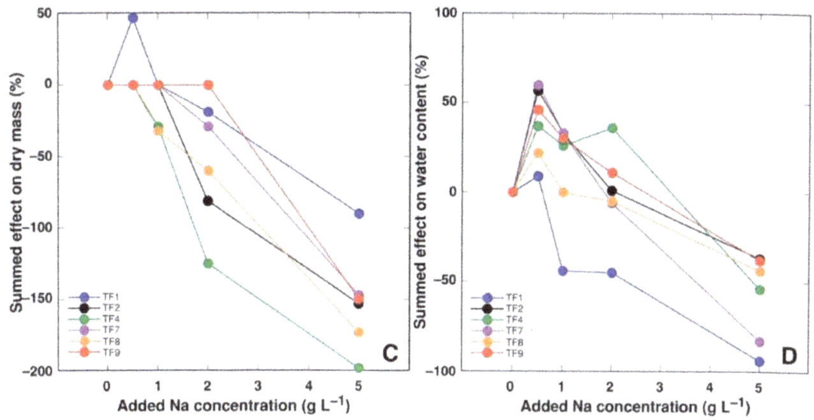

Figure 6. Summed relative effect of increasing substrate salinity on morphology (**A**), fresh mass of plant parts (**B**), dry mass of plant parts (**C**) and water content in plant parts (**D**) of *Trifolium fragiferum* plants of different accessions. Only statistically significant effects are taken into account.

In general, Na^+ accumulation capacity was organ-specific, with leaf petioles accumulating more Na^+, followed by leaf blades (Figure 7). At low salinity, there were no significant differences in accumulation of Na^+ in leaf blades (Figure 7A), leaf petioles (Figure 7B) and stolons (Figure 7C), only at the highest salinity (5 g Na^+ L^{-1}) plants from most saline habitats (TF1 and TF9) tended to accumulate more Na^+ in leaves. In contrast, differences in trend of Na^+ accumulation in dependence on increasing salinity were evident in plant roots (Figure 7). Accumulation capacity for Cl^- was also highest in leaf petioles, followed by stolons and leaf blades (Figure 8). Response of Cl^- accumulation was saturable at low substrate NaCl concentration, especially, for stolons and roots. Multivariate analysis of ion accumulation characteristics in plant parts revealed that salinity effects were rather genotype-specific, with closer similarity between TF2 and TF4, as well as TF1 and TF9 (Figure 9).

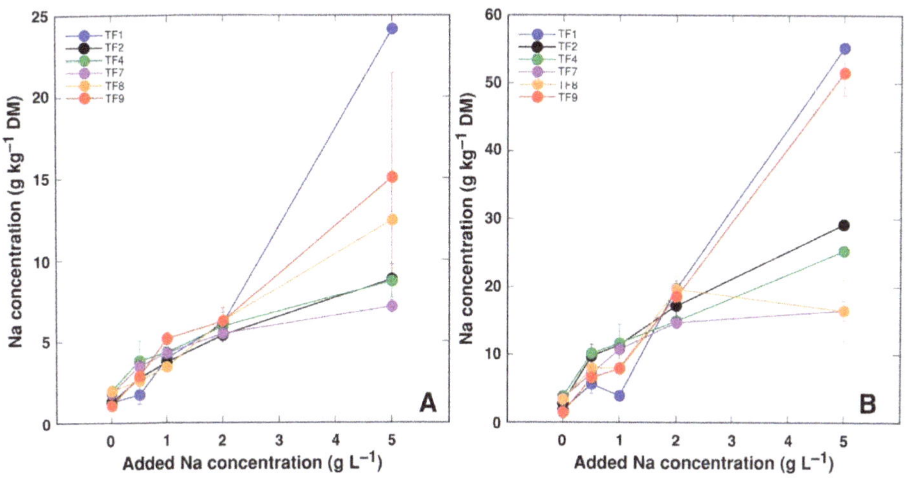

Figure 7. *Cont.*

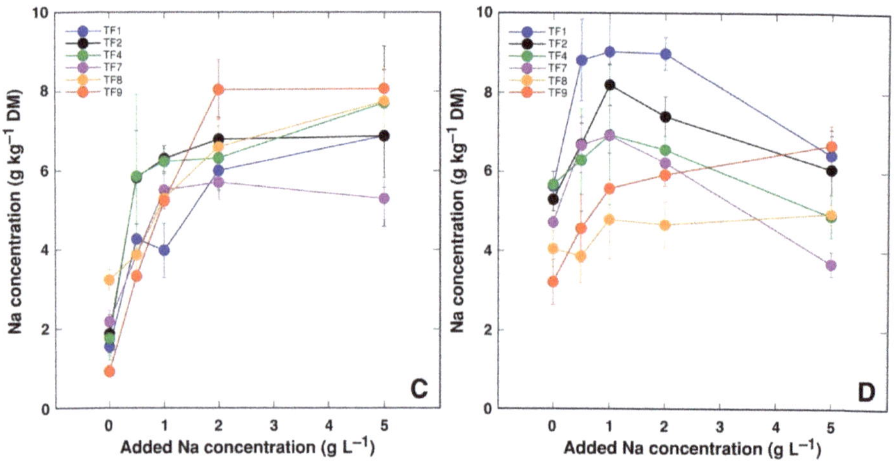

Figure 7. Effect of increasing substrate salinity on Na^+ accumulation in leaf blades (**A**), leaf petioles (**B**), stolons (**C**) and roots (**D**) of *Trifolium fragiferum* plants of different accessions.

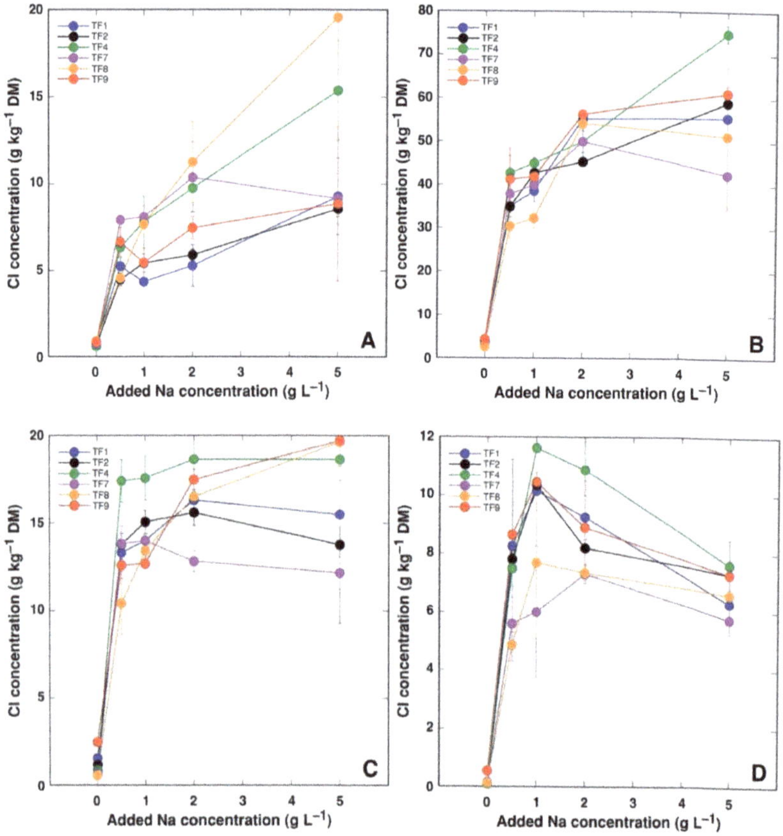

Figure 8. Effect of increasing substrate salinity on Cl^- accumulation in leaf blades (**A**), leaf petioles (**B**), stolons (**C**) and roots (**D**) of *Trifolium fragiferum* plants of different accessions.

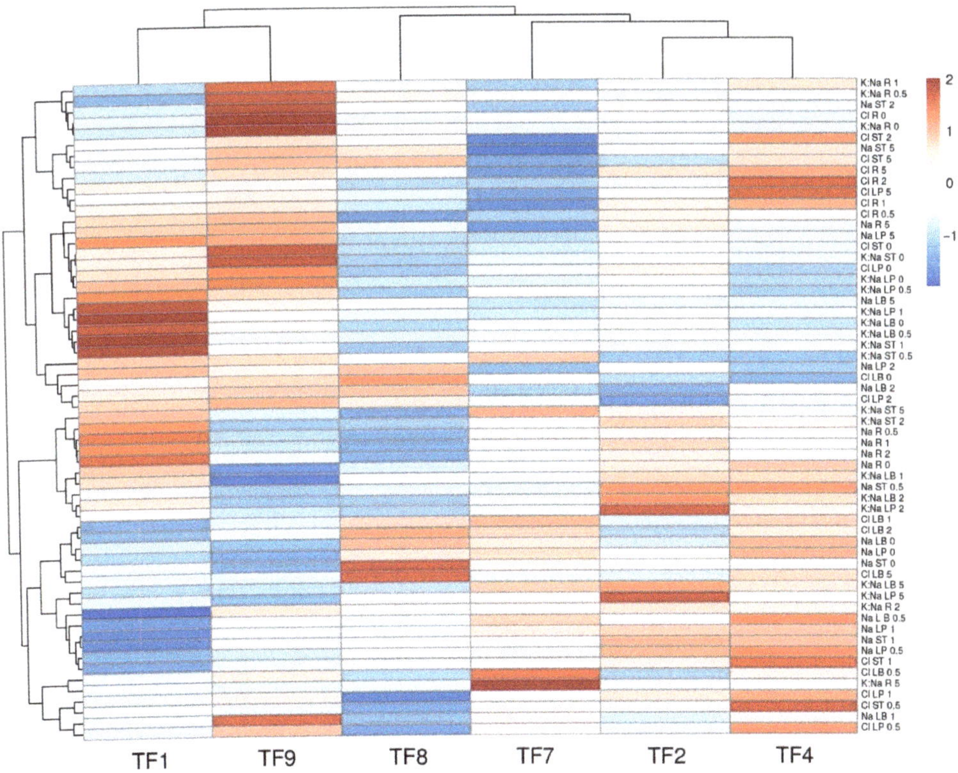

Figure 9. Generated heat map and cluster analysis on effect of increasing substrate salinity on Na^+ and Cl^- accumulation and K:Na ratio in different parts of *Trifolium fragiferum* plants of different accessions. Hierarchical clusters were generated by average linkage method with correlation distance. Color scale shows relative intensity of normalized parameter values. LB, leaf blades; LP, leaf petioles; ST, stolons; R, roots.

Effect of salinity on mineral nutrition was evaluated by comparison of relative effect of increasing substrate salinity in various plant parts for different accessions (Figure 10). The responses clearly were both genotype- and organ-specific, but some general trends were evident. Thus, Zn concentration mostly increased in all plant parts for all genotypes except TF2 and TF7, but Ca and Mg concentration decreased, except TF9. Effects on macronutrient P and K, as well as micronutrient Fe, Cu and Mn concentration were rather controversial. According to principal component analysis, diversity in mineral nutrient concentration increased with increasing salinity (Figure 11), and each genotype had rather unique mineral element response trend in different plant parts caused by salinity gradient (data not shown).

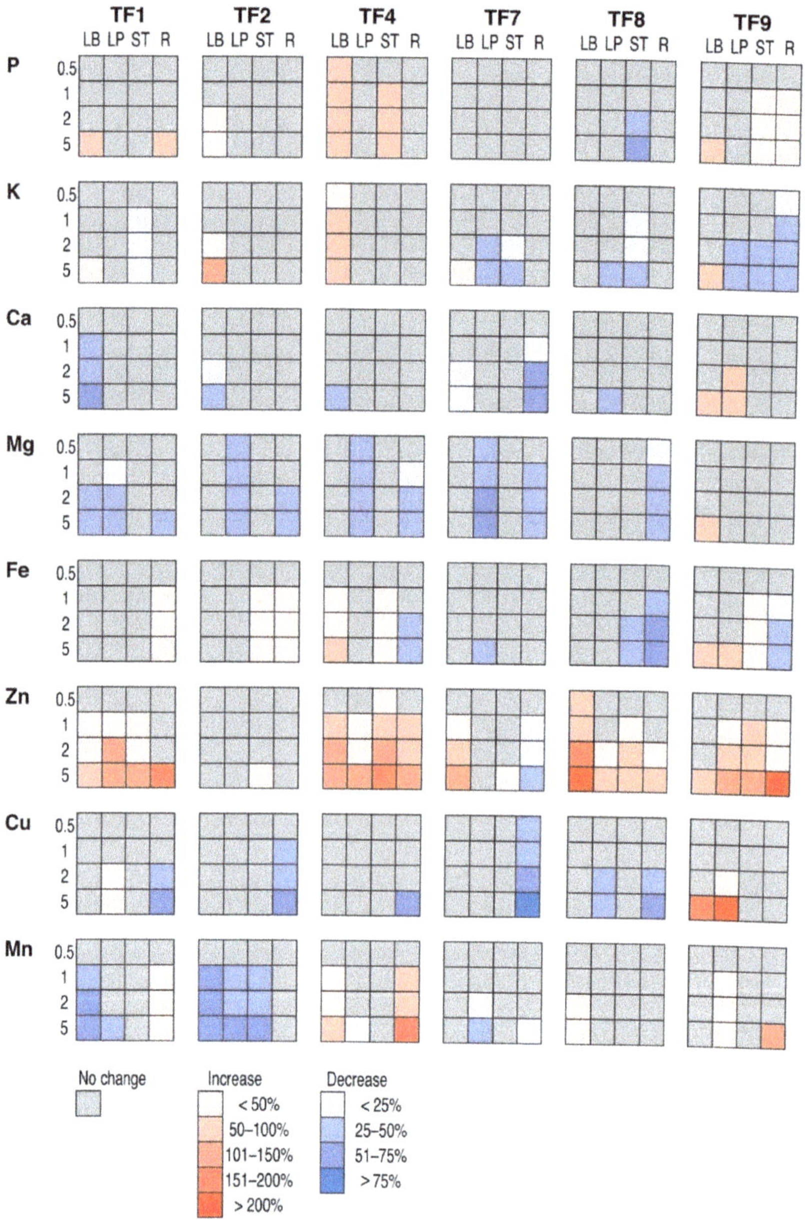

Figure 10. Relative effect of increasing substrate salinity on mineral element concentrations in leaf blades (LB), leaf petioles (LP), stolons (ST) and roots (R) of *Trifolium fragiferum* plants of different accessions. Numbers on the left side indicate added Na+ concentration (g L^{-1}). Only statistically significant effects are taken into account.

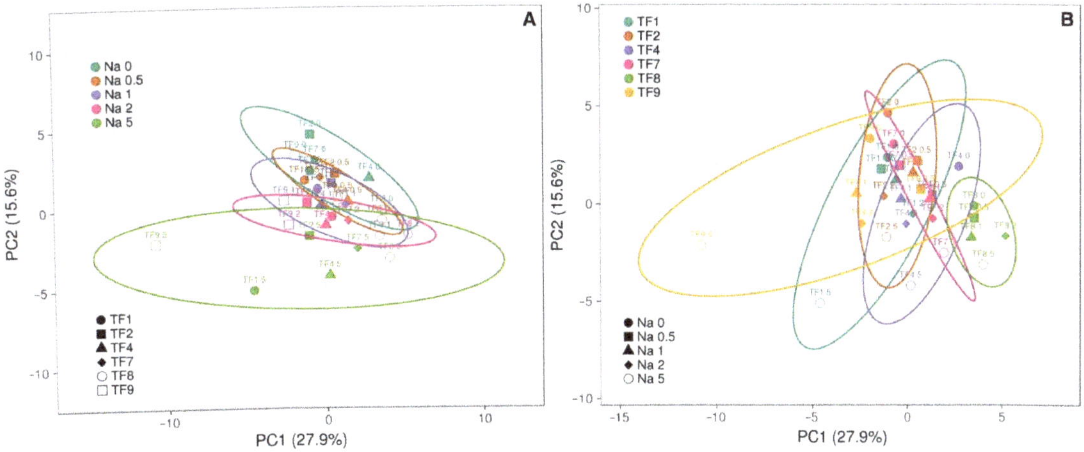

Figure 11. Principal component analysis on effect of increasing substrate salinity on mineral nutrient concentration in different parts of *Trifolium fragiferum* plants of different accessions. (**A**), grouping according salinity levels; (**B**), grouping according accessions. Prediction ellipses are such that with probability 0.95, a new observation from the same group will fall inside the ellipse. Unit variance scaling was applied to rows; singular value decomposition with imputation was used to calculate principal components. X and Y axes show principal component one and principal component two that explain 27.9% and 15.6% of the total variance, respectively.

3. Discussion

3.1. Comparison of Salinity Tolerance

T. fragiferum as a halophytic species has been included in the eHALOPH database (https://www.sussex.ac.uk/affiliates/halophytes/index.php, accessed on 2 February 2022) as based on the main criterion, tolerance to substrate EC at least 7.8 dS m^{-1} (equivalent to 7.8 mS cm^{-1}). This assumption was confirmed also by the results of the present study, with all accessions being able to grow and reproduce at 5 g Na L^{-1} with substrate $EC_{1:5}$ reaching 9.77 mS cm^{-1} (3602 mS m^{-1} by a sensor measurement). Similar salinity level was recorded also in a natural habitat of TF9 on the island of Bornholm (2749 mS m^{-1}). The species even has been defined as obligatory mesohydrohalophile, as based on its presence in salt marsh vegetation in Romania [27], but within the northern part of the distribution range it seems to be specifically associated with habitats near different water reservoirs but not with increased soil salinity [26].

Accessions of *T. fragiferum* compared in the present study were growing on soils with different salinity level (Table 2). It seems to be logical to expect that the accessions from more saline habitats (as TF1 and TF9) would show higher salinity tolerance in identical conditions of the controlled experiment in comparison to the accessions from habitats with low salinity (as TF2, TF4, TF7). However, in order to approve or reject this hypothesis, it should be understood that the degree of tolerance to changes in a particular environmental factors can be compared differently. Plants from different taxonomic groups are usually compared in a relative way, comparing percent changes of certain growth-related indices relative to control plants, in order to eliminate genotype-associated differences between the control plants. According to this approach, *T. fragiferum* plants of accession TF9 were the most tolerant to 2 g L^{-1} Na^+ treatment, but TF1 plants were the most tolerant to 5 g L^{-1} Na^+ (Figure 6C). However, in absolute terms, when looking for the accession producing the highest biomass at high salinity, *T. fragiferum* accessions TF1, TF2, TF7 and TF8 produced identically high amount of biomass at 5 g L^{-1} Na^+, with values for TF4 and TF9 being significantly lower (Figure 1B). Consequently, from a practical point of forage production, relatively sensitive cv. 'Palestine' (TF8) still would have higher yield when cultivated in saline soil because of extremely pronounced biomass production ability of control plants, when compared to relatively most tolerant accession TF9 from saline coastal habitat in Bornholm. In this respect, the most promising Latvian accession of *T. fragiferum* was TF1 from a saline wet shore meadow of Lake Liepājas: the accession showed the highest relative tolerance to 5 g L^{-1} Na^+ and also were among the accessions with the highest absolute biomass production capacity at this salinity level. Accession TF1 was also the only one used in the present study showing significant growth stimulation of both shoots and roots at 0.5 and 1 g L^{-1} Na^+. Besides, TF1 was also the accession most stable to action of several abiotic factors, with very high tolerance to soil waterlogging and repeated cutting, and high tolerance against trampling [17].

Table 2. Characterization, geographical location of accessions of *Trifolium fragiferum* used in the present study and soil electrical conductivity (EC) at the sites.

Code	Associated Water Reservoir	Habitat	Location	Coordinates	Soil EC (mS m^{-1})
TF1	Lake Liepājas	Salt-affected wet shore meadow	City of Liepāja, Latvia	56°29′29″ N, 21°1′38″ E	380 ± 124
TF2	River Lielupe	Salt-affected shore meadow	City of Jūrmala, Lielupe, River Lielupe Estuary, Latvia	57°0′11″ N, 23°55′56″ E	85 ± 5
TF4	–	Degraded urban land	City of Rīga, Vidzeme Suburb, Latvia	56°57′46″ N, 24°7′2″ E	69 ± 6
TF7	The Gulf of Riga of the Baltic Sea	Dry coastal meadow	Town of Ainaži, Latvia	57°52′8″ N, 24°21′10″ E	65 ± 4
TF8 cv. 'Palestine'	na	na	na	na	na
TF9	The Baltic Sea	Salt-affected wet coastal meadow	Hammeren, Bornholm, Denmark	55°17′54″ N 14°46′17″ E	2749 ± 209

While several studies previously have accessed salinity tolerance of *T. fragiferum* [28,29], direct comparison of the results obtained is rather difficult. The main reason is a lack of information on the precise amount of applied salts during treatments, and/or on final salinity level in substrate, measured either as substrate electrical conductivity or concentration of Na^+. When decrease in biomass accumulation is viewed as a main indication of a plant's sensitivity to salinity, a possibility that growth inhibition represents a regulated adaptive response to increased salt concentration is usually forgotten. However, changes in biomass partition within a plant with increase in substrate salinity clearly indicate that this assumption could be correct, as showed also in the present study (Figure 5).

Salinity tolerance under different flooding regimes of *T. fragiferum* cv. 'Palestine' has been compared with that of other *Trifolium* species and it was shown that the species is more sensitive to salinity than to flooding [28]. During the study with 95 *T. fragiferum* accessions and five cultivars it was concluded that within the species a wide genetic diversity exists in respect to salinity tolerance [18]. Several accessions, when grown at low salinity, even showed significant growth stimulation of their shoots. In hydroponics, dry biomass of cv. 'Palestine' decreased to 43–54% at 160 mM NaCl relative to control, but mixed plant sample from five pooled wild accessions of *T. fragiferum* had relatively better tolerance, with biomass decreasing only to 73% at the same salinity [30]. These results are similar to the ones obtained in the present study. Most importantly, it is evident that intraspecies physiological diversity of salinity responses for *T. fragiferum* exist, even from a relatively restricted territory as in the case of Latvia.

No detailed physiological mechanisms of salinity tolerance/sensitivity have been investigated for any of *Trifolium* species so far. However, some insights were made for moderately salt tolerant species *Trifolium alexandrinum*, showing that excessive accumulation of Na^+ in leaves together with inability for sequestration in vacuoles have led to inhibition of photosynthesis followed by growth inhibition [31]. Similarly, *T. repens* plants from a population accumulating lower amount of Na^+ and Cl^- in shoots, had higher shoot dry mass and lower dieback rate in saline conditions in comparison to plants from a higher-accumulating population [32]. Consequently, a relationship between Na^+ and Cl^- accumulation in plant shoot tissues and their salinity tolerance can be proposed.

3.2. Comparison of Ion Accumulation

T. fragiferum has been characterized as a species excluding Na^+ from shoots based simply on the fact that other species of the genus accumulated more Na^+ [30]. Exclusion of Na^+ and Cl^- from shoot tissues is considered as a characteristic important for salinity tolerance, but it seems to be relevant only for glycophytes [33]. In contrast, halophytic plants are able to use Na^+ for osmotic adjustment, at least, in vacuoles [34].

In the present study, both *T. fagiferum* accessions from habitats with the highest salinity (TF1 and TF9) had the highest Na^+ accumulation potential in leaf blades and petioles (Figure 7A,B). In leaf petioles of both accessions, Na^+ concentration reached >50 g kg^{-1}. Interestingly, in natural conditions, TF1 plants were not among the accessions showing highest levels of Na^+ accumulation, with Na^+ concentration in petioles reaching only 14 g kg^{-1}, while that in TF2 was 21 g kg^{-1} [26]. However, it is important that at low to moderate salinity there were no differences in Na^+ accumulation potential in leaves between different accessions in the present study in controlled conditions, but these were pronounced in stolons and, especially, in roots. Accumulation potential for the two ions in stolons and roots was less, especially, at high salinity. In comparison, Na^+ accumulation potential in shoots of *T. repens* was up to 38 g kg^{-1} Na^+ and 85 g kg^{-1} Cl^-, while in roots it was only 5 g kg^{-1} Na^+ and 8 g kg^{-1} Cl^- [35]. Consequently, relatively better salinity tolerance of *T. fragiferum* is not associated with differences in accumulation of Na^+ and Cl^-. This was also not the case when tolerance of individual *T. fragiferum* accessions were considered: the two relatively most tolerant accessions, TF1 and TF9, accumulated higher Na^+ concentration only at 5 g L^{-1} Na in leaf petioles, but no such relationship was evident in stolons and roots, or for Cl^- accumulation in all plant parts.

Plants of cv. 'Palestine' accumulated 1.63–1.86 mmol Na^+ and 1.56–1.60 mmol Cl^- per g DM in shoots (equivalent to 37.5–42.8 g kg^{-1} Na^+ and 55.4–56.8 g kg^{-1} Cl^-), but concentrations for wild accessions of *T. fragiferum* were 1.12 mmol Na^+ and 1.32 mmol Cl^- per g DM (equivalent to 25.8 g kg^{-1} Na^+ and 46.9 g kg^{-1} Cl^-) [30]. Thus, accumulation range of Na^+ and Cl^- at relatively high substrate salinity for different *T. fragiferum* genotypes is relatively similar, but significant differences in the accumulation potential between different plant parts need to be taken into account (Figures 7 and 8).

Increased water accumulation in plant tissues in response to increasing salinity can be viewed as means for dilution of soluble ions concomitantly with stimulation of vacuolar development [36]. It seems that several accessions of *T. fragiferum* employed this mechanism, especially, at low to moderate salinity (Figure 6). The relationship between salinity tolerance and salinity-induced tissue succulence has been shown for a number of species [37–40]. However, no such relationship was evident for *T. fragiferum*, as relatively salt tolerant accession TF1 showed only relatively little increase of tissue water content in comparison to that in other accessions (Figure 6).

3.3. Mineral Nutrition

Disbalance of mineral nutrition has been suggested as one of the deleterious physiological effects in plants due to high salinity [20]. However, generalization of results from mineral nutrition studies of plants under salinity seems to be rather rare in scientific literature. It has been concluded that the main nutrient-related problem during salinity could be related to mineral imbalance as a result of competition of Na^+ and Cl^- with K^+, Ca^{2+}, and NO_3^-, but micronutrient concentrations are relatively less affected [20,24,41]. It was concluded that besides rather pronounced effects on K^+ uptake and distribution, more supported is the idea of negative effect of salinity on Ca^{2+} uptake, pointing to important role of Ca^{2+} in maintenance of mineral homeostasis in saline conditions.

Increase in shoot K concentration is a commonly described tolerance-associated response of glycophytic species to salinity [33], while osmotic functions of K^+ can be taken over by Na^+ in halophytic species [42]. Shoot K^+ concentration decreased from 1.54 mmol g^{-1} DM in control plants of cv. 'Palestine' to 1.03 mmol g^{-1} DM in plants cultivated at 160 mM NaCl (equivalent to 60.1 and 40.2 g kg^{-1}), and from 1.76 mmol per g in control plants of wild accessions to 1.24 mmol per g at 160 mM NaCl (equivalent to 69.6 and 48.4 g kg^{-1}) [31]. However, no organ-specific effects on K^+ accumulation were considered so far. In the present study K^+ concentration in leaf blades significantly increased in all accessions except cv. 'Palestine' at least at the highest Na^+ concentration, but decrease in other parts was evident for several accessions (Figure 10). Interestingly, the most pronounced decrease in tissue K^+ in leaf petioles, stolons and roots was seen with increasing salinity in presumably most salinity-tolerant accession TF9, as well as for TF1 in stolons. This clearly points to ion accumulation features similar to these of halophytes.

In some species K^+:Na^+ ratio has been shown as a reliable indicator of salinity tolerance [43]. This has not been the case for *Trifolium* species, as more salinity-tolerant *T. fragiferum* had lower K^+:Na^+ ratio as less salinity-tolerant *T. repens* [44]. Also, in the present study, no relationship was found between K^+:Na^+ ratio in different organs of *T. fragiferum* plants from various accessions and their salinity tolerance.

In taxonomically and morphologically related species, *T. repens*, increasing salinity intensity induced a concomitant increase in concentration of several micronutrients in plant roots, including Fe, Mn, and Zn [35]. Proportional increase of Mn concentration with salinity has been described in another legume species, *Melilotus segetalis* [45]. Increase in concentration of Zn as a result of salinity was very noticeable, but with certain genotype-specific differences. This clearly suggests involvement in adaptations to salinity, and usually increased Zn concentration has been associated with its involvement as a component in defense-related proteins, as zinc finger proteins or antioxidative enzyme CuZn-superoxide dismutase [46,47]. It is confirmed that enhanced activity of enzymatic antioxidative system is a prerequisite for salinity tolerance [48].

As based on both literature analysis as well as the results of the present study, it seems that particular salinity-induced changes in the concentration of individual mineral elements are extremely variable even between taxonomically related species and within the species. Thus, for two closely related species, *Limonium perezii* and *Limonium sinuatum*, shoot Mg concentration either increased or decreased, respectively, as a result of increasing salinity [49]. This means that interpretation of results from mineral nutrition studies searching for general salinity effects using single species or only a few different species should be done with caution.

In the present study with *T. fragiferum*, the observed salinity-dependent changes in mineral nutrients were both genotype- and plant part-specific, with no clearly evident relationship with relative salinity tolerance of the genotype (Figure 9). An interesting general characteristic response of mineral nutrition was an increase in the diversity of distribution of concentrations of mineral nutrients at high salinity (Figure 11). Due to relatively high tolerance of all tested accessions to salinity, it seems that the recorded characteristic and genotype-specific changes in concentration of mineral nutrients in plant tissues reflect adaptive responses related to maintenance of metabolic homeostasis in plants growing in saline soil. Similarly, reallocation of mineral nutrients in all plant organs is thought to represent a whole-plant adaptive response [45]. Genotype-specific response of mineral nutrition to increased salinity has been noted also for various salt-adapted halophyte species [50].

3.4. Limitations and Benefits of the Experimental System and Future Perspectives

A study similar to this has been performed with another legume species, *Medicago ciliaris*, using seed material from seven spontaneous local populations in Tunisia, and it was concluded that this type of studies represent an efficient approach to find salt response-related genetic variation [51]. However, the problem of data interpretation in studies with legume model species is related to their symbiotic relationship with N_2-fixing bacteria. It has been shown that rhizobial symbiosis not only provides additional nitrogen substances for plant's needs but also affects tolerance to unfavorable environmental conditions, possibly through upregulation of defense-associated genes [52,53]. Consequently, different plant responses can be obtained in experiments using asymbiotic plants vs spontaneous establishment of rhizobial symbiosis vs. plants inoculated with efficient symbionts. Thus, active rhizobial symbiosis improved salinity tolerance of *Medicago sativa* plants through increased osmotic adjustment and enzymatic antioxidative capacity [54]. Similar findings have been described also for *Medicago truncatula* [55]. It has been also established that presence of rhizobial symbiosis modulates interaction between *T. fragiferum* and *T. repens* on the background of increased substrate salinity [44]. In the present study, to avoid possible problems with inadequate or/and inefficient rhizobial symbiosis when comparing plant accessions from various sites, *T. fragiferum* plants were cultivated asymbiotically. Our further studies have shown that different *T. fragiferum* accessions have highly variable degree of growth-dependence on presence of their native rhizobia (Jēkabsone et al. unpublished results), therefore, it is intended to perform future experiments on salinity responses in different genotypes of *T. fragiferum*, using their native symbiotic bacterial strains.

The novel aspects revealed by the present study concern genotype-specific effects of increased substrate salinity on biomass partitioning as well as Na^+ and Cl^- accumulation capacity between different organs of *T. fragiferum* plants from various accessions. However, salinity tolerance in plants is clearly multigenic in nature [56]. Thus, ability for sustaining ion homeostasis (including ion compartmentation), osmotic protection and antioxidative defense are listed among the most important groups of mechanisms in plant salinity tolerance [57]. In respect to osmotic adjustment under salinity, in the present study, an emphasis was put on inorganic constituents, Na^+ and K^+. However, nonionic osmotically active substances are well known for their role in maintenance of osmotic balance in plants under saline conditions, and corresponding scientific evidence has been provided from studies both in natural and controlled conditions [58–60]. Therefore, it can be proposed that accumulation of compatible osmolytes is an important constituent of salinity responses also in *T. fragiferum* plants.

At mild or moderate salinity, induction of antioxidative enzymes is an important aspect of salinity tolerance, as shown for *Beta vulgaris* spp. *vulgaris* [61]. Also, higher capacity of enzymatic antioxidative system in salt-tolerant rice landraces has been shown [43]. Our previous results have indicated that decrease of peroxidase activity in leaves of *T. fragiferum* at increased substrate salinity is a good indicator of relative salinity tolerance, but this effect seemed to be associated with salinity-induced increase in tissue water content [44]. Future studies aimed at dissecting molecular mechanisms of salinity tolerance in different *T. fragiferum* accessions clearly need to focus on enzymatic antioxidative defense system and physiological indicators of tissue integrity.

4. Materials and Methods

4.1. Plant Material

Seeds of *Trifolium fragiferum* from four accessions in Latvia (TF1, TF2, TF4, and TF7) as well as one accession from the island of Bornholm (TF9), from habitats with different salinity levels, were used in the present study (Table 2, Figure 12). *T. fragiferum* cv. 'Palestine' (TF8), obtained from Sheffield's Seeds Company (Locke, NY, USA) was used for comparison.

Figure 12. Map of the southern and central Baltic Sea region with *Trifolium fragiferum* accessions used in the present study.

4.2. Cultivation Conditions and Treatments

Plants were cultivated in asymbiotic conditions of soil culture in an automated greenhouse. All details of plant establishment and cultivation were as described previously [17]. Fully acclimatized four week-old plants were randomly divided into five treatments, five plants per treatment as biological replicates. Respective plants were treated with NaCl once a week, adding 1.27 or 2.54 g NaCl dissolved in 200 mL deionized water per container with 1 L of soil substrate until final concentration was reached within five weeks (Table 3). After achieving full treatment, soil electrical conductivity (EC) was measured in containers with HH2 meter equipped with WET-2 sensor (Delta-T Devices, Burwell, UK), and in 1:5 (v/v) soil suspension in deionized water following 15 min incubation with LAQUAtwin conductivity meter B-771 (Horiba Scientific, Kyoto, Japan). Plants were cultivated for additional three weeks then the experiment was terminated.

Table 3. Experimental treatments used in the present study. Soil EC (measured in containers with a sensor) and soil suspension EC were analyzed after reaching final treatment concentrations. EC, electrical conductivity.

Treatment	Added Salinity (mM)	Amount of Added NaCl (g)	Concentration of Added Na (g L^{-1})	Soil EC (mS m^{-1})	Soil Suspension (1:5) EC (mS cm^{-1})
Control	0	0	0	71.7 ± 7.3	0.27 ± 0.05
0.5	22	1.27	0.5	210.4 ± 10.0	0.50 ± 0.02
1	44	2.54	1.0	353.8 ± 31.0	1.96 ± 0.17
2	87	5.08	2.0	514.5 ± 36.9	3.04 ± 0.72
5	217	12.70	5.0	3602.2 ± 327.5	9.77 ± 1.18

4.3. Measurements

Plants were individually separated in different parts (roots, stolons, leaf petioles, leaf blades, flower stalks, inflorescences). Stolons, leaves and inflorescences were counted, and the length of individual stolons was measured. Plant material was weighed separately before and after drying in an oven at 60 °C for 72 h. Water content was calculated as g H_2O per g dry mass.

Mineral element analysis in dry-ashed plant material was performed as described previously [26]. After mineralization of the plant samples and dissolving the mineral fraction in either 3% HCl (P, K, Ca, Mg, Fe, Mn, Zn, Cu, Na) or deionized water (Cl), chemical analyses were done using the following methods: the levels of K, Ca, Mg, Fe, Cu, Zn and Mn were estimated by microwave plasma atomic emission spectrometer (MP-AES) Agilent 4200, these of P were analyzed by the colorimetry with ammonium molybdate in an acid-reduced medium using a spectrophotometer Jenway 6300, but values of Cl were obtained by $AgNO_3$ titration using distilled water extraction of plant ash. All analyses were performed in triplicate, using representative tissue samples from individual biological replicates.

4.4. Data Analysis

As flower-related characteristics were rather variable between individual plants, they were used only for calculation of total shoot biomass as well as for establishment of biomass partitioning [17]. The relative effect of salinity was expressed as percent changes of the parameter in comparison to the respective control plants. Comparison of the relative effect of treatments between different accessions was performed by means of summed percent changes, separately for morphological parameters (number of leaves and stolons, average and total length of stolons), fresh mass of separate plant parts, and dry mass of separate plant parts, as well as water content in plant parts. The total summed effect was calculated by combining percent effect on morphological parameters, fresh mass and dry mass. Only changes significantly statistically different from control values were taken into account for the calculation of summed effects. Effect of salinity on mineral nutrient concentration was estimated as percent increase of the respective concentration in comparison to control plants, taking into account only statistically significant changes.

Results were analyzed by KaleidaGraph (v. 5.0, Synergy Software, Reading, PA, USA). Statistical significance of differences was evaluated by one-way ANOVA using post-hoc analysis with minimum significant difference. Principal component analysis, heat map generation and cluster analysis were performed by a freely available web program ClustVis (http://biit.cs.ut.ee/clustvis/, accessed on 13 March 2022) [62]. For principal component analysis, prediction ellipses were such that with probability 0.95, a new observation from the same group will fall inside the ellipse. Unit variance scaling was applied to rows; singular value decomposition with imputation was used to calculate principal components. Hierarchical clusters were generated by average linkage method with correlation distance.

5. Conclusions

High intraspecies morphological and physiological variability is characteristic for responses of *T. fragiferum* accessions to salinity, allowing them to be described as ecotypes. While increasing salinity results in a decrease in the initial biomass differences between accessions, an expansion of morphological variability and diversity of mineral nutrient concentrations among plant parts in saline conditions is strongly pronounced. Changes in mineralome possibly reflect a reprogramming of the metabolism to adapt accordingly to changes in growth, morphology, and ion accumulation resulting from the direct effect of NaCl.

Author Contributions: U.A.-O. and G.I. proposed the research. U.A.-O., A.J., A.K., M.R. and G.I. performed the experiments and analyzed the data. G.I. drafted the manuscript. U.A.-O. and A.J. revised the manuscript. All authors have read and agreed to the published version of the manuscript.

Funding: The study was supported by the Latvian Science Council project lzp-2020/2-0349 "Molecular, physiological and ecological evaluation of Latvian genetic resources of valuable wild legume species, *Trifolium fragiferum*, in a context of sustainable agriculture".

Institutional Review Board Statement: Not applicable.

Informed Consent Statement: Not applicable.

Data Availability Statement: All data reported here is available from the authors upon request.

Acknowledgments: Participation of Lāsma Neiceniece (University of Latvia), Līva Purmale (University of Latvia), and Magnolia Garbarino (Pawling High School, NY, USA) in part of the study is sincerely acknowledged.

Conflicts of Interest: The authors declare no conflict of interest. The funders had no role in the design of the study; in the collection, analyses, or interpretation of data; in the writing of the manuscript, or in the decision to publish the results.

References

1. Maxted, N.; Ford-Lloyd, B.V.; Jury, S.; Kell, S.; Scholten, M. Towards a definition of a crop wild relative. *Biodivers. Conserv.* **2006**, *15*, 2673–2685. [CrossRef]
2. Ford-Lloyd, B.V.; Schmidt, M.; Armstrong, S.J.; Barazani, O.; Engels, J.; Hadas, R.; Hammer, K.; Kell, S.P.; Kang, D.; Khoshbakht, K.; et al. Crop wild relatives: Undervalued, underutilized and under threat? *Bioscience* **2011**, *61*, 559–565. [CrossRef]
3. Prohens, J.; Gramazio, P.; Plazas, M.; Dempewolf, H.; Kilian, B.; Diez, M.J.; Fita, A.; Herraiz, F.J.; Rodríguez-Burruezo, A.; Soler, S.; et al. Introgressiomics: A new approach for using crop wild relatives in breeding for adaptation to climate change. *Euphytica* **2017**, *213*, 158. [CrossRef]
4. Raza, A.; Razzaq, A.; Mehmood, S.S.; Zou, X.; Zhang, X.; Lv, Y.; Xu, J. Impact of Climate Change on Crops Adaptation and Strategies to Tackle Its Outcome: A Review. *Plants* **2019**, *8*, 34. [CrossRef] [PubMed]
5. Warschefsky, E.; Penmetsa, R.V.; Cook, D.R.; von Wettberg, E. Back to the wilds: Tapping evolutionary adaptations for resilient crops through systematic hybridization with crop wild relatives. *Am. J. Bot.* **2014**, *101*, 1791–1800. [CrossRef]
6. Nachshon, U. Cropland soil salinization and associated hydrology: Trends, processes and examples. *Waters* **2018**, *10*, 1030.
7. Chaudhry, S.; Sidhu, G.P.S. Climate change regulated abiotic stress mechanisms in plants: A comprehensive review. *Plant Cell Rep.* **2021**, *41*, 1–31. [CrossRef]
8. Manchanda, G.; Garg, N. Salinity and its effects on the functional biology of legumes. *Acta Physiol. Plant.* **2008**, *30*, 595–618. [CrossRef]
9. Heywood, V.H.; Zohary, D. A catalogue of the wild relatives of cultivated plants native to Europe. *Flora Mediterr.* **1995**, *5*, 375–415.
10. Zohary, M.; Heller, D. *The Genus Trifolium*; Israel Academy of Sciences and Humanities: Jerusalem, Israel, 1984; 606p.
11. Cabinet of Ministers of Latvia. Provisions on the List of Specially Protected Species and Species of restricted Use. Regulations of the Cabinet of Ministers of Latvia No. 396. 2000. Available online: https://likumi.lv/ta/id/12821-noteikumi-par-ipasi-aizsargajamo-sugu-un-ierobezoti-izmantojamo-ipasi-aizsargajamo-sugu-sarakstu (accessed on 14 March 2022).
12. Taylor, N.; Gillett, J. Crossing and morphological relationships among *Trifolium* species closely related to strawberry and Per-sian clover. *Crop Sci.* **1988**, *28*, 636–639. [CrossRef]
13. Nichols, P.G.H.; Revell, C.K.; Humphries, A.W.; Howie, J.H.; Hall, E.J.; Sandral, G.A.; Ghamkhar, K.; Harris, C.A. Temperate pasture legumes in Australia—Their history, current use, and future prospects. *Crop Pasture Sci.* **2012**, *63*, 691–725. [CrossRef]
14. Huber, H.; Wiggerman, L. Shade avoidance in the clonal herb *Trifolium fragiferum*: A field study with experimentally manip-ulated vegetation height. *Plant Ecol.* **1997**, *130*, 53–62. [CrossRef]
15. Townsend, C.E. Miscellaneous perennial clovers. In *Clover Science and Technology*; Taylor, J.L., Ed.; ASA/CSSA/SSSA: Madison, WI, USA, 1985; pp. 563–578.
16. Pederson, G.A. White clover and other perennial clovers. In *Forages—An Introduction to Grassland Agriculture*, 5th ed.; Barnes, R.F., Miller, D.A., Nelson, C.J., Eds.; Iowa State University: Ames, IA, USA, 1995; Volume 1, pp. 227–236.
17. Andersone-Ozola, U.; Jēkabsone, A.; Purmale, L.; Romanovs, M.; Ievinsh, G. Abiotic Stress Tolerance of Coastal Accessions of a Promising Forage Species. *Trifolium fragiferum*. *Plants* **2021**, *10*, 1552. [CrossRef]
18. Rumbaugh, M.D.; Pendery, B.M.; James, D.W. Variation in the salinity tolerance of strawberry clover (*Trifolium fragiferum* L.). *Plant Soil* **1993**, *153*, 265–271. [CrossRef]
19. Janssen, J.A.M.; Rodwell, J.S. *European Red List of Habitats: Part 2. Terrestrial and Freshwater Habitats*; European Union: Brussels, Belgium, 2016.
20. Hu, Y.; Schmidhalter, U. Drought and salinity: A comparison of their effects on mineral nutrition of plants. *J. Plant Nutr. Soil Sci.* **2005**, *168*, 541–549. [CrossRef]
21. Chen, M.; Yang, Z.; Liu, J.; Zhu, T.; Wei, X.; Fan, H.; Wang, B. Adaptation mechanisms of salt excluders under saline conditions and its applications. *Int. J. Mol. Sci.* **2018**, *19*, 3668. [CrossRef]
22. Singh, M.; Kumar, J.; Singh, S.; Singh, V.P.; Prasad, S.M. Roles of osmoprotectants in improving salinity and drought tolerance in plants: A review. *Res. Environ. Sci. Biotechnol.* **2015**, *14*, 407–426. [CrossRef]
23. Wu, H. Plant salt tolerance and Na^+ sensing and transport. *Crop J.* **2018**, *6*, 215–225. [CrossRef]
24. Grattan, S.R.; Grieve, C.M. Salinity–mineral nutrient relations in horticultural crops. *Sci. Hortic.* **1998**, *78*, 127–157. [CrossRef]

25. Maxted, N.; Scholten, M.; Codd, R.; Ford-Lloyd, B. Creation and use of a national inventory of crop wild relatives. *Biol. Conserv.* **2007**, *140*, 142–159. [CrossRef]
26. Andersone-Ozola, U.; Jēkabsone, A.; Karlsons, A.; Romanovs, M.; Ievinsh, G. Soil chemical properties and mineral nutrition of Latvian accessions of *Trifolium fragiferum*, a crop wild relative plant species. *Environ. Exp. Biol.* **2021**, *19*, 245–254.
27. Ciocârlan, V.; Georgescu, M.I.; Săvulescu, E.; Anastasiu, P. Plopul salt marshes (Tulcea County)—An unique area for halophytes in Romania. *Acta Horti Bot. Bucurest.* **2013**, *40*, 27–32. [CrossRef]
28. Rogers, M.E.; West, D.W. The Effects of Rootzone Salinity and Hypoxia on Shoot and Root Growth in *Trifolium* Species. *Ann. Bot.* **1993**, *72*, 503–509. [CrossRef]
29. Can, E.; Arslan, M.; Sener, O.; Daghan, H. Response of strawberry clover (*Trifolium fragiferum* L.) to salinity stress. *Res. Crops* **2013**, *14*, 576–584.
30. Rogers, M.E.; Colmer, T.D.; Frost, K.; Henry, D.; Cornwall, D.; Hulm, E.; Hughes, S.; Nichols, P.G.H.; Craig, A.D. The influence of NaCl salinity and hypoxia on aspects of growth in *Trifolium* species. *Crop Pasture Sci.* **2009**, *60*, 71–82. [CrossRef]
31. Abogadallah, G.M. Sensitivity of *Trifolium alexandrinum* L. to salt stress is related to the lack of long-term stress-induced gene expression. *Plant Sci.* **2010**, *178*, 491–500. [CrossRef]
32. Rogers, M.E.; Noble, C.L. Variation in growth and ion accumulation between two selected populations of *Trifolium repens* L. differing in salt tolerance. *Plant Soil* **1992**, *146*, 131–136. [CrossRef]
33. Ashraf, M. Some important physiological selection criteria for salt tolerance in plants. *Flora-Morphol. Distrib. Funct. Ecol. Plants* **2004**, *199*, 361–376. [CrossRef]
34. Barros, N.L.F.; Marques, D.N.; Tadaiesky, L.B.A.; de Souza, C.R.B. Halophytes and other molecular strategies for the genera-tion of salt-tolerant crops. *Plant Physiol. Biochem.* **2021**, *162*, 581–591. [CrossRef]
35. Cekstere, G.; Karlsons, A.; Grauda, D. Salinity-induced responses and resistance in *Trifolium repens* L. *Urban For. Urban Green.* **2015**, *14*, 225–236. [CrossRef]
36. Ogburn, R.M.; Edwards, E.J. The ecological water-use strategies of succulent plants. In *Advances in Botanical Research*; Kader, J.-C., Delseny, M., Eds.; Academic Press: Burlington, MA, USA, 2010; Volume 55, pp. 179–225.
37. Reimann, C.; Breckle, S. Salt tolerance and ion relations of *Salsola kali* L.: Differences between ssp. tragus (L.) Nyman and ssp. *ruthenica* (Iljin) Soó. *New Phytol.* **1995**, *130*, 37–45. [CrossRef]
38. Hajiboland, R.; Bahrami-Rad, S.; Zeinalzade, N.; Atazadeh, E.; Akhani, H.; Poschenrieder, C. Differential functional traits underlying the contrasting salt tolerance in *Lepidium* species. *Plant Soil* **2020**, *448*, 315–334. [CrossRef]
39. Palchetti, M.V.; Reginato, M.; Llanes, A.; Hornbacher, J.; Papenbrock, J.; Barboza, G.E.; Luna, V.; Cantero, J.J. New insights into the salt tolerance of the extreme halophytic species *Lycium humile* (Lycieae, Solanaceae). *Plant Physiol. Biochem.* **2021**, *163*, 166–177. [CrossRef] [PubMed]
40. Belghith, I.; Senkler, J.; Abdelly, C.; Braun, H.-P.; Debez, A. Changes in leaf ecophysiological traits and proteome profile provide new insights into variability of salt response in the succulent halophyte *Cakile maritima*. *Funct. Plant Biol.* **2022**. [CrossRef] [PubMed]
41. Grattan, S.R.; Grieve, C.M. Mineral element acquisition and growth response of plants grown in saline environments. *Agric. Ecosyst. Environ.* **1992**, *38*, 275–300. [CrossRef]
42. Pan, Y.-Q.; Guo, H.; Wang, S.-M.; Zhao, B.; Zhang, J.-L.; Ma, Q.; Yin, H.-J.; Bao, A.-K. The photosynthesis, Na^+/K^+ homeostasis and osmotic adjustment of *Atriplex canescens* in response to salinity. *Front. Plant Sci.* **2016**, *7*, 848. [CrossRef]
43. Rasel, M.; Tahjib-Ul-Arif, M.; Hossain, M.A.; Hassan, L.; Farzana, S.; Brestic, M. Screening of Salt-Tolerant Rice Landraces by Seedling Stage Phenotyping and Dissecting Biochemical Determinants of Tolerance Mechanism. *J. Plant Growth Regul.* **2021**, *40*, 1853–1868. [CrossRef]
44. Dūmiņš, K.; Andersone-Ozola, U.; Samsone, I.; Elferts, D.; Ievinsh, G. Growth and physiological performance of a coastal species *Trifolium fragiferum* as asffected by a coexistence with *Trifolium repens*, NaCl treatment and inoculation with rhizobia. *Plants* **2021**, *10*, 2196. [CrossRef]
45. Romero, J.M.; Marañón, T. Allocation of biomass and mineral elements in *Melilotus segetalis* (annual sweetclover): Effects of NaCl salinity and plant age. *New Phytol.* **1996**, *132*, 565–573. [CrossRef]
46. Han, G.; Liu, C.; Guo, J.; Qiao, Z.; Sui, N.; Qiu, N.; Wang, B. C_2H_2 zinc finger proteins: Master regulators of abiotic stress responses in plants. *Front. Plant Sci.* **2020**, *11*, 115. [CrossRef]
47. Sofy, M.R.; Elhindi, K.M.; Farouk, S.; Alotaibi, M.A. Zinc and Paclobutrazol Mediated Regulation of Growth, Upregulating Antioxidant Aptitude and Plant Productivity of Pea Plants under Salinity. *Plants* **2020**, *9*, 1197. [CrossRef] [PubMed]
48. Canalejo, A.; Martínez-Domínguez, D.; Córdoba, F.; Torronteras, R. Salt tolerance is related to a specific antioxidant response in the halophyte cordgrass, *Spartina densiflora*. *Estuar. Coast. Shelf Sci.* **2014**, *146*, 68–75. [CrossRef]
49. Grieve, C.M.; Poss, J.; Grattan, S.; Shouse, P.; Lieth, J.; Zeng, L. Productivity and Mineral Nutrition of *Limonium* Species Irrigated with Saline Wastewaters. *HortScience* **2005**, *40*, 654–658. [CrossRef]
50. Yepes, L.; Chelbi, N.; Vivo, J.-M.; Franco, M.; Agudelo, A.; Carvajal, M.; Martínez-Ballesta, M.D.C. Analysis of physiological traits in the response of Chenopodiaceae, Amaranthaceae, and Brassicaceae plants to salinity stress. *Plant Physiol. Biochem.* **2018**, *132*, 145–155. [CrossRef]

51. Mbarki, S.; Skalicky, M.; Vachova, P.; Hajihashemi, S.; Jouini, L.; Zivcak, M.; Tlustos, P.; Brestic, M.; Hejnak, V.; Khelil, A.Z. Comparing Salt Tolerance at Seedling and Germination Stages in Local Populations of *Medicago ciliaris* L. to *Medicago intertexta* L. and *Medicago scutellata* L. *Plants* **2020**, *9*, 526. [CrossRef] [PubMed]
52. Deakin, W.J.; Broughton, W.J. Symbiotic use of pathogenic strategies: Rhizobial protein secretion systems. *Nat. Rev. Genet.* **2009**, *7*, 312–320. [CrossRef]
53. Sharma, M.P.; Grover, M.; Chourasiya, D.; Bharti, A.; Agnihotri, R.; Maheshwari, H.S.; Pareek, A.; Buyer, J.S.; Sharma, S.K.; Schütz, L.; et al. Deciphering the Role of Trehalose in Tripartite Symbiosis Among Rhizobia, Arbuscular Mycorrhizal Fungi, and Legumes for Enhancing Abiotic Stress Tolerance in Crop Plants. *Front. Microbiol.* **2020**, *11*, 509919. [CrossRef]
54. Wang, Y.; Zhang, Z.; Zhang, P.; Cao, Y.; Hu, T.; Yang, P. Rhizobium symbiosis contribution to short-term salt stress tolerance in alfalfa (*Medicago sativa* L.). *Plant Soil* **2016**, *402*, 247–261. [CrossRef]
55. Irshad, A.; Rehman, R.N.U.; Abrar, M.; Saeed, Q.; Sharif, R.; Hu, T. Contribution of rhizobium-legume symbiosis in salt stress tolerance in *Medicago truncatula* evaluated through photosynthesis, antioxidant enzymes, and compatible solutes accumulation. *Sustainability* **2021**, *13*, 3369. [CrossRef]
56. Agarwal, P.K.; Shukla, P.S.; Gupta, K.; Jha, B. Bioengineering for Salinity Tolerance in Plants: State of the Art. *Mol. Biotechnol.* **2013**, *54*, 102–123. [CrossRef]
57. Arif, Y.; Singh, P.; Siddiqui, H.; Bajguz, A.; Hayat, S. Salinity induced physiological and biochemical changes in plants: An omic approach towards salt stress tolerance. *Plant Physiol. Biochem.* **2020**, *156*, 64–77. [CrossRef] [PubMed]
58. Tipirdamaz, R.; Gagneul, D.; Duhazé, C.; Aïnouche, A.; Monnier, C.; Özkum, D.; Larher, F. Clustering of halophytes from an inland salt marsh in Turkey according to their ability to accumulate sodium and nitrogenous osmolytes. *Environ. Exp. Bot.* **2006**, *57*, 139–153. [CrossRef]
59. Grigore, M.N.; Boscaiu, M.; Vicente, O. Assessment of the relevance of osmolyte biosynthesis for salt tolerance of halophytes under natural conditions. *Eur. J. Plant Sci. Biotechnol.* **2011**, *5*, 12–19.
60. Slama, I.; Abdelly, C.; Bouchereau, A.; Flowers, T.; Savouré, A. Diversity, distribution and roles of osmoprotective compounds accumulated in halophytes under abiotic stress. *Ann. Bot.* **2015**, *115*, 433–447. [CrossRef] [PubMed]
61. Tahjib-Ul-Arif, M.; Sohag, A.A.M.; Afrin, S.; Bashar, K.K.; Afrin, T.; Mahamud, A.G.M.S.U.; Polash, M.A.S.; Hossain, M.T.; Sohel, M.A.T.; Brestic, M.; et al. Differential response of sugar beet to long-term mild to severe salinity in a soil–pot culture. *Agriculture* **2019**, *9*, 223. [CrossRef]
62. Metsalu, T.; Vilo, J. ClustVis: A web tool for visualizing clustering of multivariate data using Principal Component Analysis and heatmap. *Nucleic Acids Res.* **2015**, *43*, W566–W570. [CrossRef] [PubMed]

plants

MDPI

Article

The Halophyte Species *Solanum chilense* Dun. Maintains Its Reproduction despite Sodium Accumulation in Its Floral Organs

Servane Bigot [1,*], Paula Pongrac [2], Martin Šala [3], Johannes T. van Elteren [3], Juan-Pablo Martínez [4], Stanley Lutts [1] and Muriel Quinet [1]

1 Groupe de Recherche en Physiologie Végétale (GRPV), Earth and Life Institute-Agronomy (ELI-A), Université Catholique de Louvain, Croix du Sud 4-5, 1348 Louvain-la-Neuve, Belgium; stanley.lutts@uclouvain.be (S.L.); muriel.quinet@uclouvain.be (M.Q.)

2 Department of Biology, Biotechnical Faculty, University of Ljubljana, Večna Pot 111, 1000 Ljubljana, Slovenia; paula.pongrac@bf.uni-lj.si

3 Department of Analytical Chemistry, National Institute of Chemistry, Hajdrihova 19, 1000 Ljubljana, Slovenia; Martin.Sala@ki.si (M.Š.); elteren@ki.si (J.T.v.E.)

4 Instituto de Investigaciones Agropecuarias (INIA-La Cruz), Chorrillos 86, La Cruz 2280454, Chile; jpmartinez@inia.cl

* Correspondence: servane.bigot@uclouvain.be

Citation: Bigot, S.; Pongrac, P.; Šala, M.; van Elteren, J.T.; Martínez, J.-P.; Lutts, S.; Quinet, M. The Halophyte Species *Solanum chilense* Dun. Maintains Its Reproduction despite Sodium Accumulation in Its Floral Organs. *Plants* **2022**, *11*, 672. https://doi.org/10.3390/plants11050672

Academic Editors: Oscar Vicente and Marius-Nicusor Grigore

Received: 7 February 2022
Accepted: 25 February 2022
Published: 28 February 2022

Publisher's Note: MDPI stays neutral with regard to jurisdictional claims in published maps and institutional affiliations.

Abstract: Salinity is a growing global concern that affects the yield of crop species, including tomato (*Solanum lycopersicum*). Its wild relative *Solanum chilense* was reported to have halophyte properties. We compared salt resistance of both species during the reproductive phase, with a special focus on sodium localization in the flowers. Plants were exposed to NaCl from the seedling stage. Salinity decreased the number of inflorescences in both species but the number of flowers per inflorescence and sepal length only in *S. lycopersicum*. External salt supply decreased the stamen length in *S. chilense*, and it was associated with a decrease in pollen production and an increase in pollen viability. Although the fruit set was not affected by salinity, fruit weight and size decreased in *S. lycopersicum*. Concentrations and localization of Na, K, Mg, and Ca differed in reproductive structures of both species. Inflorescences and fruits of *S. chilense* accumulated more Na than *S. lycopersicum*. Sodium was mainly located in male floral organs of *S. chilense* but in non-reproductive floral organs in *S. lycopersicum*. The expression of Na transporter genes differed in flowers of both species. Overall, our results indicated that *S. chilense* was more salt-resistant than *S. lycopersicum* during the reproductive phase and that differences could be partly related to dissimilarities in element distribution and transport in flowers.

Keywords: salinity; inflorescences; ion localization

1. Introduction

Tomato (*Solanum lycopersicum*) is cultivated worldwide and is of great economic importance. In 2020, more than 6 Mha of tomato plants was cultivated and 252 Mt of fruits was harvested [1]. Plant breeding increased tomato yields, and the world average yield in 2020 was 598 t ha^{-1} with values ranging from 14 t to 5 kt ha^{-1}, depending on the region and the cultural mode [1]. However, tomato is sensitive to abiotic stresses, including salinity, because of its glycophytic nature [2]. Salinity is a growing global concern, and it is estimated that salinity is present in 900 million ha of soils worldwide [3]. Sodium chloride (NaCl) is the most common of salts and represents more than 90% of salt in the world [4]. Tomato is cultivated in many countries affected by salinity (e.g., East Asia, the Middle East, and North Africa), and salinity decreases tomato yield by on average 50% for an electrical conductivity of 5 dS m^{-1} [5].

Despite decades of tomato breeding programs, resistance to abiotic stress has been neglected [6]. Indeed, since the 1960s, tomato improvement has mainly focused on fruit yield, shelf-life, and taste [7,8]. Because of the self-pollination of cultivated tomato and varietal selection, genetic diversity has been considerably lost in this species [9]. Miller and Tanksley [10] estimated that the *S. lycopersicum* genome contained less than 5% of the genetic variation of its wild relatives and, according to Bretó et al. [11], this species is considered to have the lowest genetic diversity in the tomato clade (Clade II of *Solanum*, consisting of *S. lycopersicum*, *S. tuberosum*, and *S. muricatum*, [12]). *Solanum lycopersicum* has many wild relatives including a few originating from harsh environments [13]. The use of resistant wild relatives in breeding is a common practice to improve the resistance of crop species to abiotic stresses [14]. *Solanum chilense* is a wild tomato relative native from the Atacama desert, one of the most salty and arid areas in the world [15,16]. Due to its high level of genetic variability, *S. chilense* is considered one of the most promising sources of genes for selection of tomato genotypes resistant to abiotic and biotic stress [11,17,18]. Like some tomato relatives, *S. chilense* is self-incompatible and requires cross-pollination, while *S. lycopersicum* is self-compatible and self-pollinates [16]. The resistance of *S. chilense* to biotic stress has been largely investigated, and this species has been used in breeding programs for resistance to viruses such as the tomato yellow leaf curl virus [19] or the cucumber mosaic virus [20]. However, despite a great interest in improving the abiotic stress resistance of tomato, investigation into the resistance of *S. chilense* to abiotic stress such as salinity is rarely studied [2,21].

The effects of NaCl stress on *S. lycopersicum* culture have been explored for a long time, and studies have mainly focused on vegetative growth or yield parameters [8,21–23]. Even if fruit formation is a direct function of reproduction efficiency, the flowering stage is a necessary process before fructification and is consequently impacted by salinity stress before fruit formation. However, the effect of salt on reproductive structures has been little explored in tomato, although abiotic stresses and more specifically salinity may have an impact on the flowering stage. The reproductive phase is indeed considered one of the most sensitive plant developmental stages toward salinity [24]. Ultimately, salinity leads to a decrease in fruit yield and fruit weight and modification of sugar concentration and antioxidant compounds [25,26]. However, earlier in the reproductive development, it can lead to decrease of flower production or decrease of pollen germination and pollen tube growth and even modifications of flower morphology [24,27,28]. In tomato, salinity was shown to induce inflorescence failure and fertility decrease [29,30]. Nevertheless, how salinity affects the flowering and reproductive stage of the halophyte *S. chilense* remains largely unknown.

Solanum chilense has been shown to accumulate more Na in the vegetative aerial parts than *S. lycopersicum* in response to salt [2] but Na accumulation in the reproductive parts has not been investigated as yet. Sodium transport and storage play key roles in the plant response to salinity [31]. Transporters of mineral elements involved in salinity resistance have been widely studied in several plant species, including tomato [32,33]. Several families of transporters are indeed involved in salinity resistance at different stages, especially to maintain Na and potassium (K) homeostasis [34,35]. Briefly, sodium can enter the cell via class I-HKT (High Affinity K^+) transporters and non-selective cation channels. Other transporters, such as the SOS (salt overly sensitive) pathway genes are involved in Na exclusion [34–36]. NHX (vacuolar Na^+/H^+ antiporters) transporters are believed to be Na^+/H^+ exchangers implied in vacuolar Na^+ sequestration [37,38]. Other transporters may play a role in salinity resistance in other ways. HAK (High Affinity K^+) transporters are involved in potassium nutrition and so could help against salt stress [39]. AKT2/3 (inward-rectifying K^+ channel) is a potassium transporter involved in sucrose import in the phloem, which is also activated in response to salt stress [40,41]. In inflorescences of tomato, silencing of HKT1;2 was shown to increase the Na^+/K^+ ratio [25]. However, involvement of transporters activity in salinity resistance in the reproductive structures remains largely unknown in tomato.

In this paper, we compared the Na and K concentrations and localization in the reproductive structures of the halophyte *S. chilense* and the glycophyte *S. lycopersicum* as affected by salt stress and investigated responses of the reproduction of *S. chilense* to salt stress. We aimed to answer the following questions: (1) How does salinity affect flowering, flower development and fertility, and fruit production in these species? (2) Does salinity affect Na and mineral accumulation and partitioning in the reproductive structures of the two species? (3) Does a different Na partitioning in flowers affect flower fertility? (4) What are the responses of putative Na transporters and their contribution to Na accumulation and partitioning in the reproductive structures?

2. Results

2.1. Impact of Salinity on Reproductive Growth

Salt stress was applied before floral transition up to fruit maturation. Throughout the experiment, *S. lycopersicum* produced more leaves on the main stem than *S. chilense*, even under salt stress conditions (Figure 1a,b, Table S2). At 113 days after stress imposition (DASt), the average number of leaves on the main stem was 34.06 ± 5.72 in *S. lycopersicum* and 29.47 ± 4.77 in *S. chilense* (Figure 1a,b). Salt decreased the leaf production in both species (Figure 1a,b): leaf production decreased gradually with stress intensity in *S. lycopersicum* while it was similar in plants treated with 60 and 120 mM NaCl in *S. chilense*. As *S. chilense* had a bushier appearance than *S. lycopersicum*, the total number of leaves produced at 85 DASt was higher in *S. chilense* (80.22 ± 46.63) than in *S. lycopersicum* (47.67 ± 29.97) but it also decreased by 71% and 65% with salt stress, respectively.

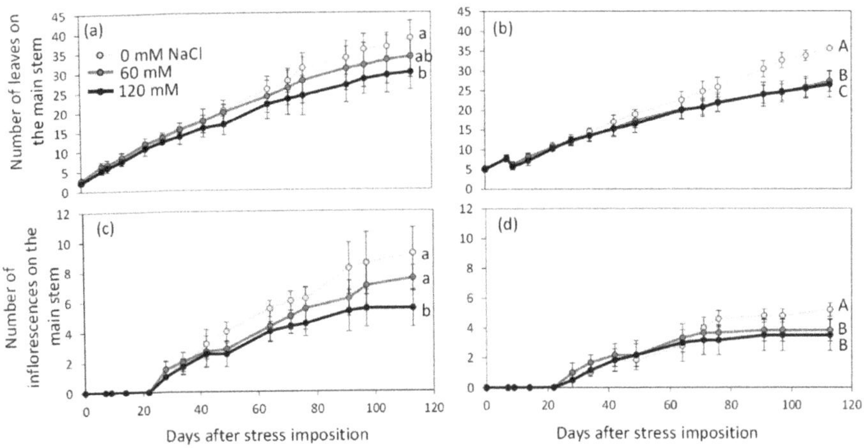

Figure 1. Number of leaves (**a**,**b**) and of inflorescences (**c**,**d**) on the main stem of *Solanum lycopersicum* (**a**,**c**) and *Solanum chilense* (**b**,**d**) grown in perlite:vermiculite mixture supplied with 0, 60, and 120 mM NaCl from 0 to 113 days after stress imposition. Data are means ± SD, treatments followed by different letters are significantly different (lowercase, *S. lycopersicum*, uppercase, *S. chilense*) at $p < 0.05$ for a same species at 113 DASt.

Regarding reproductive growth, flowering times of the initial and sympodial segments were similar between species and salt treatments (Tables 1 and S2). However, as observed for leaf production, *S. lycopersicum* produced more inflorescences on the main stem than *S. chilense* (Figure 1c,d): at 113 DASt, 7.39 ± 1.97 and 4.12 ± 1.05 inflorescences were observed on the main stem of *S. lycopersicum* and *S. chilense*, respectively. Taking into account the ramifications, the total number of inflorescences per plant was similar in both species (Tables 1 and S2). NaCl decreased the number of inflorescences on the main stem and the total number of inflorescences per plant in both species (Tables 1 and S2); the effect was dose-dependent in *S. lycopersicum* but not in *S. chilense* (Figure 1c,d, Table 1). The number of floral buds per inflorescence was always higher in *S. chilense* than in *S. lycopersicum* (Tables 1 and S2). This number decreased with salt stress in *S. lycopersicum* but not in *S. chilense*. In the same way, salinity decreased the percentage of flower buds reaching anthesis only in the cultivated tomato (Tables 1 and S2).

Table 1. Effects of salt stress on flowering parameters of *Solanum lycopersicum* and *Solanum chilense* grown at 0, 60, and 120 mM NaCl.

Flowering Parameters	*S. lycopersicum*			*S. chilense*		
	0 mM NaCl	60 mM NaCl	120 mM NaCl	0 mM NaCl	60 mM NaCl	120 mM NaCl
FT initial segment [1]	11.2 ± 1.3 [a]	10.8 ± 1.2 [a]	11.0 ± 0.9 [a]	12 ± 1.1 [A]	10.0 ± 1.5 [A]	11.8 ± 1.7 [A]
FT sympodial segment [1]	2.7 ± 0.3 [a]	3.0 ± 0.3 [a]	3.0 ± 0.4 [a]	3.8 ± 1.2 [A]	3.1 ± 0.5 [A]	4.2 ± 2.4 [A]
inflorescences per plant	96.0 ± 61.9 [a]	28.8 ± 5.3 [a]	10.0 ± 1.4 [b]	160.3 ± 58.9 A	53.5 ± 17.9 [A]	7.8 ± 4.9 [B]
floral buds per inflorescence	8.32 ± 2.29 [a]	6.56 ± 1.19 [b]	6.14 ± 1.39 [b]	12.5 ± 7.18 A	11.38 ± 9.40 [A]	10.40 ± 5.58 [A]
open flowers per inflorescence (%)	74.6 ± 19.4 [a]	55.6 ± 25.5 [b]	50.7 ± 27.1 [b]	71.3 ± 29.6 A	54.5 ± 34.0 [A]	47.5 ± 30.8 [A]

[1] FT: flowering time; expressed in number of leaves; data are means ± standard deviation, different letters indicate significant difference for each species (lowercase, *S. lycopersicum*, uppercase, *S. chilense*) at $p < 0.05$.

2.2. Impact of Salinity on Flower Morphology and Fertility

Flower morphology differed among tomato species (Tables 2 and S2): sepals, petals, and stamens were always longer in *S. lycopersicum* than in *S. chilense*, while pistils were longer in *S. chilense* than in *S. lycopersicum* and style exertion was only observed in *S. chilense*. Salt affected flower morphology by decreasing the length of sepals in *S. lycopersicum* and modifying the length of stamens in *S. chilense*.

Table 2. Effects of salt stress on flowering morphology and fertility of *Solanum lycopersicum* and *Solanum chilense* grown at 0, 60, and 120 mM NaCl.

Flower Parameters	*S. lycopersicum*			*S. chilense*		
	0 mM NaCl	60 mM NaCl	120 mM NaCl	0 mM NaCl	60 mM NaCl	120 mM NaCl
Sepal length (cm)	1.18 ± 0.23 [a]	0.89 ± 0.16 [b]	0.91 ± 0.15 [b]	0.62 ± 0.1 [A]	0.61 ± 0.08 [A]	0.63 ± 0.15 [A]
Petal length (cm)	1.36 ± 0.18 [a]	1.24 ± 0.23 [a]	1.34 ± 0.17 [a]	1.12 ± 0.2 [A]	1.25 ± 0.22 [A]	1.18 ± 0.17 [A]
Stamen length (cm)	0.84 ± 0.09 [a]	0.8 ± 0.08 [a]	0.85 ± 0.07 [a]	0.80 ± 0.05 [AB]	0.82 ± 0.08 [A]	0.74 ± 0.06 [B]
Style + ovary length (cm)	0.94 ± 0.07 [a]	0.86 ± 0.1 [a]	0.94 ± 0.09 [a]	1.18 ± 0.11 [A]	1.11 ± 0.12 [A]	1.08 ± 0.09 [A]
Style exsertion (cm)	ND	ND	ND	0.38 ± 0.12 [A]	0.29 ± 0.15 [A]	0.33 ± 0.09 [A]
Stigma receptivity (%)	88.6 ± 26.4 [a]	81 ± 29.5 [a]	84.4 ± 30.1 [a]	96.4 ± 13.4 [A]	100 ± 0 [A]	100 ± 0 [A]
Pollen viability (%)	84.7 ± 13.5 [a]	82.5 ± 21.2 [a]	81.6 ± 14.2 [a]	58.3 ± 26.1 [B]	68.9 ± 25 [A]	63 ± 14.3 [AB]
Pollen grains per anther (×1000)	19.2 ± 14.2 [a]	13.9 ± 15.2 [a]	16.3 ± 10.2 [a]	68.0 ± 35.1 [A]	48.5 ± 23.8 [AB]	37.4 ± 19.3 [B]

ND, no style exsertion. Data are means ± standard deviation, different letters indicate significant difference for each species (lowercase, *S. lycopersicum*, uppercase, *S. chilense*) at $p < 0.05$.

Flower fertility was assessed by stigma receptivity, pollen production, and viability (Tables 2 and S2). Overall, stigma receptivity was slightly lower in *S. lycopersicum* than in *S. chilense*. *S. lycopersicum* also produced fewer pollen grains per anther than *S. chilense*. However, pollen viability was 23% higher in *S. lycopersicum* than in *S. chilense*. Salt did not affect stigma receptivity, pollen viability, or the number of pollen grains per anther in *S. lycopersicum*. However, in *S. chilense*, the number of pollen grains per anther decreased with salt while pollen viability increased gradually with salt concentration.

2.3. Impact of Salinity on Fruit Production and Quality

Fruit set was higher in *S. chilense* than in *S. lycopersicum* and was not affected by salt stress whatever the species (Tables 3 and S2).

Table 3. Effects of salt stress on fructification parameters of *Solanum lycopersicum* and *Solanum chilense* grown at 0, 60, and 120 mM NaCl.

Fruit Parameters	*S. lycopersicum*			*S. chilense*		
	0 mM NaCl	**60 mM NaCl**	**120 mM NaCl**	**0 mM NaCl**	**60 mM NaCl**	**120 mM NaCl**
Fruit set (%)	47.9 ± 15.1 [a]	43.5 ± 22.5 [a]	38.9 ± 21.1 [a]	51.7 ± 40.6 [A]	60 ± 37.7 [A]	44.3 ± 33.2 [A]
FW (g)	47.7 ± 10.3 [a]	22.5 ± 6 [b]	14.5 ± 5.3 [c]	0.65 ± 0.2 [A]	0.79 ± 0.26 [A]	0.84 ± 0.21 [A]
DW (g)	3.42 ± 1.98 [a]	2.01 ± 0.56 [b]	1.44 ± 0.62 [b]	0.12 ± 0.03 [A]	0.10 ± 0.02 [A]	0.10 ± 0.01 [A]
WC (%)	91.91 ± 3.77 [a]	89.67 ± 0.42 [b]	88.19 ± 0.9 [c]	80.72 ± 2.87 [C]	82.49 ± 7.17 [B]	87.49 ± 1.6 [A]
Circumference (cm)	14.4 ± 0.92 [a]	11.62 ± 0.74 [b]	10.1 ± 1.13 [c]	3.15 ± 0.24 [A]	3.48 ± 0.73 [A]	3.57 ± 0.62 [A]
Number of seeds/fruit	91.17 ± 46.02 [a]	72.77 ± 33.19 [ab]	50.08 ± 16.04 [b]	21.22 ± 4.47 [A]	22.00 ± 5.28 [A]	24.88 ± 10.21 [A]
Number of seeds/fruit FW (g)	2.11 ± 0.73 [b]	3.31 ± 0.91 [a]	3.76 ± 2.1 [a]	34.37 ± 13.44 [A]	30.72 ± 11.83 [A]	25.93 ± 7.18 [A]
Sugar concentration (°Brix)	5.54 ± 0.52 [c]	7.95 ± 0.44 [b]	9.2 ± 0.95 [a]	18.4 ± 3.2 [A]	11.95 ± 4.12 [B]	10.15 ± 2.35 [B]
pH	4.52 ± 0.09 [a]	4.4 ± 0.11 [b]	4.31 ± 0.11 [b]	4.67 ± 0.26 [A]	4.07 ± 0.49 [B]	3.8 ± 0.26 [B]

Data are means ± standard deviation; different letters indicate significant difference for each species (lowercase, *S. lycopersicum*, uppercase, *S. chilense*) at $p < 0.05$. DW, FW, dry and fresh weights; WC, water content.

S. lycopersicum produced bigger fruits than *S. chilense*. Indeed, fruit FW, DW, WC, and size were higher in *S. lycopersicum* than in *S. chilense* (Table 3). Following the fruit size, the number of seeds per fruit was 69% higher in *S. lycopersicum* than in *S. chilense* (Table 3), although, when expressed per gram of fruit FW, the number of seeds was 90% higher in *S. chilense* than in *S. lycopersicum*. Salinity mainly affected fruit growth in *S. lycopersicum* as fruit DW, FW, WC, and size decreased with a higher salt concentration in *S. lycopersicum* while salinity modified only fruit WC in *S. chilense*, which increased with salt concentration (Table 3). However, the number of seeds per fruit or per gram of fruit FW were not affected by salinity whatever the species (Table 3).

Concerning fruit quality, fruits of *S. lycopersicum* were less sweet and less acidic than those of *S. chilense* (Table 3): sugar content and pH were, respectively, 3.3 and 1.1 times lower in fruits of *S. lycopersicum* than in the ones of *S. chilense* under control conditions. Salinity affected fruit quality in both species (Table S2). The fruit sugar content was modified in different ways according to the species: sugar concentration increased in *S. lycopersicum* but decreased in *S. chilense* with salt concentration (Table 3). However, fruit pH decreased with salinity in both species (Table 3).

2.4. Impact of Salinity on Mineral Concentration and Distribution in Reproductive Organs

2.4.1. Inflorescences and Flowers

Inflorescences of *S. chilense* accumulated more Na than the ones of *S. lycopersicum* (Figure 2a), even under control conditions. Salinity induced a significant increase in Na concentration in the inflorescences of both species (Figure 2a, Table S2), although it was larger in *S. chilense* than in *S. lycopersicum*. Indeed, Na concentration increased by 223% and 465%, given as the percentual difference between control and 120 mM NaCl treated plants in *S. lycopersicum* and *S. chilense*, respectively. Moreover, Na distribution mapping showed that, in addition to the Na concentration, there was a difference in Na location inside the flowers in the two species (Figures 3 and S1). In *S. chilense*, Na mainly accumulated in the male organs (Figure 3), whereas in *S. lycopersicum*, most of the Na was located in the receptacle and pedicel (Figure 3). Moreover, the ratio between the number of counts of Na in the floral receptacle and reproductive (stamens + pistil) floral whorls was higher in *S. lycopersicum* than in *S. chilense* and increased with salt stress, mainly in *S. lycopersicum* (Table 4). The ovary had the lowest Na signal compared to the rest of the flower in both species (Figure 3). As a result, the ratio between the Na signal in the stamens and the pistil was higher in *S. chilense* than in *S. lycopersicum* (Table 4). This ratio decreased with salt stress in both species.

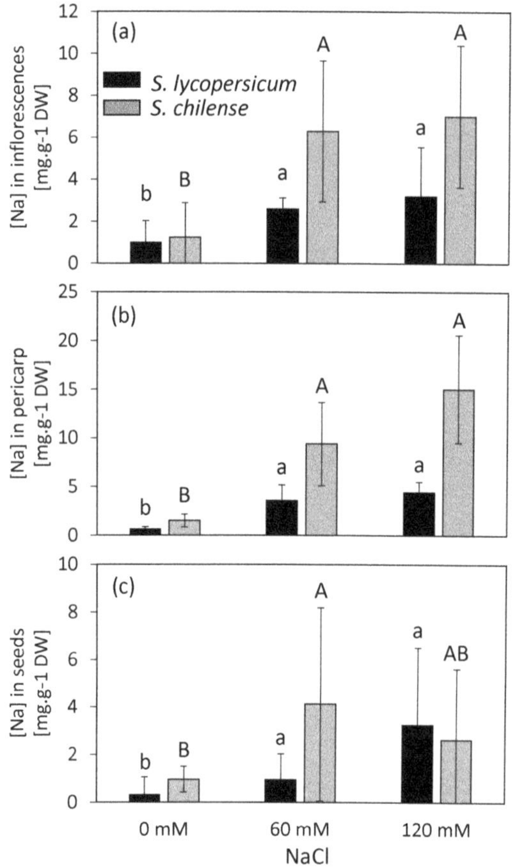

Figure 2. Sodium (Na) concentration in inflorescences (**a**), pericarp of fruits (**b**), and seeds (**c**) of *Solanum lycopersicum* and *Solanum chilense* grown in perlite:vermiculite mixture supplied with 0, 60, and 120 mM NaCl. Data are means ± SD; treatments followed by different letters are significantly different (lowercase, *S. lycopersicum*, uppercase, *S. chilense*) at $p < 0.05$ for a same species.

As for Na, inflorescences of *S. chilense* accumulated more K than those of *S. lycopersicum* (Tables 5 and S2). Salinity did not affect the K concentration in the inflorescences whatever the species (Tables 5 and S2). The K/Na ratio was, however, higher in the inflorescences of *S. lycopersicum* than in those of *S. chilense* and decreased with salt stress in both species (Tables 5 and S2). In flowers of *S. chilense*, K mainly accumulated in male organs with no accumulation in female organs (Figures 4 and S2). In contrast, K accumulated mainly in female organs in *S. lycopersicum* (Figures 4 and S2). As a result, the ratio of the number of counts of K in stamens and pistil was higher in *S. chilense* than in *S. lycopersicum* (Table 4). However, the ratio between the K signals in floral receptacle and reproductive floral organs was similar in both species under control conditions but decreased with salt in *S. chilense* and not in *S. lycopersicum* (Table 4).

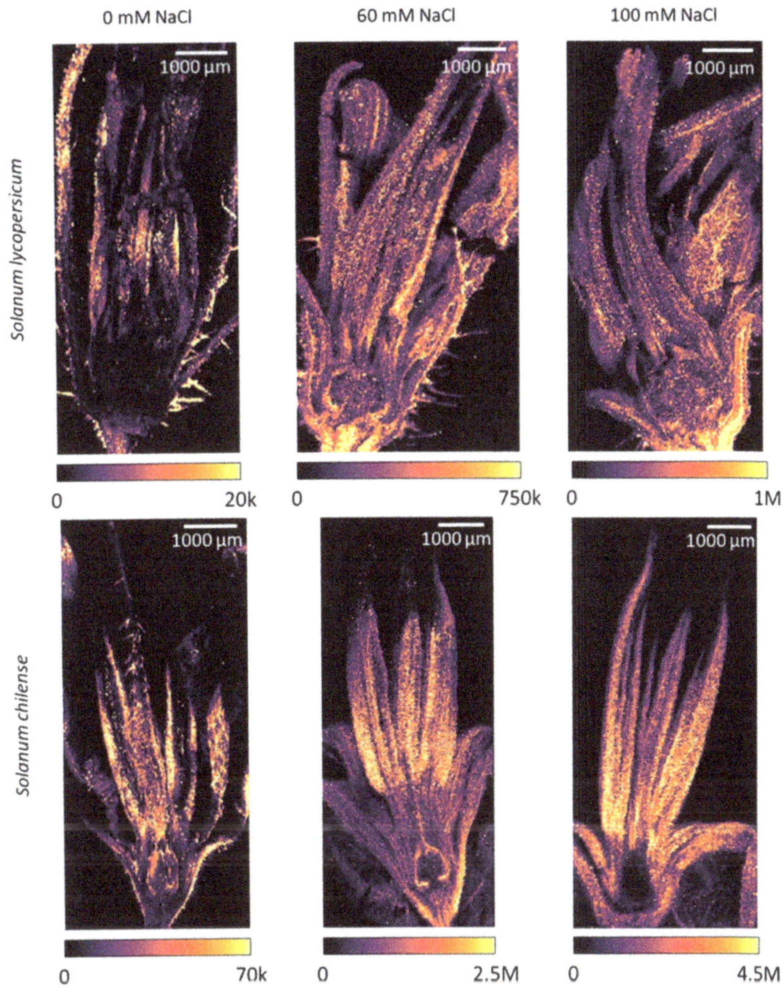

Figure 3. Sodium (Na) distribution in flowers of *Solanum lycopersicum* (top row) *and Solanum chilense* (bottom row) grown in perlite:vermiculite mixture supplied with 0, 60, and 100 mM NaCl as revealed by LA-ICP-MS (laser ablation inductively coupled plasma mass spectroscopy) and visualized using ImageJ (version 1.53a). Color legends represent the number of counts per pixel ($20 \times 20\ \mu m^2$) of each analysis.

Inflorescences of *S. lycopersicum* accumulated about 10 times more Ca than those of *S. chilense*, and their Ca concentrations were not affected by salinity (Tables 5 and S2). Ca mainly accumulated in floral receptacle of *S. lycopersicum* and mainly in reproductive floral organs of *S. chilense* (Figures 4 and S3). Indeed, the ratio between the Ca signals in floral receptacle and reproductive floral organs was higher in *S. lycopersicum* than in *S. chilense* (Table 4). Ca was particularly visible in the ovary of salt-treated *S. lycopersicum* flowers (Figures 4 and S3), explaining the lower ratio of Ca signal between stamens and pistil in salt-treated flowers (Table 4).

The concentration of Mg in inflorescences of *S. lycopersicum* was more important than in those of *S. chilense* (Tables 5 and S2). However, only the former was affected by salinity (Tables 5 and S2). Mg mainly accumulated in the stamens and ovary of *S. chilense* and in the ovary of *S. lycopersicum* (Figures 4 and S4). The ratio of Mg signals between floral receptacle and reproductive floral organs and between stamens and pistil decreased with salt stress in *S. chilense* and *S. lycopersicum*, respectively (Table 4).

Table 4. Effects of salt stress on ratio (vegetative/reproductive organs and male/female organs) of mineral elements signals in flowers of *Solanum lycopersicum* and *Solanum chilense* grown at 0, 60, and 100 mM NaCl.

Mineral	***S. lycopersicum***			***S. chilense***		
	0 mM NaCl	**60 mM NaCl**	**100 mM NaCl**	**0 mM NaCl**	**60 mM NaCl**	**100 mM NaCl**
vegetative/reproductive floral organs						
Na	0.36 ± 0.12	0.65 ± 0.11	1 ± 0.26	0.18 ± 0.07	0.43 ± 0.04	0.31 ± 0.02
K	0.45 ± 0.1	0.43 ± 0.05	0.5 ± 0.03	0.43 ± 0.19	0.22 ± 0.09	0.14 ± 0.02
Ca	0.54 ± 0.07	0.42 ± 0.05	0.69 ± 0.24	0.34 ± 0.29	0.29 ± 0.12	0.26 ± 0.08
Mg	0.86 ± 0.16	0.34 ± 0.1	0.77 ± 0.19	0.67 ± 0.46	0.23 ± 0.02	0.3 ± 0.08
male/female floral organs						
Na	0.82 ± 0.05	0.55 ± 0.16	0.5 ± 0.07	1.24 ± 0.68	0.59 ± 0.27	0.7 ± 0.49
K	0.63 ± 0.03	0.41 ± 0	0.34 ± 0.2	0.83 ± 0.06	0.6 ± 0.16	0.68 ± 0.54
Ca	0.64 ± 0.24	0.26 ± 0.02	0.3 ± 0.14	1.32 ± 0.07	0.35 ± 0.11	0.68 ± 0.29
Mg	0.49 ± 0.05	0.35 ± 0.12	0.3 ± 0.09	0.62 ± 0.2	0.38 ± 0.04	0.52 ± 0.37

Relative signal intensities obtained by LA-ICP-MS (laser ablation inductively coupled plasma mass spectroscopy) are expressed in counts. Signal intensities are correlated with the concentrations of a particular element (comparisons could be performed per element but not between elements).

Table 5. Effects of salt stress K, Ca, and Mg concentrations of different organs of *Solanum lycopersicum* and *Solanum chilense* grown at 0, 60, and 120 mM NaCl.

Mineral	***S. lycopersicum***			***S. chilense***		
	0 mM NaCl	**60 mM NaCl**	**120 mM NaCl**	**0 mM NaCl**	**60 mM NaCl**	**120 mM NaCl**
Inflorescences						
K (mg g^{-1} DW)	27.63 ± 2.62 ^{a}	26.09 ± 6 ^{a}	23.97 ± 4.36 ^{a}	30.76 ± 5.06 ^{A}	31.97 ± 4.81 ^{A}	26.47 ± 5.21 ^{A}
K/Na	43.01 ± 23.32 ^{a}	10.34 ± 2.87 ^{b}	10.15 ± 5.7 ^{b}	10.66 ± 17.26 A	4.74 ± 4.43 ^{B}	1.26 ± 1.14 ^{B}
Ca (mg g^{-1} DW)	1.03 ± 0.87 ^{a}	1.22 ± 0.82 ^{a}	1.09 ± 1.02 ^{a}	0.18 ± 0.14 ^{A}	0.08 ± 0.08 ^{A}	0.29 ± 0.01 ^{A}
Mg (mg g^{-1} DW)	4.92 ± 1.22 ^{a}	5.62 ± 1.96 ^{a}	3.64 ± 0.62 ^{b}	3.45 ± 1.04 ^{A}	2.94 ± 0.42 ^{A}	3.04 ± 0.94 ^{A}
Pericarp						
K (mg g^{-1} DW)	38.98 ± 6.36 ^{a}	32.14 ± 9.24 ^{b}	26.20 ± 5.88 ^{b}	39.22 ± 4.66 ^{A}	27.75 ± 4.47 ^{B}	27.99 ± 6.74 ^{B}
Ca (mg g^{-1} DW)	0.73 ± 0.23 ^{a}	0.49 ± 0.23 ^{b}	0.64 ± 0.24 ^{ab}	1.49 ± 0.56 ^{A}	1.27 ± 0.23 ^{A}	1.54 ± 0.56 ^{A}
Mg (mg g^{-1} DW)	1.36 ± 0.26 ^{a}	1.09 ± 0.43 ^{b}	1.08 ± 0.12 ^{b}	2.10 ± 0.33 ^{A}	2.14 ± 0.58 ^{A}	2.46 ± 0.40 ^{A}
seeds						
K (mg g^{-1} DW)	8.09 ± 5.09 ^{a}	9.79 ± 7.28 ^{a}	14.42 ± 9.2 ^{a}	20.43 ± 7.6 ^{A}	9.85 ± 5.68 ^{B}	8.71 ± 7.21 ^{B}
Ca (mg g^{-1} DW)	0.77 ± 0.53 ^{a}	0.71 ± 0.54 ^{a}	0.52 ± 0.13 ^{a}	0.89 ± 0.23 ^{A}	0.85 ± 0.24 ^{A}	0.85 ± 0.39 ^{A}
Mg (mg g^{-1} DW)	3.60 ± 0.93 ^{a}	3.57 ± 0.92 ^{a}	2.40 ± 0.75 ^{a}	2.74 ± 0.27 ^{A}	2.38 ± 0.26 ^{B}	2.69 ± 0.37 ^{AB}

Concentrations (mg g^{-1} DW) are measured by AAS (atomic absorption spectrometry). Data are means ± standard deviation, different letters indicate significant difference for each species (lowercase, *S. lycopersicum*, uppercase, *S. chilense*) at $p < 0.05$. DW, dry weight.

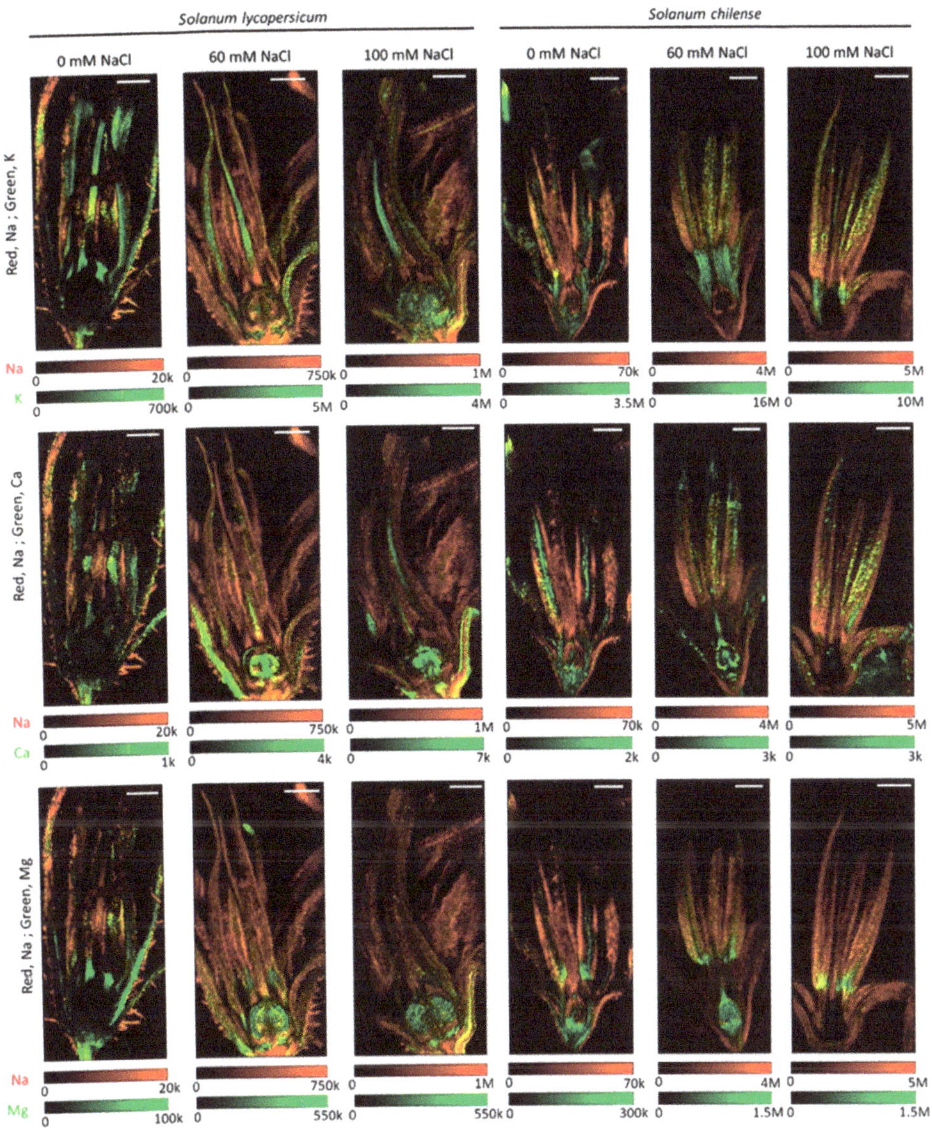

Figure 4. Distribution of sodium (Na) shown in red and potassium (K), calcium (Ca) and magnesium (Mg) shown in green and their co-localization (yellow) in flowers of *Solanum lycopersicum* and *Solanum chilense* grown in perlite:vermiculite mixture supplied with 0, 60, and 100 mM NaCl. Distribution of individual element was determined using LA-ICP-MS (laser ablation inductively coupled plasma mass spectroscopy) and visualized using ImageJ (version 1.53a). Color legends represent the number of counts per pixel (20 × 20 μm^2) for each analysis and each element. Signal intensities are correlated with the concentrations of a particular element. Scale bar = 1000 μm.

2.4.2. Fruits and Seeds

The Na concentration in fruit pericarp was similar to in the inflorescences for the same species (*S. lycopersicum*, $t_{101} = -0.161$, $p = 0.872$, *S. chilense*, $t_{49} = -0.818$, $p = 0.417$, Figure 2a,b).

Nevertheless, as observed in the inflorescences, the pericarp of *S. lycopersicum* fruits were less concentrated in Na than the pericarp of *S. chilense* fruits (Figure 2b, Table S2): the difference was about 2.4 times that of control plants, 2.6 times that of 60 mM NaCl treated plants, and 3.4 times that of 120 mM NaCl treated plants. Salinity indeed increased the Na concentration in the pericarp of both species but to a higher extent in *S. chilense*. For both species, the Na concentration was 0.6 and 0.4 times lower in seeds than in pericarp for *S. lycopersicum* and *S. chilense*, respectively, but again, seeds of *S. chilense* contained more Na that the ones of *S. lycopersicum* (Figure 2c, Table S2). However, the Na concentration increased with salt stress in the seeds of *S. lycopersicum*, but only slightly in those of *S. chilense* (Figure 2b,c).

The concentrateion of K in the pericarp was similar in both species and decreased significantly with salt stress in both species (Tables 5 and S2). However, the K concentration in the seeds was higher under control conditions in *S. chilense* than in *S. lycopersicum*, and it decreased with salinity only in the former so that the K concentration was similar in the seeds of stressed plants of both species (Tables 5 and S2).

The concentration of Ca was higher in the pericarp of *S. chilense* than in the one of *S. lycopersicum*, but there was no clear difference under salinity (Tables 5 and S2). However, the Ca concentration in seeds did not differ between species (Tables 5 and S2). The concentration of Mg was higher in the pericarp of *S. chilense* than in the one of *S. lycopersicum*, but it was higher in the seeds of *S. lycopersicum* than in the ones of *S. chilense* (Tables 5 and S2).

2.5. Impact of Salinity on the Expression of Mineral Transporters in Flowers

To improve our understanding of Na accumulation and its distribution in flowers, we investigated the expression of genes coding for transporters involved in Na transport in flowers at anthesis. We particularly focused on the SOS pathway, and the NHX, HKT and HAK transporters.

Concerning the SOS pathway, *SOS1* expression was higher in *S. lycopersicum* than in *S. chilense*, while the opposite trend was observed for *SOS3* expression (Figure 5a,c, Table S2). However, there was no difference of expression for *SOS2* between species (Figure 5b, Table S2). Salt stress increased *SOS1* expression in both species but more significantly and at a lower salt concentration in *S. lycopersicum* than in *S. chilense* (Figure 5a). Expression of *SOS2* and *SOS3*, respectively, increased and decreased with salt in *S. lycopersicum* only; nevertheless, a decrease of *SOS3* expression was observed in *S. chilense* at 60 mM NaCl (Figure 5b,c).

The gene *NHX3*, which encodes a tonoplast transporter, had similar expression levels in both species regardless of treatment (Figure 5d, Table S2), contrary to *NHX4*, which was more expressed in *S. lycopersicum* than in *S. chilense* at least in salt-treated flowers (Figure 5e, Table S2). Salt stress decreased the expression of *NHX3* and increased the expression of *NHX4* in *S. lycopersicum* but did not affect their expression in *S. chilense* (Figure 5d,e).

The expression of *HKT1;2* was slightly higher in *S. lycopersicum* than in *S. chilense* and decreased with salt treatment in both species from 60 mM NaCl (Figure 5f, Table S2).

The expression of *SlHAK14* and *SlHAK3* was higher in *S. chilense* than in *S. lycopersicum*, while the expression of *SlAKT2/3* and *CNGC10* was similar in both species (Figure 5g–j, Table S2). Salinity affected these genes differently, depending on the species. The expression of *SlHAK14* gradually increased with salt in *S. lycopersicum* but decreased in *S. chilense* at 60 mM NaCl only (Figure 5g). The expression of *SlAKT2/3* increased in *S. lycopersicum* from 60 mM NaCl but was unchanged in *S. chilense* (Figure 5h). The expression of *SlHAK3* was stable in *S. lycopersicum* but decreased at 60 mM NaCl in *S. chilense* (Figure 5i). The expression of *CNGC10* was stable in *S. lycopersicum* but increased at 120 mM NaCl in *S. chilense* (Figure 5j).

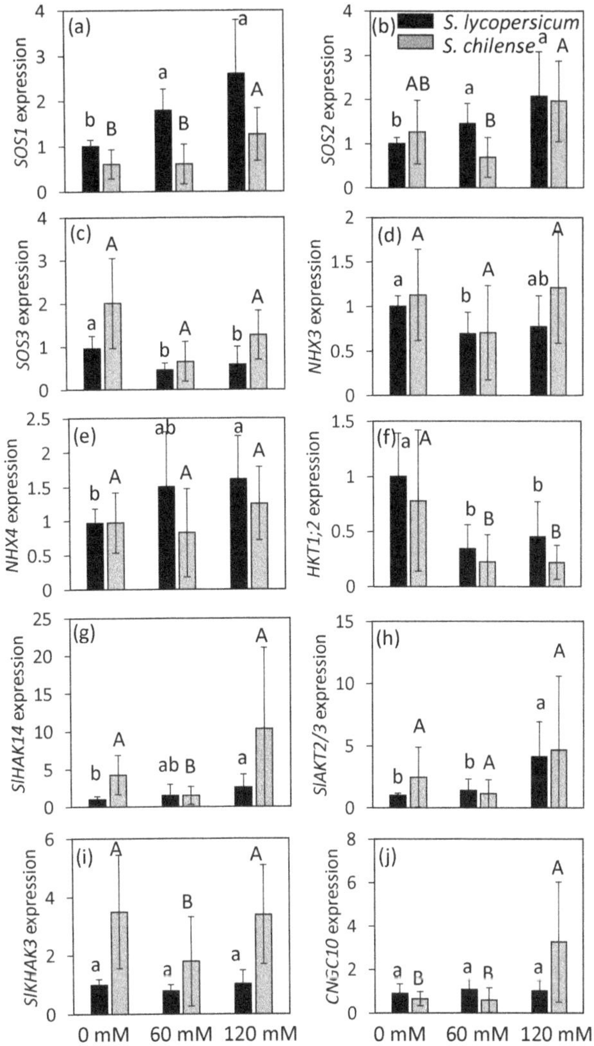

Figure 5. Expression of 10 genes involved in minerals transport analyzed by qRT-PCR on flowers of *Solanum lycopersicum* and *Solanum chilense* growing at 0, 60, and 120 mM NaCl. (**a**) *SOS1* (Salt Overly Sensitive 1, Solyc01g005020); (**b**) *SOS2* (Salt Overly Sensitive 2, Solyc12g009570); (**c**) *SOS3* (Salt Overly Sensitive 3, Solyc06g051970); (**d**) *NHX3* (vacuolar Na^+/H^+ antiporter 3, Solyc01g067710); (**e**) *NHX4* (vacuolar Na^+/H^+ antiporter 4, Solyc01g098190); (**f**) *HKT1;2* (class I—High affinity K^+ transporter 2, Solyc07g014680); (**g**) *SlHAK14* (High Affinity K^+ transporter 14, Solyc09g074820); (**h**) *SlAKT2/3* (inward-rectifying K^+ channel, Solyc10g024360); (**i**) *SlHAK3* (High Affinity K^+ transporter 3, Solyc12g096580); (**j**) *CNGC10* (Cyclic Nucleotide Gated Channel 10, Solyc05g050350). The tomato elongation factor gene (*LeEF-1α*, Solyc06g005060) and TIP41-like protein (*TIP41*, Solyc10g04985) were used as the reference genes. Expressions are given based on *S. lycopersicum* grown at 0 mM NaCl, to which a value of 1 was assigned. Data are means ± SD, treatments followed by different letters are significantly different at $p < 0.05$ for the same species (lowercase, *S. lycopersicum*, uppercase, *S. chilense*).

2.6. Correlations among Flower Morphology, Mineral Concentrations, and Gene Expression

Analysis of correlations among flower fertility parameters, concentrations of elements in inflorescences and flowers, and expression of mineral transporters in flowers showed a different behavior between both species (Figure 6). Overall, few correlations were observed between flower fertility parameters and mineral concentrations in the flowers, mainly in *S. lycopersicum* (Figure 6a,b). In *S. chilense*, the number of pollen grains per stamen was negatively correlated with the concentration of Na in inflorescences, although this correlation was not observed in *S. lycopersicum*. Some correlations were observed between floral organ size and elements signals in the reproductive structures in both species (Figure 6a,b). In *S. lycopersicum*, sepal length was negatively correlated with the ratio of Na signal between vegetative and reproductive floral organs and positively correlated with the ratio of elements signals between male and female reproductive organs and with the K/Na ratio in the inflorescences (Figure 6a). Moreover, in *S. chilense*, the pistil length and the style exertion were negatively correlated with, respectively, the Na concentration in the inflorescence and the ratio of Na signal between vegetative and reproductive organs (Figure 6b). Stamen and pistil lengths were also negatively correlated with, respectively, the Ca and Mg concentrations in inflorescences and positively correlated with the ratio of Mg signals between vegetative and reproductive floral organs in *S. lycopersicum*. Correlations between Na signals in reproductive structures and Na transporter gene expression also differed among species (Figure 6c,d). The Na concentration in inflorescences was negatively correlated with the expression of *SOS3* and positively correlated with the expression of *SOS2* and *SlHAK14* in *S. lycopersicum* while it was negatively correlated with the expression of *HKT1;2* in *S. chilense* (Figure 6c,d). The ratio of Na concentrations in male and female floral organs was negatively correlated with the expression of *SOS1*, *SOS2*, and *SlAKT2/3* and positively correlated with the expression of *NHX3* and *HKT1;2* in *S. lycopersicum* while it was positively correlated with the expression of *SOS3* and *HKT1;2* in *S. chilense*. The ratio between Na signals in vegetative and reproductive floral organs was negatively correlated with the expression of *NHX3* and *HKT1;2* in both species; it was also negatively correlated with the expression of *SlHAK3* and *SOS3* in *S. chilense* and positively correlated with the expression of *SOS1*, *SOS2*, *SlHAK14*, and *SlAKT2/3* in *S. lycopersicum* (Figure 6c,d).

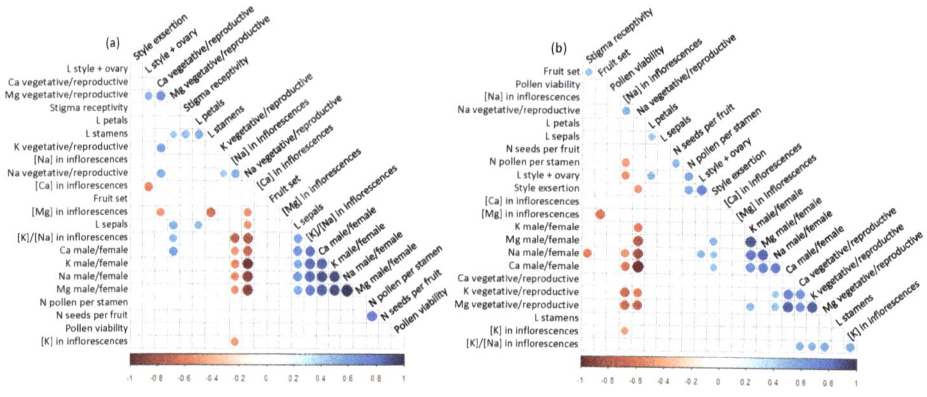

Figure 6. *Cont.*

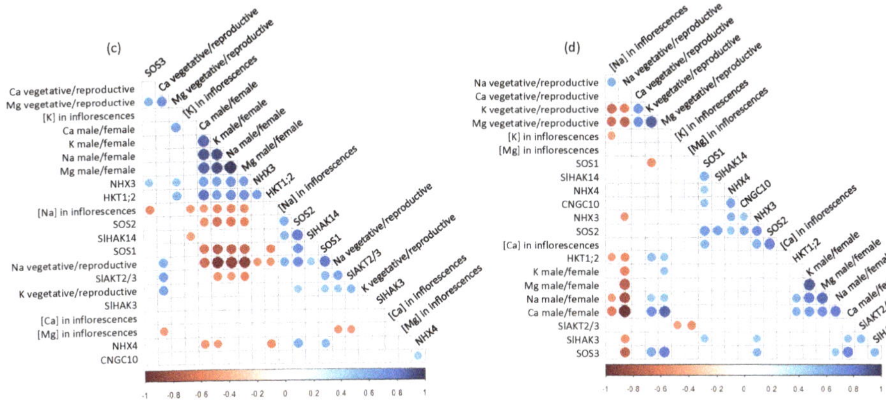

Figure 6. (**a**,**b**) Correlation graphs of concentrations of elements in inflorescences, ratios of element signals in the vegetative/reproductive organs and male/female organs on flowers and fertility parameters of flowers of *Solanum lycopersicum* (**a**) and *Solanum chilense* (**b**). (**c**,**d**) Correlation graphs of concentrations of elements in inflorescences, ratios of element signals in the vegetative/reproductive organs and male/female organs on flowers and expression of mineral transporters in flowers of *S. lycopersicum* (**c**) and *S. chilense* (**d**). Only significant correlations ($p < 0.05$) are indicated with circles. Negative correlations are highlighted in red and positive correlations in blue. *CNGC10*, Cyclic Nucleotide Gated Channel 10; *HKT1;2*, class I—High affinity K^+ transporter 2; L petals, sepals, style + ovary: length of, respectively, petals, sepals, and the sum of style and the ovary; *NHX3, 4*, vacuolar Na^+/H^+ antiporter 3, 4; N pollen per stamen, seeds per fruit: number of, respectively, pollen grains per stamen and seeds per fruit; *SlAKT2/3*, inward-rectifying K^+ channel; *SlHAK3, 14*, High Affinity K^+ transporter 3,14; *SOS1, 2, 3*, Salt Overly Sensitive 1, 2, 3.

3. Discussion

3.1. Salinity Affects Reproductive Structures in Both Species

Flowering and reproduction differed between *S. lycopersicum* and *S. chilense*. The former is considered as an autonomous flowering plant [42] while the latter is a short-day plant [16,43]. Moreover, *S. lycopersicum* is self-compatible and self-pollinates while *S. chilense* is self-incompatible and requires insect pollination [13]. We observed that salinity affected the reproductive phase in both species but in different ways. Salt stress decreased the number of inflorescences in both species but the number of floral buds and opened flowers per inflorescence was only reduced in *S. lycopersicum*. *Solanum chilense* produced more flowers per inflorescence than *S. lycopersicum* like most wild tomato relatives, which could be an advantage for breeding [44], but this parameter was not affected by salt stress in *S. chilense*. Inflorescence and flower production seemed thus more affected by salinity in *S. lycopersicum* than in *S. chilense*, and the effect was more dose-dependent in the former than in the latter. Flower abortion was previously observed under salt conditions in cultivated tomato [29]. A decrease in inflorescence and flower production and an increase in flower abortion are common phenomena observed in response to stress; abortion of spikelets was, for instance, observed in rice under salinity treatments [45].

Salinity also affected flower morphology and fertility. Flower morphology differed between species: the ratio between corolla and calyx area was higher in *S. chilense* than in *S. lycopersicum*, and style exertion was observed only in the former. These differences could be related to the self-incompatibility of *S. chilense* [16] that needs to attract pollinators for cross-pollination. Concerning floral organs, salt decreased sepal length in *S. lycopersicum* and decreased stamen length in *S. chilense*. Modification of flower morphology due to salinity was reported in *Spergularia maritima* (petal size increased in salinity treatments) [27]. In tomato, other environmental constraints such as temperature also affect flower morphology [46,47]. Those modifications could have an impact on flower attractivity for pollinators, as it has been shown in *Raphanus sativus* [48] or *Borago officinalis* [49,50]. Flower and petal size are indeed important floral signals for pollinators [49]. The decrease of stamen length observed in salt-treated *S. chilense* was associated with a decrease in the number of pollen grain per anther and an increase in pollen viability. However, in our study, pollen production and viability were not affected by salinity in *S. lycopersicum*, and stigma receptivity was not affected by salt stress whatever the species. Anther development and microsporogenesis are generally considered the most sensitive reproductive stages to abiotic stresses, which could explain the more important effect on male organs than on female organs [51]. Gynoecium fertility is not often affected by abiotic stress in tomato or is affected as a consequence of male development failure [52,53].

In accordance with the low impact of salinity on flower fertility, fruit set was not affected by salt treatment in our study whatever the species. However, fruit weight, size, and water content decreased with salinity in *S. lycopersicum* while these parameters were not affected or even increased (for WC) under salt treatment in *S. chilense*. Moreover, the seed set decreased with salinity in *S. lycopersicum* but not in *S. chilense*. The effect of salinity on flower fertility is thus not sufficient to explain the salt-induced modifications of fruit parameters despite the positive correlation between pollen per anther and seeds per fruit. Pollen tube growth, fertilization, and seed development may be affected by abiotic stress such as salinity [54]. Moreover, the decrease of sepal length observed in salt-treated *S. lycopersicum* may limit sepal photosynthesis and reduce the supply of carbohydrates for fruit and seed growth as observed in hellebore [55]. It was indeed reported that photosynthesis of green reproductive organs contribute in a significant way to fruit growth [56,57]. A decrease of yield in *S. lycopersicum* subjected to salinity has frequently been described and was explained by a decrease in fruit size rather than by a decrease in fruit number [21,58], which corroborates our observations. Martínez et al. [21] compared fruit yield in *S. lycopersicum* and *S. chilense* in response to NaCl (0–80 mM) and observed that, although salt decreased fruit production and fruit weight in *S. lycopersicum*, it did not affect these parameters in *S. chilense*. *Solanum chilense* seems thus able to maintain its fruit production in salt conditions. Maintenance of fruit size and seed set under salt stress could be of great interest for tomato improvement. However, salinity affected fruit quality in both species. We observed that salinity increased fruit sugar concentrations in *S. lycopersicum* but decreased it in *S. chilense*; salt also decreased fruit pH in both species. Martínez et al. [21,26] also observed a change in fruit quality in both species as a response to salt. For example, they observed that both species differed regarding their main antioxidant compounds and that salinity increased the antioxidant capacity in *S. chilense* while it decreased it in *S. lycopersicum* [26].

3.2. Salinity Affects Mineral Accumulation and Distribution Which May Affect Fertility

The decrease of inflorescence and flower production and of flower fertility as well as the increase of flower abortion in response to abiotic stress is often explained in terms of competition for assimilates or alteration of carbohydrates metabolism [29,30,59]. However, in response to salinity, we may not exclude that the negative impact on flower production and fertility could be due to an accumulation of toxic ions in the reproductive structures [60,61].

The sodium concentration increased in the inflorescences and the fruits of salt treated plants of both species as soon as they were exposed to 60 mM NaCl, but final concentrations in *S. chilense* were higher than in *S. lycopersicum*. However, Na concentrations were lower in the seeds than in the pericarp, suggesting that the plant protect the next generation. A limitation of toxic ions in the seeds has indeed been reported in other plant species such as rice [62] and *Kosteletzkya pentacarpos* [63]. It was previously shown that *S. chilense* accumulated more Na in the vegetative aerial parts than *S. lycopersicum* during vegetative growth [64]. Our results showed that a similar situation occurred in the reproductive organs. The higher salinity resistance of *S. chilense* compared to *S. lycopersicum* regarding flower and fruit production can therefore not be explained by Na exclusion in the reproductive parts.

However, the Na distribution in the flowers differed in the species. In *S. lycopersicum*, Na was mostly accumulated in the non-reproductive parts of the flowers and especially in the pedicel and receptacle. This suggests that *S. lycopersicum* protects the reproductive organs by limiting Na accumulation in this sensitive tissue. Ghanem et al. [29] previously reported that *S. lycopersicum* limited Na accumulation in the reproductive organs and particularly in pollen grains. However, we may not exclude that the higher Na accumulation in the non-reproductive floral organs contributed to the decrease of sepal length. The sepal length was indeed negatively correlated with the Na signal ratio between floral receptacle and reproductive floral organs in *S. lycopersicum*. It is known that Na accumulation reduced vegetative growth in *S. lycopersicum* [64,65]. In *S. chilense*, Na accumulated more in reproductive floral organs and mainly in stamens. This could explain the decrease of stamen length and pollen production observed in salt-treated *S. chilense*. The number of pollen grains per stamen was indeed negatively correlated with the concentration of Na in the inflorescences in this species. In *S. lycopersicum*, the exclusion of Na in the male floral organs probably led to the protection of pollen because neither pollen viability nor the number of pollen grains per stamen were affected by salt stress in our study. It is often reported that male reproductive floral organs are more affected by abiotic stress than female floral organs in tomato [29,66], suggesting that the latter is more protected than the former. However, we observed that the ratio of the Na signals between male and female floral organs decreased with salt in both species. Regarding female floral organs, Na accumulated in the external tissues over the ovary but not in the ovules in *S. chilense*, whereas in *S. lycopersicum*, Na signal was low in female organs but was distributed in the whole ovary. Such differences in Na localization between species may explain the effects of salinity on fruit development in both species. Fruit and seed development were indeed more affected in *S. lycopersicum* than in *S. chilense*.

In addition to the accumulation of Na, modification of the concentration or localization of other key minerals may also affect flower development and fertility. Indeed, K is an essential macronutrient in flower development, particularly for stamen and pollen grains [67]. We observed that the concentrations of K and Na in inflorescences were negatively correlated and that the K/Na ratio decreased with salt stress in both species although K concentrations in inflorescences were not affected by salinity. In vegetative organs, a decrease of K is often observed in response to NaCl [2,29,64,68], which negatively affects C/N nutrition and the activity of several enzymes [69,70]. The maintenance of sufficient K concentration in inflorescences despite salt stress can be explained by the importance of this element for reproductive development and especially for elongation of filaments and release of pollen [67]. For example, K contributed to anther dehiscence and pollen imbibition in rice [71,72]. Decrease of the K/Na ratio is commonly reported as symptomatic of salinity stress [73]. Surprisingly, we observed that the K/Na ratio is more important in the inflorescences of *S. lycopersicum* than in those of *S. chilense*, even at high NaCl concentration. Albaladejo et al. [74] observed also a more significant decrease in K concentration with salinity in the halophyte *S. pennellii* than in *S. lycopersicum*. They hypothesized that this wild tomato species is able to withstand K deficiency by using Na in osmoregulation: K may indeed be replaced by Na in non-specific activities in a few species [69], notably in enzyme activities [75]. This could be a resistance strategy also shared by *S. chilense* to withstand the Na accumulation. Magnesium is also required for pollen development since mutants in the Mg transporter family genes, AtMGT, showed pollen-abortive phenotypes [76]. We observed that Mg accumulated in the stamens and the ovary of *S. chilense* and in the ovary of *S. lycopersicum*, suggesting also a potential role for ovary and fruit development. Because of its fundamental role in phloem export of carbohydrates, Mg is of critical importance during the reproductive growth stage of plants to maintain and maximize carbohydrates transport to sink organs [77]. Calcium is known to play a key role in pollination and pollen tube growth [78] as well as in fruit development [79]. We observed that Ca concentration and localization also differed between both species and Ca accumulated in ovaries in response to salt. Concentrations of Mg and Ca were higher in the inflorescences of *S. lycopersicum* than in the ones of *S. chilense*. However, more research is required to understand their role in flower and fruit development.

3.3. Mineral Transporters Are Involved in Na Accumulation and Partitioning in the Reproductive Structures

To better understand the localization of Na in tomato reproductive structures, we investigated the expression of genes coding for Na transporters. SOS1 is a Na^+/H^+ exchanger activated by the complex formed by SOS2 and SOS3 [80–82]. The SOS pathway is involved in Na exclusion out of the cell [83–85]. We observed that expression of *SOS1* and *SOS2* increased with salt stress in flowers of *S. lycopersicum* and to a lesser extent in the ones of *S. chilense*. Moreover, their expression was positively correlated with the Na concentration ratio between non-reproductive and reproductive floral organs in *S. lycopersicum*. Surprisingly, we found a decrease of *SOS3* expression with salt in *S. lycopersicum*, despite its role in activation of SOS1 [83]. However, pathways other than the SOS2–SOS3 complex are involved in Na^+ activation of SOS1 [83]. Induction of the expression of *SOS1* and *SOS2* is commonly reported in response to salt stress in vegetative parts, and their overexpression induces a better salt resistance [81,86,87]. By contrast, knock-out mutants of these genes lead to a decrease in salt resistance [88,89]. Our results suggested that the SOS pathway is also activated in reproductive organs in response to salt stress. In contrast to our results, Romero-Aranda et al. [25] did not observe any induction of *SOS1* in inflorescences of tomato near-isogenic lines homozygous for *S. cheesmaniae SOS1* allele under salinity conditions. The involvement of *SOS1* in inflorescences thus seems species-dependent in the tomato clade and may differ among halophyte and glycophyte species. Based on those results, *SOS1* expression is induced in salt response in a higher extent in the glycophyte *S. lycopersicum* than in the halophytes *S. cheesmaniae* and *S. chilense* at the reproductive level. We indeed observed that the expression of *SOS* genes was correlated with Na concentrations in the inflorescences of *S. lycopersicum* but not of *S. chilense*.

Other genes involved in the Na transport at the cell level are *NHX3* and *NHX4*, which encode tonoplast transporters involved in the import of Na to the vacuole [90–92]. In our study, *NHX3* expression decreased and *NHX4* expression increased with salt stress in *S. lycopersicum* flowers while their expression was not affected by salinity in *S. chilense* flowers. This differs with the results of Gálvez et al. [91], who compared the response of *S. lycopersicum* and *S. pimpinellifolium* to salinity. They indeed observed that *NHX3* and *NHX4* were upregulated by salinity, especially in the wild halophyte *S. pimpinellifolium* [91]. However, they analyzed plants at the vegetative stage and did not investigate expression in the reproductive organs. We may thus not exclude that the involvement of *NHX* genes differ in vegetative and reproductive organs in tomato species subjected to salinity. Nevertheless, Bassil et al. [67] have shown that, in Arabidopsis, *AtNHX1* and *AtNHX2* are involved in flower development by regulating vacuolar pH and K^+ homeostasis and that Na^+ could partially substitute K^+ in presence of salt. *AtNHX1* and *AtNHX2* are the closest *AtNHX* homologs of *SlNHX4* [91]. We could hypothesize that, in *S. lycopersicum*, under salt stress conditions, the increase of *NHX4* expression would be related to an attempt to increase the K concentration in the anthers, whereas the fact that *S. chilense* could use Na instead of K for flower development and therefore would not require high *NHX4* expression remains an open question.

Other transporters are involved in Na and K transport. HKT1;2 belongs to HKT1-like transporters whose role is to remove Na from the xylem in the roots [93]. However, it has been shown that this gene family is important in salinity resistance during the reproductive stage [25,94]. In our study, *HKT1;2* expression decreased with salinity in both species, possibly explaining the accumulation of Na in the inflorescences. This gene seems to be involved in the partitioning of Na in the flowers as its expression was positively correlated with the Na ratio between male and female floral organs and negatively correlated with the Na ratio between vegetative and reproductive floral organs in both species. *SlAKT2/3* is a phloem K transporter involved in long-distance transport of sucrose [40]. This gene is expressed in tomato flowers and especially sepals [95,96]. We observed that the expression of *SlAKT2/3* increased with NaCl in *S. lycopersicum* but not in *S. chilense*. In the same way, the expression of *SlHAK14* increased with salt stress in *S. lycopersicum* only. SlHAK14 and SlHAK3 are K transporters belonging to the KT/KUP/HAK family [97], and they are both very highly expressed in pollen [96]. In *S. chilense*, the expression of both *SlHAK14* and *SlHAK3* decreased at a concentration of 60 mM NaCl compared to the other treatments. The expression of *SlHAK14* negatively correlated with the K concentration in inflorescences in *S. lycopersicum* only, suggesting a different role in element regulation in *S. lycopersicum* and in *S. chilense*. The expression of *CNGC10* also differed among tomato species. It increased with salinity in *S. chilense* but not in *S. lycopersicum*. This gene is linked to the import of Na and K in flowers, and its expression is inhibited by salinity in Arabidopsis [98]. Its higher expression in *S. chilense* could partly explain the higher Na concentration in inflorescences and flowers of *S. chilense* compared to *S. lycopersicum*. Our results suggest that Na and K transport could be differently regulated in flowers of *S. lycopersicum* and *S. chilense*. Moreover, correlations between transporters expression and mineral concentrations in flowers differed in both species, mainly for the SOS pathway. Further studies are required to decipher the role of transporters in Na and K localization in flowers of both species.

4. Materials and Methods

4.1. Plant Material and Growth Conditions

Seeds of *Solanum lycopersicum* L. cv Ailsa Craig (accession LA2838A) and of *Solanum chilense* Dunal (accession LA4107) were obtained from the Tomato Genetics Resource Center (TGRC, University of California, Davis, CA, USA) and INIA-La Cruz (La Cruz, Chile), respectively. *S. chilense* was subjected to 6 days pre-germination in Petri dishes on humid filter paper at 25 °C and 12 h photoperiod before sowing in peat compost (DCM, Amsterdam, The Netherlands) and transferred to a temperate greenhouse. Sowing of *S. lycopersicum* was performed in the same peat compost and in the same greenhouse 13 days after the sowing of *S. chilense* so they would be of the same developmental stage at the start of stress application. When the two-leaf stage was reached, the plants were individually transplanted in pots (2.5 L) on perlite/vermiculite (50% *v*/*v*) and were grown under the same temperate greenhouse conditions (24 ± 1.5 °C, 63 ± 8% RH day, 21 ± 0.8 °C, 67 ± 5% RH night, 16 h-photoperiod). In addition to natural light, supplementary lighting was provided by LED LumiGrow lights (650 W, red-blue) to maintain a minimum light intensity (mean light in the middle of a cloudy day 181.33 ± 63.42 µmol m^{-2} s^{-1}). Plants were watered three times a week with modified Hoagland solution (5 mM KNO_3, 5.5 mM $Ca(NO_3)_2$, 1 mM $NH_4H_2PO_4$, 0.5 mM $MgSO_4$, 25 µM KCl, 10 µM H_3BO_4, 1 µM $MnSO_4$, 0.25 µM CuSO4, 1 µM $ZnSO_4$, 10 µM $(NH_4)_6Mo_7O$ and 1.87 g L^{-1} Fe-EDTA, and pH 5.5–6). After four days of acclimation, plants were randomly divided into four groups (25 plants per group) receiving 0, 60, 100, or 120 mM NaCl (respectively, 0.86, 7.07, 10.82, and 12.72 mS cm^{-1}). Salt solutions were applied three times a week at the same time that the Hoagland solution, with volumes depending on the physiological stage of the plant.

4.2. Growth

Vegetative growth was assessed by counting the number of leaves on the main stem on 10 plants per condition and species, once a week. Reproductive growth was also assessed on the same 10 plants per condition and species. Flowering time of the initial and the sympodial segments were assessed by counting the number of leaves below the first inflorescence and between inflorescences, respectively. The number of inflorescences on the main stem was counted once a week from 20 days after stress imposition (DASt). The number of flower buds and flowers at anthesis per inflorescence was followed on the second and third inflorescences.

Per condition and species, 11 to 20 flowers at anthesis from the second inflorescence of the main stem were harvested to evaluate the length of sepals, petals, stamens, pistil, and ovary. The style exertion was also assessed for *S. chilense* by measuring the length of the pistil outside the stamen cone. Organs were dissected, flattened, and measured using ImageJ (version 1.53a).

4.3. Flower Fertility

To detect stigma receptivity, peroxidase activity was tested at the stigma's surface according to Dafni and Maués [99]. At anthesis, 14 to 22 flowers per condition were harvested. Stigmas were dissected and immersed for 5 min in acetate buffer with 112.2 mM $CaCl_2 \cdot 2H_2O$, 2.3 mM 3-amino-9-ethylcarbazole diluted in N-N-dimethylformamide, and 0.014% H_2O_2 (v/v). The reddish-brown color developed on the surface was scored by 0 (no receptive stigma) or 1 (receptive stigma). Pollen viability was assessed on two stamens of the same flowers using Alexander dye [100]. Pollen was considered viable when a red coloration appeared, whereas it was considered non-viable when its coloration was green. A minimum of 100 pollen grains was counted by anther. The number of pollen grains per anther was determined by crushing an anther in 40 µL of Alexander's dye and counting using ImageJ as described by Ayenan et al. [101], showing a pollen size of 5–800 pixel^2 and a circularity of 0.3–1.0. Six pictures were taken by anther, and two anthers per flower and 10 flowers per condition and species were analyzed.

4.4. Fruit Parameters

For fruit production, flowers of *S. lycopersicum* were self-pollinated, and flowers of the self-incompatible *S. chilense* were hand pollinated with pollen from the same condition. The fruit set was assessed by the ratio between the number of obtained fruits and the number of pollinated flowers. Fruits were collected at the maturity stage. The number of seeds per fruit, circumference, and fresh weight (FW) were measured for 10 to 15 fruits per condition and species. For the same fruits, sugar concentration was estimated in degrees Brix by refractometry (Eclipse, Bellingham + Stanley, Tunbridge Wells, UK), and the pH of the juice was evaluated by pH paper (Dosatest pH test strips pH 3.6–6.1, VWR).

4.5. Mineral Elements Concentrations and Element Distribution

Sodium (Na), potassium (K), calcium (Ca), and magnesium (Mg) were quantified in inflorescences, pericarp, and seeds of fruits. Material was oven-dried at 70 °C for 72 h, and 50 to 100 mg dry weight (DW) was weighted and digested in 4 mL of warm 68% (v/v) HNO_3. After complete dissolution, minerals were dissolved in aqua regia (HCl 37%:HNO_3 68% 3:1), filtered (Whatman, 11 µm), and quantified by flame atomic absorption spectrophotometry (ICE 3300, Thermo Scientific, Waltham, MA, USA) using suitable standards (Spectracer-CPACHEM; accredited through ISO/IEC17025). Quantification was performed on at least nine samples per condition and species.

Flowers of both species growing at 0, 60, and 100 mM were longitudinally cut using a platinum coated razor blade and sandwiched between two aluminum foils, flattened, frozen in liquid nitrogen, and freeze-dried (−30 °C, 0.210 mbar, Alpha 2–4, Christ, Osterode am Harz, Germany) for 72 h. Two flowers per condition and species were placed on double sided Scotch® tape on glass slides, and the distribution of Na, Mg, K, and Ca was evaluated by laser ablation inductively coupled plasma mass spectrometry (LA-ICP-MS, Agilent 7900×, Agilent Technologies, Palo Alto, CA and Analyte G2, Teledyne Photon Machines Inc., Bozeman, MT, USA). The laser ablation system contains a HelEx II 2-volume ablation cell with integrated Aerosol Rapid Introduction System [102]. The imaging parameters for best image quality were set according to van Elteren et al. [103] (LA settings: square 20 µm beam size, 275 Hz, dosage 11, 1 J/cm^2; ICP-MS: acquisition time 40 ms, dwell times Mg, Na, K, 7 ms and Ca 12 ms). Distribution of elements was visualized using ImageJ [104] by adjusting contrasts and using Look Up Table (LUT) menu. Colocalisation maps (Na with K, Mg, or Ca) were generated by merging channels in ImageJ. Number of counts in specific organs was estimated in two flowers per condition and per species using ROI (Regio Of Interest) manager by selecting an ovary, a style, one stamen, and a floral receptacle. The ratio between the number of counts of each element in the male part (one stamen) and female parts (ovary and style) and the ratio between vegetative (floral receptacle) and reproductive (stamen, ovary, and style) parts were determined.

4.6. Transporters Expression Analysis by qRT-PCR

The expression of 10 genes coding for mineral transporters was analyzed. Genes were selected according to the literature [25,37,40,98,105–107] and on transcriptome profiling of inflorescences of tomato during salt stress imposition [96]. When the sequences were not described in tomato, sequences of tomato homologs were identified using nucleotide BLAST again National Center for Biotechnology Information (NCBI) and Sol Genomics Network (SGN) databases and alignment with BioEdit. A first bioinformatics study of the expression of these genes was analyzed via available databases (TomExpress, SGN, [96]). The obtained full-length tomato sequences were used for primer design using Primer3Plus [108]. The analyzed genes and primer sequences are described in Table S1.

Flowers at anthesis were collected at 35 DASt and stored in liquid nitrogen. RNA extraction was performed on three samples of 100 mg of flowers per condition and species using TRI Reagent Solution (Ambion, Austin, TX, USA) with DNase treatment (RQ1 DNase 1 U/µg Promega, Leiden, The Netherlands) according to the manufacturer's instructions. First-strand cDNA was synthesized from 1 µg RNA using the Revertaid H Minus First Strand cDNA Synthesis Kit (ThermoFisher, Waltham, MA, USA). The concentration and purity of the RNA were measured using a NanoDrop ND-1000 spectrophotometer (Thermo Scientific, Villebon-sur-Yvette, France). Transcript levels were quantified in two independent qPCR (in triplicates for each of the three biological replicates) using the GoTaq qPCR Master Mix (Promega) in StepOnePlus Real-Time PCR systems (Applied Biosystems, Foster City, CA, USA). Cycling conditions were initial denaturation 10 min at 95 °C, then 40 cycles of 15 s at 95 °C, and 1 s at 60 °C. The tomato housekeeping genes *LeEF1-α* (Elongation factor 1-alpha, Solyc06g005060) and *TIP41* (TIP41-like protein, Solyc10g049850) were used as reference genes [109]. Results were expressed using the $\Delta\Delta$Ct calculation method in arbitrary units by comparison to the expression of *S. lycopersicum* under control conditions, and normalization was carried out with *LeEF1-α* and *TIP41*. A melt-curve analysis was performed to check the specific amplifications.

4.7. Statistical Analysis

All statistical analyses were performed in RStudio (R Development Core Team, 2017). Normality distribution and homoscedasticity were verified using the Shapiro–Wilk and Levene's tests, respectively, and data were transformed when required. When possible, two-way analysis of variance (ANOVA II) was used to compare species, salinity, and their interactions. Comparisons between the two species were analyzed using the Student's test, the permutation Student's *t*-test (if normality was not met), or the Wilcoxon test (if homoscedasticity was not met). For a single species, comparisons between NaCl treatments were made using one-way analysis of variance (ANOVA I), ANOVA I using the permutation test (if normality was not met), or the Kruskal–Wallis test (if homoscedasticity was not met), followed by appropriate post-hoc tests. Data are shown as means ± standard deviation. For results obtained by LA-ICP-MS, no statistical treatment was applied because of the lack of repetitions (two repetitions per condition and species). Statistical results are presented in Table S2.

Supplementary Materials: The following supporting information can be downloaded at: https://www.mdpi.com/article/10.3390/plants11050672/s1, Figure S1: Distribution of sodium (Na) in flowers of Solanum lycopersicum (top row) and Solanum chilense (bottom row) grown in perlite:vermiculite mixture supplied with 0, 60 and 100 mM NaCl as revealed by LA-ICP-MS (Laser ablation inductively coupled plasma mass spectroscopy) and visualized using ImageJ (version 1.53a) by using the same scale for all treatments. Colour legend represents the number of counts per pixel (20 × 20 µm^2), the number of counts is linearly proportional to the Na concentration. Flowers are the same than in Figure 3, Figure S2: Distribution of potassium (K) in flowers of Solanum lycopersicum (top row) and Solanum chilense (bottom row). For details, see the legend of Figure S1, Figure S3: Distribution of calcium (Ca) in flowers of Solanum lycopersicum (top row) and Solanum chilense (bottom row). For details, see the legend of Figure S1, Figure S4: Distribution of magnesium (Mg) in flowers of Solanum lycopersicum (top row) and Solanum chilense (bottom row). For details, see the legend of Figure S1, Table S1: List of genes and their primers used for qRT-PCR and their efficiency, Table S2: Statistical results for the analyzed parameters.

Author Contributions: Conceptualization, S.B., S.L., J.-P.M. and M.Q.; methodology, S.B., P.P., J.T.v.E. and M.Q.; formal analysis, S.B. and M.Q.; investigation, S.B., P.P. and M.Š.; resources, M.Q., S.L., J.-P.M. and J.T.v.E.; writing—original draft preparation, S.B. and M.Q.; writing—review and editing, S.B., P.P., M.Š., J.T.v.E., J.-P.M., S.L. and M.Q.; supervision, M.Q.; project administration, M.Q., S.L. and J.-P.M.; funding acquisition, M.Q., S.L., J.-P.M. and J.T.v.E. All authors have read and agreed to the published version of the manuscript.

Funding: This research was funded by the Belgian "Fonds National de la Recherche" Scientifique (FRS-FNRS), grant number CDR J.0136.19, the FSR-UCLouvain 2018–2020, and by the WBI/Chili project number 17. P.P., M.Š. and J.T.v.E. acknowledge the support from the Slovenian Research Agency (ARRS), contracts numbers P1-0034, P1-0212, N1-0090 and J7-9418.

Data Availability Statement: The datasets generated during and/or analyzed during the current study are available from the corresponding author on reasonable request.

Acknowledgments: The authors are grateful to Brigitte Van Pee, Marie-Eve Renard, and Katarina Vogel-Mikuš for technical support. They also would like to thank the Fondation Universitaire de Belgique for its contribution to publication fees.

Conflicts of Interest: The authors declare no conflict of interest. The funders had no role in the design of the study; in the collection, analyses, or interpretation of data; in the writing of the manuscript; or in the decision to publish the results.

References

1. FAOStats Tomato Yields and Cultivated Area of Tomato in the World. Available online: https://www.fao.org/faostat/en/#data/QCL (accessed on 28 January 2022).
2. Gharbi, E.; Martínez, J.-P.; Benahmed, H.; Fauconnier, M.-L.; Lutts, S.; Quinet, M. Salicylic Acid Differently Impacts Ethylene and Polyamine Synthesis in the Glycophyte *Solanum lycopersicum* and the Wild-Related Halophyte *Solanum chilense* Exposed to Mild Salt Stress. *Physiol. Plant.* **2016**, *158*, 152–167. [CrossRef] [PubMed]
3. Singh, R.P.; Prasad, P.V.V.; Reddy, K.R. Impacts of Changing Climate and Climate Variability on Seed Production and Seed Industry. In *Advances in Agronomy*; Elsevier: Rome, Italy, 2013; Volume 118, pp. 49–110, ISBN 978-0-12-405942-9.
4. Dudley, L. Salinity in the Soil Environment. In *Handbook of Plant and Crop Stress*; Marcel Dekker, Inc.: New York, NY, USA, 1994; pp. 13–30.
5. Campos, C.A.B.; Fernandes, P.D.; Gheyi, H.R.; Blanco, F.F.; Gonçalves, C.B.; Campos, S.A.F. Yield and Fruit Quality of Industrial Tomato under Saline Irrigation. *Sci. Agric.* **2006**, *63*, 146–152. [CrossRef]
6. Gerszberg, A.; Hnatuszko-Konka, K. Tomato Tolerance to Abiotic Stress: A Review of Most Often Engineered Target Sequences. *Plant Growth Regul.* **2017**, *83*, 175–198. [CrossRef]
7. Bai, Y.; Lindhout, P. Domestication and Breeding of Tomatoes: What Have We Gained and What Can We Gain in the Future? *Ann. Bot.* **2007**, *100*, 1085–1094. [CrossRef]
8. Diouf, I.A.; Derivot, L.; Bitton, F.; Pascual, L.; Causse, M. Water Deficit and Salinity Stress Reveal Many Specific QTL for Plant Growth and Fruit Quality Traits in Tomato. *Front. Plant Sci.* **2018**, *9*, 279. [CrossRef]
9. Rick, C.M.; Chetelat, R.T. Utilization of Related Wild Species for Tomato Improvement. *Acta Hortic.* **1995**, *412*, 21–38. [CrossRef]
10. Miller, J.C.; Tanksley, S.D. RFLP Analysis of Phylogenetic Relationships and Genetic Variation in the Genus *Lycopersicon*. *Theor. Appl. Genet.* **1990**, *80*, 437–448. [CrossRef]
11. Bretó, M.P.; Asins, M.J.; Carbonell, E.A. Genetic Variability in *Lycopersicon* Species and Their Genetic Relationships. *Theor. Appl. Genet.* **1993**, *86*, 113–120. [CrossRef]
12. Asamizu, E.; Ezura, H. Inclusion of Tomato in the Genus Solanum as "*Solanum lycopersicum*" Is Evident from Phylogenetic Studies. *J. Jpn. Soc. Hortic. Sci.* **2009**, *78*, 3–5. [CrossRef]
13. Peralta, I.E.; Spooner, D.M.; Knapp, S. Taxonomy of Wild Tomatoes and Their Relatives (*Solanum* Sect. Lycopersicoides, Sect. Juglandifolia, Sect. Lycopersicon; Solanaceae). *Syst. Bot. Monogr.* **2008**, *84*, 186.
14. Dwivedi, S.L.; Scheben, A.; Edwards, D.; Spillane, C.; Ortiz, R. Assessing and Exploiting Functional Diversity in Germplasm Pools to Enhance Abiotic Stress Adaptation and Yield in Cereals and Food Legumes. *Front. Plant Sci.* **2017**, *8*. [CrossRef] [PubMed]
15. Houston, J. Variability of Precipitation in the Atacama Desert: Its Causes and Hydrological Impact. *Int. J. Climatol.* **2006**, *26*, 2181–2198. [CrossRef]
16. Chetelat, R.T.; Pertuzé, R.A.; Faúndez, L.; Graham, E.B.; Jones, C.M. Distribution, Ecology and Reproductive Biology of Wild Tomatoes and Related Nightshades from the Atacama Desert Region of Northern Chile. *Euphytica* **2009**, *167*, 77–93. [CrossRef]
17. Nakazato, T.; Warren, D.L.; Moyle, L.C. Ecological and Geographic Modes of Species Divergence in Wild Tomatoes. *Am. J. Bot.* **2010**, *97*, 680–693. [CrossRef] [PubMed]

18. Böndel, K.B.; Nosenko, T.; Stephan, W. Signatures of Natural Selection in Abiotic Stress-Responsive Genes of *Solanum chilense*. *R. Soc. Open Sci.* **2018**, *5*, 171198. [CrossRef] [PubMed]
19. Ji, Y.; Scott, J.W.; Schuster, D.J.; Maxwell, D.P. Molecular Mapping of Ty-4, a New Tomato Yellow Leaf Curl Virus Resistance Locus on Chromosome 3 of Tomato. *J. Am. Soc. Hortic. Sci.* **2009**, *134*, 281–288. [CrossRef]
20. Stamova, B.S.; Chetelat, R.T. Inheritance and Genetic Mapping of Cucumber Mosaic Virus Resistance Introgressed from *Lycopersicon chilense* into Tomato. *Theor. Appl. Genet.* **2000**, *101*, 527–537. [CrossRef]
21. Martínez, J.-P.; Antúnez, A.; Pertuzé, R.; Acosta, M.D.P.; Palma, X.; Fuentes, L.; Ayala, A.; Araya, H.; Lutts, S. Effects of Saline Water on Water Status, Yield and Fruit Quality of Wild (*Solanum chilense*) and Domesticated (*Solanum lycopersicum* Var. Cerasiforme) Tomatoes. *Exp. Agric.* **2012**, *48*, 573–586. [CrossRef]
22. Bolarín, M.C.; Pérez-Alfocea, F.; Cano, E.A.; Estañ, M.T.; Caro, M. Growth, Fruit Yield, and Ion Concentration in Tomato Genotypes after Pre- and Post-Emergence Salt Treatments. *J. Am. Soc. Hortic. Sci.* **1993**, *118*, 655–660. [CrossRef]
23. Tapia, G.; Méndez, J.; Inostroza, L. Different Combinations of Morpho-Physiological Traits Are Responsible for Tolerance to Drought in Wild Tomatoes *Solanum chilense* and *Solanum peruvianum*. *Plant Biol.* **2016**, *18*, 406–416. [CrossRef]
24. Samineni, S.; Siddique, K.H.M.; Gaur, P.M.; Colmer, T.D. Salt Sensitivity of the Vegetative and Reproductive Stages in Chickpea (*Cicer arietinum* L.): Podding Is a Particularly Sensitive Stage. *Environ. Exp. Bot.* **2011**, *71*, 260–268. [CrossRef]
25. Romero-Aranda, M.R.; González-Fernández, P.; Pérez-Tienda, J.R.; López-Diaz, M.R.; Espinosa, J.; Granum, E.; Traverso, J.Á.; Pineda, B.; Garcia-Sogo, B.; Moreno, V.; et al. Na^+ Transporter HKT1;2 Reduces Flower Na^+ Content and Considerably Mitigates the Decline in Tomato Fruit Yields under Saline Conditions. *Plant Physiol. Biochem.* **2020**, *154*, 341–352. [CrossRef] [PubMed]
26. Martínez, J.P.; Fuentes, R.; Farías, K.; Lizana, C.; Alfaro, J.F.; Fuentes, L.; Calabrese, N.; Bigot, S.; Quinet, M.; Lutts, S. Effects of Salt Stress on Fruit Antioxidant Capacity of Wild (*Solanum chilense*) and Domesticated (*Solanum lycopersicum* Var. Cerasiforme) Tomatoes. *Agronomy* **2020**, *10*, 1481. [CrossRef]
27. Delesalle, V.A.; Mazer, S.J. Nutrient Levels and Salinity Affect Gender and Floral Traits in the Autogamous *Spergularia marina*. *Int. J. Plant Sci.* **1996**, *157*, 621–631. [CrossRef]
28. White, A.C.; Colmer, T.D.; Cawthray, G.R.; Hanley, M.E. Variable Response of Three *Trifolium repens* Ecotypes to Soil Flooding by Seawater. *Ann. Bot.* **2014**, *114*, 347–355. [CrossRef]
29. Ghanem, M.E.; van Elteren, J.; Albacete, A.; Quinet, M.; Martínez-Andújar, C.; Kinet, J.-M.; Pérez-Alfocea, F.; Lutts, S. Impact of Salinity on Early Reproductive Physiology of Tomato (*Solanum lycopersicum*) in Relation to a Heterogeneous Distribution of Toxic Ions in Flower Organs. *Funct. Plant Biol.* **2009**, *36*, 125. [CrossRef]
30. Liu, H.; Wang, C.; Chen, H.; Zhou, B. Genome-Wide Transcriptome Analysis Reveals the Molecular Mechanism of High Temperature-Induced Floral Abortion in *Litchi chinensis*. *BMC Genom.* **2019**, *20*, 127. [CrossRef]
31. Shabala, S. Learning from Halophytes: Physiological Basis and Strategies to Improve Abiotic Stress Tolerance in Crops. *Ann. Bot.* **2013**, *112*, 1209–1221. [CrossRef]
32. Wu, H. Plant Salt Tolerance and Na^+ Sensing and Transport. *Crop J.* **2018**, *6*, 215–225. [CrossRef]
33. Wang, Z.; Hong, Y.; Zhu, G.; Li, Y.; Niu, Q.; Yao, J.; Hua, K.; Bai, J.; Zhu, Y.; Shi, H.; et al. Loss of Salt Tolerance during Tomato Domestication Conferred by Variation in a Na^+/K^+ Transporter. *EMBO J.* **2020**, *39*, e103256. [CrossRef]
34. Assaha, D.V.M.; Ueda, A.; Saneoka, H.; Al-Yahyai, R.; Yaish, M.W. The Role of Na^+ and K^+ Transporters in Salt Stress Adaptation in Glycophytes. *Front. Physiol.* **2017**, *8*, 509. [CrossRef] [PubMed]
35. Basu, S.; Kumar, A.; Benazir, I.; Kumar, G. Reassessing the Role of Ion Homeostasis for Improving Salinity Tolerance in Crop Plants. *Physiol. Plant.* **2020**, *174*, 502–519. [CrossRef]
36. Asins, M.J.; Raga, V.; Roca, D.; Belver, A.; Carbonell, E.A. Genetic Dissection of Tomato Rootstock Effects on Scion Traits under Moderate Salinity. *Theor. Appl. Genet.* **2015**, *128*, 667–679. [CrossRef] [PubMed]
37. McCubbin, T.; Bassil, E.; Zhang, S.; Blumwald, E. Vacuolar Na^+/H^+ NHX-Type Antiporters Are Required for Cellular K^+ Homeostasis, Microtubule Organization and Directional Root Growth. *Plants* **2014**, *3*, 409–426. [CrossRef] [PubMed]
38. Bassil, E.; Zhang, S.; Gong, H.; Tajima, H.; Blumwald, E. Cation Specificity of Vacuolar NHX-Type Cation/H^+ Antiporters. *Plant Physiol.* **2019**, *179*, 616–629. [CrossRef] [PubMed]
39. Rubio, F.; Fon, M.; Ródenas, R.; Nieves-Cordones, M.; Alemán, F.; Rivero, R.M.; Martínez, V. A Low K^+ Signal Is Required for Functional High-Affinity K^+ Uptake through HAK5 Transporters. *Physiol. Plant.* **2014**, *152*, 558–570. [CrossRef]
40. Deeken, R.; Geiger, D.; Fromm, J.; Koroleva, O.; Ache, P.; Langenfeld-Heyser, R.; Sauer, N.; May, S.; Hedrich, R. Loss of the AKT2/3 Potassium Channel Affects Sugar Loading into the Phloem of Arabidopsis. *Planta* **2002**, *216*, 334–344. [CrossRef]
41. Maathuis, F.J.M. The Role of Monovalent Cation Transporters in Plant Responses to Salinity. *J. Exp. Bot.* **2006**, *57*, 1137–1147. [CrossRef]
42. Quinet, M.; Kinet, J.-M. Transition to Flowering and Morphogenesis of Reproductive Structures in Tomato. *Int. J. Plant Dev. Biol.* **2007**, *1*, 64–74.
43. Soyk, S.; Müller, N.A.; Park, S.J.; Schmalenbach, I.; Jiang, K.; Hayama, R.; Zhang, L.; Van Eck, J.; Jiménez-Gómez, J.M.; Lippman, Z.B. Variation in the Flowering Gene SELF PRUNING 5G Promotes Day-Neutrality and Early Yield in Tomato. *Nat. Genet.* **2017**, *49*, 162–168. [CrossRef]
44. Welty, N.; Radovich, C.; Meulia, T.; van der Knaap, E. Inflorescence Development in Two Tomato Species. *Can. J. Bot.* **2007**, *85*, 111–118. [CrossRef]

45. Aref, F.; Rad, H.E. Physiological Characterization of Rice under Salinity Stress during Vegetative and Reproductive Stages. *Indian J. Sci. Technol.* **2012**, *5*, 10. [CrossRef]
46. Lozano, R.; Angosto, T.; Gómez, P.; Payán, C.; Capel, J.; Huijser, P.; Salinas, J.; Martínez-Zapater, J.M. Tomato Flower Abnormalities Induced by Low Temperatures Are Associated with Changes of Expression of MADS-Box Genes. *Plant Physiol.* **1998**, *117*, 91–100. [CrossRef] [PubMed]
47. Giorno, F.; Wolters-Arts, M.; Mariani, C.; Rieu, I. Ensuring Reproduction at High Temperatures: The Heat Stress Response during Anther and Pollen Development. *Plants* **2013**, *2*, 489–506. [CrossRef]
48. Stanton, M.L.; Preston, R.E. Ecological Consequences and Phenotypic Correlates of Petal Size Variation in Wild Radish, *Raphanus sativus* (Brassicaceae). *Am. J. Bot.* **1988**, *75*, 528–539. [CrossRef]
49. Descamps, C.; Boubnan, N.; Jacquemart, A.-L.; Quinet, M. Growing and Flowering in a Changing Climate: Effects of Higher Temperatures and Drought Stress on the Bee-Pollinated Species *Impatiens glandulifera* Royle. *Plants* **2021**, *10*, 988. [CrossRef]
50. Descamps, C.; Quinet, M.; Baijot, A.; Jacquemart, A.-L. Temperature and Water Stress Affect Plant-Pollinator Interactions in *Borago officinalis* (Boraginaceae). *Ecol. Evol.* **2018**, *8*, 3443–3456. [CrossRef]
51. Zhang, Z.; Hu, M.; Xu, W.; Wang, Y.; Huang, K.; Zhang, C.; Wen, J. Understanding the Molecular Mechanism of Anther Development under Abiotic Stresses. *Plant Mol. Biol.* **2021**, *105*, 1–10. [CrossRef]
52. Xu, J.; Wolters-Arts, M.; Mariani, C.; Huber, H.; Rieu, I. Heat Stress Affects Vegetative and Reproductive Performance and Trait Correlations in Tomato (*Solanum lycopersicum*). *Euphytica* **2017**, *213*, 156. [CrossRef]
53. Pan, C.; Yang, D.; Zhao, X.; Jiao, C.; Yan, Y.; Lamin-Samu, A.T.; Wang, Q.; Xu, X.; Fei, Z.; Lu, G. Tomato Stigma Exsertion Induced by High Temperature Is Associated with the Jasmonate Signalling Pathway. *Plant Cell Environ.* **2019**, *42*, 1205–1221. [CrossRef]
54. Baby, T.; Collins, C.; Tyerman, S.D.; Gilliham, M. Salinity Negatively Affects Pollen Tube Growth and Fruit Set in Grapevines and Is Not Mitigated by Silicon. *Am. J. Enol. Vitic.* **2016**, *67*, 218–228. [CrossRef]
55. Herrera, C.M. Post-Floral Perianth Functionality: Contribution of Persistent Sepals to Seed Development in *Helleborus foetidus* (Ranunculaceae). *Am. J. Bot.* **2005**, *92*, 1486–1491. [CrossRef] [PubMed]
56. Hetherington, S.E.; Smillie, R.M.; Davies, W.J. Photosynthetic Activities of Vegetative and Fruiting Tissues of Tomato. *J. Exp. Bot.* **1998**, *49*, 1173–1181. [CrossRef]
57. Lytovchenko, A.; Eickmeier, I.; Pons, C.; Osorio, S.; Szecowka, M.; Lehmberg, K.; Arrivault, S.; Tohge, T.; Pineda, B.; Anton, M.T.; et al. Tomato Fruit Photosynthesis Is Seemingly Unimportant in Primary Metabolism and Ripening but Plays a Considerable Role in Seed Development. *Plant Physiol.* **2011**, *157*, 1650–1663. [CrossRef] [PubMed]
58. El-Mogy, M.M.; Garchery, C.; Stevens, R. Irrigation with Salt Water Affects Growth, Yield, Fruit Quality, Storability and Marker-Gene Expression in Cherry Tomato. *Acta Agric. Scand. Sect. B Soil Plant Sci.* **2018**, *68*, 727–737. [CrossRef]
59. Karapanos, I.C.; Mahmood, S.; Thanopoulos, C. Fruit Set in Solanaceous Vegetable Crops as Affected by Floral and Environmental Factors. *Eur. J. Plant Sci. Biotechnol.* **2008**, *1*, 88–105.
60. Khatun, S.; Flowers, T.J. Effects of Salinity on Seed Set in Rice. *Plant Cell Environ.* **1995**, *18*, 61–67. [CrossRef]
61. An, D.; Chen, J.-G.; Gao, Y.-Q.; Li, X.; Chao, Z.-F.; Chen, Z.-R.; Li, Q.-Q.; Han, M.-L.; Wang, Y.-L.; Wang, Y.-F.; et al. AtHKT1 Drives Adaptation of *Arabidopsis thaliana* to Salinity by Reducing Floral Sodium Content. *PLoS Genet.* **2017**, *13*, e1007086. [CrossRef]
62. Vromman, D.; Lutts, S.; Lefèvre, I.; Somer, L.; De Vreese, O.; Šlejkovec, Z.; Quinet, M. Effects of Simultaneous Arsenic and Iron Toxicities on Rice (*Oryza sativa* L.) Development, Yield-Related Parameters and As and Fe Accumulation in Relation to As Speciation in the Grains. *Plant Soil* **2013**, *371*, 199–217. [CrossRef]
63. Han, R.; Quinet, M.; André, E.; van Elteren, J.T.; Destrebecq, F.; Vogel-Mikuš, K.; Cui, G.; Debeljak, M.; Lefèvre, I.; Lutts, S. Accumulation and Distribution of Zn in the Shoots and Reproductive Structures of the Halophyte Plant Species *Kosteletzkya virginica* as a Function of Salinity. *Planta* **2013**, *238*, 441–457. [CrossRef]
64. Gharbi, E.; Martínez, J.-P.; Benahmed, H.; Lepoint, G.; Vanpee, B.; Quinet, M.; Lutts, S. Inhibition of Ethylene Synthesis Reduces Salt-Tolerance in Tomato Wild Relative Species *Solanum chilense*. *J. Plant Physiol.* **2017**, *210*, 24–37. [CrossRef] [PubMed]
65. Bacha, H.; Tekaya, M.; Drine, S.; Guasmi, F.; Touil, L.; Enneb, H.; Triki, T.; Cheour, F.; Ferchichi, A. Impact of Salt Stress on Morpho-Physiological and Biochemical Parameters of *Solanum lycopersicum* Cv. *Microtom Leaves. S. Afr. J. Bot.* **2017**, *108*, 364–369. [CrossRef]
66. Müller, F.; Xu, J.; Kristensen, L.; Wolters-Arts, M.; de Groot, P.F.M.; Jansma, S.Y.; Mariani, C.; Park, S.; Rieu, I. High-Temperature-Induced Defects in Tomato (*Solanum lycopersicum*) Anther and Pollen Development Are Associated with Reduced Expression of B-Class Floral Patterning Genes. *PLoS ONE* **2016**, *11*, e0167614. [CrossRef]
67. Bassil, E.; Tajima, H.; Liang, Y.-C.; Ohto, M.; Ushijima, K.; Nakano, R.; Esumi, T.; Coku, A.; Belmonte, M.; Blumwald, E. The *Arabidopsis* Na^+/H^+ Antiporters NHX1 and NHX2 Control Vacuolar PH and K^+ Homeostasis to Regulate Growth, Flower Development, and Reproduction. *Plant Cell* **2011**, *23*, 3482–3497. [CrossRef]
68. Gharbi, E.; Martínez, J.-P.; Benahmed, H.; Hichri, I.; Dobrev, P.I.; Motyka, V.; Quinet, M.; Lutts, S. Phytohormone Profiling in Relation to Osmotic Adjustment in NaCl-Treated Plants of the Halophyte Tomato Wild Relative Species *Solanum chilense* Comparatively to the Cultivated Glycophyte *Solanum lycopersicum*. *Plant Sci.* **2017**, *258*, 77–89. [CrossRef]
69. Wakeel, A.; Farooq, M.; Qadir, M.; Schubert, S. Potassium Substitution by Sodium in Plants. *Crit. Rev. Plant Sci.* **2011**, *30*, 401–413. [CrossRef]
70. Hu, W.; Coomer, T.D.; Loka, D.A.; Oosterhuis, D.M.; Zhou, Z. Potassium Deficiency Affects the Carbon-Nitrogen Balance in Cotton Leaves. *Plant Physiol. Biochem.* **2017**, *115*, 408–417. [CrossRef]

71. Rehman, S.; Yun, S.J. Developmental Regulation of K Accumulation in Pollen, Anthers, and Papillae: Are Anther Dehiscence, Papillae Hydration, and Pollen Swelling Leading to Pollination and Fertilization in Barley (*Hordeum vulgare* L.) Regulated by Changes in K Concentration? *J. Exp. Bot.* **2006**, *57*, 1315–1321. [CrossRef]
72. Chen, Y.; Ma, J.; Miller, A.J.; Luo, B.; Wang, M.; Zhu, Z.; Ouwerkerk, P.B.F. OsCHX14 Is Involved in the K^+ Homeostasis in Rice (*Oryza sativa*) Flowers. *Plant Cell Physiol.* **2016**, *57*, 1530–1543. [CrossRef]
73. Mel, V.C.; Bado, V.B.; Ndiaye, S.; Djaman, K.; Aissata Bama Nati, D.; Manneh, B.; Futakuchi, K. Predicting Rice Yield under Salinity Stress Using K/Na Ratio Variable in Plant Tissue. *Commun. Soil Sci. Plant Anal.* **2019**, *50*, 1321–1329. [CrossRef]
74. Albaladejo, I.; Meco, V.; Plasencia, F.; Flores, F.B.; Bolarin, M.C.; Egea, I. Unravelling the Strategies Used by the Wild Tomato Species *Solanum pennellii* to Confront Salt Stress: From Leaf Anatomical Adaptations to Molecular Responses. *Environ. Exp. Bot.* **2017**, *135*, 1–12. [CrossRef]
75. Nieves-Cordones, M.; Al Shiblawi, F.R.; Sentenac, H. Roles and Transport of Sodium and Potassium in Plants. In *The Alkali Metal Ions: Their Role for Life*; Sigel, A., Sigel, H., Sigel, R.K.O., Eds.; Springer International Publishing: Cham, Switzerland, 2016; Volume 16, pp. 291–324, ISBN 978-3-319-21755-0.
76. Li, J.; Huang, Y.; Tan, H.; Yang, X.; Tian, L.; Luan, S.; Chen, L.; Li, D. An Endoplasmic Reticulum Magnesium Transporter Is Essential for Pollen Development in *Arabidopsis*. *Plant Sci.* **2015**, *231*, 212–220. [CrossRef] [PubMed]
77. Cakmak, I.; Kirkby, E.A. Role of Magnesium in Carbon Partitioning and Alleviating Photooxidative Damage. *Physiol. Plant.* **2008**, *133*, 692–704. [CrossRef] [PubMed]
78. Steinhorst, L.; Kudla, J. Calcium—A Central Regulator of Pollen Germination and Tube Growth. *Biochim. Biophys. Acta BBA Mol. Cell Res.* **2013**, *1833*, 1573–1581. [CrossRef]
79. Hocking, B.; Tyerman, S.D.; Burton, R.A.; Gilliham, M. Fruit Calcium: Transport and Physiology. *Front. Plant Sci.* **2016**, *7*, 569. [CrossRef]
80. Ishitani, M.; Liu, J.; Halfter, U.; Kim, C.-S.; Shi, W.; Zhu, J.-K. SOS3 Function in Plant Salt Tolerance Requires N-Myristoylation and Calcium Binding. *Plant Cell* **2000**, *12*, 1667–1677. [CrossRef]
81. Belver, A.; Olías, R.; Huertas, R.; Rodríguez-Rosales, M.P. Involvement of SlSOS2 in Tomato Salt Tolerance. *Bioengineered* **2012**, *3*, 298–302. [CrossRef]
82. Li, D.; Ma, N.-N.; Wang, J.-R.; Yang, D.-Y.; Zhao, S.-J.; Meng, Q.-W. Overexpression of Tomato Enhancer of SOS3-1 (LeENH1) in Tobacco Enhanced Salinity Tolerance by Excluding Na^+ from the Cytosol. *Plant Physiol. Biochem.* **2013**, *70*, 150–158. [CrossRef]
83. Ji, H.; Pardo, J.M.; Batelli, G.; Van Oosten, M.J.; Bressan, R.A.; Li, X. The Salt Overly Sensitive (SOS) Pathway: Established and Emerging Roles. *Mol. Plant* **2013**, *6*, 275–286. [CrossRef]
84. Álvarez-Aragón, R.; Haro, R.; Benito, B.; Rodríguez-Navarro, A. Salt Intolerance in Arabidopsis: Shoot and Root Sodium Toxicity, and Inhibition by Sodium-plus-Potassium Overaccumulation. *Planta* **2016**, *243*, 97–114. [CrossRef]
85. Almeida, D.M.; Oliveira, M.M.; Saibo, N.J.M. Regulation of Na^+ and K^+ Homeostasis in Plants: Towards Improved Salt Stress Tolerance in Crop Plants. *Genet. Mol. Biol.* **2017**, *40*, 326–345. [CrossRef]
86. Shi, H.; Lee, B.; Wu, S.-J.; Zhu, J.-K. Overexpression of a Plasma Membrane Na^+/H^+ Antiporter Gene Improves Salt Tolerance in *Arabidopsis thaliana*. *Nat. Biotechnol.* **2003**, *21*, 81–85. [CrossRef]
87. Huertas, R.; Olías, R.; Eljakaoui, Z.; Gálvez, F.J.; Li, J.; De Morales, P.A.; Belver, A.; Rodríguez-Rosales, M.P. Overexpression of SlSOS2 (SlCIPK24) Confers Salt Tolerance to Transgenic Tomato: SlSOS2 and Tomato Salt Tolerance. *Plant Cell Environ.* **2012**, *35*, 1467–1482. [CrossRef]
88. Batelli, G.; Verslues, P.E.; Agius, F.; Qiu, Q.; Fujii, H.; Pan, S.; Schumaker, K.S.; Grillo, S.; Zhu, J.-K. SOS2 Promotes Salt Tolerance in Part by Interacting with the Vacuolar H^+-ATPase and Upregulating Its Transport Activity. *Mol. Cell. Biol.* **2007**, *27*, 7781–7790. [CrossRef] [PubMed]
89. Ariga, H.; Katori, T.; Yoshihara, R.; Hase, Y.; Nozawa, S.; Narumi, I.; Iuchi, S.; Kobayashi, M.; Tezuka, K.; Sakata, Y.; et al. *Arabidopsis* Sos1 Mutant in a Salt-Tolerant Accession Revealed an Importance of Salt Acclimation Ability in Plant Salt Tolerance. *Plant Signal. Behav.* **2013**, *8*, e24779. [CrossRef] [PubMed]
90. Leidi, E.O.; Barragán, V.; Rubio, L.; El-Hamdaoui, A.; Ruiz, M.T.; Cubero, B.; Fernández, J.A.; Bressan, R.A.; Hasegawa, P.M.; Quintero, F.J.; et al. The AtNHX1 Exchanger Mediates Potassium Compartmentation in Vacuoles of Transgenic Tomato. *Plant J.* **2010**, *61*, 495–506. [CrossRef] [PubMed]
91. Gálvez, F.J.; Baghour, M.; Hao, G.; Cagnac, O.; Rodríguez-Rosales, M.P.; Venema, K. Expression of LeNHX Isoforms in Response to Salt Stress in Salt Sensitive and Salt Tolerant Tomato Species. *Plant Physiol. Biochem.* **2012**, *51*, 109–115. [CrossRef]
92. Villalta, I.; Reina-Sánchez, A.; Bolarín, M.C.; Cuartero, J.; Belver, A.; Venema, K.; Carbonell, E.A.; Asins, M.J. Genetic Analysis of Na^+ and K^+ Concentrations in Leaf and Stem as Physiological Components of Salt Tolerance in Tomato. *Theor. Appl. Genet.* **2008**, *116*, 869–880. [CrossRef]
93. Almeida, P.; de Boer, G.-J.; de Boer, A.H. Differences in Shoot Na^+ Accumulation between Two Tomato Species Are Due to Differences in Ion Affinity of HKT1;2. *J. Plant Physiol.* **2014**, *171*, 438–447. [CrossRef]
94. Suzuki, K.; Yamaji, N.; Costa, A.; Okuma, E.; Kobayashi, N.I.; Kashiwagi, T.; Katsuhara, M.; Wang, C.; Tanoi, K.; Murata, Y.; et al. OsHKT1;4-Mediated Na^+ Transport in Stems Contributes to Na^+ Exclusion from Leaf Blades of Rice at the Reproductive Growth Stage upon Salt Stress. *BMC Plant Biol.* **2016**, *16*, 22. [CrossRef]
95. Deeken, R.; Sanders, C.; Ache, P.; Hedrich, R. Developmental and Light-Dependent Regulation of a Phloem-Localised K^+ Channel of *Arabidopsis thaliana*: Regulation of Akt2/3-MRNA Expression. *Plant J.* **2000**, *23*, 285–290. [CrossRef] [PubMed]

96. Zouine, M.; Maza, E.; Djari, A.; Lauvernier, M.; Frasse, P.; Smouni, A.; Pirrello, J.; Bouzayen, M. TomExpress, a Unified Tomato RNA-Seq Platform for Visualization of Expression Data, Clustering and Correlation Networks. *Plant J.* **2017**, *92*, 727–735. [CrossRef] [PubMed]
97. Liu, J.; Liu, J.; Liu, J.; Cui, M.; Huang, Y.; Tian, Y.; Chen, A.; Xu, G. The Potassium Transporter SlHAK10 Is Involved in Mycorrhizal Potassium Uptake. *Plant Physiol.* **2019**, *180*, 465–479. [CrossRef]
98. Jin, Y.; Jing, W.; Zhang, Q.; Zhang, W. Cyclic Nucleotide Gated Channel 10 Negatively Regulates Salt Tolerance by Mediating Na^+ Transport in Arabidopsis. *J. Plant Res.* **2015**, *128*, 211–220. [CrossRef] [PubMed]
99. Dafni, A.; Maués, M.M. A Rapid and Simple Procedure to Determine Stigma Receptivity. *Sex. Plant Reprod.* **1998**, *11*, 177–180. [CrossRef]
100. Alexander, M.P. Differential Staining of Aborted and Nonaborted Pollen. *Stain Technol.* **1969**, *44*, 117–122. [CrossRef]
101. Ayenan, M.A.T.; Danquah, A.; Ampomah-Dwamena, C.; Hanson, P.; Asante, I.K.; Danquah, E.Y. Optimizing Pollencounter for High Throughput Phenotyping of Pollen Quality in Tomatoes. *MethodsX* **2020**, *7*, 100977. [CrossRef]
102. van Acker, T.; Van Malderen, S.J.M.; Van Helden, T.; Stremtan, C.; Šala, M.; van Elteren, J.T.; Vanhaecke, F. Analytical Figures of Merit of a Low-Dispersion Aerosol Transport System for High-Throughput LA-ICP-MS Analysis. *J. Anal. At. Spectrom.* **2021**, *36*, 1201–1209. [CrossRef]
103. van Elteren, J.T.; Šelih, V.S.; Šala, M. Insights into the Selection of 2D LA-ICP-MS (Multi)Elemental Mapping Conditions. *J. Anal. At. Spectrom.* **2019**, *34*, 1919–1931. [CrossRef]
104. Schindelin, J.; Arganda-Carreras, I.; Frise, E.; Kaynig, V.; Longair, M.; Pietzsch, T.; Preibisch, S.; Rueden, C.; Saalfeld, S.; Schmid, B.; et al. Fiji: An Open-Source Platform for Biological-Image Analysis. *Nat. Methods* **2012**, *9*, 676–682. [CrossRef]
105. Olías, R.; Eljakaoui, Z.; Li, J.; De Morales, P.A.; Marín-Manzano, M.C.; Pardo, J.M.; Belver, A. The Plasma Membrane Na^+/H^+ Antiporter SOS1 Is Essential for Salt Tolerance in Tomato and Affects the Partitioning of Na^+ between Plant Organs. *Plant Cell Environ.* **2009**, *32*, 904–916. [CrossRef] [PubMed]
106. Asins, M.J.; Villalta, I.; Aly, M.M.; Olías, R.; Álvarez De Morales, P.; Huertas, R.; Li, J.; Jaime-Pérez, N.; Haro, R.; Raga, V.; et al. Two Closely Linked Tomato HKT Coding Genes Are Positional Candidates for the Major Tomato QTL Involved in Na^+/K^+ Homeostasis: HKT Genes Likely to Underlie a Major Tomato QTL. *Plant Cell Environ.* **2013**, *36*, 1171–1191. [CrossRef] [PubMed]
107. Jaime-Pérez, N.; Pineda, B.; García-Sogo, B.; Atares, A.; Athman, A.; Byrt, C.S.; Olías, R.; Asins, M.J.; Gilliham, M.; Moreno, V.; et al. The Sodium Transporter Encoded by the *HKT1;2* Gene Modulates Sodium/Potassium Homeostasis in Tomato Shoots under Salinity. *Plant Cell Environ.* **2017**, *40*, 658–671. [CrossRef]
108. Untergasser, A.; Nijveen, H.; Rao, X.; Bisseling, T.; Geurts, R.; Leunissen, J.A.M. Primer3Plus, an Enhanced Web Interface to Primer3. *Nucleic Acids Res.* **2007**, *35*, W71–W74. [CrossRef]
109. Leelatanawit, R.; Saetung, T.; Phuengwas, S.; Karoonuthaisiri, N.; Devahastin, S. Selection of Reference Genes for Quantitative Real-Time PCR in Postharvest Tomatoes (*Lycopersicon esculentum*) Treated by Continuous Low-Voltage Direct Current Electricity to Increase Secondary Metabolites. *Int. J. Food Sci. Technol.* **2017**, *52*, 1942–1950. [CrossRef]

plants

MDPI

Article

Comparative Analysis of Root Na^+ Relation under Salinity between *Oryza sativa* and *Oryza coarctata*

Tetsuya Ishikawa [1], Lana Shabala [1], Meixue Zhou [1], Gayatri Venkataraman [2], Min Yu [3], Gothandapani Sellamuthu [2,4], Zhong-Hua Chen [5] and Sergey Shabala [1,3,*]

1 Tasmanian Institute of Agriculture, College of Science and Engineering, University of Tasmania, Hobart, TAS 7005, Australia; tetsuya.ishikawa@utas.edu.au (T.I.); l.shabala@utas.edu.au (L.S.); meixue.zhou@utas.edu.au (M.Z.)
2 Plant Molecular Biology Laboratory, M. S. Swaminathan Research Foundation, III Cross Street, Taramani Institutional Area, Chennai 600113, India; gayatri@mssrf.res.in (G.V.); sellamuthu@fld.czu.cz (G.S.)
3 International Research Centre for Environmental Membrane Biology, Foshan University, Foshan 528000, China; yumin0820@hotmail.com
4 Forest Molecular Entomology Lab, Excellent Team for Mitigation (ETM), Faculty of Forestry and Wood Sciences, Czech University of Life Sciences Prague, 16500 Prague, Czech Republic
5 School of Science, Hawkesbury Institute for the Environment, Western Sydney University, Penrith, NSW 2751, Australia; z.chen@westernsydney.edu.au
* Correspondence: sergey.shabala@utas.edu.au

Abstract: Na^+ toxicity is one of the major physiological constraints imposed by salinity on plant performance. At the same time, Na^+ uptake may be beneficial under some circumstances as an easily accessible inorganic ion that can be used for increasing solute concentrations and maintaining cell turgor. Two rice species, *Oryza sativa* (cultivated rice, salt-sensitive) and *Oryza coarctata* (wild rice, salt-tolerant), demonstrated different strategies in controlling Na^+ uptake. Glasshouse experiments and gene expression analysis suggested that salt-treated wild rice quickly increased xylem Na^+ loading for osmotic adjustment but maintained a non-toxic level of stable shoot Na^+ concentration by increased activity of a high affinity K^+ transporter HKT1;5 (essential for xylem Na^+ unloading) and a Na^+/H^+ exchanger NHX (for sequestering Na^+ and K^+ into root vacuoles). Cultivated rice prevented Na^+ uptake and transport to the shoot at the beginning of salt treatment but failed to maintain it in the long term. While electrophysiological assays revealed greater net Na^+ uptake upon salt application in cultivated rice, *O. sativa* plants showed much stronger activation of the root plasma membrane Na^+/H^+ Salt Overly Sensitive 1 (SOS1) exchanger. Thus, it appears that wild rice limits passive Na^+ entry into root cells while cultivated rice relies heavily on SOS1-mediating Na^+ exclusion, with major penalties imposed by the existence of the "futile cycle" at the plasma membrane.

Keywords: rice; salinity; halophyte; root; microelectrode ion flux; MIFE; transporters

Citation: Ishikawa, T.; Shabala, L.; Zhou, M.; Venkataraman, G.; Yu, M.; Sellamuthu, G.; Chen, Z.-H.; Shabala, S. Comparative Analysis of Root Na^+ Relation under Salinity between *Oryza sativa* and *Oryza coarctata*. *Plants* **2022**, *11*, 656. https://doi.org/10.3390/plants11050656

Academic Editors: Oscar Vicente and Marius-Nicusor Grigore

Received: 6 February 2022
Accepted: 23 February 2022
Published: 28 February 2022

Publisher's Note: MDPI stays neutral with regard to jurisdictional claims in published maps and institutional affiliations.

1. Introduction

Sodium toxicity is considered to be a major constraint affecting plant performance caused by salt stress in the long term. As a result of selective breeding, salinity-tolerant rice cultivars accumulate less Na^+ in the shoot compared with sensitive cultivars [1–5]. A number of previous studies focused on the mechanism of Na^+ retrieval back from xylem operated by a high-affinity K^+ transporter OsHKT1;5 that reduces shoot Na^+ accumulation in rice [5–9]. Once unloaded from the xylem, Na^+ needs to be extruded into external medium. Root Na^+ exclusion is known to be operated by a Na^+/H^+ exchanger Salt Overly Sensitive 1 (SOS1) at the root epidermis [10,11], and beneficial effects of enhancement of SOS1 operation fuelled by H^+-ATPase on salt tolerance in rice was demonstrated [12,13]. In addition to Na^+ exclusion, increasing biosynthesis of organic osmolytes has been also targeted to improve salinity-induced osmotic stress tolerance in rice [14–16]. However, despite numerous attempts, achievements in breeding salinity-tolerant rice are still quite modest [17,18].

If one enhances Na^+ exclusion by SOS1, then plants need to rely on de novo synthesis of organic osmolytes (compatible solutes) for osmotic adjustment, which comes with a considerable energy cost, leading a depletion of the ATP pool [19,20]. Therefore, SOS1-mediated root Na^+ exclusion activity did not correlate with overall salinity tolerance in barley [21] and rice varieties [22] when assessed by direct Na^+ flux measurements using electrophysiological techniques. The efficacy of Na^+ exclusion strategy mediated by SOS1 in rice is further complicated by the presence of the apoplastic pathway of Na^+ entry, named as bypass flow. Despite anatomical barriers, bypass flow causes a significant amount of passive Na^+ entry from sites of lateral root emergence, areas of weak Casparian strip barrier formation, and cell walls near the root apices that has long been considered as a major component of high salt sensitivity in rice [7,23–26]. Due to this passive Na^+ entry pathway, SOS1 transporters in rice may operate in a "futile cycle", depleting energy but not achieving a significant reduction in Na^+ content. Thus, selection of inappropriate traits (i.e., Na^+ exclusion and de novo synthesis of compatible solutes) can be the reason of the failure to produce salinity tolerant rice over the past decades.

Instead of excluding Na^+ and synthesising organic osmolytes, the ability of utilising Na^+ can be considered to be a more effective trait in conferring salinity tolerance. Although accumulating excessive amount of Na^+ can become toxic for plants, Na^+ uptake is desirable because this element is highly soluble and easily available (especially under salinity) for plants to increase osmotic pressure, absorb water, and sustain turgor [27,28]. A sharp increase of xylem Na^+ loading and shoot Na^+ accumulation can be an efficient means of osmotic adjustment, and this Na^+ utilisation mechanism has been reported from halophytes and salinity-tolerant barley genotypes [29–32]. Excessive Na^+ elevation in the cytosol causes toxicity; therefore, effective Na^+ sequestration into vacuoles mediated by tonoplast $Na^+(K^+)/H^+$ exchanger (NHX) has to be accompanied with the above mechanisms of Na^+ utilisation. Recently, a need for a shift from crop breeding for Na^+ exclusion towards conferring superior traits benefitting from Na^+ called "halophytism" was suggested [33].

The only halophytic relative of wild rice species, *O. coarctata*, is known to grow under high level of salinity (20–40 ds m^{-1}) that is lethal for cultivated rice (*O. sativa*) species [34–36]. *O. coarctata* has long been known to maintain low leaf Na^+/K^+ ratio [37], showing greater Na^+ accumulation in the root rather than the shoot under salinity [38]. Secretion of Na^+ via external microhairs [39], efficient performance of NHX [40], and a high transport capacity of HKT1;5 [41] are considered to contribute to superior ionic homeostasis under salinity in this species. Due to high salinity tolerance within the genus of *Oryza*, *O. coarctata* has been considered as an important resource of gene pools to improve salinity tolerance in rice cultivars [35,41]. However, detailed mechanisms of maintaining Na^+ homeostasis in this species have been less understood due to the limited number of studies at the cellular level.

We hypothesise that *O. coarctata* possesses mechanisms, wherein Na^+ is utilised rather than excluded, for adapting to a saline environment. To test this hypothesis, we compared a range of physiological variables (e.g., biomass change, relative water content, chlorophyll content, and stomatal conductance) between salt-grown cultivated (*O. sativa*) and wild (*O. coarctata*) rice species and linked them with kinetics of Na^+ transport in plant roots; Na^+ concentrations in root, leaf, and xylem sap; and expression of SOS1, NHX, and HKT1 transporter genes. The overall research aim was to explore the mechanisms of Na^+ uptake, exclusion, and translocation differentiating Na^+ homeostasis between these two rice species.

2. Results

2.1. Biomass Change, Relative Water Content, and Physiological Responses

After four weeks of salinity treatment, prominent differences were observed in plant biomass (FW) and relative water content (RWC) between cultivated and wild rice species (Figure 1). Cultivated rice significantly ($p < 0.05$) declined in biomass and RWC in response to the increase of salinity levels (Figure 1A,C). Notably, cultivated rice treated with 100 mM NaCl showed an eightfold decline in biomass compared with its controls (Figure 1C). In contrast, both 50 and 100 mM NaCl treatments did not significantly decrease both biomass and RWC compared with the control in wild rice (Figure 1B,D).

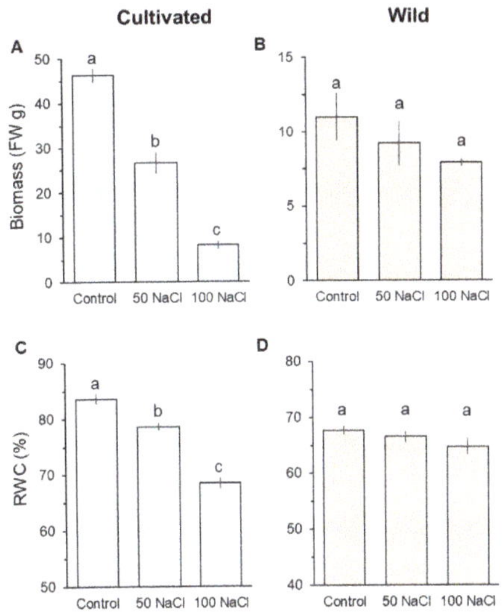

Figure 1. Whole-plant biomass change in fresh weight measured from cultivated (**A**) and wild (**B**) rice species. Shoot relative water content (RWC) of cultivated (**C**) and wild rice species (**D**). Seedlings were exposed to 0 (control), 50, or 100 mM NaCl for four weeks. Different letters indicate significant differences ($p < 0.05$, one-way ANOVA followed by LSD tests). Mean ± SE ($n = 6$).

Physiological characteristics were also affected only in cultivated rice in response to salinity (Figure 2). Two weeks of 100 mM NaCl treatment significantly ($p < 0.001$) reduced chlorophyll content (Figure 2A) and stomatal conductance (Figure 2B) in cultivated rice, while wild rice showed almost the same values between control and salt-treated plants (Figure 2A,B). The above observations suggest that wild rice is considerably more salinity-tolerant at the whole-plant level compared with cultivated rice.

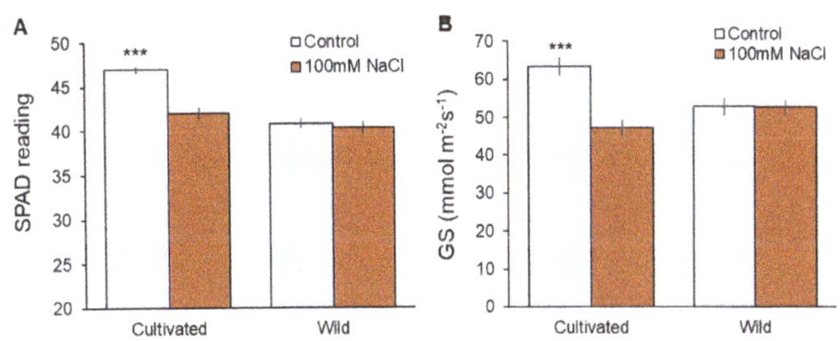

Figure 2. Physiological responses to two weeks of salinity (100 mM NaCl) treatment in cultivated and wild rice species. (**A**) SPAD value (chlorophyll content); (**B**) Gs (stomatal conductance). Asterisks indicate significant differences within the plant species (*** significant at $p < 0.001$, Student's t-tests). Mean ± SE ($n = 10$).

2.2. Root, Leaf, and Xylem Sap Na^+ Concentrations

Under non-saline conditions, wild rice showed about two- and fourfold higher Na^+ concentrations in root and leaf sap, respectively, compared with cultivated rice (Figure 3A,B).

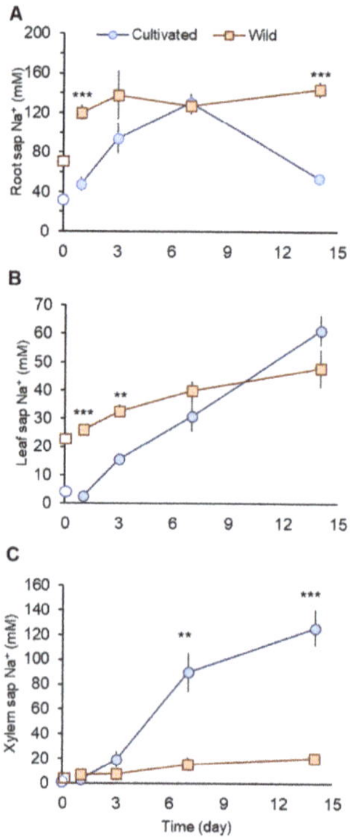

Figure 3. Na^+ concentrations in root (**A**), leaf (**B**), and xylem sap (**C**) in cultivated and wild rice species under 100 mM NaCl treatments at different time points after the commencement of salinity. Open symbols describe Na^+ concentrations before salinity onset. Asterisks indicate significant differences between plant species within the same harvest day. (** $p < 0.01$; *** $p < 0.001$, Student's t-tests). Mean ± SE (n = 3–6).

One day after the commencement of salinity treatment (day 1), cultivated rice showed significantly ($p < 0.001$) smaller Na^+ concentrations in root and leaf sap than those in wild rice (Figure 3A,B). At the same time (day 1), Na^+ concentration in xylem sap in cultivated rice showed only a very marginal increase (not significant at $p < 0.05$), while in wild rice, this increase was substantial (threefold; from 2.55± 0.49 in control plant to 7.28 ± 1.19 mM in salt-treated plant; significant at $p < 0.05$). Thus, at the beginning of a salinity event, cultivated rice may have mechanisms operative to prevent root Na^+ uptake and xylem Na^+ loading. In contrast, wild rice showed increased xylem Na^+ loading and Na^+ transport to the shoot.

When salinity stress was prolonged, cultivated rice Na^+ concentration in the root sap increased until day 7, but dropped sharply at day 14 to become significantly ($p < 0.001$) lower than that seen in wild rice (Figure 3A). This sharp drop of root sap Na^+ in cultivated rice can be accounted for by increased Na^+ transport to the shoot. Leaf sap Na^+ concentration in cultivated rice progressively increased over the period of salinity treatment, with a sharp increase after day 7 (Figure 3B). Xylem sap Na^+ concentration in cultivated rice was not significantly different compared with wild rice until day 3. However, a sharp and substantial increase in the xylem sap Na^+ concentration in cultivated rice was observed at day 7, with values being significantly (fourfold, $p < 0.01$) higher than for wild rice (Figure 3C). The increase in xylem sap Na^+ concentration in cultivated rice was observed until day 14 (Figure 3C). In contrast, although salt-treated wild rice showed approximately twofold higher root sap Na^+ concentrations over the period of salinity treatment compared with control, the variation in root sap Na^+ concentrations were not as large (Figure 3A). Therefore, wild rice may possess superior ability to retain Na^+ in the root under prolonged salinity compared with cultivated rice. Further, wild rice also maintained rather stable Na^+ concentrations in leaf and xylem sap over the period of salinity treatment (Figure 3B,C).

2.3. Transcriptional Analysis of Genes Related to Na^+ Transport

HKT1;4 and HKT1;5 are known to mediate retrieval back of Na^+ from the xylem that contribute to reduce shoot Na^+ accumulation in rice [6,42]. Salinity treatment did not significantly ($p < 0.05$) change expression of *HKT1;4* in the elongation root zone (EZ), but significantly downregulated it in the mature zone (MZ) in both species (Figure 4A,B). It is reported that HKT1;4 mediates Na^+ unloading in a range of conditions (submillimolar Na^+ to high salinity) in cultivated rice [42]. Further, OsHKT1;4 has been suggested to have a more prominent role in mediating Na^+ unloading in the leaf sheath at the reproductive stage, preventing over-accumulation of Na^+ in the leaf blade under salinity [43]. Therefore, it may be considered that HKT1;4 has a very minor or no role in Na^+ transport into xylem parenchyma cells under saline conditions tested here. *HKT1;5* expression was downregulated in cultivated rice but upregulated in wild rice by salinity (Figure 4C,D). In MZ, cultivated rice showed 59.2% decrease in *HKT1;5* expression in response to salinity (significant at $p < 0.05$), in contrast to wild rice that showed an 85.6% increase in *HKT1;5* expression (significant at $p < 0.05$, Figure 4C,D). Tonoplast Na^+/H^+ antiporter (NHX1) mediates Na^+ sequestration into vacuoles to reduce excessive increase of cytosolic Na^+ concentration [44]. Having a dual affinity for both Na^+ and K^+ NHX1 also catalyses K^+/H^+ exchange at the tonoplast membrane [45]. *NHX1* expression was significantly increased in both root zones of wild rice (sevenfold and twofold in EZ and MZ, respectively; Figure 4F). In contrast, cultivated rice showed downregulated *NHX1* expression in response to salinity (Figure 4E)—for example, there was a threefold reduction in *NHX1* expression (significant at $p < 0.05$) in EZ under salinity in the cultivated rice. Thus, wild rice showed greater expressions of *HKT1;5* and *NHX* under salinity compared with cultivated rice.

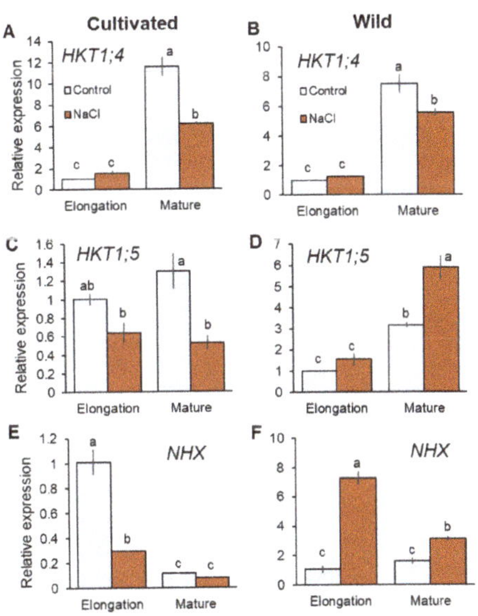

Figure 4. RT-qPCR analysis of the gene expressions of *HKT1;4* (**A**,**B**), *HKT1;5* (**C**,**D**), and *NHX1* (**E**,**F**) in mature and elongation root zones of cultivated and wild rice species under control and salinity (100 mM NaCl, 48 h) conditions. Different letters indicate significant differences ($p < 0.05$, one-way ANOVA followed by LSD tests). Mean ± SE ($n = 3$).

2.4. Ion Flux Measurements on the Root Epidermis in Response to NaCl

2.4.1. NaCl-Induced Na^+ Influx

Na^+ influx into elongation (EZ) and mature root zones (MZ) of both two rice species were induced by salt (100 mM NaCl) application. However, cultivated rice showed higher influx than wild rice (Figure 5A,B). The peak value of net Na^+ flux in MZ in the cultivated rice was much higher (about twofold, significant at $p < 0.05$) than that in wild rice (marked as "no inhibitor" in Figure 6G).

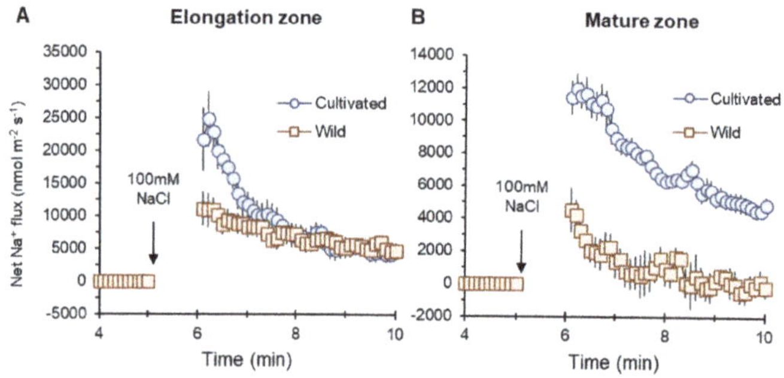

Figure 5. Transient net Na^+ flux measured from elongation (**A**) and mature (**B**) root zones of cultivated and wild rice species in response to 100 mM NaCl application. Mean ± SE (n = 7–10).

2.4.2. SOS1 Operations in Reducing Net Na^+ Influx

Cellular Na^+ exclusion in plants is mediated by Na^+/H^+ exchanger (SOS1) fuelled by H^+-ATPase activity at the root plasma membrane [12,22]. Thus, net Na^+ influx into the root can be explained by the difference between unidirectional Na^+ entry into the root and SOS1-mediated Na^+ efflux from the root. Pharmacological experiments revealed that net Na^+ efflux was decreased by both amiloride (an inhibitor of Na^+/H^+ exchanger: SOS1) and sodium orthovanadate (vanadate: H^+-ATPase blocker) pre-treatments in both species (Figure 6A–D). The peak Na^+ flux values were significantly ($p < 0.05$) increased by amiloride and vanadate pre-treatments compared with no-inhibitor within the same species, except vanadate pre-treatment in wild rice (Figure 6G). This suggests activity of SOS1 fuelled by H^+-ATPase at the root plasma membrane in both species. Compared with wild rice, cultivated rice showed a greater shift towards net Na^+ influx caused by SOS1 inhibition. The increases in peak Na^+ flux caused by amiloride pre-treatment (relative to no-inhibitor) were 15,346 and 6685 nmol m^{-2} s^{-1} in cultivated and wild rice, respectively (Figure 6G). Likewise, cultivated rice also showed greater increase of peak Na^+ influx by vanadate pre-treatment than wild rice (7414 vs. 4636 nmol m^{-2} s^{-1}, Figure 6G). This suggests that cultivated rice relies more on SOS1 activity for cellular Na^+ extrusion at the root epidermis under salinity than wild rice.

2.4.3. Na^+ Influx through NSCC

Non-selective cation channels (NSCC) have been known as a major pathway of Na^+ entry into the root [46]. Although $GdCl_3$ (Gd^{3+}; NSCC blocker) pre-treatment did not significantly change peak Na^+ influx values in two species, it reduced the peak of Na^+ influx in cultivated rice by 24.6%, while in wild rice, this reduction was only 10.3% (Figure 6G). Moreover, kinetics of net Na^+ influx was always smaller in the root treated with Gd^{3+} relative to "no-inhibitor" control in cultivated rice after salt application (Figure 6E), while Gd^{3+}-induced difference in net Na^+ flux in wild rice was less obvious (Figure 6F). These observations suggest that NSCCs may play a smaller role in Na^+ uptake in wild rice compared with cultivated rice.

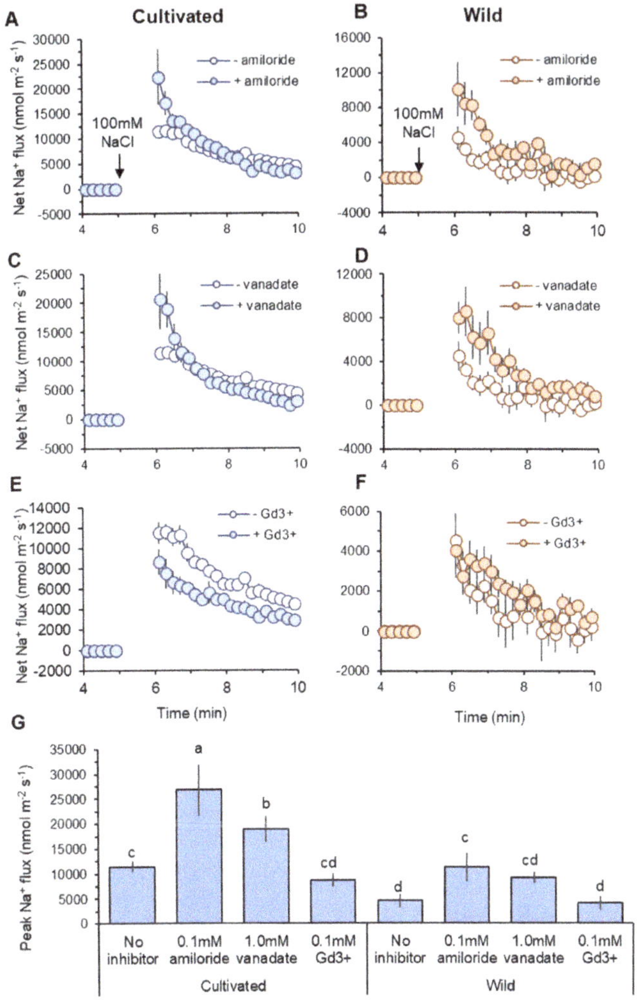

Figure 6. Pharmacological analysis of transient net Na^+ flux measured from mature root zones of cultivated and wild rice species in response to 100 mM NaCl application. Roots were pre-treated for 1 h with one of the following known inhibitors: 0.1 mM amiloride, an inhibitor of Na^+/H^+ exchanger (**A**,**B**); 1 mM sodium orthovanadate (vanadate), H^+-ATPase blocker (**C**,**D**); 0.1 mM $GdCl_3$ (Gd^{3+}), non-selective cation channel NSCC blocker (**E**,**F**). Peak Na^+ flux identified as maximum flux value during measurements (**G**). Different letters indicate significant differences ($p < 0.05$, one-way ANOVA followed by LSD tests). Mean ± SE (n = 6–8).

2.4.4. H^+ Flux in SOS1 Operations

Vanadate pre-treatment induced only slight reductions in net H^+ efflux in both species (Figure 7A). However, amiloride pre-treatment induced a much more prominent increase in net H^+ efflux (reduction in the amount of H^+ exchanged by Na^+ in SOS1 operations) in cultivated rice compared with wild rice (Figure 7B). These data can be taken as evidence for higher SOS1 activity in cultivated rice to reduce Na^+ uptake compared with wild rice.

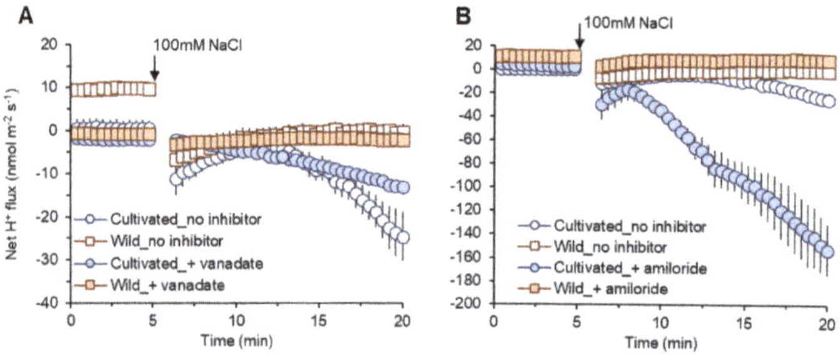

Figure 7. Transient net H^+ flux measured from mature root zones of cultivated and wild rice species in response to 100 mM NaCl application with 1 h of pre-treatment of known inhibitors: 1 mM sodium orthovanadate (vanadate), H^+-ATPase blocker (**A**); 0.1 mM amiloride, an inhibitor of Na^+/H^+ exchanger (**B**). Mean ± SE (*n* = 5–6).

2.5. Analysis of SOS1 Functional Activity

To assess functional activity of SOS1 in the root plasma membrane, we used the so-called "recovery protocol" [47]. The idea behind it is that the root is exposed to salinity and allowed to accumulate Na^+ for some time and induce expression of SOS1 genes required for its extrusion. The roots are then transferred to Na^+-free media and, after transient processes in the apoplast (Donnan system) are over, the magnitude of net Na^+ efflux reflects the functional activity of SOS1-like exchanger.

Consistent with reported expression of SOS1 genes, both rice species showed dramatically higher net Na^+ efflux in the root elongation zone (EZ) than mature zone (MZ) (Figure 8A). In EZ, about 80% greater net Na^+ efflux was observed from cultivated rice root without inhibitor than those in wild rice (cultivated rice: -651 ± 48 vs. wild rice: -359 ± 53 nmol m^{-2} s^{-1}, Figure 8A). Amiloride (SOS1 inhibitor) pre-treatment significantly ($p < 0.05$) reduced net Na^+ efflux in cultivated rice in both root zones, while amiloride-induced decrease of Na^+ efflux in wild rice was not significant in both root zones (Figure 8A,B). Wild rice showed a significant decrease in net Na^+ efflux by Gd^{3+} (NSCC blocker) pre-treatment in both root zones, while cultivated rice showed Gd^{3+}-induced decrease of Na^+ efflux (with significance, $p < 0.05$) in only EZ (Figure 8A,B). The above observations suggest that cultivated rice mediates greater Na^+ efflux than wild rice, and therefore cultivated more relies on SOS1 activity for Na^+ exclusion compared with wild rice. On the other hand, passive Na^+ leakage through NSCC (rather than active Na^+ exclusion by SOS1) largely contributed to Na^+ efflux from the root of wild rice.

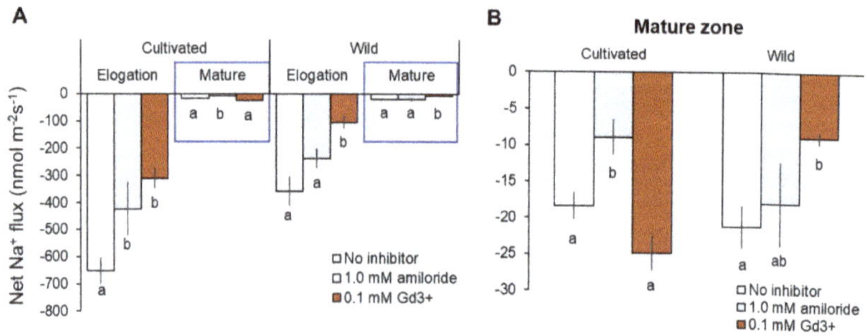

Figure 8. Na^+ efflux from elongation and mature root zones (**A**) and only mature root zone (**B**) of cultivated and wild rice after the removal of 100 mM NaCl (48 h treatment) with known inhibitors: 0.1 mM amiloride, an inhibitor of Na^+/H^+ exchanger; 0.1 mM $GdCl_3$ (Gd^{3+}), NSCC blocker. Steady-state net Na^+ flux was measured 20 min after NaCl removal. Different letters indicate significant differences within the same root zone in the same species ($p < 0.05$, one-way ANOVA followed by LSD tests). Mean ± SE (*n* = 5–6).

Transcriptional analysis showed that *SOS1* expressions were significantly higher in wild rice than cultivated rice under both control and salinity (100 mM NaCl for 48 h) conditions (Figure 9A,B). Salinity-induced changes in *SOS1* expressions were not significant in both root zones of cultivated rice, while those in wild rice were significant downregulation and upregulation in EZ and MZ, respectively (Figure 9A,B). As the above differences in SOS1 transcriptions can hardly explain actual SOS1 operations observed from Na^+ flux measurements, it appears that SOS1 activities might be regulated at the post-translational rather than transcriptional level.

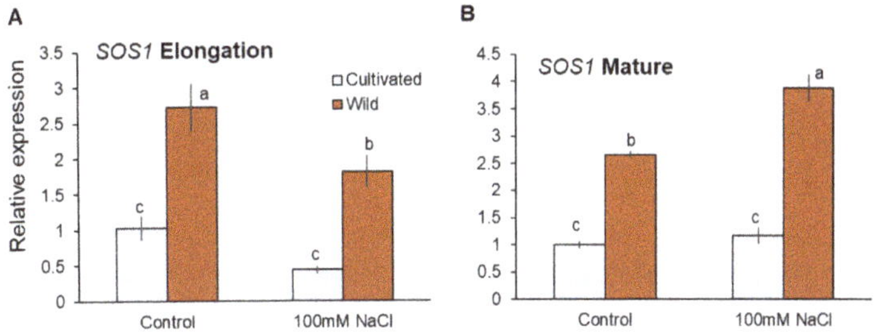

Figure 9. RT-qPCR analysis of the gene expressions of SOS1 in mature (**A**) and elongation (**B**) root zones of cultivated and wild rice species under control and salinity (100 mM NaCl, 48 h) conditions. Different letters indicate significant differences ($p < 0.05$, one-way ANOVA followed by LSD tests). Mean ± SE ($n = 3$).

3. Discussion

3.1. Leaf Tissue Na^+ Tolerance Observed in Wild Rice Is Highly Important for the Overall Salinity Tolerance in this Species

Wild rice possesses fourfold higher leaf sap Na^+ concentration than cultivated rice under un-salinised (0 mM NaCl) conditions (Figure 3B). Halophytes typically possess a greater amount of Na^+ in their leaf tissues compared with glycophytes [29], and this phenomenon is attributed to a likely role of Na^+ in maintenance of cell turgor [28]. Two weeks of salinity (100 mM NaCl) treatment increased leaf sap Na^+ concentration in both rice species, with no significant difference between them (Figure 3B). However, a significant reduction in chlorophyll content was observed in cultivated rice (Figure 2A) that is a typical symptom of Na^+ toxicity [48]. This was not observed in wild rice (Figure 2B), indicating its higher tissue tolerance to Na^+ [49] that may be conferred by a superior sequestration of Na^+ into vacuoles [40,50]. Superior tissue tolerance has been shown to confer salinity tolerance in the wild rice species *O. rufipogon* [50], and this trait has been suggested to be targeted for rice breeding instead of Na^+ exclusion [49,50]. Here, leaf tissue Na^+ tolerance was reported as being a hallmark for one of the most salt tolerant rice species, *O. coarctata*, validating the above conclusion.

3.2. Different Means of Osmotic Adjustment Differentiated Stress Tolerance between Two Species

Onset of salinity treatment also triggers osmotic stress, causing plant dehydration. Plants increase osmotic pressure in the cells and regain turgor in response to osmotic stress in a process called osmotic adjustment. There are two means of osmotic adjustment, namely, synthesis of organic osmolytes (compatible solutes) and accumulation of inorganic ions within cells [30,51]. At the early stage of salinity (one day after the stress onset), a significant ($p < 0.05$) increase in xylem sap Na^+ concentration was observed in wild rice, but not in cultivated rice. Thus, it is reasonable to suggest that wild rice relies on Na^+ transfer to the shoot for osmotic adjustment, while cultivated rice heavily relies on de novo synthesis of organic osmolytes and tries to minimise xylem Na^+ loading.

As biosynthesis of organic osmolytes is a highly energy-consuming process, it leads to growth penalties under prolonged salinity [20,52]. In addition to osmotic adjustment, stomatal operation is also a critical factor under osmotic stress conditions. In response to drought or salinity stress, stomatal closure is induced by ABA accumulation to conserve water in plants [53], and only cultivated rice significantly ($p < 0.001$) reduced stomatal conductance under salinity (Figure 2B). Reduced stomatal conductance results in a decrease of the ability of a plant to assimilate CO_2, thus limiting photosynthesis and plant growth [50,54,55]. Despite energy cost by de novo synthesis of organic osmolytes and reduced CO_2 assimilation due to stomatal closure, cultivated rice showed symptoms of dehydration (significant decrease of RWC at $p < 0.05$, Figure 1C). Osmotic adjustment by means of Na^+ accumulation is quick, energy-saving, and more efficient compared with de novo synthesis of organic osmolytes [20,30,56] and may be the reason that wild rice did not show a significant dehydration (Figure 1D) and stomatal closure (Figure 2B). Thus, utilisation of Na^+ for osmotic adjustment is a significant trait differentiating tolerance to salinity induced osmotic stress between rice species.

3.3. *Maintenance of Na^+ Homeostasis under Long-Term Salinity Is the Key Determinant of Overall Salinity Tolerance in Wild Rice*

Na^+ toxicity is considered as a main constraint imposed by the long-term salinity stress [57]. In addition to the response to salinity induced osmotic stress (explained in the above section), the two rice species differently controlled Na^+ uptake and transport to avoid Na^+ toxicity.

Cultivated rice showed significantly ($p < 0.001$) lower Na^+ concentrations in the root sap than wild rice at the beginning (day 1) of the salinity treatment (Figure 3A) that may be explained by greater activities of Na^+/H^+ exchanger (SOS1) located at the root plasma membrane for mediating Na^+ exclusion in this species (Figures 6 and 8; see also the next section for more discussion). Cellular Na^+ exclusion mediated by SOS1 activity fuelled by H^+-ATPase is an energy-consuming process as well as de novo synthesis of organic osmolytes for osmotic adjustment [19]. This suggests that cultivated rice expends a substantial amount of energy, leading to depletion of ATP pool when salt stress is prolonged. At a later stage of salinity imposition, cultivated rice may be low on available energy and therefore unable to control significant Na^+ entry into the root and thus Na^+ transport to the shoot, resulting in severe Na^+ toxicity leading to significant biomass reductions (Figure 1A). The above pattern of Na^+ transport under salinity observed in cultivated rice is also typically observed in salinity sensitive glycophytic species [29,30].

Compared with cultivated rice, wild rice has more efficiently control over Na^+ transport during the imposition of salinity. Once osmotic adjustment by means of Na^+ accumulation is achieved, wild rice maintains rather stable leaf and xylem sap Na^+ concentrations (from day 3; Figure 3B,C), and this is coupled with root Na^+ accumulation that is significantly greater than in cultivated rice, two weeks after the onset of salinity stress (Figure 3A). This pattern of Na^+ transport is effective to avoid shoot Na^+ toxicity. Although a functional role of OsHKT1;5 is still questioned due to direct measurements of Na^+ flux from root stele [58], OsHKT1;5 within the *SKC1* locus is suggested to mediate xylem Na^+ unloading that reduces shoot Na^+ accumulation under salinity [59]. In addition, effective Na^+ sequestration into root vacuoles through NHX was found to be a key determinant of salinity tolerance in barley and wheat [21,60]. As only wild rice shows significantly upregulated *HKT1;5* and *NHX1* expressions in the root in response to salinity (Figure 4C–F), it is plausible to suggest that this species may transfer an excessive amount of Na^+ from the xylem to root vacuoles. For Na^+ in root vacuoles to be retained for avoiding Na^+ toxicity, effective control of Na^+ back-leak into cytosol [61] is required and is suggested to operate in wild rice. The above mechanisms of root Na^+ sequestration rather than exclusion observed in wild rice may be a critical determinant of salinity tolerance, allowing this species to maintain normal metabolism and plant growth under long-term salinity (Figure 1B).

3.4. *Smaller Net Na^+ Entry in Wild Rice Root Is Not Attributable to SOS1 Activity*

In response to salt (100 mM NaCl) application, net Na^+ influx into the roots of both rice species was observed in electrophysiological experiments (Figure 5). Na^+ enters into the root through major two pathways, namely, NSCC and HKT [46]. Sensing Na^+ entry results in the elevation of cytosolic Ca^{2+}, cGMP, and H_2O_2 production [62,63] that activates H^+-ATPase-mediating H^+ efflux fuelling SOS1 activity to exclude Na^+ from the cytosol [64,65]. Due to the above Na^+ efflux system, the difference between Na^+ entry and Na^+ exclusion can explain the observed net Na^+ influx into the roots of both rice species. In response to transient application of salt stress, wild rice showed much lesser net Na^+ influx compared with cultivated rice (Figure 5A,B). However, this smaller net Na^+ influx was not due to greater activity of SOS1-mediated Na^+ extrusion. Both amiloride (a blocker of SOS1 exchanger) and vanadate (H^+-ATPase inhibitor) pre-treatments suggested greater involvement of SOS1 activity in cultivated rice than in wild rice (Figures 6A–D,G and 7A,B). Na^+ influx into root cells under salinity is a passive process [20], as both low cytosolic Na^+ concentrations and the negative electrical difference at the plasma membrane readily mediate Na^+ movement into cells [66]. Under non-saline conditions, wild rice possesses much greater root sap Na^+ concentration (Figure 3A) and less negative membrane potential (data are not shown) compared with cultivated rice. Therefore, wild rice may be able to allow smaller degree of salinity-induced Na^+ gradient moving into the plasma membrane of root cells than cultivated rice (Figure 6E,F), thus showing smaller net Na^+ influx in response to a sudden increase of external salt concentration.

3.5. Limiting Na^+ Exclusion by SOS1 Activity under Long-Term Salinity Is Crucial to Improve Salinity Tolerance in Rice Species

Na^+ efflux measurements in Na^+-free solution after the removal of salt (the so called "recovery protocol"; [22,47] revealed the difference of root Na^+ efflux mechanisms under long-term salinity (100 mM NaCl for 48 h) between cultivated and wild rice. A considerable degree of Na^+ efflux from cultivated rice root was mediated by amiloride-sensitive SOS1 activity, while measured Na^+ efflux from wild rice root was mostly due to passive Na^+ movement through Gd^{3+}-sensitive NSCC (Figure 8A,B). Moreover, in the root elongation zone (where SOS1 is predominantly located; [60,67], significantly larger Na^+ efflux was observed from cultivated rice than wild rice. Therefore, greater activity of SOS1 for Na^+ extrusion in cultivated rice under salinity was clearly observed from electrophysiological experiments. However, transcriptional changes in *SOS1* did not correlate with the observations at a functional level (Figure 9A,B). This is consistent with previous observations that the actual operation of SOS1 protein activity at a functional level does not always correlate with changes in transcript levels [22]. Further posttranslational mechanisms have been shown to be a major control point determining SOS1 activity in plants [68].

As mentioned in the previous sections, the issue with cellular Na^+ exclusion mediated by SOS1 activity fuelled by sharp H^+ gradient comes with a high energy cost. For this transporter to operate efficiently, for each Na^+ ion expelled across the plasma membrane, one H^+ ion needs to be extruded via H^+-ATPase (every H^+ extrusion hydrolyses one ATP; [19,20]). Therefore, a considerable energy penalty is imposed on cultivated rice due to the existence of the above-mentioned "futile cycle" at the root plasma membrane. Moreover, a passive apoplastic pathway of Na^+ entry into the root named bypass flow in *O. sativa* species [7,23–26] may impose further detrimental effects, due to a futile cycle. It is therefore plausible to suggest that wild rice may possesses a superior ability to sequester Na^+ into root vacuoles and limit passive Na^+ entry through Na^+-permeable channels/transporters under salinity, thus making SOS1-mediated Na^+ extrusion playing only a small role in its overall salinity stress tolerance.

4. Materials and Methods

4.1. Plant Materials and Growth Conditions

Seeds of cultivated rice (*O. sativa* cv. Koshihikari) were obtained from Western Sydney University and then multiplied using glasshouse facilities at Tasmanian Institute of Agriculture, University of Tasmania, Hobart, Australia. Seeds were pre-germinated in an incubator (30 °C for three days) and sown into plastic cell trays filled with the standard potting mix containing 70% perlite and 30% sand using half strength Hoagland solution (see [69] for details). Two weeks after the sowing of pre-germinated seeds, young seedlings (three to four leaves stage) of cultivated rice were transplanted into the pots with the mixture of soil collected from University Farm, University of Tasmania, Cambridge, Tasmania, Australia (Chromosol, see [70] for details) and potting mix (65/35% *w*/*w*). The pot volume was 1.5 L, and each pot contained one plant. Wild rice (*O. coarctata*) seedlings were obtained from the Swaminathan Research Foundation (Chennai, India) and were propagated vegetatively. For vegetative propagation, *O. coarctata* seedlings were grown in 15 L plastic tubs with the mixture of soil and potting mix (see details above), which were filled with tap water up to the soil surface. Newly developed three to four leaf stages of wild rice seedlings (about one month after the emergence of the new plantlets) were carefully separated and transplanted to the pots under the same condition as for cultivated rice. Pots with transplanted cultivated or wild rice seedlings were placed in a 15 L plastic tub filled with tap water up to the soil surface (4× pots per tub). One week after transplanting, salinity stress was imposed for four weeks (0, 50, or 100 mM NaCl) by the replacement of tap water in the tubs by appropriate NaCl solutions. Plants were grown in the greenhouse (temperature: 25 ± 2 °C; 12 h light/12 h dark photoperiod).

4.2. Biomass Measurement and Relative Water Content

Whole-plant biomass (fresh weight: FW) was measured before transplanting to the pots, and all transplanted seedlings were labelled to be identified. After four weeks of salinity treatments (when specific effects of Na^+ toxicity dominate), whole-plant FW was measured from labelled seedlings again, and the changes of FW (a biomass gain or loss) over the exposure to salinity were calculated. For measuring FW, seedlings were carefully removed from the growing medium, and their roots were then gently washed with a tap water to remove soil and quickly blotted. Roots and shoots were separated and weighted. Plant tissues were oven-dried, and shoot relative water content (RWC) was calculated.

4.3. Physiological Responses (Chlorophyll Content and Stomatal Conductance)

Chlorophyll content and stomatal conductance were measured from the second youngest fully expanded leaves two weeks after the commencement of salinity using SPAD-502 m (Konica Minolta, Osaka, Japan) and Decagon Leaf Porometer (Decagon Devices Inc., Pullman, WA, USA), respectively, as described in [71].

4.4. Root, Shoot, and Xylem Sap Na^+ Analysis

Root, leaf, and xylem saps were collected at different time points (1, 3, 7, and 14 days) after the commencement of salinity. Leaf sap was taken from the second youngest fully expanded leaves. Harvested roots were washed with 10 mM $CaCl_2$ to remove apoplastic Na^+ and quickly blotted. Harvested leaf and root samples were put into Eppendorf tubes and stored in the −20 °C freezer. Frozen samples were then thawed under the room temperature, and sap was obtained by hand-squeezing, as described in [71]. Xylem sap was collected using Scholander pressure chamber (Plant Moisture Systems, Santa Barbara, CA, USA). Each sample of xylem sap was collected from one to three plants per pot. Collected samples were weighed with 0.1 mg accuracy, then diluted and kept in the fridge. The content of Na^+ and K^+ in all sap samples was then measured using a flame photometry (model: PFP7, Jenway, Felsted, Dunmow, Essex, United Kingdom).

4.5. RNA Isolation and Real-Time Quantitative PCR Analysis

Excised root segments (3–5 cm long) were taken from the seedlings one month after transplanting of young seedlings into the pots (see the details of growing condition in Section 4.1). Seedlings were treated with 100 mM NaCl for 48 h before root harvest. Root segments were gently washed and blotted, and then cut into elongation and mature zone segments (1.0–2.0 and 12–15 mm from the root tip, respectively). Total RNA was isolated using RNAiso Plus (Takara, Shiga, Japan) as per the manufacturer's protocol. First-strand cDNA synthesis was performed in a 20 μL reaction volume with 1 μg of total RNA and Superscript III (Invitrogen, Carlsbad CA, USA) at 42 °C for 60 min, followed by heat inactivation at 70 °C for 10 min. Real-time PCR (Quant Studio 6; Thermo Fisher, Waltham, MA, USA) was carried out in a reaction volume of 10 μL (1 μL of cDNA, 5 μL of Takara TB GreenTM Primix Ex TaqTM II (2×), 0.5 μL each of a given primer pair (final concentration of 250 nM each)) under the following cycling conditions: 95 °C (30 s), 40 cycles of denaturation at 95 °C (5 s), annealing and extension at 60 °C (30 s) in a 96-well optical reaction plate (Thermo Fisher, Waltham, MA, USA). The primer pairs listed in Appendix A were used to amplify fragments of indicated sizes for each target gene. Each real-time PCR reaction was performed in triplicate. Amplicon specificity was verified by melt curve analysis (60–95 °C at 40 cycles) and subsequent agarose gel electrophoresis. Gene expression was quantified using the comparative CT ($2^{-\Delta\Delta CT}$) quantitation method. Three biological replicates were used in all cases.

4.6. Ion Flux Measurements

For electrophysiological experiments, newly developed underground adventitious roots from rhizomes of wild rice or crown of cultivated rice (crown root) were cut and taken one month after transplanting of young seedlings (three to four leaves stage) to the pots (see the details of growing condition in Section 4.1). Seedlings were grown in tap water until root harvest. Net ion fluxes were measured by using non-invasive ion-selective microelectrode (MIFE) technique (University of Tasmania, Hobart, Australia). Complete description of the theory of MIFE measurements, fabrication of ion-selective microelectrodes, and calibration processes have been previously explained in our past studies [72,73]. For preparation of H^+-selective microelectrodes, commercially available ionophore cocktail (Merck, Germany; catalogue number 95291) was front-filled on the tips of electrodes. For Na^+ measurement, an improved calixarene-based Na^+ ionophore cocktail [74,75] was used.

Excised root segments (3–5 cm long) were carefully washed in a basic salt medium (BSM; 0.5 mM KCl, 0.1 mM $CaCl_2$; pH 5.7, unbuffered) solution and then immobilised in the measuring chamber containing BSM. Fluxes of Na^+ and H^+ were measured from epidermal cells of elongation and mature root zones (1.0–2.0 and 12–15 mm from the root tip, respectively). Steady net ion fluxes were measured for five minutes in BSM solution, then salt treatment was applied to bring the final NaCl concentration (100 mM NaCl) in the measuring chamber. The resulting transient ion flux was recorded for up to 25 min. For the pharmacological experiment, excised root segments were pre-treated with known inhibitors (Appendix B) for one hour before ion flux measurement. Measurements were conducted from at least five individual plants

4.7. Measuring Na^+/H^+ Exchanger Activity

To quantify activity of the plasma membrane Na^+/H^+ exchangers mediating Na^+ extrusion from plant roots, we used a so called "recovery protocol" as described in the previous study [47]. Root segments (3–5 cm long) were cut and taken from the seedlings (see the details of growing condition and root harvest in Sections 4.1 and 4.6) treated with 0 or 100 mM NaCl for 48 h before root harvest. Excised root segments were thoroughly and quickly washed with 10 mM $CaCl_2$ to remove apoplastic NaCl and rinsed with double-distilled water. The roots were then transferred into Na^+-free BSM solution and kept for 20 min, and net Na^+ flux was measured for 3–5 min.

Author Contributions: Conceptualisation—S.S., G.V., Z.-H.C., M.Z., L.S.; methodology- L.S., S.S.; data acquisition—T.I., G.S.; formal analysis—T.I., S.S., L.S.; writing—original draft preparation, T.I.; writing—review and editing, S.S., L.S., M.Z., G.V., Z.-H.C., M.Y.; supervision S.S., L.S., M.Z.; funding acquisition—S.S., G.V., M.Y. All authors have read and agreed to the published version of the manuscript.

Funding: This research was funded by Grant-Australia-India Strategic Research Fund, Department of Industry, Innovation and Science (project AISRF48490); Indo-Australian Biotechnology Fund (BT/Indo-Aus/09/03/2015) grant provided by the Department of Biotechnology, Government of India; and China National Natural Science Foundation (projects 31961143001 and 31870249). S.S. also acknowledges the support provided by the National Biodiversity Authority (NBA), Indian Government, in assisting with *O. coarctata* material transfer (Appl.no.1712).

Data Availability Statement: Not applicable.

Conflicts of Interest: The authors declare no conflict of interest.

Appendix A. List of Primers for RT-qPCR Analysis

Oligos Name	Oligos (5′ → 3′)
OsHKT1; 4	GACAGATCAATCCAGACCATCTC
	AGCCTcCCAAAGAACATCAc
OcNHX1	GAGAGGAGCTGTGTCGATTGc
	GGTAGCAGCAGCCTGATCAATG
OcHKT1; 5	ATTCTGGcTCCAACTGCTGIACT
	GTGAAGATCAGGTCCAAGTCCAT
OcSOS1	AGAAGTTCAAGAGGAATCCACCAT
	GGATCGTGCcATGTCCTTT

Appendix B. List of Inhibitors for Pharmacological Experiments

Name	Mode of Action	Concentration
Amiloride	Na^+/H^+ exchanger inhibitor	0.1 mM
Sodium orthovanadate (vanadate)	H^+ - A1Pase blocker	1 mM
GdC1$_3$ (Gd^{3+})	NSCC blocker	0.1 mM

References

1. Golldack, D.; Quigley, F.; Michalowski, C.B.; Kamasani, U.R.; Bohnert, H.J. Salinity stress-tolerant and -sensitive rice (*Oryza sativa* L.) regulate AKT1-type potassium channel transcripts differently. *Plant Mol. Biol.* **2003**, *51*, 71–81. [CrossRef] [PubMed]
2. Lee, K.S.; Choi, W.Y.; Ko, J.C.; Kim, T.S.; Gregorio, G.B. Salinity tolerance of japonica and indica rice (*Oryza sativa* L.) at the seedling stage. *Planta* **2003**, *216*, 1043–1046. [CrossRef] [PubMed]
3. Lin, H.X.; Zhu, M.Z.; Yano, M.; Gao, J.P.; Liang, Z.W.; Su, W.A.; Hu, X.H.; Ren, Z.H.; Chao, D.Y. QTLs for Na^+ and K^+ uptake of the shoots and roots controlling rice salt tolerance. *Theor. Appl. Genet.* **2004**, *108*, 253–260. [CrossRef]
4. Ren, Z.; Gao, J.-P.; Li, L.-G.; Cai, X.-L.; Huang, W.; Chao, D.-Y.; Zhu, M.-Z.; Wang, Z.-Y.; Luan, S.; Lin, H.-X. A rice quantitative trait locus for salt tolerance encodes a sodium transporter. *Nat. Gen.* **2005**, *37*, 1141–1146. [CrossRef]
5. Cotsaftis, O.; Plett, D.; Shirley, N.; Tester, M.; Hrmova, M. A two-staged model of Na^+ exclusion in rice explained by 3D modeling of HKT transporters and alternative splicing. *PLoS ONE* **2012**, *7*, e39865. [CrossRef] [PubMed]
6. Kobayashi, N.I.; Yamaji, N.; Yamamoto, H.; Okubo, K.; Ueno, H.; Costa, A.; Tanoi, K.; Matsumura, H.; Fujii-Kashino, M.; Horiuchi, T.; et al. OsHKT1;5 mediates Na^+ exclusion in the vasculature to protect leaf blades and reproductive tissues from salt toxicity in rice. *Plant J.* **2017**, *91*, 657–670. [CrossRef]
7. Horie, T.; Karahara, I.; Katsuhara, M. Salinity tolerance mechanisms in glycophytes: An overview with the central focus on rice plants. *Rice* **2012**, *5*, 11. [CrossRef]
8. Platten, J.D.; Egdane, J.A.; Ismail, A.M. Salinity tolerance, Na^+ exclusion and allele mining of HKT1;5 in *Oryza sativa* and *O. glaberrima*: Many sources, many genes, one mechanism? *BMC Plant Biol.* **2013**, *13*, 32. [CrossRef]
9. Shohan, M.U.S.; Sinha, S.; Nabila, F.H.; Dastidar, S.G.; Seraj, Z.I. HKT1;5 transporter gene expression and association of amino acid substitutions with salt tolerance across rice genotypes. *Front. Plant Sci.* **2019**, *10*, 18. [CrossRef]
10. Shi, H.Z.; Ishitani, M.; Kim, C.S.; Zhu, J.K. The *Arabidopsis thaliana* salt tolerance gene *SOS1* encodes a putative Na^+/H^+ antiporter. *Proc. Natl. Acad. Sci. USA* **2000**, *97*, 6896–6901. [CrossRef]
11. Shi, H.; Lee, B.H.; Wu, S.J.; Zhu, J.K. Overexpression of a plasma membrane Na^+/H^+ antiporter gene improves salt tolerance in Arabidopsis thaliana. *Nat. Biotechnol.* **2003**, *21*, 81–85. [CrossRef] [PubMed]

12. Gupta, A.; Shaw, B.P. Biochemical and molecular characterisations of salt tolerance components in rice varieties tolerant and sensitive to NaCl: The relevance of Na^+ exclusion in salt tolerance in the species. *Funct. Plant Biol.* **2021**, *48*, 72. [CrossRef] [PubMed]
13. Do, P.T.; Pham, H.Q.; Nguyen, H.; Le, D.H. Assessment of sequence polymorphism and gene expression of OsSOS1 gene in two contrasting rice genotypes. *Acta Biol. Crac. Ser. Bot.* **2018**, *60*, 25–34.
14. Bhatt, T.; Sharma, A.; Puri, S.; Minhas, A.P. Salt tolerance mechanisms and approaches: Future scope of halotolerant genes and rice landraces. *Rice Sci.* **2020**, *27*, 368–383. [CrossRef]
15. Luo, D.; Niu, X.; Yu, J.; Yan, J.; Gou, X.; Lu, B.-R.; Liu, Y. Rice choline monooxygenase (OsCMO) protein functions in enhancing glycine betaine biosynthesis in transgenic tobacco but does not accumulate in rice (*Oryza sativa* L. ssp. *japonica*). *Plant Cell Rep.* **2012**, *31*, 1625–1635. [CrossRef] [PubMed]
16. Karthikeyan, A.; Pandian, S.K.; Ramesh, M. Transgenic indica rice cv. ADT 43 expressing a Δ1-pyrroline-5-carboxylate synthetase (P5CS) gene from *Vigna aconitifolia* demonstrates salt tolerance. *Plant Cell Tissue Organ Cult.* **2011**, *107*, 383–395. [CrossRef]
17. Solis, C.A.; Yong, M.T.; Vinarao, R.; Jena, K.; Holford, P.; Shabala, L.; Zhou, M.; Shabala, S.; Chen, Z.-H. Back to the wild: On a quest for donors toward salinity tolerant rice. *Front. Plant Sci.* **2020**, *11*, 323. [CrossRef]
18. Hoang, T.M.L.; Tran, T.N.; Nguyen, T.K.T.; Williams, B.; Wurm, P.; Bellairs, S.; Mundree, S. Improvement of salinity stress tolerance in rice: Challenges and opportunities. *Agronomy* **2016**, *6*, 54. [CrossRef]
19. Munns, R.; Day, D.A.; Fricke, W.; Watt, M.; Arsova, B.; Barkla, B.J.; Bose, J.; Byrt, C.S.; Chen, Z.; Foster, K.J.; et al. Energy costs of salt tolerance in crop plants. *New Phytol.* **2020**, *225*, 1072–1090. [CrossRef]
20. Munns, R.; Passioura, J.B.; Colmer, T.D.; Byrt, C.S. Osmotic adjustment and energy limitations to plant growth in saline soil. *New Phytol.* **2020**, *225*, 1091–1096. [CrossRef]
21. Wu, H.; Shabala, L.; Zhou, M.; Su, N.; Wu, Q.; Ul-Haq, T.; Zhu, J.; Mancuso, S.; Azzarello, E.; Shabala, S. Root vacuolar Na^+ sequestration but not exclusion from uptake correlates with barley salt tolerance. *Plant J.* **2019**, *100*, 55–67. [CrossRef] [PubMed]
22. Liu, J.; Shabala, S.; Shabala, L.; Zhou, M.; Meinke, H.; Venkataraman, G.; Chen, Z.-H.; Zeng, F.; Zhao, Q. Tissue-specific regulation of Na^+ and K^+ transporters explains genotypic differences in salinity stress tolerance in rice. *Front. Plant Sci.* **2019**, *10*, 1361. [CrossRef]
23. Yeo, A.R.; Yeo, M.E.; Flowers, T.J. The contribution of an apoplastic pathway to sodium uptake by rice roots in saline conditions. *J. Exp. Bot.* **1987**, *38*, 1141–1153. [CrossRef]
24. Garcia, A.; Rizzo, C.; Ud-Din, J.; Bartos, S.; Senadhira, D.; Flowers, T.; Yeo, A. Sodium and potassium transport to the xylem are inherited independently in rice, and the mechanism of sodium: Potassium selectivity differs between rice and wheat. *Plant Cell Environ.* **1997**, *20*, 1167–1174. [CrossRef]
25. Krishnamurthy, P.; Ranathunge, K.; Nayak, S.; Schreiber, L.; Mathew, M.K. Root apoplastic barriers block Na^+ transport to shoots in rice (*Oryza sativa* L.). *J. Exp. Bot.* **2011**, *62*, 4215–4228. [CrossRef]
26. Flam-Shepherd, R.; Huynh, W.Q.; Coskun, D.; Hamam, A.M.; Britto, D.T.; Kronzucker, H.J. Membrane fluxes, bypass flows, and sodium stress in rice: The influence of silicon. *J. Exp. Bot.* **2018**, *69*, 1679–1692. [CrossRef]
27. Pardo, J.M.; Quintero, F.J. Plants and sodium ions: Keeping company with the enemy. *Genome Biol.* **2002**, *3*, reviews1017.1. [CrossRef]
28. Wu, H. Plant salt tolerance and Na^+ sensing and transport. *Crop J.* **2018**, *6*, 215–225. [CrossRef]
29. Zarei, M.; Shabala, S.; Zeng, F.; Chen, X.; Zhang, S.; Azizi, M.; Rahemi, M.; Davarpanah, S.; Yu, M.; Shabala, L. Comparing kinetics of xylem ion loading and its regulation in halophytes and glycophytes. *Plant Cell Physiol.* **2019**, *61*, 403–415. [CrossRef]
30. Bose, J.; Shabala, L.; Pottosin, I.; Zeng, F.; Velarde-Buendia, A.M.; Massart, A.; Poschenrieder, C.; Hariadi, Y.; Shabala, S. Kinetics of xylem loading, membrane potential maintenance, and sensitivity of K^+-permeable channels to reactive oxygen species: Physiological traits that differentiate salinity tolerance between pea and barley. *Plant Cell Environ.* **2014**, *37*, 589–600. [CrossRef]
31. Zhu, M.; Zhou, M.; Shabala, L.; Shabala, S. Physiological and molecular mechanisms mediating xylem Na^+ loading in barley in the context of salinity stress tolerance. *Plant Cell Environ.* **2017**, *40*, 1009–1020. [CrossRef] [PubMed]
32. Sun, Y.; Lindberg, S.; Shabala, L.; Morgan, S.; Shabala, S.; Jacobsen, S.E. A comparative analysis of cytosolic Na^+ changes under salinity between halophyte quinoa (*Chenopodium quinoa*) and glycophyte pea (*Piswn sativwn*). *Environ. Exp. Bot.* **2017**, *141*, 154–160. [CrossRef]
33. Liu, M.; Pan, T.; Allakhverdiev, S.I.; Yu, M.; Shabala, S. Crop halophytism: An environmentally sustainable solution for global food security. *Trends Plant Sci.* **2020**, *25*, 630–634. [CrossRef] [PubMed]
34. Sengupta, S.; Majumder, A.L. Insight into the salt tolerance factors of a wild halophytic rice, *Porteresia coarctata*: A physiological and proteomic approach. *Planta* **2009**, *229*, 911–929. [CrossRef] [PubMed]
35. Garg, R.; Verma, M.; Agrawal, S.; Shankar, R.; Majee, M.; Jain, M. Deep Transcriptome sequencing of wild halophyte rice, *Porteresia coarctata*, provides novel insights into the salinity and submergence tolerance factors. *DNA Res.* **2014**, *21*, 69–84. [CrossRef] [PubMed]
36. Mondal, T.; Rawal, H.; Gaikwad, K.; Sharma, T.; Singh, N. First de novo draft genome sequence of *Oryza coarctata*, the only halophytic species in the genus *Oryza*. *F1000Research* **2017**, *6*, 1750. [CrossRef]
37. Flowers, T.J.; Flowers, S.A.; Hajibagheri, M.A.; Yeo, A.R. Salt tolerance in the halophytic wild rice, *Porteresia coarctata* Tateoka. *New Phytol.* **1990**, *114*, 675–684. [CrossRef]

38. Prusty, M.R.; Kim, S.-R.; Vinarao, R.; Entila, F.; Egdane, J.; Diaz, M.G.Q.; Jena, K.K. Newly identified wild rice accessions conferring high salt tolerance might use a tissue tolerance mechanism in leaf. *Front. Plant Sci.* **2018**, *9*, 15. [CrossRef]
39. Rajakani, R.; Sellamuthu, G.; Saravanakumar, V.; Kannappan, S.; Shabala, L.; Meinke, H.; Chen, Z.-H.; Zhou, M.; Parida, A.; Shabala, S.; et al. Microhair on the adaxial leaf surface of salt secreting halophytic *Oryza coarctata* Roxb. show distinct morphotypes: Isolation for molecular and functional analysis. *Plant Sci.* **2019**, *285*, 248–257. [CrossRef]
40. Kizhakkedath, P.; Jegadeeson, V.; Venkataraman, G.; Parida, A. A vacuolar antiporter is differentially regulated in leaves and roots of the halophytic wild rice *Porteresia coarctata* (Roxb.) Tateoka. *Mol. Biol. Rep.* **2015**, *42*, 1091–1105. [CrossRef]
41. Somasundaram, S.; Véry, A.-A.; Vinekar, R.S.; Ishikawa, T.; Kumari, K.; Pulipati, S.; Kumaresan, K.; Corratgé-Faillie, C.; Sowdhamini, R.; Parida, A.; et al. Homology modeling identifies crucial amino-acid residues that confer higher Na^+ transport capacity of OcHKT1;5 from *Oryza coarctata* Roxb. *Plant Cell Physiol.* **2020**, *61*, 1321–1334. [CrossRef] [PubMed]
42. Khan, I.; Mohamed, S.; Regnault, T.; Mieulet, D.; Guiderdoni, E.; Sentenac, H.; Véry, A.-A. constitutive contribution by the rice oSHKT1;4 Na^+ transporter to xylem sap desalinization and low Na^+ accumulation in young leaves under low as high external Na^+ conditions. *Front. Plant Sci.* **2020**, *11*, 1130. [CrossRef] [PubMed]
43. Suzuki, K.; Yamaji, N.; Costa, A.; Okuma, E.; Kobayashi, N.I.; Kashiwagi, T.; Katsuhara, M.; Wang, C.; Tanoi, K.; Murata, Y.; et al. OsHKT1;4-mediated Na^+ transport in stems contributes to Na^+ exclusion from leaf blades of rice at the reproductive growth stage upon salt stress. *BMC Plant Biol.* **2016**, *16*, 22. [CrossRef]
44. Bassil, E.; Blumwald, E. The ins and outs of intracellular ion homeostasis: NHX-type cation/H^+ transporters. *Curr. Opin. Plant Biol.* **2014**, *22*, 1–6. [CrossRef] [PubMed]
45. Leidi, E.O.; Barragán, V.; Rubio, L.; El-Hamdaoui, A.; Ruiz, M.T.; Cubero, B.; Fernández, J.A.; Bressan, R.A.; Hasegawa, P.M.; Quintero, F.J.; et al. The AtNHX1 exchanger mediates potassium compartmentation in vacuoles of transgenic tomato. *Plant J.* **2010**, *61*, 495–506. [CrossRef] [PubMed]
46. Munns, R.; Tester, M. Mechanisms of salinity tolerance. *Ann. Rev. Plant Biol.* **2008**, *59*, 651–681. [CrossRef]
47. Cuin, T.A.; Bose, J.; Stefano, G.; Jha, D.; Tester, M.; Mancuso, S.; Shabala, S. Assessing the role of root plasma membrane and tonoplast Na^+/H^+ exchangers in salinity tolerance in wheat: In planta quantification methods. *Plant Cell Environ.* **2011**, *34*, 947–961. [CrossRef] [PubMed]
48. Ashraf, M.; Harris, P.J.C. Photosynthesis under stressful environments: An overview. *Photosynthetica* **2013**, *51*, 163–190. [CrossRef]
49. Chakraborty, K.; Mondal, S.; Ray, S.; Samal, P.; Pradhan, B.; Chattopadhyay, K.; Kar, M.K.; Swain, P.; Sarkar, R.K. Tissue tolerance coupled with ionic discrimination can potentially minimize the energy cost of salinity tolerance in rice. *Front. Plant Sci.* **2020**, *11*, 265. [CrossRef]
50. Solis, C.A.; Yong, M.; Venkataraman, G.; Milham, P.; Zhou, M.; Shabala, L.; Holford, P.; Shabala, S.; Chen, Z. Sodium sequestration confers salinity tolerance in an ancestral wild rice. *Physiol. Plant.* **2021**, *172*, 1594–1608. [CrossRef]
51. Hariharan, G.; Purvaja, R.; Robin, R.; Ramesh, R. Evaluation of the multiple biomarkers on identification of the vulnerable coastal pollution hotspots. *Environ. Sci. Pollut. Res.* **2016**, *23*, 23281–23290. [CrossRef] [PubMed]
52. Hariadi, Y.; Marandon, K.; Tian, Y.; Jacobsen, S.-E.; Shabala, S. Ionic and osmotic relations in quinoa (*Chenopodium quinoa* willd.) plants grown at various salinity levels. *J. Exp. Bot.* **2011**, *62*, 185–193. [CrossRef]
53. Bharath, P.; Gahir, S.; Raghavendra, A.S. Abscisic acid-induced stomatal closure: An important component of plant defense against abiotic and biotic stress. *Front. Plant Sci.* **2021**, *12*, 65114. [CrossRef] [PubMed]
54. Munns, R.; Sharp, R.E. Involvement of abscisic-acid in controlling plant-growth in soils of low water potential. *Aust. J. Plant Physiol.* **1993**, *20*, 425–437. [CrossRef]
55. Adem, G.D.; Roy, S.J.; Huang, Y.; Chen, Z.-H.; Wang, F.; Zhou, M.; Bowman, J.P.; Holford, P.; Shabala, S. Expressing Arabidopsis thaliana V-ATPase subunit C in barley (*Hordeum vulgare*) improves plant performance under saline condition by enabling better osmotic adjustment. *Funct. Plant Biol.* **2017**, *44*, 1147. [CrossRef]
56. Shabala, S.; Shabala, L. Ion transport and osmotic adjustment in plants and bacteria. *Biomol. Concepts* **2011**, *2*, 407–419. [CrossRef]
57. Munns, R. Genes and salt tolerance: Bringing them together. *New Phytol.* **2005**, *167*, 645–663. [CrossRef]
58. Al Nayef, M.; Solis, C.; Shabala, L.; Ogura, T.; Chen, Z.; Bose, J.; Maathuis, F.J.M.; Venkataraman, G.; Tanoi, K.; Yu, M.; et al. Changes in expression level of OsHKT1;5 alters activity of membrane transporters involved in K^+ and Ca^{2+} acquisition and homeostasis in salinized rice roots. *Int. J. Mol. Sci.* **2020**, *21*, 4882. [CrossRef]
59. Ismail, A.M.; Horie, T. Genomics, physiology, and molecular breeding approaches for improving salt tolerance. *Annu. Rev. Plant Biol.* **2017**, *68*, 405–434. [CrossRef]
60. Wu, H.; Shabala, L.; Azzarello, E.; Huang, Y.; Pandolfi, C.; Su, N.; Wu, Q.; Cai, S.; Bazihizina, N.; Wang, L.; et al. Na^+ extrusion from the cytosol and tissue-specific Na^+ sequestration in roots confer differential salt stress tolerance between durum and bread wheat. *J. Exp. Bot.* **2018**, *69*, 3987–4001. [CrossRef]
61. Shabala, S.; Chen, G.; Chen, Z.-H.; Pottosin, I. The energy cost of the tonoplast futile sodium leak. *New Phytol.* **2020**, *225*, 1105–1110. [CrossRef] [PubMed]
62. Bose, J.; Pottosin, I.I.; Shabala, S.S.; Palmgren, M.G.; Shabala, S. Calcium efflux systems in stress signaling and adaptation in plants. *Front. Plant Sci.* **2011**, *2*, 85. [CrossRef] [PubMed]
63. Maathuis, F.J.M. Sodium in plants: Perception, signalling, and regulation of sodium fluxes. *J. Exp. Bot.* **2014**, *65*, 849–858. [CrossRef] [PubMed]

64. Maughan, P.J.; Turner, T.; Coleman, C.E.; Elzinga, D.B.; Jellen, E.; Morales, J.A.; Udall, J.A.; Fairbanks, D.J.; Bonifacio, A. Characterization of salt overly sensitive 1 (SOS1) gene homoeologs in quinoa (*Chenopodium quinoa* Willd.). *Genome* **2009**, *52*, 647–657. [CrossRef]
65. Bose, J.; Rodrigo-Moreno, A.; Lai, D.; Xie, Y.; Shen, W.; Shabala, S. Rapid regulation of the plasma membrane H+-ATPase activity is essential to salinity tolerance in two halophyte species, *Atriplex lentiformis* and *Chenopodium quinoa*. *Ann. Bot.* **2015**, *115*, 481–494. [CrossRef]
66. Blumwald, E.; Aharon, G.S.; Apse, M.P. Sodium transport in plant cells. *Biochim. Biophys. Acta* **2000**, *1465*, 140–151. [CrossRef]
67. Shi, H.Z.; Quintero, F.J.; Pardo, J.M.; Zhu, J.K. The putative plasma membrane Na^+/H^+ antiporter SOS1 controls long-distance Na^+ transport in plants. *Plant Cell* **2002**, *14*, 465–477. [CrossRef]
68. Gupta, A.; Shaw, B.P.; Sahu, B.B. Post-translational regulation of the membrane transporters contributing to salt tolerance in plants. *Funct. Plant Biol.* **2021**, *48*, 1199. [CrossRef]
69. Chen, Z.; Zhou, M.; Newman, I.A.; Mendham, N.J.; Zhang, G.; Shabala, S. Potassium and sodium relations in salinised barley tissues as a basis of differential salt tolerance. *Funct. Plant Biol.* **2007**, *34*, 150–162. [CrossRef]
70. DPIPWE. *Reconnaissance Soil Map Series of Tasmania Hobart*; Department of Primary Industries, Parks, Water and Environment, Tasmanian Government: Hobart, Australia, 2021. Available online: https://dpipwe.tas.gov.au/Documents/hobasoil.pdf (accessed on 10 August 2021).
71. Ishikawa, T.; Shabala, S. Control of xylem Na^+ loading and transport to the shoot in rice and barley as a determinant of differential salinity stress tolerance. *Physiol. Plant.* **2019**, *165*, 619–631. [CrossRef]
72. Shabala, S.N.; Newman, I.A.; Morris, J. Oscillations in H^+ and Ca^{2+} ion fluxes around the elongation region of corn roots and effects of external pH. *Plant Physiol.* **1997**, *113*, 111–118. [CrossRef] [PubMed]
73. Shabala, S.; Demidchik, V.; Shabala, L.; Cuin, T.A.; Smith, S.J.; Miller, A.J.; Davies, J.M.; Newman, I.A. Extracellular Ca^{2+} ameliorates NaCl-induced K^+ loss from Arabidopsis root and leaf cells by controlling plasma membrane K^+-permeable channels. *Plant Physiol.* **2006**, *141*, 1653–1665. [CrossRef]
74. Jayakannan, M.; Babourina, O.; Rengel, Z. Improved measurements of Na^+ fluxes in plants using calixarene-based microelectrodes. *J. Plant Physiol.* **2011**, *168*, 1045–1051. [CrossRef] [PubMed]
75. Jayakannan, M.; Bose, J.; Babourina, O.; Rengel, Z.; Shabala, S. Salicylic acid improves salinity tolerance in Arabidopsis by restoring membrane potential and preventing salt-induced K^+ loss via a GORK channel. *J. Exp. Bot.* **2013**, *64*, 2255–2268. [CrossRef] [PubMed]

plants

Article

Physiological Adaptation of Three Wild Halophytic *Suaeda* Species: Salt Tolerance Strategies and Metal Accumulation Capacity

Farag Ibraheem [1,2,*], Ateeq Al-Zahrani [1] and Ahmed Mosa [3,*]

1 Biology and Chemistry Department, Al Qunfodah University College, Umm Al-Qura University, Al Qunfodah 21912, Saudi Arabia; aaalzahrani@uqu.edu.sa
2 Botany Department, Faculty of Science, Mansoura University, Mansoura 35516, Egypt
3 Soils Department, Faculty of Agriculture, Mansoura University, Mansoura 35516, Egypt
* Correspondence: fllbraheem@uqu.edu.sa (F.I.); ahmedmosa@mans.edu.eg (A.M.)

Abstract: Understanding salt tolerance mechanisms in halophytes is critical for improving the world's agriculture under climate change scenarios. Herein, the physiological and metabolic responses of *Suaeda monoica*, *Suaeda vermiculata*, and *Suaeda schimperi* against abiotic stress in their natural saline environment on the east coast of the Red Sea were investigated. The tested species are exposed to different levels of salinity along with elemental disorders, including deficiency in essential nutrients (N&P in particular) and/or elevated levels of potentially toxic elements. The tested species employed common and species-specific tolerance mechanisms that are driven by the level of salinity and the genetic constitution of *Suaeda* species. These mechanisms include: (i) utilization of inorganic elements as cheap osmotica (Na+ in particular), (ii) lowering C/N ratio (*S. monoica* and *S. schimperi*) that benefits growth priority, (iii) efficient utilization of low soil N (*S. vermiculata*) that ensures survival priority, (v) biosynthesis of betacyanin (*S. schimperi* and *S. vermiculata*) and (vi) downregulation of overall metabolism (*S. vermiculata*) to avoid oxidative stress. Based on their cellular metal accumulation, *S. monoica* is an efficient phytoextractor of Cr, Co, Cu, Ni, and Zn, whereas *S. vermiculata* is a hyper-accumulator of Hg and Pb. *S. schimperi* is an effective phytoextractor of Fe, Hg, and Cr. These results highlight the significance of *Suaeda* species as a promising model halophyte and as phytoremediators of their hostile environments.

Keywords: *Suaeda*; salinity; physiology; oxidative stress; potential toxic elements; betacyanin; carbon; nitrogen; phytoremediation

Citation: Ibraheem, F.; Al-Zahrani, A.; Mosa, A. Physiological Adaptation of Three Wild Halophytic *Suaeda* Species: Salt Tolerance Strategies and Metal Accumulation Capacity. *Plants* **2022**, *11*, 537. https://doi.org/10.3390/plants11040537

Academic Editors: Oscar Vicente and Marius-Nicusor Grigore

Received: 4 January 2022
Accepted: 15 February 2022
Published: 17 February 2022

1. Introduction

Coastal salt marshes are transition zones between land and sea and act as natural buffers against deteriorative impacts of saltwater intrusion, coastal erosion, and contaminants release [1]. These regions usually contain large levels of salinity along with substantial amounts of potentially toxic elements (PTEs) as a result of various anthropogenic activities (e.g., rapid urbanization, marine construction, oil spilling, domestic waste dumping, landfilling due to the advancement of a seaside framework, brine discharge from desalination plants and agricultural practices [2]. Climate change is expected to increase temperature and evapotranspiration and thus can aggravate salinity and PTE-induced stress, particularly in arid and semi-arid regions [3]. Such harsh conditions of salinity and PTEs in arid salt marshes restrict plant vegetation to halophytic plants, which evolved exceptional ability to grow and reproduce in a highly saline environment [1,4]. Interestingly, the ongoing increase in atmospheric CO_2 can improve the salinity tolerance of C3 and C4 halophytes [5]. Therefore, these unique plants can contribute significantly to carbon sequestration and thus can reduce the impact of global climate change [1]. Halophytes can further be used as intercropping and rotating species to improve crops' productivity, given their high potential to desalinize the high salt accumulations [6]. These features highlight the potentiality of halophytes as promising biological resources for improving the world's agriculture in the climate change scenario via genetic and biotechnological approaches. Intensive research has been undertaken for a better understanding of the salt tolerance mechanisms in glycophytes and halophytes. However, the full picture, particularly in halophytes, is far from clear.

To cope with salinity-induced challenges, halophytes employ common as well as species-specific mechanisms to minimize their detrimental effects. Common salt tolerance mechanisms include (i) regulation/compartmentalization of ions (Na^+ and Cl^-) uptake and localization in vacuoles [7], (ii) accumulation of organic osmolytes in the cytoplasm to balance the osmotic effects in vacuoles [8,9], and (iii) maintaining a balance between the production of reactive oxygen species (ROS) and the total quenching activity of antioxidative system [10]. Species-specific mechanisms may include succulence, extrusion of toxic ions, special anatomical structures (hairs, salt glands), redistribution of excessive ions to senescent leaves [11,12], and synthesis of stress-related pigments with specific physiological functions such as betacyanin [13]. Activation of the above tolerance mechanisms involves the diversion of a significant portion of essential plant metabolites such as carbohydrates and nitrogenous compounds away from biomass production [14]. Such metabolic shunting drives a trade-off between halophyte growth and survival: responses that differ among taxa and are not fully understood [15]. Along with their excess salt ions, halophytic habitats are enriched with PTEs. To cope with the adverse effects of such PTEs, halophytes employ various mechanisms, including metals stabilization in the root zone, complexation with root exudates, changing the metal ions, precipitation as insoluble deposits inside vacuoles, and establishing a partnership with heavy metal tolerant soil microorganisms [16,17].

Halophytes belong to different angiosperms plant families, suggesting a polyphyletic origin of salt tolerance [18]. Among these families, Amaranthaceae (previously known as Chenopodiaceae) is an interesting example as it contains the largest number of known halophytes with high capabilities of salt tolerance [18]. *Suaeda* is one of the extreme obligate halophytic chenopods and has been proposed as a model system for the dissection of salt tolerance in halophytes [18]. *Suaeda* species are generally perennials chamaephyte (dwarf-shrub) with succulent leaves. They are distributed in various saline habitats with different salinity levels and exhibit differential capabilities of withstanding high salinity levels ranging from 200 mM to 400 mM or even more [19]. Along with their significance as model plants for dissection of salt tolerance, they have been suggested as promising biological tools for desalinization of hypersaline lands because of their high capacity of salt uptake and accumulation [20]. In fact, some *Suaeda* species can remove more than two tons of salt/hectare in a single harvest [21,22]. In addition, *Suaeda* species have been acknowledged for their high efficiency as phytoremediators with the ability to uptake substantial amounts of PTEs [23]. These halophytic species are adapted to overcome PTEs accumulation similar to glycophytes [24]. The salt stress tolerance mechanisms of *Suaeda* species, similar to most other dicotyledonous halophytes, mainly depend on the accumulation of Na^+ and Cl^- in leaves [25,26], where their succulence enables the dilution of ions concentration and thus alleviate ions toxicity [27]. Furthermore, osmolytes such as glycine betaine, proline, and sugars play key roles in osmotic adjustment in some *Suaeda* species [28,29]. Further, non-enzymatic antioxidants such as flavonoids and phenolics, along with antioxidant enzymes, contribute to their adaptation against salinity-induced oxidative stress [10,30].

The east coast of the Red Sea is a typical hyper-arid saline region with high temperature, limited precipitation, high salinity, and vulnerability to contamination derived from oil trading and other anthropogenic activities developing along the Red Sea coast. It is particularly rich in *Suaeda* species. Examples include *S. egyptiaca*, *S. fruticosa*, *S. monoica*, *S. vermiculata*, *S. pruinosa*, and *S. schimperi* [31]. The current harsh climatic conditions in the region, as well as the predicted climate change-induced increase in temperature and evapotranspiration, are expected to exacerbate salinity-induced deleterious effects on *Suaeda* growth and physiology in the region [1,32]. Up to our knowledge, the differences in the activity and the relative contribution of the above individual mechanisms to salt tolerance in *Suaeda* species in the area have not been reported. In addition, the species/habitat variations in *Suaeda* species seem to depend on differential efficiency of salt tolerance mechanisms among these species, which may be associated with different adaptive physiological and molecular mechanisms in response to different levels of soil salinity [33].

In the current study, three species of the genus *Suaeda* including *Suaeda monoica* Forssk. ex J.F.Gmel., *Suaeda vermiculata* Forssk. ex J.F.Gmel., and *Suaeda schimperi* Moq. that dominate three separate salt marshes at different vicinity to the east coast of the Red Sea were selected. These species are genetically related but differ in their leaf reddening phenotype, which has been reported as an adaptive strategy for salt tolerance [34]. The aim of the current study was to assess the impact of the interaction between different levels of soil salinity and the differential physiological responses of the selected *Suaeda* species, if any, on their successful adaptation in particular salt marshes. Herein, the hypothesis is that the successful adaptation of different *Suaeda* species in the selected salt marshes is shaped by the interplay between the magnitude of salinity in their rhizospheric soil and their relative salt tolerance evolutionary strategies. In addition, common and species-specific physiological responses may operate among species. Therefore, performing a comparative analysis of the physiological responses of these genetically related species against the physicochemical properties of their rhizospheric soil would be a useful approach to gain insights on possible common and species-specific tolerance mechanisms in these species.

The specific objectives of this study are to (i) monitor the levels of soil salinity, nutrients status, and PTEs concentration in the rhizospheric soil of the tested *Suaeda* species, (ii) determine water and nutrient status of *Suaeda* species as affected by physicochemical properties of soil, (iii) explore critical biochemical indicators of *Suaeda* species relevant to their physiological adaptation, and (v) evaluate the bioaccumulation capacity of PTEs by *Sauuda* species for the future phytoremediation planning.

2. Results

2.1. Soil Physicochemical Properties

Sand was the dominant component among soil fractions with a higher silt and clay content in soil supporting *S. vermiculata* (Table 1). The texture was sandy in soils supporting *S. monoica* and *S. schimperi*; however, it was sandy clay loam in the soil supporting *S. vermiculata*. The soil supporting *S. vermiculata* had higher porosity (49.93%) than both *S. monoica* (39.98%) and *S. schimperi* (43.89%). Water holding capacity in all soils was generally low (31.98–38.15%), with a relative superiority of soil supporting *S. vermiculata* (Table 1).

Table 1. Physicochemical analyses of the investigated soils. Shown are the means of three biological replicates ± standard deviation.

	Physicochemical Parameters		*S. monoica*	*S. vermiculata*	*S. schimperi*
Soil physical properties	Particle size distribution (%)	Sand	92.6	76.0	95.8
		Silt and clay	6.9	23.3	3.7
	Texture		Sandy	Sandy clay loam	Sandy
	Water holding capacity (%)		31.98 ± 3.05	38.15 ± 3.52	35.02 ± 3.21
	Porosity (%)		39.98 ± 3.53	49.93 ± 4.57	43.89 ± 3.93
Soil chemical properties	EC (dS m^{-1})		5.04 ± 0.55	18.37 ± 1.91	16.25 ± 1.04
	pH		8.10 ± 0.44	8.65 ± 0.38	7.78 ± 0.24
	$CaCO_3$ (%)		0.55 ± 060	0.79 ± 0.66	0.79 ± 0.30
	Water soluble anions (Cmol/100 g)	HCO_3^-	5.57 ± 0.60	6.35 ± 0.66	5.33 ± 0.30
		Cl^-	6.46 ± 0.67	13.76 ± 3.28	14.29 ± 2.29
	Water soluble cations (Cmol/100 g)	Na^+	2.51 ± 0.15	5.49 ± 0.23	4.89 ± 0.30
		K^+	0.46 ± 0.09	1.10 ± 0.12	0.98 ± 0.10
		Ca^{2+}	7.20 ± 0.39	6.15 ± 0.41	7.60 ± 0.44
		Mg^{2+}	1.79 ± 0.08	5.66 ± 0.69	10.42 ± 0.89

The tested soils were all alkaline, with pH values between 7.78 and 8.65 (Table 1). The EC values in soils supporting *S. vermiculata* (18.37 dSm^{-1}) and *S. schimperi* (16.25 dSm^{-1}) were significantly higher than soil supporting *S. monoica* (5.04 dSm^{-1}) (Table 1). Furthermore, soil supporting *S. vermiculata* and *S. schimperi* had comparable levels of calcium carbonate (0.79%), which were higher than soil supporting *S. monoica* (0.55%). The significantly high EC values of soils supporting *S. vermiculata* and *S. schimperi* were associated with higher water-soluble Cl^-, Na^+, and K^+ ions. In addition, the high pH value of soil supporting *S. vermiculata* was correlated with high soluble HCO_3^- concentration (6.35 Cmol/100 g). The soil supporting *S. schimperi* had about two- and six-fold greater Mg^{2+} concentration (10.42 Cmol/100 g) than soils supporting *S. vermiculata* and *S. monoica*, respectively (Table 1).

Regarding the nutrient status of the tested soils, available phosphorus in soil supporting *S. vermiculata* (9.40 mg kg^{-1}) was higher than those supporting *S. schimperi* (6.53 mg kg^{-1}) and *S. monoica* (8.17 mg kg^{-1}) (Table 2). The average concentrations of K^+ in rhizospheric soils of *S. schimperi*, *S. vermiculata*, and *S. monoica* were 383.8, 427.8, 178.8 mg kg^{-1}; whereas the corresponding values were 1519.1, 1229.9 and 1441.0 mg kg^{-1} for Ca^{2+} and 1250.2, 679.4, and 215.3 mg kg^{-1} for Mg^{2+}. Soil organic elements were generally low. Carbon was not detected in soil supporting *S. monoica*, recorded low value in soil supporting *S. vermiculata* (0.23%), and was relatively high in soil supporting *S. schimperi* (1.95%). Nitrogen and sulfur were not detected in soils supporting *S. vermiculata* and *S. monoica* and showed low values in soil supporting *S. schimperi* (0.18 and 0.47%, respectively).

Table 2. Total organic elements, available inorganic nutrients, and available sodium in the soil surface layer of the studied locations. Shown are the means of three biological replicates ± standard deviation. ND indicates that the element was under its detection limit.

Elemental Concentrations		*S. monoica*	*S. vermiculata*	*S. schimperi*
Available nutrients ($mg\ kg^{-1}$)	P	8.17 ^{a} ± 0.38	9.40 ^{a} ± 0.44	6.53 ^{b} ± 31
	K	178.8 ^{b} ± 6.2	427.8 ^{a} ± 15.4	383.8 ^{a} ± 19.0
	Ca	1441.0 ^{a} ± 127.0	1229.9 ^{b} ± 114.9	1519.1 ^{a} ± 145.9
	Mg	215.3 ^{c} ± 6.3	679.4 ^{b} ± 3.7	1250.2 ^{a} ± 13.3
Available Na^+ ($mg\ kg^{-1}$)		578.3 ^{c} ± 45	1262.1 ^{a} ± 124.3	1103.5 ^{b} ± 108.2
Total organic elements (%)	C	N.D.	0.23 ^{b} ± 0.07	1.95 ^{a} ± 0.19
	N	N.D.	N.D.	0.18 ± 0.13
	H	0.23 ^{b} ± 0.03	0.51 ^{a} ± 0.04	0.593 ^{a} ± 0.04
	S	N.D.	N.D.	0.47 ± 0.03

Means followed by the same letter are not significantly different at the probability level of 5% according to LSD.

2.2. PTEs Concentration in the Tested Soils and Their Ecological Risk Assessment

Soil available contents of PTEs varied among locations and were ranked based on their average values (mg kg^{-1}) as Mn (18.21), Fe (7.94), Pb (4.71), Zn (3.62), Cu (2.26), Hg (1.55), Ni (0.71), Cr (0.333), Co (0.326), and Cd (0.07) (Table 3). Soil supporting *S. schimperi* had the highest levels of Cd, Co, Cu, Fe, and Mn. Meanwhile, the highest values of Cr, Hg, and Ni were recorded in soil supporting *S. monoica*. The soil supporting *S. vermiculata*, however, exhibited the highest Pb and Zn values. These results were greatly higher than other reported values in *Typic Torripsamment* such as [35] (Cd, Cu, Fe, Mn, Ni, Pb, and Zn), [36] (Cd, Cr, Co, Mn, and Ni), and [37] (Cd, Cu, Fe, Mn, Ni, Pb, and Zn) (Table 3).

Ecological risk assessments of PTEs in different locations are shown in Figure 1 and are categorized into six classes to interpret the obtained contamination levels of PTEs (Table S1, Supplementary Data). Geo-accumulation index indicated an uncontaminated effect of Cd, Cu, Fe, Pb, and Zn in all locations ($0 < I_{geo} < 1$). Other toxic elements (Cr, Hg, Mn, and Ni) showed slight contamination ($1 < I_{geo} < 2$). Meanwhile, Co exhibited moderate-to-high contamination ($3 < I_{geo} < 4$) effect in the studied locations with a higher risk in soil supporting *S. schimperi*.

Table 3. PTEs concentration in the soil surface layer of the studied locations and in leaves of the tested *Suaeda* species. Shown are the means of three biological replicates ± standard deviation. ND indicates that the element was under its detection limit.

Toxic Elements	*S. monoica*		*S. vermiculata*		*S. schimperi*	
	Concentration (mg kg^{-1})					
	Soil	Plant	Soil	Plant	Soil	Plant
Cd	ND	ND	$0.024^{b} \pm 0.002$	$0.496^{a} \pm 0.055$	$0.112^{a} \pm 0.011$	$0.499^{a} \pm 0.058$
Cr	$0.380^{a} \pm 0.038$	$177.108^{a} \pm 15.591$	$0.286^{b} \pm 0.029$	$18.668^{c} \pm 2.272$	ND	$50.166^{b} \pm 5.262$
Co	$0.279^{b} \pm 0.028$	$6.734^{a} \pm 0.662$	$0.121^{c} \pm 0.012$	$5.032^{b} \pm 0.526$	$0.577^{a} \pm 0.057$	$3.516^{c} \pm 0.377$
Cu	$2.070^{b} \pm 0.391$	$75.125^{a} \pm 7.873$	$2.160^{b} \pm 0.743$	$17.040^{b} \pm 1.635$	$2.566^{a} \pm 0.056$	$19.488^{b} \pm 2.002$
Fe	$5.019^{c} \pm 0.441$	$902.942^{c} \pm 85.291$	$8.147^{b} \pm 0.762$	$1286.0^{b} \pm 113.4$	$10.643^{a} \pm 1.082$	$2038.4^{a} \pm 188.6$
Hg	$1.813^{a} \pm 0.155$	ND	$1.367^{b} \pm 0.123$	$29.229^{a} \pm 2.545$	$1.456^{b} \pm 0.147$	$16.871^{b} \pm 1.132$
Mn	$11.884^{b} \pm 0.295$	$146.304^{b} \pm 15.319$	$13.511^{b} \pm 0.576$	$102.29^{c} \pm 11.52$	$29.237^{a} \pm 1.2211$	$166.053^{a} \pm 14.973$
Ni	$1.623^{a} \pm 0.056$	$16.852^{a} \pm 1.981$	$0.441^{b} \pm 0.020$	$4.518^{b} \pm 0.503$	$0.066^{c} \pm 0.013$	$5.964^{b} \pm 0.654$
Pb	$3.812^{b} \pm 0.147$	ND	$5.617^{a} \pm 0.216$	$18.651^{a} \pm 2.342$	ND	$9.918^{b} \pm 1.119$
Zn	$1.992^{c} \pm 0.069$	$113.735^{a} \pm 11.109$	$5.633^{a} \pm 0.071$	$74.080^{c} \pm 8.148$	$3.233^{b} \pm 0.063$	$93.904^{b} \pm 10.077$

Means followed by the same letter are not significantly different at the probability level of 5% according to LSD.

According to enrichment factor (E_f), some PTEs (Cd, Cu, Fe, Pb, and Zn) exhibited no enrichment in all locations ($0 < E_f < 1$). Other elements (Cr and Mn) recorded minor enrichment ($1 < E_f < 3$). Mercury (Hg) and Ni showed a moderate enrichment ($3 < E_f < 5$) in soil supporting *S. monoica*. However, cobalt showed moderate to severe enrichment ($5 < E_f < 10$) in *S. monoica* and *S. schimperi* supporting soils.

The contamination factor (C_f) index showed minor contamination ($C_f < 2$) of Cd, Cu, Fe, Pb, and Zn. Moderate contamination ($2 \leq C_F < 5$) was observed with Cr (*S. monoica* and *S. vermiculata* supporting soils), Co (*S. vermiculata* supporting soil), Hg (*S. vermiculata* and *S. schimperi* supporting soils), and Mn (*S. monoica* and *S. vermiculata* supporting soils). Some PTEs reached significant contamination ($5 \leq C_F < 20$) including Co (*S. monoica* and *S. schimperi* supporting soils), Hg (*S. monoica* supporting soil), Mn (*S. schimperi* supporting soil), and Ni (*S. monoica* supporting soil).

Ecological risk index (E_r^i) pointed to a low risk of most PTEs (Cr, Cu, Fe, Mn, Ni, Pb, and Zn) in all locations ($E_r^i < 40$). In contrast, Hg showed a high risk ($160 \leq E_r^i < 320$) in all locations given its high toxicity coefficient (40). Meanwhile, Cd and Co exhibited moderate to considerable risk with higher values in *S. schimperi* supporting soils. The combined effect of PTEs hazard was explored using a modified degree of contamination and pollution load index. The modified degree of contamination showed low contamination ($1.5 \leq mCD < 2.2$) with *S. vermiculata* supporting soil. However, *S. monoica* and *S. schimperi* supporting soils recorded moderate contamination ($2.2 \leq mCD < 4.4$). Conversely, the pollution load index revealed an unpolluted effect ($0 < P_{LI} \leq 1$) of soils in the studied locations (Figure 1).

2.3. PTEs Concentration in the Tested Suaeda Species

To gain insights into the PTEs transport to the aerial parts of *Suaeda* plants, the cellular concentrations of PTEs in leaves of the tested *Suaeda* species were analyzed and compared (Table 3). Values of PTEs (mg kg^{-1}) averaged as: Cd (0.50), Cr (81.98), Co (5.09), Cu (37.22), Fe (1409.15), Hg (23.05), Mn (138.22), Ni (9.11), Pb (14.28), and Zn (93.91). *Suaeda. monoica* showed the highest concentrations of Cr, Co, Cu, Ni, and Zn. *Suaeda schimperi*; however, had the highest values of Cd, Fe, and Mn. Meanwhile, *S. vermiculata* exhibited the highest values of Hg and Pb.

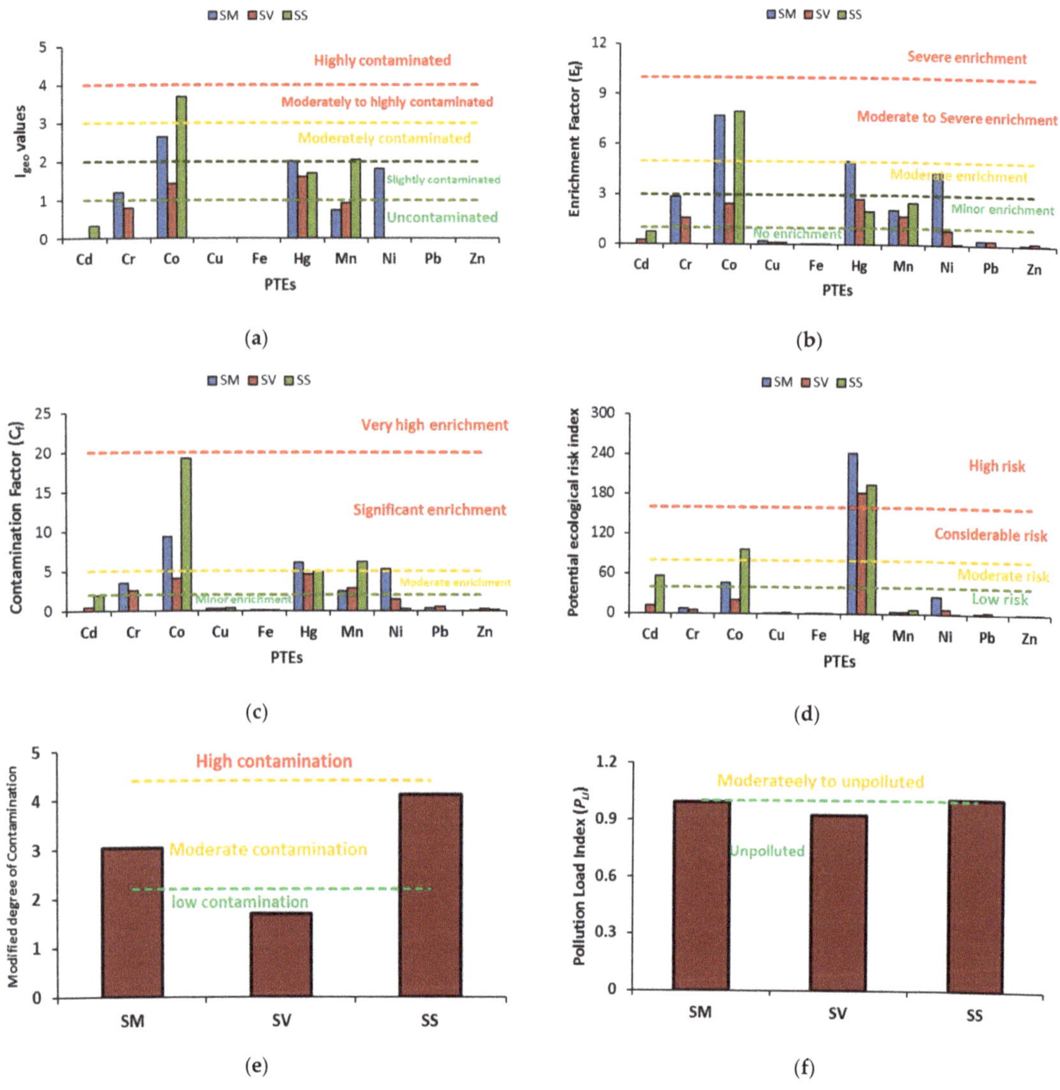

Figure 1. Ecological risk assessment indices of TEs in different soils supporting the tested *Suaeda* species: (**a**) geo-accumulation index; (**b**) enrichment factor; (**c**) contamination factor; (**d**) potential ecological risk index; (**e**) modified degree of contamination; and (**f**) pollution Load Index. SM: *S. monoica*, SV: *S. vermiculata*, and SS: *S. schimperi*.

Our calculations of the bioaccumulation factors (BCR) of PTEs of the tested *Suaeda* species revealed high BCR values that ranged between 3.3 and 466.1 (Figure 2). These findings illustrated the potential utilization of these plants as phytoextractors since values greater than 1.0 pointed to hyperaccumulating plants; however, values below 1.0 are indicative of excluder plants. Chromium (Cr) showed the highest bioaccumulation among PTEs, but Pb showed the lowest value.

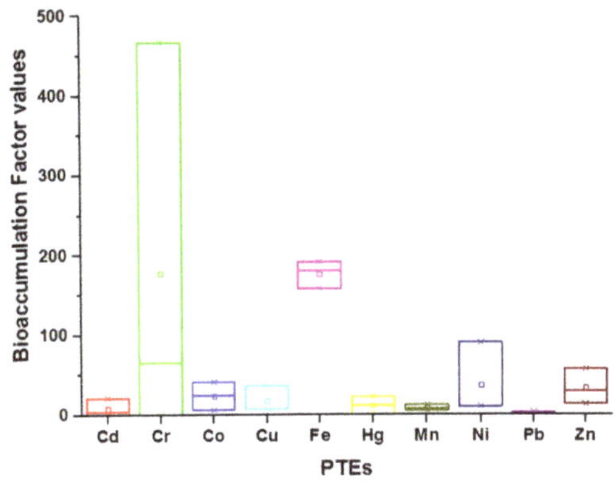

Figure 2. Bioaccumulation index of the tested *Suaeda* species. Values were calculated by dividing the metal concentration in *Suaeda* species by its corresponding concentration in rhizospheric soil. Plants with values higher than 1.0 are considered phytoextractors.

2.4. Levels of Inorganic and Organic Nutrients in Leaves of the Investigated Suaeda Species

Leaves of the tested *Suaeda* species varied significantly in their content of inorganic minerals (Table 4). Both *S. monoica* (1.485 mg g^{-1}) and *S. schimperi* (1.275 mg g^{-1}) contained significantly higher P than *S. vermiculata* (0.452 mg g^{-1}). Meanwhile, Ca^{2+} concentrations in *Suaeda* species were comparable (10.38–12.63 mg g^{-1}). Likewise, Mg^{2+} concentrations were relatively similar in *S. schimperi* and *S. vermiculata* (6.63 mg g^{-1}), and both were significantly higher than *S. monoica* (4.99 mg g^{-1}). The concentrations of K^+ were 10.04, 12.69, 10.55 mg g^{-1} DWT in leaves of *S. monoica, S. vermiculata*, and *S. schimperi*, respectively, whereas their corresponding concentrations of Na^+ were 11.01, 17.84, and 16.19 mg g^{-1} DWT. Given the accumulation patterns of both Na^+ and K^+ in the tested species, *S. schimperi* had the highest Na^+/K^+ ratio (1.536), whereas *S. monoica* had the lowest ratio (1.097). *S. vermiculata* had intermediate Na^+/K^+ ratio (1.407). *Suaeda monoica* had greater leaf C content (332.6 mg g^{-1}) than *S. schimperi* (265.16 mg g^{-1}) and *S. vermiculata* (245.76 mg g^{-1}). Similarly, *S. monoica* and *S. schimperi* had total leaf N of 32.04 mg g^{-1} and 29.79 mg g^{-1}, respectively, and both were significantly higher than *S. vermiculata* (11.96 mg g^{-1}) (Table 4). Given the observed differences in their total leaf C and N, the tested *Suaeda* species differed significantly in their C/N ratio. *Suaeda vermiculata* had the highest C/N ratio (20.99), whereas both *S. monoica* and *S. schimperi* had C/N values of 10.24 and 9.09, respectively. Leaves of *S. monoica* and *S. vermiculata* had relatively similar levels of Sulfur (S), whereas its level in *S. schimperi* was below the detection limit.

Table 4. Elemental concentration in leaves of the tested *Suaeda* species (mg g^{-1} DWT). Shown are the mean values of three biological replicates. ND indicates that the element was under its detection limit.

Elements Concentration (mg g^{-1} DWT)	*S. monoica*	*S. vermiculata*	*S. schimperi*
P	1.485 ^{a} ± 0.156	0.452 ^{c} ± 0.126	1.275 ^{b} ± 0.130
Ca	10.38 ^{b} ± 1.15	10.82 ^{ab} ± 1.22	12.63 ^{a} ± 1.27
Mg	4.99 ^{b} ± 0.51	6.51 ^{a} ± 0.71	6.74 ^{a} ± 0.63
K	10.04 ^{b} ± 0.87	12.69 ^{a} ± 1.33	10.55 ^{b} ± 1.19
Na	11.01 ^{b} ± 1.15	17.84 ^{a} ± 1.85	16.19 ^{a} ± 1.75
Na/K ratio	1.097 ± 0.045	1.407 ± 0.002	1.536 ± 0.009
C	322.56 ^{a} ± 33.22	245.76 ^{b} ± 24.58	265.16 ^{b} ± 24.25
N	32.04 ^{a} ± 3.65	11.96 ^{c} ± 1.60	29.79 ^{b} ± 4.03
S	0.377 ^{a} ± 0.031	0.357 ^{a} ± 0.025	N.D.
C/N ratio	10.07	20.55	9.09

Means followed by the same letter are not significantly different at the probability level of 5% according to LSD.

2.5. Variation in Photosynthetic Pigments and Carbohydrate Synthesis

Because of their importance as functional and responsive traits to salinity conditions, photosynthetic pigments were measured and compared among species (Figure 3). *Suaeda schimperi* leaves contained the highest concentration of Chl a (0.84 mg g^{-1} FWT), whereas *S. vermiculata* had the lowest (0.30 mg g^{-1} FWT) among species. *Suaeda monoica* had intermediate Chl a concentration (0.54 mg g^{-1} FWT). Relatively comparable statistical relations were obtained for Chl b and total Chl (Figure 3A). Our analysis of carbohydrate residues revealed consistently higher mean values of total soluble sugars (TSS), sucrose, starch, and total carbohydrates in *S. monoica* than the other two species. *Suaeda schimperi* maintained the lowest mean values of carbohydrate residues among species (Figure 3B).

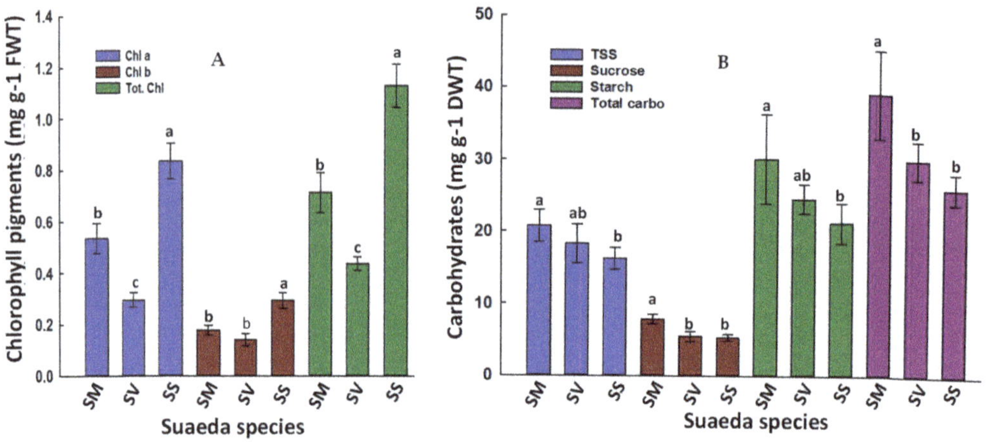

Figure 3. Leaf photosynthetic pigments and C assimilation: (**A**) Photosynthetic pigments: Chlorophyll a (Chl a), Chlorophyll b (Chl b), Total Chlorophyll (Tot. Chl); (**B**) Carbohydrates: Total soluble sugars (TSS), sucrose, starch, total carbohydrates. Shown are the mean values of three biological replicates. Means with the same letter are not significantly different at the probability level of 5% according to LSD. SM: *S. monoica*, SV: *S. vermiculata*, and SS: *S. schimperi*.

2.6. Changes in Total Soluble Proteins and Amino Acid Profiles

The tested species varied significantly in their soluble proteins and amino acids content. *Suaeda monoica* had greater leaf soluble protein (12.61 mg g^{-1} DWT) than *S. schimperi* (8.30 mg g^{-1} DWT) and *S. vermiculata* (6.65 mg g^{-1} DWT) (Figure 4A). Clear differences in leaf total amino acids were noted among species. *Suaeda monoica* (83.10 mg g^{-1} DWT) and *S. schimperi* (73.32 mg g^{-1} DWT) had more than two-fold greater total amino acids than *S. vermiculata* (30.29 mg g^{-1} DWT) (Figure 4A). No qualitative differences were observed in amino acid profiles among species, yet significant differences in the relative contribution of each amino acid to the amino acids pool of the tested species were noted (Figure 4B). *Suaeda vermiculata* consistently had the lowest concentrations of individual amino acids among species. Compared to its individual amino acid concentrations, the fold change of the corresponding amino acids ranged from 2.2 to 6.0 in *S. monoica* and from 2.05 to 4.0 in *S. schimperi*. Glutamic and aspartic acids dominated the amino acid pool across species. *Suaeda monoica* and *S. schimperi* had a relatively similar ranking pattern of amino acids within the pool; however, such a pattern was significantly disturbed in *S. vermiculata*. Interestingly, *S monoica* accumulated significantly higher proline (4.71 mg g^{-1} DWT) than *S. schimperi* (4.15 mg g^{-1} DWT) and *S. vermiculata* (1.68 mg g^{-1} DWT). These findings were coordinated with similar results of phenylalanine in *S. monoica* (4.62 mg g^{-1} DWT), *S. schimperi* (3.95 mg g^{-1} DWT), and *S. vermiculata* (1.49 mg g^{-1} DWT) (Figure 4B).

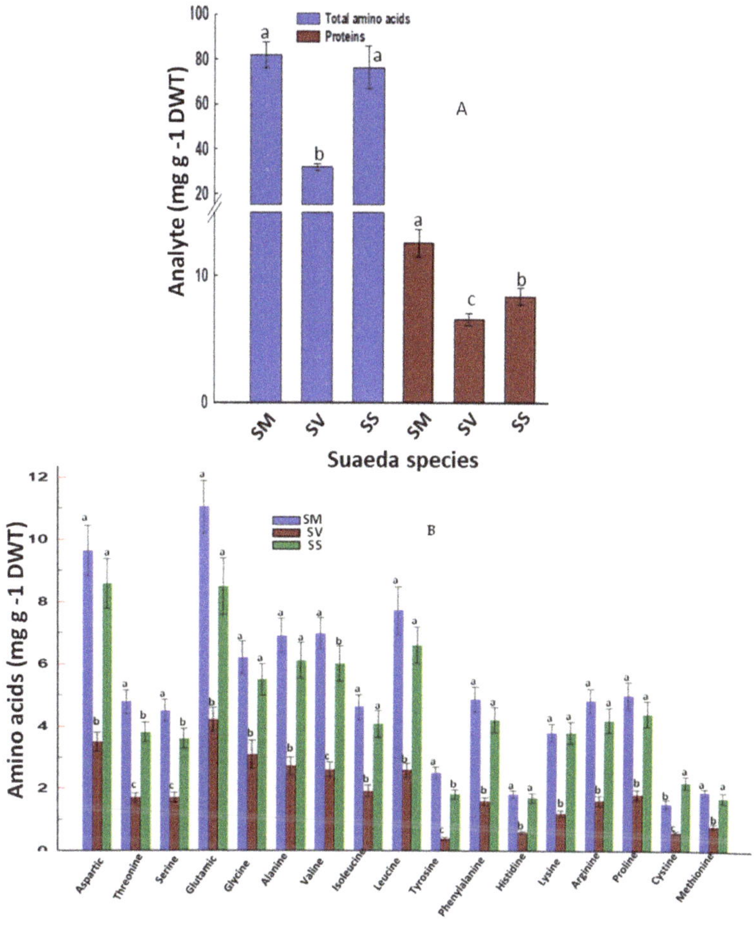

Figure 4. Leaf amino acids and total soluble proteins; (**A**) Total amino acids and total soluble proteins; (**B**) Amino acid profiles. Shown are the mean values of three biological replicates. Means with the same letter are not significantly different at the probability level of 5% according to LSD. SM: *S. monoica*, SV: *S. vermiculata*, and SS: *S. schimperi*.

2.7. Alterations in Oxidative Stress and Antioxidants Secondary Metabolites

Lipid peroxidation (Malondialdehyde, MDA) and H_2O_2 were monitored to test salinity- and PTEs-induced oxidative stress in *Suaeda* species (Figure 5). Measurements of leaf MDA revealed unexpectedly greater MDA concentration (15.55 nmol g^{-1} FWT) in *S. schimperi* than *S. monoica* (4.20 nmol g^{-1} FWT) and *S. vermiculata* (2.42 nmol g^{-1} FWT) (Figure 5A). Such responses were associated with significantly higher H_2O_2 values in *S. schimperi* (1.16 µmol g^{-1} FWT) than *S. monoica* (0.23 µmol g^{-1} FWT) and *S. vermiculata* (0.12 µmol g^{-1} FWT) (Figure 5A). The corresponding differences in the cellular levels of antioxidants such as total phenolics, flavonoids, carotenes, betacyanin, and reduced glutathione, were also measured and compared. Our analysis revealed great differences among *Suaeda* species, with *S. vermiculata* being the lowest in flavonoids and phenolic levels (Figure 5B). Compared to *S. vermiculata*, *S. monoica* and *S. schimperi* had about 2.69- and 7.25-fold higher flavonoids and 3.0- and 8.3 -fold higher phenolics, respectively. Similarly, *S. vermiculata* had significantly lower carotenoids (0.09 mg g^{-1} FWT) than *S. monoica* (0.14 mg g^{-1} FWT) and *S. schimperi* (0.25 mg g^{-1} FWT) (Figure 5C). It also had significantly lower reduced glutathione (0.67 mmol g^{-1} FWT) than *S. schimperi* (0.98 mmol g^{-1} FWT) and *S. monoica* (1.28 mmol g^{-1} FWT). On the other hand, *S. monoica* had significantly lower betacyanin (0.69 mg g^{-1} FWT) than both *S. vermiculata* (4.42 mg g^{-1} FWT) and *S. schimperi* (17.40 mg g^{-1} FWT) (Figure 5C).

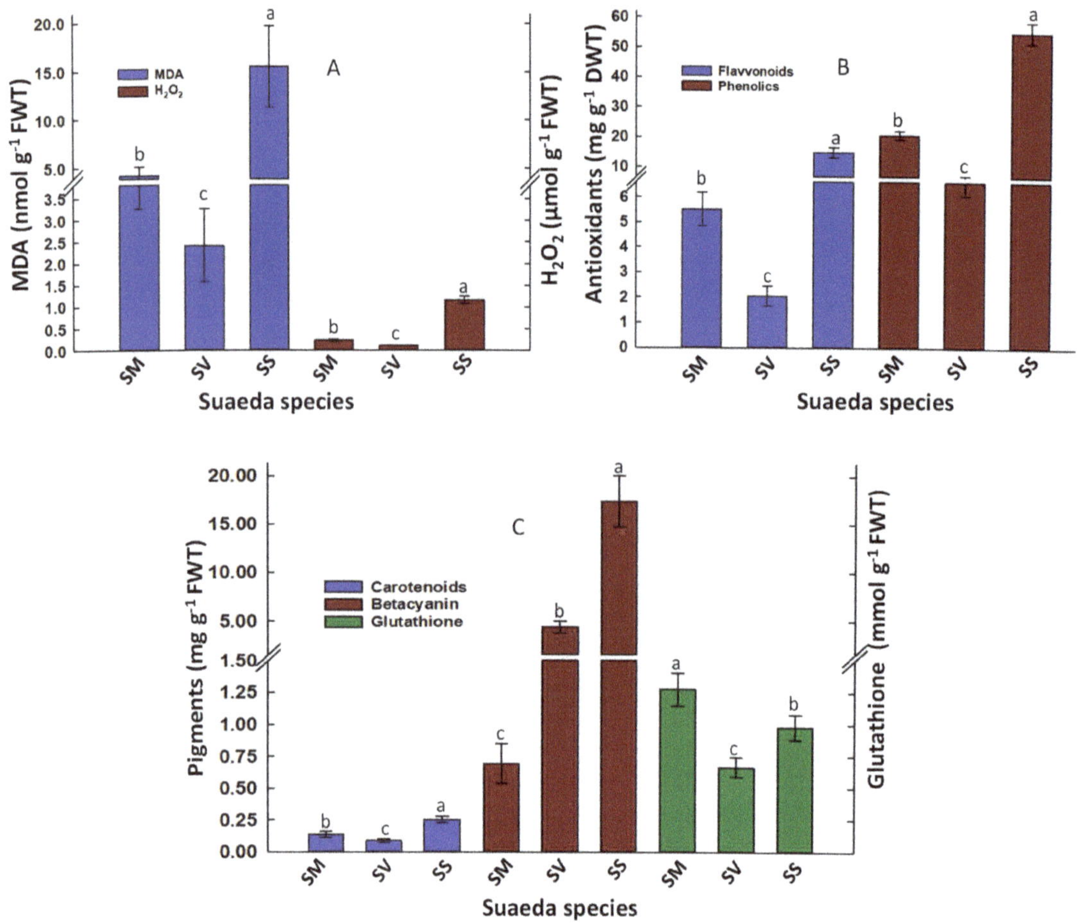

Figure 5. Oxidative stress and antioxidants indicators: (**A**) lipid peroxidation (Malondialdehyde; MDA) and hydrogen peroxide (H_2O_2); (**B**) Total flavonoids, total phenolic, and reduced glutathione; (**C**) antioxidant pigments (carotenoids and betacyanin). Shown are the mean values of three biological replicates. Means with the same letter are not significantly different at the probability level of 5% according to LSD. SM: *S. monoica*, SV: *S. vermiculata*, and SS: *S. schimperi*.

2.8. Correlation among the Tested Physiological Responses

In order to assess the correlation among the various physiological attributes with the studied *Suaeda* species, we subjected a matrix of the determined parameters to principal component analysis (PCA). *Suaeda schimperi* was segregated in the upper right side of the PCA plot, where it showed a correlation to H_2O_2, MDA, total flavonoids, total phenolics, and total chlorophyll. However, *S. monoica* was separated on the lower right side and revealed a close correlation to sucrose, starch, TSS, leaf C, and total soluble proteins. On the other hand, *S. vermiculata* was segregated in the lower left side of the PCA plot and showed a close correlation to leaf K and leaf C/N ratio (Figure 6).

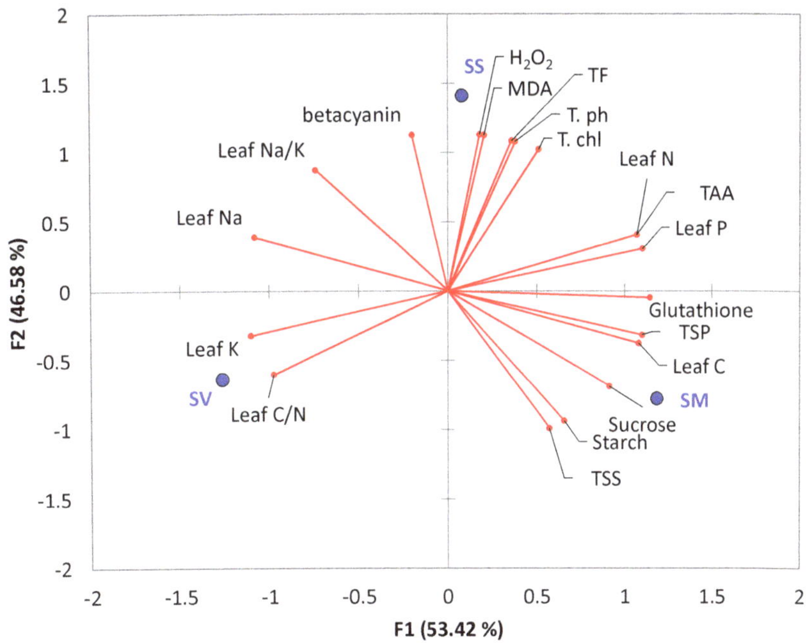

Figure 6. Biplot of principal component analysis of the monitored physiological attributes in the tested *Suaeda* species. SM: *S. monoica*, SV: *S. vermiculata*, and SS: *S. schimperi*. Abbreviated parameters are TSS: total soluble sugars, T. chl: total chlorophyll, T.AA: total amino acids, MDA: Malonaldehyde, TF: total flavonoids, Tph: total phenolics, and TSP: total soluble protein.

3. Discussion

The hyper-arid climate of the eastern coast of the Red Sea aggravates the high salinity- and PTE-induced deteriorative effects on the physiology of halophytes in the region, including *Suaeda* species. Such harmful effects are predicted to be more intense under climate change scenarios. The successful adaptation of *Suaeda* plants in such harsh environments reflects unique physiological, biochemical, and cellular adjustments that enable them to overcome salinity-induced constraints such as severe ionic, osmotic, and oxidative stresses [38]. Therefore, targeting these plants in their saline natural habitats is critical for a better understanding of their tolerance against salt stress. Herein, three *Suaeda* species that are genetically related but naturally distributed in different salt marshes in the region were selected to investigate their possible distinctive physiological adaptation against physicochemical properties of the rhizospheric soil in their natural habitats.

3.1. Soil Physicochemical Properties and the Relative Magnitudes of Salt Stress Imposed on the Investigated Suaeda Species

Soils supporting the investigated *Suaeda* species had relatively similar physical features (Table 1), which may not impose significant restrictions on their root growth. The tested soils also had low WHC values, which were correlated with high sand content that usually exhibits small cohesion forces to hold water molecules against gravity [39]. The relatively high WHC in soils supporting *S. vermiculata* and *S. schimperi* is attributed mainly to their high silt and clay and organic matter content, respectively (Table 1).

The rhizospheric soils of *S. vermiculata* and *S. schimperi* had significantly higher EC than that of *S. monoica* (Table 1). The high EC values in the tested arid saline habitats are attributed mainly to the synergistic interplay among the harsh metrological data in the region (high temperature, low rainfall, and increased evapotranspiration). The increased rate of evapotranspiration enhances the upward movement of water and its dissolved salts, which accumulate in the soil surface layers leading to hypersaline conditions and the observed white crust during summer [40]. The higher alkaline value of soil supporting *S. vermiculata* (8.65) is mainly correlated with the relatively higher silt and clay content of the soil, which encourages binding of metal ions (Na^+ in particular) with a high potential of NaH_2CO_3. Signs of that are the higher water-soluble Na^+ and HCO_3^- concentrations in soil supporting *S. vermiculata.* The obtained EC values of the rhizospheric soil are equivalent to NaCl concentrations that are beyond the stimulatory concentrations of most *Suaeda* species [41]. This higher EC value of soil supporting *S. vermiculata* is also attributed to its higher silt and clay content, which retains higher amounts of salts onto the soil matrix. The strong variations in EC, Na^+, and Cl^- among the tested rhizospheric soils indicate that the tested species experience different magnitudes of salt stress. The high soluble Mg^{2+} concentration in soil supporting *S. schimperi* is mainly due to its closeness to the Red Sea coast. The high colloidal content of soil supporting *S. vermiculata* maintained its sorption capacity of soluble ions (HCO_3^-, Na^+ and K^+).

Most soils under the tested *Suaeda* species are deficient in C, (<5%) N (<280 kg ha^{-1}), and S (<10 mg kg^{-1}) [42] (Table 2). Alike, phosphorus concentration showed low values according to soil fertility standards. Such soil nutrient deficiency could reduce leaf nutrient contents and thus may exacerbate the deleterious physiological effects of salinity. Along with the above differences in soil salinity, the rhizospheric soils exhibited qualitative and quantitative differences in the composition of the PTEs pool (Table 3). The soil supporting *S. schimperi* had the highest Co values, whereas soil supporting *S. monoica* showed the highest Hg content. The high levels of these elements might be derived from various anthropogenic activities developed along the east coast of the Red Sea.

The complex interplay between these PTEs and salinity interferes with the uptake and transport of essential elements (nutrients) and/or PTEs [38,43,44] and thus affects their cellular levels in the tested species (Table 3). Values of most PTEs showed higher levels (mg kg^{-1}) than those justified by standard regulatory bodies (WHO, FAO, and EPA): Cd (0.01), Cr (1.3), Cu (10), Fe (425), Ni (10), Pb (2), and Zn (100) [37]. Such high values of PTEs suggest a potential use of these plants for phytoremediation purposes rather than their utilization in human/animal nutrition [24,38]. Signs of that are the high bioaccumulation indices of PTEs since the studied halophytes exhibited outstanding hyperaccumulating potentials with BCR values ranging between 3.3 and 466.1 (Figure 2). These results suggest that the investigated *Suaeda* species can be used as efficient biological tools for PTEs phytoremediation in contaminated soils. The higher bioaccumulation factor of Cr could be due to the negative charge of chromate ions, which are weakly bound onto soil colloids and are easily taken up by plants.

3.2. Utilization of Inorganic Ions as "Cheap" Osmotoica in the Tested Suaeda Species

Accumulation of inorganic salts as energy cost-effective osmotica is a relevant strategy of salt tolerance in many halophytes [45,46]. In the current study, the tested species accumulated high levels of cations, particularly Na^+ and K^+ (Table 4). The active transport of such high levels of inorganic salts to leaves of the tested species may indicate a constitutive mechanism for osmotic adjustment [46]. Consistent with that, a significant contribution of the transported ions in the aerial parts to salt tolerance in *S. fruticosa* has been reported [47]. Although dicots are known for their active transport of Na^+ to their aerial shoot, maintaining a balanced Na^+/K^+ ratio in the cytoplasm is an important mechanism for salt tolerance because of the interference of Na with K uptake/accumulation and its disruptive effects on protein synthesis and activities of several cytoplasmic proteins [45,48]. In the current study, *S. monoica* had the lowest Na^+/K^+ ratio among species (Table 4), suggesting a pronounced role of such a low Na/K ratio in salt tolerance in this species similar to other halophytes such as *Limoium* species [46]. *Suaeda schimperi* and *S. vermiculata* accumulated higher Mg concentration than *S. monoica* suggesting a role of Mg in salt tolerance in these two species, possibly via interference with mRNA translation and consequently protein synthesis. Similar findings have been reported in other halophytes such as *Atriplex isatidea* and *Inula crithmoides* [49]. Interestingly, leaves of *S. schimperi* and *S. monoica* had significantly higher phosphorus than *S. vermiculata*. Along with being a major component of many primary and secondary cellular molecules [46], P has been implicated in improving salt tolerance via improving physiological mechanisms that promoted the full recovery of stressed plants [50]. It is worth mentioning that the tested species maintained relatively similar levels of Ca^{2+}, which is known for its role in maintaining Na^+ and K^+ homeostasis via the Salt Overly Sensitive pathway [51].

The observed variation in elemental concentration in leaves of the tested species was associated with their known succulence, particularly in *S. vermiculata*, which had more succulent leaves than the other two species (Figure S1). The succulence phenotype in *S. vermiculata* was associated with its higher foliar Na^+ level. These results are consistent with those of [52], who indicated that Na^+ contributes significantly to succulence phenotype. In addition, its higher foliar ionic content can contribute to building up a gradient in the osmotic potential, which allows *S. vermiculata* to take up more water to secure efficient osmotic adjustment to overcome the low external water potential. The differences in leaf succulence among the tested species may reflect comparable differences in cell size at the tissue level as well as leaf anatomy [52]. Many succulents are described as salt accumulators and are considered the most salt-tolerant because of their succulence, which enables them to have higher vacuolar concentrations of Na^+ and Cl^- than external concentrations and to avoid desiccation in dry soil [53]. Succulent species with high leaf C may improve the energy returns from carbon investment for cellular components favoring salt tolerance.

3.3. Relative Physiological Responses of the Tested Suaeda Species in Response to Their Soil Microenvironment

The variation in soil salinity and PTEs in the current study area are expected to negatively impact most aspects of plant growth and physiology. The tested species differed significantly in their C assimilation and its related physiological traits. Despite the superiority of *S. schimperi* in photosynthetic pigments (Chl a, Chl, b, and total Chl.; Figure 3A), *S. monoica* tended to have consistently higher averages TSS, sucrose, starch, total carbohydrates, and total leaf C than the other two species (Figure 3B, Table 4). The higher carbohydrates and C accumulation in *S. monoica* may reflect its higher efficiency in C assimilation/sequestration and can be attributed to several reasons: its higher leaf N and leaf P (Table 4), which are critical nutrients for gas exchange and C assimilation as well as other cellular fundamental activities [53]. A positive correlation between the foliar leaf N and/or P and photosynthetic efficiency as well as the greater stochiometric homeostasis of leaf N in N-deficient soils have been reported [53,54]. According to the PCA and correlation analyses, a significant positive correlation ($r \approx 0.99$) was recoded between leaf N with leaf P and total amino acids (Table S3, Supplementary Data). Likewise, leaf C/N exhibited a significant positive correlation with leaf P and total amino acids (0.96 and 0.90, respectively). Furthermore, the inappreciable levels of Cd, Hg, and Pb in leaves of *S. monoica* (Table 3) can have a positive cumulative effect on the efficiency of carbohydrates accumulation. In contrast, the high Cd and Hg in *S. vermiculata* and *S. schimperi* can significantly reduce their photosynthetic activity. For example, Cd forms mercaptide with the thiol group of RUBISCO protein and thus hinders its activity and consequently suppresses their photosynthetic efficiency, which explains their reduced carbohydrates accumulation [55]. It is worth mentioning that *S. monoica* had high levels of other PTEs such as Cr, Cu, and Ni; however, its higher carbohydrate content may suggest that the toxicity of these minerals was tolerated internally in *S. monoica* either by compartmentalization or binding these PTEs in less toxic forms [56,57]. The relatively low carbohydrate values in *S. vermiculata* and *S. schimperi* can be also attributed to the high salinity of their rhizospheric soil (Table 1) as well as their relatively higher leaf Na^+ (Table 4), which seems to be beyond their optimum salt range and their ability of compartmentalization into vacuoles [29] and thus accumulates in the cytoplasm and exerts its inhibitory biological effects [58–60]. Other possible reasons for the reduced carbohydrates in *S. vermiculata* are the reduced content of photosynthetic pigments (Figure 3A), low leaf-N (Table 4), reduced carbon reductive tissue area, and high foliar ABA content [61]. In fact, down regulation of carbohydrates and other biochemical and physiological processes may be a strategy that *S. vemiculata* employs to avoid the oxidative stress it encounters in its natural habitat. It is worth mentioning that, compared to both *S. monoica* and *S. schimperi*, *S. vermmiculata* had significantly higher foliar C/N ratio (Table 4), which is an indicator on improved nitrogen use efficiency (NUE) in N-deficient soil [62]. These results fit nicely with the adaptive growth hypothesis, which suggests that "plants with higher C/N ratio promote NUE under strong N-limited conditions to ensure survival priority, whereas plants with a lower C/N ratio under less N-limited environments benefit growth priority" [15]. On the other hand, the extensive consumption of carbohydrates for building up carbon skeletons of amino acids in *S schimperi* (Figure 4) and the severe oxidative stress (Figure 5A) it encounters partially explain its reduced levels of carbohydrate.

The tested species exhibited significant variation in amino acid biosynthesis, which showed a significant positive correlation with leaf P (r = 0.97) (Table S2, Supplementary Data). *Suaeda monoica* and *S. schimperi* were more efficient in amino acids biosynthesis than *S. vermiculata* (Figure 4). *Suaeda monoica* and *S. schimperi* had a relatively similar ranking of amino acids within the pool; however, such ranking was significantly disturbed in *S. vermiculata* (Figure 4B). Glutamic and aspartic acids dominated the amino acid pool across species, reflecting their "housekeeping" functions in the three species. Interestingly, *S monoica* and *S. schimperi* accumulated significantly higher foliar proline and phenylalanine contents than *S. vermiculata* suggesting a potential role of proline as an important compatible osmolyte in both *S. monoica* and *S. schimperi* but not in *S. vermiculata*. This is consistent with the recently reported low proline level in shoots of *S. vermiculata* [28]. Proline and phenylalanine are two critical amino acids under stress. The former is an important osmolyte in many halophytes, whereas the latter is the main entry point to the phenylpropanoid pathway through which flavonoid and phenolic compounds are synthesized [63]. Consistent with that, our measurements of these secondary metabolites revealed significantly higher flavonoids and phenolic in *S. schimperi* and *S. monoica* than *S. vermiculata* (Figure 5B). In addition, our analysis suggests a significant positive correlation between flavonoids and total phenolics (r = 0.97) and betacyanin (r = 0.89) (Table S3, Supplementary Data).

3.4. Relative Oxidative Stress and Antioxidants Synthesis in the Tested Suaeda Species

The tested species are exposed to oxidative stress of different severity. *Suaeda schimperi* suffers the highest stress as indicated by its highest levels of MDA (Figure 5A). Such responses revealed an imbalance between the production of ROS and radical quenchers in *S. schimperi* at the cellular level [64]. Such high MDA level in *S. schimperi* is attributed to the high salinity of its rhizospheric soils (Table 1), its higher foliar level of H_2O_2 (Figure 5A), and the simultaneous toxicity of PTEs in their leaves, particularly Hg, Cr, and Fe (Table 3). High salinity induces ROS generation via disruption of electron transport in chloroplasts and mitochondria and thus induces lipid peroxidation [65,66]. A higher level of H_2O_2 can directly induce oxidative stress because of its oxidation potential or indirectly via the generation of highly reactive hydroxyl radicals via Fenton's reaction in the presence of increased levels of transient PTEs [67]. Unfortunately, cells do not have an enzymatic system to detoxify such hydroxyl radicals [67]. Generation of leaf H_2O_2 correlates positively with leaf contents of total flavonoids and total phenolics (r = 0.97). Further, the level of Hg in *S. schimperi* leaves exceeded the toxic threshold of Hg in plants [68], thereby interfering with mitochondrial activity and triggering ROS generation and MDA accumulation [17,38]. Mercury (Hg) can also hinder water flow in *S. schimperi* via binding to water channel proteins and induction of stomatal closure [69]. Such a high level of MDA was reflected in its relatively low TSS, sucrose, starch, and total carbohydrates but not in the level of photosynthetic pigments suggesting that the salinity-induced oxidative stress affects photosynthetic activity rather than chlorophyll synthesis in *S. schimperi*. A positive significant correlation was recorded between total chlorophyll with H_2O_2 (r = 0.92), total flavonoids (r = 0.97), and total phenolics (r = 0.96) (Table S3, Supplementary Data). Unlike *S. schimperi*, the relatively low levels of MDA in leaves of *S. vermiculata* and *S. monoica* are attributed, in part, to their low levels of H_2O_2 (Figure 5A) and to their relatively high Zn and Pb (Table 3), which might induce antioxidant enzymes (CAT, SOD GPX) [70]. In addition, the high succulence in these two species (Figure S1) may also minimize the detrimental effects of salinity and PTEs and thus reduce their-induced oxidative stress. Further, the reduced oxidative stress in *S. vermiculata* leaves may also be attributed to salinity-induced guaiacol peroxidase activity and to its overall downregulated biochemical and physiological processes (Figures 3 and 4) [28].

Suaeda vermiculata and *S. schimperi* accumulated 6- and 25- fold higher betacyanin than *S. monoica*, respectively (Figure 5C), suggesting a possible role of this stress-related pigment in salt tolerance in these two species. This hypothesis is supported by the reported positive correlation between *Suaeda* leaf betacyanin content and both soil salinity and leaf H_2O_2 [71,72]. Our results are thus consistent with the physiological roles of betacyanin in the prevention of salt toxicity [72], acting as an osmotic pigment, antioxidant, and protecting halophytes against the H_2O_2-induced protein oxidation [72]. In fact, a trade-off between leaf chlorophyll and betacyanin for maintaining growth and survival in saline environments has been reported in *S. salsa* (Li et al., 2018; Wang et al., 2007) and *S. japonica* (Hayakawa and Agarie, 2010). In the current study, the high level of betacyanin (Figure 5C) and the reduced level of all chlorophyll fractions (Figure 3A) in *S. vermiculata* suggest that the betacyanin/chlorophyll trade-off scenario operates in this species to minimize the salinity-induced ROS (Figure 5A). The obtained results also pointed to a significant negative correlation between leaf Na^+ and glutathione (r = −0.93) (Table S3, Supplementary Data). In *S. schimperi*, despite its high levels of betacyanin and other measured antioxidants (carotenoids, phenolics, flavonoids, and reduced glutathione) (Figure 5B,C), it suffered from the highest magnitude of oxidative stress (Figure 5A). These results do not necessarily minimize the biological significance of these compounds in salt tolerance in *S. schimperi* but may rather indicate that the levels of these compounds are not sufficient to completely neutralize the severe salinity- and PTEs-induced ROS production in *S. schimperi* because of its very high saline rhizospheric soil (Table 1). In addition, these compounds may also be involved in other salt tolerance mechanisms such as the protection of photosynthetic pigments against degradation by salt stress in this species [73], which partially explains the high photosynthetic pigments in *S. schimperi* in the current study (Figure 3A). Therefore, *S. vermiculata* and *S. schimperi* seem to invest in betacyanin synthesis as an adaptive strategy against the severe salinity stress they encounter in their harsh environment. In fact, the leaf reddening phenotypes because of betacyanin accumulation can be easily recognized in these two species. In addition, the reduced oxidative stress in *S. vermiculata*, regardless of the high EC values and PTEs content in its rhizospheric soil, suggests that this halophyte may have optimized its growth, biochemical processes, and antioxidant defense to minimize the oxidative damage it encounters in its natural environment. This hypothesis is consistent with a recent study, which indicated that *S. vermiculata* might downregulate its biochemical and physiological processes to avoid oxidative stress [28].

4. Materials and Methods

4.1. Study Site and the Selected Suaeda Species

The current study was carried out on three *Suaeda* species (*Suaeda monoica* Forssk. ex J.F.Gmel., *Suaeda vermiculata* Forssk. ex J.F.Gmel., and *Suaeda schimperi* Moq) that naturally grow in salt marsh habitat on the eastern coast of the Red Sea at Al-Qunfudah Governorate (19°7′35.1″ N, 41°4′43.9″ E), southwest of Saudi Arabia. The habitat has typical characteristics of not flooded salt marshes. The region has a typical arid dry climate with a maximum temperature of 42.6 °C and a minimum temperature of 21 °C. In addition, the tested salt marshes receive erratic and irregular precipitation in time and quantity. Some metrological data in the study area are illustrated in Table S2 (Supplementary Data). *Suaeda monoica* is a leaf succulent 1 m height bushy shrub, whereas both *S. vermiculata* and *S. schimperi* are woody shrubs with tiny oval and cylindrical succulent leaves, respectively [53]. The *Suaeda* species were identified according to [31].

4.2. Plants and Soil Sampling

Six homogenous medium size representative plants from each of the dominant *Suaeda* species at the flowering stage in each of the selected salt marshes were marked, guarded, and used for collection of plant material for downstream analysis. From each individual plant, three batches of leaves (100 each) were collected, washed thoroughly with deionized water, blotted dry, and divided into two groups. The first group was frozen immediately in liquid N and transferred to −80 °C. The second group was collected in plastic bags and brought on ice to the laboratory, where their fresh weight was recorded and then dried in an electric oven at 70 °C until reaching constant dry weights. The dried leaves were then ground into homogeneous powder using a stainless-steel grinder and used for elemental analyses.

Soil samples were simultaneously collected with plant materials as described by [38]. The selected plants were uprooted; their roots were separated, put into sterile plastic bags, and brought to the laboratory. Roots were gently shaken to remove loosely attaching soil particles. Firmly root-adhering soil particles (rhizospheric soil) were air-dried at room temperature, passed through 2 mm sieve, and stored in sterile polyethylene bags for soil physicochemical analyses.

4.3. Soil Physicochemical and Plant Elemental Analyses

4.3.1. Soil PHYSICOCHEMICAL ANALYSES

Particle size distribution of soil was determined using sieve methods according to [74]. Water holding capacity of soil was determined using the gravimetric method, hydraulic conductivity by Darcy's law, and the soil porosity from the measured values of soil particle and bulk densities calculations [75]. Soil electrical conductivity (EC) was measured using HANNA (HI9835) EC meter in 1:2.5 soil/water extract, soil pH (1:2.5 DI water suspension) by Jenway 3505 pH/mV/Temperature Meter and the total carbonate content (expressed as $CaCO_3$) using the gasometric determination following 6.0 M HCl application [76]. Water-soluble cations and anions (1:2.5 DI water extract) were determined using standard methods [76]: Na^+ using a Sherwood, flame photometer (MODEL 360), Ca^{2+}, Mg^{2+}, and K^+ using ICP-OES Thermo Scientific™ iCAP™ 7000 Plus Series, $CO_3{}^{2-}$ and $HCO_3{}^-$ by titration with a standardized H_2SO_4 solution and Cl^- by $AgNO_3$ titration. Total organic elements concentration in soil (C, N, H, and S) was determined using dry combustion method by a Thermo Scientific Flash 2000 analyzer. Available concentrations of PTEs were determined using ICP-OES after extraction by diethylene tri-amine Penta acetic acid (DTPA).

4.3.2. Plant Analyses

Plant Elemental Analysis: Fine powdered dried leaves were used for determination of elements such as C, N, H, and S using CHNS analyzer (Thermo Scientific Flash 2000) following dry combustion technique. Other inorganic elements were determined using ICP-OES in the acid-digested leaf samples. Plant samples were digested using HCl/HNO_3 mixture (3:1 v/v) in a microwave digester (Milestone MLS 1200 Mega).

Leaf Water Content and Succulence: Leaf water content (LWC), relative to fresh weight, was calculated using the equation: LWC = [100 × (leaves fresh weight-leaves dry weight)/(leaves fresh weight)] [77]. Leaf succulence was calculated using the equation [succulence = (leaves fresh weight − leaves dry weight)/leaves dry weight] as described previously [78].

Chlorophyll Pigments: Photosynthetic pigments including chlorophyll a (Chl a), chlorophyll b (Chl b), and carotenoids in 50 mg frozen leaves were extracted in 10 mL cold aqueous acetone (80%) and their concentrations were measured spectrophotometrically at 663.2, 646.8, and 470.0 nm according to [79] and were expressed as mg g^{-1} FWT.

Determination of Betacyanin: Quantities of 500 mg powdered, frozen leaves were extracted by grinding in 20 mL cold methanol at 4 °C for 30 min. The homogenates were centrifuged at 4 °C at 10,000 rpm for 10 min. The supernatants were discarded, and the pellets were re-extracted in distilled water at 4 °C. The concentrations of betacyanin were then measured spectrophotometrically at 538 nm and calculated using the molar extraction coefficient of betacyanin of 5.66×10^4 [80].

Determination of Carbohydrate Fractions: Carbohydrate residues in 100 mg powdered dry leaves were extracted in aqueous ethanol (80%). The ethanolic extracts were collected and completed to specific volumes and used for spectrophotometric determination of total soluble sugars (TSS) and sucrose using anthrone reagent at 620 nm [81]. Starch in sugar-free plant residue was extracted in perchloric acid: water (6.5:1), and the liberated sugars were then estimated using anthrone method as described previously [82]. Carbohydrate fractions were calculated using standard curves of pure glucose and sucrose and expressed as mg g^{-1} DWT.

Determination of Total Soluble Proteins and Amino Acids: Total soluble proteins (TSP) were extracted by grinding 500 mg of frozen leaves in chilled acetone to remove pigments [83]. TSP in dry precipitates was then extracted in Tris-HCl buffer (0.05 mM, pH 9.0) and then determined using coomassie brilliant blue G 250 spectrophotometrically at 595 nm [84]. TSP concentrations were calculated using bovine serum albumin standard curve and expressed as mg g^{-1} DWT. Amino acid analysis was carried out using amino acid analyzer (Biochrom 30; Biochrom Ltd., Cambridge Science Park, Cambridge, England) as described by AOAC (2012).

Assessment of Oxidative Stress in Leaves: Lipid peroxidation and hydrogen peroxide (H_2O_2) were monitored as key indicators of oxidative damage in leaves. Lipid peroxidation was monitored as the level of malondialdehyde (MDA) using 2-thiobarbituric acid method spectrophotometrically as described previously [85] with minor modification. Frozen leaf tissues (500 mg) were extracted in 5 mL 10% trichloroacetic acid (w/v), and the homogenates were centrifuged at 4 °C for 10 min at 4000 rpm. A total of 0.5 mL of the supernatants were mixed with 0.5 mL of thiobarbituric acid (0.6%; w/v), and the mixtures were incubated at 95 °C for 15 min, cooled on ice, and centrifuged at 4 °C for 10 min at 4000 rpm. The absorbance of the pink color was measured at 450, 532, and 600 nm. The MDA concentration was calculated using the extinction coefficient of 155 and expressed as nmol g^{-1} FWT. For H_2O_2, 500 mg of powdered, frozen leaf tissues were homogenized in 5 mL cold phosphate buffer (50 mM potassium phosphate, 1 mM EDTA, pH 7.5) and centrifuged at 4 °C for 15 min at 4000 rpm. The supernatants were then collected and used for the measurement of H_2O_2 spectrophotometrically using a hydrogen peroxide assay kit (Biodiagonistic, HP 25, Giza, Egypt) according to the manufacturer's instructions.

Estimation of Antioxidant Substances: Powdered dry leaves (100 mg) were extracted in acetone to remove chlorophyll, dried, and resuspended in distilled water. Total flavonoids were measured spectrophotometrically using $AlCl_3$ reagent and quercetin as a standard at 410 nm [86]. Total phenolics in the same aqueous extracts were determined spectrophotometrically using the Folin–Ciocalteu method and gallic acid as a standard at 760 nm [87]. The concentrations of both total flavonoids and phenolics were expressed as mg g^{-1} DWT. Reduced glutathione was extracted by homogenizing 500 mg of frozen leaf tissues in 5 mL cold potassium phosphate buffer (50 mM potassium phosphate, pH 7.5, 1 mM EDTA). The homogenates were centrifuged for 15 min at 4000 rpm and 4 °C. Supernatants were then collected and used for spectrophotometric determination of reduced glutathione using Biodiagonistic kit (GR 2511) following the manufacturer's instructions.

4.4. Quality Control

Plant and soil measurements were carried out by an ISO/IEC 17025 accredited laboratory to ensure data accuracy and verification. Analytical measurements were conducted under constant temperature (25 ± 0.5 °C) with standardized protocols of replication controls and blanks. Deionized water (18.2 MΩ) (Nanopure water, Barnstead) was used for chemical solutions preparation, and all analytical grade chemical reagents (Merck-Darmstadt, Germany) were used without further purification procedures. A certified soil reference material (BIPEA, France) was used for optimization of soil analysis accuracy and verification. In addition, the accuracy of inorganic element determination was verified using Thermo Fisher Scientific standard solutions ($R^2 \geq 0.99$). Organic elements measurement was optimized by a BBOT standard ($C_{26}H_{26}N_2SO_2$): carbon (72.52%), hydrogen (6.09%), nitrogen (6.51%), and sulfur (7.44%). The recovery values of organic and inorganic elements oscillated in the range 93.6–103.4, and the data precision was justified at maximum relative standard deviation (RSD) value $\leq 5\%$. Values of limit of detection (LOD) for inorganic elements ($\mu g\ L^{-1}$) were Al (52.6), Cd (53.4), Cr (31.8), Co (37.2), Cu (20.8), Fe (46.8), Hg (23.5), Mn (29.7), Ni (40.2), Pb (54.9) and Zn (40.9).

4.5. Statistical Analysis

ANOVA analysis for the studied parameters was performed using COHORT/COSTAT software (798 Lighthouse Ave. PMB 329, Monterey, CA, USA) using the least significant difference test (LSD) at the significance level of 95%. In addition, data (Figure 2) were also presented in box and whisker plots using OriginPro 9.1: mean (dot), median (center line), lower quartile (lower border of the box), and upper quartile (upper border of the box). Pearson's correlations coefficients were applied to study the relationships among different physiological responses. Principal component analysis (PCA) was carried out using XLSTAT statistical computer software package, version 14 (Addinsoft, New York, NY, USA).

5. Conclusions

The unique genetic and physiological characteristics of halophytes support their high potential utilization as promising biological resources for improving the world's agriculture under climate change scenarios. This investigation deliberates on the premise that studying biochemical and physiological features of *Suaeda* species in their natural environments will support our planning for the future management of agricultural practices, especially in arid climate conditions. The key findings of the current investigation can be summarized as:

- *Suaeda* species are exposed to varying levels of salinity stress along with nutrient stress either as deficiency of essential nutrients such as N, K, P, or as elevated levels of PTEs.
- *Suaeda* species employ different and efficient adaptive strategies to maintain cellular homeostasis against increased levels of salinity in their rhizospheric soils.
- The high accumulation potential of PTEs, based on the bioaccumulation index of the tested *Suaeda* species, highlights their potentiality as efficient phytoextractors of soil pollutants.
- The obtained differences among the tested *Suaeda* species in the current study are driven mainly by species-specific tolerance strategies, and such specificity is shaped by the level of salinity and the genetic constitution of halophytic species.
- In essence, the obtained results of this investigation fulfill the proposed specific objectives and support the set hypotheses of the current study.

Supplementary Materials: The following are available online at https://www.mdpi.com/article/10.3390/plants11040537/s1, Table S1: Ecological Risk Assessment Parameters and Their Contamination Levels. Table S2: Mean values of Metrological data in the study area. Table S3: Correlation values among physiological attributes in the tested *Suaeda species*.

Author Contributions: Conceptualization, F.I. and A.M.; Methodology, F.I. and A.A.-Z.; software, F.I. and A.M.; validation, F.I. and A.M.; formal analysis, F.I. and A.M.; resources, F.I. and A.A.-Z.; data curation, F.I. and A.M.; writing—original draft preparation, F.I., A.A.-Z. and A.M.; writing review and editing, F.I., A.A.-Z. and A.M.; visualization, A.A.-Z.; supervision, F.I.; project administration, F.I.; funding acquisition, F.I. and A.A.-Z. All authors have read and agreed to the published version of the manuscript.

Funding: The authors would like to thank the Deanship of Scientific Research at Umm Al-Qura University for supporting this work by Grant Code: 17-SCI-1-01-0015.

Institutional Review Board Statement: Not applicable.

Informed Consent Statement: Not applicable.

Data Availability Statement: Not applicable.

Acknowledgments: The research team would like to thank Mohamed Al-Helasi and Hassan El-Shehaby for their help during samples collection and processing. The authors are also thankful to Eman Al-Ghareeb and Ahmed Abd-ElGawad for their help with statistical analysis.

Conflicts of Interest: The authors declare no conflict of interest.

References

1. Szymansky, C.-M.; Muscolo, A.; Yeo, M.; Colville, L.; Clatworthy, I.; Salge, T.; Seal, C.E. Elemental localisation and a reduced glutathione redox state protect seeds of the halophyte Suaeda maritima from salinity during over-wintering and germination. *Environ. Exp. Bot.* **2021**, *190*, 104569. [CrossRef]
2. Mahboob, S.; Ahmed, Z.; Khan, M.F.; Virik, P.; Al-Mulhm, N.; Baabbad, A.A. Assessment of heavy metals pollution in seawater and sediments in the Arabian Gulf, near Dammam, Saudi Arabia. *J. King Saud Univ. Sci.* **2021**, *34*, 101677. [CrossRef]
3. El-Sorogy, A.; Al-Kahtany, K.; Youssef, M.; Al-Kahtany, F.; Al-Malky, M. Distribution and metal contamination in the coastal sediments of Dammam Al-Jubail area, Arabian Gulf, Saudi Arabia. *Mar. Pollut. Bull.* **2018**, *128*, 8–16. [CrossRef] [PubMed]
4. Seal, C.E.; Flowers, T.J.; Esnaud, C.; Yeo, M.; Muscolo, A. Is chloride toxic to seed germination in mixed-salt environments? A case study with the coastal halophyte Suaeda maritima in the presence of seawater. *Plant Stress* **2021**, *2*, 100030. [CrossRef]
5. Geissler, N.; Hussin, S.; El-Far, M.M.; Koyro, H.-W. Elevated atmospheric CO2 concentration leads to different salt resistance mechanisms in a C3 (Chenopodium quinoa) and a C4 (Atriplex nummularia) halophyte. *Environ. Exp. Bot.* **2015**, *118*, 67–77. [CrossRef]
6. Ben Hamed, K.; Castagna, A.; Ranieri, A.; García-Caparrós, P.; Santin, M.; Hernandez, J.A.; Espin, G.B. Halophyte based Mediterranean agriculture in the contexts of food insecurity and global climate change. *Environ. Exp. Bot.* **2021**, *191*, 104601. [CrossRef]
7. Isayenkov, S.V.; Maathuis, F.J.M. Plant Salinity Stress: Many Unanswered Questions Remain. *Front. Plant Sci.* **2019**, *10*, 80. [CrossRef]
8. Benjamin, J.J.; Lucini, L.; Jothiramshekar, S.; Parida, A. Metabolomic insights into the mechanisms underlying tolerance to salinity in different halophytes. *Plant Physiol. Biochem.* **2019**, *135*, 528–545. [CrossRef]
9. Tipirdamaz, R.; Gagneul, D.; Duhazé, C.; Aïnouche, A.; Monnier, C.; Özkum, D.; Larher, F. Clustering of halophytes from an inland salt marsh in Turkey according to their ability to accumulate sodium and nitrogenous osmolytes. *Environ. Exp. Bot.* **2006**, *57*, 139–153. [CrossRef]
10. Ahmed, H.A.I.; Shabala, L.; Shabala, S. Tissue-specificity of ROS-induced K+ and Ca2+ fluxes in succulent stems of the perennial halophyte Sarcocornia quinqueflora in the context of salinity stress tolerance. *Plant Physiol. Biochem.* **2021**, *166*, 1022–1031. [CrossRef]
11. Bueno, M.; Lendínez, M.L.; Calero, J.; del Pilar Cordovilla, M. Salinity responses of three halophytes from inland saltmarshes of Jaén (southern Spain). *Flora-Morphol. Distrib. Funct. Ecol. Plants* **2020**, *266*, 151589. [CrossRef]
12. Matinzadeh, Z.; Akhani, H.; Abedi, M.; Palacio, S. The elemental composition of halophytes correlates with key morphological adaptations and taxonomic groups. *Plant Physiol. Biochem.* **2019**, *141*, 259–278. [CrossRef] [PubMed]
13. Saleem, H.; Khurshid, U.; Sarfraz, M.; Tousif, M.I.; Alamri, A.; Anwar, S.; Alamri, A.; Ahmad, I.; Abdallah, H.H.; Mahomoodally, F.M.; et al. A comprehensive phytochemical, biological, toxicological and molecular docking evaluation of Suaeda fruticosa (L.) Forssk.: An edible halophyte medicinal plant. *Food Chem. Toxicol.* **2021**, *154*, 112348. [CrossRef] [PubMed]
14. Ravi, S.; Young, T.; Macinnis-Ng, C.; Nyugen, T.V.; Duxbury, M.; Alfaro, A.C.; Leuzinger, S. Untargeted metabolomics in halophytes: The role of different metabolites in New Zealand mangroves under multi-factorial abiotic stress conditions. *Environ. Exp. Bot.* **2020**, *173*, 103993. [CrossRef]
15. Zhang, J.; He, N.; Liu, C.; Xu, L.; Chen, Z.; Li, Y.; Wang, R.; Yu, G.; Sun, W.; Xiao, C.; et al. Variation and evolution of C:N ratio among different organs enable plants to adapt to N-limited environments. *Glob. Chang. Biol.* **2020**, *26*, 2534–2543. [CrossRef]
16. Ding, Z.; Alharbi, S.; Almaroai, Y.A.; Eissa, M.A. Improving quality of metal-contaminated soils by some halophyte and non-halophyte forage plants. *Sci. Total Environ.* **2021**, *764*, 142885. [CrossRef]
17. El-Banna, M.F.; Mosa, A.; Gao, B.; Yin, X.; Wang, H.; Ahmad, Z. Scavenging effect of oxidized biochar against the phytotoxicity of lead ions on hydroponically grown chicory: An anatomical and ultrastructural investigation. *Ecotoxicol. Environ. Saf.* **2019**, *170*, 363–374. [CrossRef]
18. Flowers, T.J.; Colmer, T.D. Salinity tolerance in halophytes. *New Phytol.* **2008**, *179*, 945–963. [CrossRef]
19. Liu, R.; Cui, B.; Lu, X.; Song, J. The positive effect of salinity on nitrate uptake in Suaeda salsa. *Plant Physiol. Biochem.* **2021**, *166*, 958–963. [CrossRef]
20. Capó, X.; Tejada, S.; Ferriol, P.; Pinya, S.; Mateu-Vicens, G.; Montero-González, I.; Box, A.; Sureda, A. Hypersaline water from desalinization plants causes oxidative damage in Posidonia oceanica meadows. *Sci. Total Environ.* **2020**, *736*, 139601. [CrossRef]

21. Mu, Y.; Tang, D.; Mao, L.; Zhang, D.; Zhou, P.; Zhi, Y.; Zhang, J. Phytoremediation of secondary saline soil by halophytes with the enhancement of γ-polyglutamic acid. *Chemosphere* **2021**, *285*, 131450. [CrossRef] [PubMed]
22. Wang, L.; Wang, X.; Jiang, L.; Zhang, K.; Tanveer, M.; Tian, C.; Zhao, Z. Reclamation of saline soil by planting annual euhalophyte Suaeda salsa with drip irrigation: A three-year field experiment in arid northwestern China. *Ecol. Eng.* **2021**, *159*, 106090. [CrossRef]
23. Samundeeswari, S.; Lakshmi, S. Heavy metals phytoremediation potentiality of Suaeda maritima (L.) Dumort. and Sesuvium portulacastrum L. from influence of tannery effluents. *Int. J. Curr. Res. Life Sci.* **2018**, *7*, 1934–1941.
24. Caparrós, P.G.; Ozturk, M.; Gul, A.; Batool, T.S.; Pirasteh-Anosheh, H.; Unal, B.T.; Altay, V.; Toderich, K.N. Halophytes have potential as heavy metal phytoremediators: A comprehensive review. *Environ. Exp. Bot.* **2021**, *193*, 104666. [CrossRef]
25. Guo, J.; Du, M.; Lu, C.; Wang, B. NaCl improves reproduction by enhancing starch accumulation in the ovules of the euhalophyte Suaeda salsa. *BMC Plant Biol.* **2020**, *20*, 262. [CrossRef]
26. Wang, S.-M.; Zhang, J.; Flowers, T.J. Low-Affinity Na+ Uptake in the Halophyte Suaeda maritima. *Plant Physiol.* **2007**, *145*, 559–571. [CrossRef]
27. Zeng, Y.; Li, L.; Yang, R.; Yi, X.; Zhang, B. Contribution and distribution of inorganic ions and organic compounds to the osmotic adjustment in Halostachys caspica response to salt stress. *Sci. Rep.* **2015**, *5*, 13639. [CrossRef]
28. Al-Shamsi, N.; Hussain, M.I.; El-Keblawy, A. Physiological responses of the xerohalophyte Suaeda vermiculata to salinity in its hyper-arid environment. *Flora Morphol. Distrib. Funct. Ecol. Plants* **2020**, *273*, 151705. [CrossRef]
29. Behr, J.H.; Bouchereau, A.; Berardocco, S.; Seal, C.E.; Flowers, T.J.; Zörb, C. Metabolic and physiological adjustment of Suaeda maritima to combined salinity and hypoxia. *Ann. Bot.* **2017**, *119*, 965–976.
30. Li, Q.; Song, J. Analysis of widely targeted metabolites of the euhalophyte Suaeda salsa under saline conditions provides new insights into salt tolerance and nutritional value in halophytic species. *BMC Plant Biol.* **2019**, *19*, 388. [CrossRef]
31. Collenette, S. An Illustrated guide to the flowers of Saudi Arabia. *Brittonia* **1985**, *38*, 430–432.
32. Guo, J.; Chen, Y.; Lu, P.; Liu, M.; Sun, P.; Zhang, Z. Roles of endophytic bacteria in Suaeda salsa grown in coastal wetlands: Plant growth characteristics and salt tolerance mechanisms. *Environ. Pollut.* **2021**, *287*, 117641. [CrossRef] [PubMed]
33. Khan, M.A.; Ungar, I.A.; Showalter, A.M. The effect of salinity on the growth, water status, and ion content of a leaf succulent perennial halophyte, Suaeda fruticosa (L.) Forssk. *J. Arid. Environ.* **2000**, *45*, 73–84. [CrossRef]
34. Hayakawa, K.; Agarie, S. Physiological Roles of Betacyanin in a Halophyte, Suaeda japonicaMakino. *Plant Prod. Sci.* **2010**, *13*, 351–359. [CrossRef]
35. El-Arby, A.M.; Elbordiny, M.M. Impact of reused wastewater for irrigation on availability of heavy metals in sandy soils and their uptake by plants. *J. Appl. Sci. Res.* **2006**, *2*, 106–111.
36. Shaheen, S.M.; Shams, M.S.; Khalifa, M.R.; El-Dali, M.A.; Rinklebe, J. Various soil amendments and environmental wastes affect the (im)mobilization and phytoavailability of potentially toxic elements in a sewage effluent irrigated sandy soil. *Ecotoxicol. Environ. Saf.* **2017**, *142*, 375–387. [CrossRef]
37. Alnaimy, M.A.; Shahin, S.A.; Vranayova, Z.; Zelenakova, M.; Abdel-Hamed, E.M.W. Long-Term Impact of Wastewater Irrigation on Soil Pollution and Degradation: A Case Study from Egypt. *Water* **2021**, *13*, 2245. [CrossRef]
38. Ibraheem, F.; Al-Hazmi, N.; El-Morsy, M.; Mosa, A. Ecological Risk Assessment of Potential Toxic Elements in Salt Marshes on the East Coast of the Red Sea: Differential Physiological Responses and Adaptation Capacities of Dominant Halophytes. *Sustainability* **2021**, *13*, 11282. [CrossRef]
39. Mosa, A.; Taha, A.A.; Elsaeid, M. In-situ and ex-situ remediation of potentially toxic elements by humic acid extracted from different feedstocks: Experimental observations on a contaminated soil subjected to long-term irrigation with sewage effluents. *Environ. Technol. Innov.* **2021**, *23*, 101599. [CrossRef]
40. Hussain, M.; Al-Shaibani, A.; Al-Ramadan, K.; Wood, W.W. Geochemistry and isotopic analysis of brines in the coastal sabkhas, Eastern region, Kingdom of Saudi Arabia. *J. Arid Environ.* **2020**, *178*, 104142. [CrossRef]
41. Wang, F.; Xu, Y.-G.; Wang, S.; Shi, W.; Liu, R.; Feng, G.; Song, J. Salinity affects production and salt tolerance of dimorphic seeds of Suaeda salsa. *Plant Physiol. Biochem.* **2015**, *95*, 41–48. [CrossRef] [PubMed]
42. Patil, P.L.; Bidari, B.I.; Hebbara, M.; Katti, J.; Dilvaranaik, S.; Vishwanatha, S.; Geetanjali, H.M.; Dasog, G.S. Identification of soil fertility constraints by GIS in Bedwatti sub watershed under Northern dry zone of Karnataka for site specific recommendations. *J. Farm Sci.* **2017**, *30*, 206–211.
43. El-Naggar, A.; Ahmed, N.; Mosa, A.; Niazi, N.K.; Yousaf, B.; Sharma, A.; Sarkar, B.; Cai, Y.; Chang, S.X. Nickel in soil and water: Sources, biogeochemistry, and remediation using biochar. *J. Hazard. Mater.* **2021**, *419*, 126421. [CrossRef] [PubMed]
44. Moghanm, F.S.; El-Banna, A.; El-Esawi, M.A.; Abdel-Daim, M.M.; Mosa, A.; Abdelaal, K.A. Genotoxic and Anatomical Deteriorations Associated with Potentially Toxic Elements Accumulation in Water Hyacinth Grown in Drainage Water Resources. *Sustainability* **2020**, *12*, 2147. [CrossRef]
45. Al Hassan, M.; Estrelles, E.; Soriano, P.; López-Gresa, M.P.; Bellés, J.M.; Boscaiu, M.; Vicente, O. Unraveling Salt Tolerance Mechanisms in Halophytes: A Comparative Study on Four Mediterranean Limonium Species with Different Geographic Distribution Patterns. *Front. Plant Sci.* **2017**, *8*, 1438. [CrossRef]
46. González-Orenga, S.; Ferrer-Gallego, P.P.; Laguna, E.; López-Gresa, M.P.; Donat-Torres, M.P.; Verdeguer, M.; Vicente, O.; Boscaiu, M. Insights on Salt Tolerance of Two Endemic Limonium Species from Spain. *Metabolites* **2019**, *9*, 294. [CrossRef]

47. Gil, R.; Bautista, I.; Boscaiu, M.; Lidón, A.L.; Wankhade, S.; Sánchez, H.; Llinares, J.; Vicente, O. Responses of five Mediterranean halophytes to seasonal changes in environmental conditions. *AoB Plants* **2014**, *6*, plu049. [CrossRef]
48. Assaha, D.V.M.; Ueda, A.; Saneoka, H.; Al-Yahyai, R.; Yaish, M.W. The role of Na+ and K+ transporters in salt stress adaptation in glycophytes. *Front. Physiol.* **2017**, *8*, 509. [CrossRef]
49. Flowers, T.J.; Dalmond, D. Protein synthesis in halophytes: The influence of potassium, sodium and magnesium in vitro. *Genet. Asp. Plant Miner. Nutr.* **1993**, *146*, 195–203.
50. Cocozza, C.; Brilli, F.; Pignattelli, S.; Pollastri, S.; Brunetti, C.; Gonnelli, C.; Tognetti, R.; Centritto, M.; Loreto, F. The excess of phosphorus in soil reduces physiological performances over time but enhances prompt recovery of salt-stressed Arundo donax plants. *Plant Physiol. Biochem.* **2020**, *151*, 556–565. [CrossRef]
51. Mahajan, S.; Pandey, G.K.; Tuteja, N. Calcium- and salt-stress signaling in plants: Shedding light on SOS pathway. *Arch. Biochem. Biophys.* **2008**, *471*, 146–158. [CrossRef] [PubMed]
52. Zaier, M.M.; Ciudad-Mulero, M.; Cámara, M.; Pereira, C.; Ferreira, I.C.; Achour, L.; Kacem, A.; Morales, P. Revalorization of Tunisian wild Amaranthaceae halophytes: Nutritional composition variation at two different phenotypes stages. *J. Food Compos. Anal.* **2020**, *89*, 103463. [CrossRef]
53. Wang, L.; Zhao, G.; Lilong, W.; Zhang, M.; Zhang, L.; Zhang, X.; Guanxiang, Z.; Xu, S. C:N:P Stoichiometry and Leaf Traits of Halophytes in an Arid Saline Environment, Northwest China. *PLoS ONE* **2015**, *10*, e0119935. [CrossRef] [PubMed]
54. Alshaal, T.; El-Ramady, H.; Al-Saeedi, A.H.; Shalaby, T.; Elsakhawy, T.; Omara, A.E.D.; Gad, A.; Hamad, E.; El-Ghamry, A.; Mosa, A. The rhizosphere and plant nutrition under climate change. In *Essential Plant Nutrients*; Springer: Berlin, Germany, 2017; pp. 275–308.
55. Bagheri, R.; Ahmad, J.; Bashir, H.; Iqbal, M.; Qureshi, M.I. Changes in rubisco, cysteine-rich proteins and antioxidant system of spinach (Spinacia oleracea L.) due to sulphur deficiency, cadmium stress and their combination. *Protoplasma* **2017**, *254*, 1031–1043. [CrossRef] [PubMed]
56. Mosa, A.; El-Banna, M.F.; Gao, B. Biochar filters reduced the toxic effects of nickel on tomato (Lycopersicon esculentum L.) grown in nutrient film technique hydroponic system. *Chemosphere* **2016**, *149*, 254–262. [CrossRef]
57. El-Banna, M.F.; Mosa, A.; Gao, B.; Yin, X.; Ahmad, Z.; Wang, H. Sorption of lead ions onto oxidized bagasse-biochar mitigates Pb-induced oxidative stress on hydroponically grown chicory: Experimental observations and mechanisms. *Chemosphere* **2018**, *208*, 887–898. [CrossRef]
58. Flowers, T.; Yeo, A. Ion Relations of Plants Under Drought and Salinity. *Funct. Plant Biol.* **1986**, *13*, 75–91. [CrossRef]
59. Munns, R. Comparative physiology of salt and water stress. *Plant, Cell Environ.* **2002**, *25*, 239–250. [CrossRef]
60. Flowers, T.J.; Colmer, T.D. Plant salt tolerance: Adaptations in halophytes. *Ann. Bot.* **2015**, *115*, 327–331. [CrossRef]
61. Thalmann, M.; Santelia, D. Starch as a determinant of plant fitness under abiotic stress. *New Phytol.* **2017**, *214*, 943–951. [CrossRef]
62. Xu, S.; Sardans, J.; Zhang, J.; Peñuelas, J. Variations in foliar carbon:nitrogen and nitrogen:phosphorus ratios under global change: A meta-analysis of experimental field studies. *Sci. Rep.* **2020**, *10*, 12156. [CrossRef]
63. El Moukhtari, A.; Cabassa-Hourton, C.; Farissi, M.; Savouré, A. How Does Proline Treatment Promote Salt Stress Tolerance During Crop Plant Development? *Front. Plant Sci.* **2020**, *11*, 1127. [CrossRef]
64. Bhaduri, A.M.; Fulekar, M.H. Antioxidant enzyme responses of plants to heavy metal stress. *Rev. Environ. Sci. Bio/Technol.* **2012**, *11*, 55–69. [CrossRef]
65. Nxele, X.; Klein, A.; Ndimba, B. Drought and salinity stress alters ROS accumulation, water retention, and osmolyte content in sorghum plants. *South Afr. J. Bot.* **2017**, *108*, 261–266. [CrossRef]
66. Hasanuzzaman, M.; Raihan, R.H.; Masud, A.A.C., Rahman, K.; Nowroz, F.; Rahman, M.; Nahar, K.; Fujita, M. Regulation of Reactive Oxygen Species and Antioxidant Defense in Plants under Salinity. *Int. J. Mol. Sci.* **2021**, *22*, 9326. [CrossRef]
67. Bose, J.; Rodrigo-Moreno, A.; Shabala, S. ROS homeostasis in halophytes in the context of salinity stress tolerance. *J. Exp. Bot.* **2014**, *65*, 1241–1257. [CrossRef] [PubMed]
68. Zhou, Z.S.; Huang, S.Q.; Guo, K.; Mehta, S.K.; Zhang, P.C.; Yang, Z.M. Metabolic adaptations to mercury-induced oxidative stress in roots of Medicago sativa L. *J. Inorg. Biochem.* **2007**, *101*, 1–9. [CrossRef]
69. Tang, B.; Chen, J.; Wang, Z.; Qin, P.; Zhang, X. Mercury accumulation response of rice plant (Oryza sativa L.) to elevated atmospheric mercury and carbon dioxide. *Ecotoxicol. Environ. Saf.* **2021**, *224*, 112628. [CrossRef] [PubMed]
70. Wu, H.; Liu, X.; Zhao, J.; Yu, J. Regulation of Metabolites, Gene Expression, and Antioxidant Enzymes to Environmentally Relevant Lead and Zinc in the Halophyte Suaeda salsa. *J. Plant Growth Regul.* **2013**, *32*, 353–361. [CrossRef]
71. Wang, C.-Q.; Chen, M.; Wang, B.-S. Betacyanin accumulation in the leaves of C3 halophyte Suaeda salsa L. is induced by watering roots with H_2O_2. *Plant Sci.* **2007**, *172*, 1–7. [CrossRef]
72. Li, Y.; Cui, L.; Yao, X.; Ding, X.; Pan, X.; Zhang, M.; Li, W.; Kang, X. Trade-off between leaf chlorophyll and betacyanins in Suaeda salsa in the Liaohe estuary wetland in northeast China. *J. Plant Ecol.* **2018**, *11*, 569–575. [CrossRef]
73. Mane, A.V.; Karadge, B.A.; Samant, J.S. Salinity induced changes in photosynthetic pigments and polyphenols of Cymbopogon nardus (L.) Rendle. *J. Chem. Pharm. Res* **2010**, *2*, 338–347.
74. Gee, G.W.; Or, D. 2.4 Particle-size analysis. *Methods Soil Analysis. Part* **2002**, *4*, 255–293.
75. Dewis, J.; Freitas, F. Physical and chemical methods of soil and water analysis. *FAO Soils Bull.* **1970**, *10*, 275.
76. Jackson, M.L. *Soil Chemical Analysis: Advanced Course*; UW-Madison Libraries Parallel Press: Madison, WI, USA, 2005.

77. Ullah, S.; Skidmore, A.K.; Groen, T.A.; Schlerf, M. Evaluation of three proposed indices for the retrieval of leaf water content from the mid-wave infrared (2–6 μm) spectra. *Agric. For. Meteorol.* **2013**, *171–172*, 65–71. [CrossRef]
78. Dehan, K.; Tal, M. Salt tolerance in the wild relatives of the cultivated tomato: Responses of Solanum pennellii to high salinity. *Irrig. Sci.* **1978**, *1*, 71–76. [CrossRef]
79. Vernon, L.P. Spectrophotometric determination of chlorophylls and pheophytins in plant extracts. *Anal. Chem.* **1960**, *32*, 1144–1150. [CrossRef]
80. Skalicky, M.; Kubes, J.; Shokoofeh, H.; Tahjib-Ul-Arif, M.; Vachova, P.; Hejnak, V. Betacyanins and Betaxanthins in Cultivated Varieties of Beta vulgaris L. Compared to Weed Beets. *Molecules* **2020**, *25*, 5395. [CrossRef]
81. Hansen, J.; Møller, I. Percolation of starch and soluble carbohydrates from plant tissue for quantitative determination with anthrone. *Anal. Biochem.* **1975**, *68*, 87–94. [CrossRef]
82. Sadasivam, S.; Manickam, A. Determination of total carbohydrate by anthrone method. In *Bochemical Methods*; New age International (P) Limited: New Delhi, India, 1996; pp. 8–9.
83. Yadav, S.; Srivastava, A.; Biswas, S.; Chaurasia, N.; Singh, S.K.; Kumar, S.; Srivastava, V.; Mishra, Y. Comparison and optimization of protein extraction and two-dimensional gel electrophoresis protocols for liverworts. *BMC Res. Notes* **2020**, *13*, 60. [CrossRef]
84. Bradford, M.M. A rapid and sensitive method for the quantitation of microgram quantities of protein utilizing the principle of protein-dye binding. *Anal. Biochem.* **1976**, *72*, 248–254. [CrossRef]
85. Ouyang, S.-Q.; Liu, Y.-F.; Liu, P.; Lei, G.; He, S.-J.; Ma, B.; Zhang, W.-K.; Zhang, J.-S.; Chen, S.-Y. Receptor-like kinase OsSIK1 improves drought and salt stress tolerance in rice (Oryza sativa) plants. *Plant J.* **2010**, *62*, 316–329. [CrossRef] [PubMed]
86. Zhishen, J.; Mengcheng, T.; Jianming, W. The determination of flavonoid contents in mulberry and their scavenging effects on superoxide radicals. *Food Chem.* **1999**, *64*, 555–559. [CrossRef]
87. Alonsoborbalan, A.; Zorro, L.; Guillen, D.; Barroso, C.G. Study of the polyphenol content of red and white grape varieties by liquid chromatography–mass spectrometry and its relationship to antioxidant power. *J. Chromatogr. A* **2003**, *1012*, 31–38. [CrossRef]

plants

MDPI

Article

Bioactive Compounds in *Sarcocornia* and *Arthrocnemum*, Two Wild Halophilic Genera from the Iberian Peninsula

Irene Sánchez-Gavilán, Esteban Ramírez Chueca and Vicenta de la Fuente García *

Departamento de Biología, Facultad de Ciencias, Universidad Autónoma de Madrid, Cantoblanco, 28049 Madrid, Spain; irene.sanchezgavilan@estudiante.uam.es (I.S.-G.); esteban.ramirez@uam.es (E.R.C.)
* Correspondence: vicenta.fuente@uam.es

Abstract: (1) Background: this study describes bioactive compounds in the following halophytes: *Sarcocornia* (*S. alpini*, *S. pruinosa*, and *S. perennis*) and *Arthrocnemum* (*A. macrostachyum*). The material comes from: coastal marshes in Tinto River, Guadiana River, and some interior provinces from the Iberian Peninsula. (2) Methods: the techniques used were Folin–Ciocalteu, GC-MS, and ESI-MS/MS. (3) Results: Five phenolic acids were found in *Sarcocornia*: trans-cinnamic, salicylic, veratric, coumaric, and caffeic acids. In addition, in *Arthronemum*, ferulic acid was also detected. The obtained flavonoids were cyanidin-3-O-arabinoside, luteolin-7-glucoside, dihydroquercetin, and p-coumaroyl-glucoside. They also presented fatty acids, such as palmitic, linoleic, and oleic acids in *Sarcocornia*, while palmitic, linolenic, and stearic acids were the main fatty acids in *A. macrostachyum*. (4) Conclusions: the high diversity of the compounds identified confirms the relation between nutritional interest and salt tolerance in halophytes.

Keywords: halophytes; salt tolerance; bioactive compounds; flavonoids; fatty acids

Citation: Sánchez-Gavilán, I.; Ramírez Chueca, E.; de la Fuente García, V. Bioactive Compounds in *Sarcocornia* and *Arthrocnemum*, Two Wild Halophilic Genera from the Iberian Peninsula. *Plants* **2021**, *10*, 2218. https://doi.org/10.3390/plants10102218

Academic Editors: Oscar Vicente and Marius-Nicusor Grigore

Received: 22 September 2021
Accepted: 15 October 2021
Published: 19 October 2021

1. Introduction

The genera *Arthrocnemum* Moq. and *Sarcocornia* A.J. Scott (Chenopodiaceae/Amaranthaceae) include succulent chamaephytes that are specialized in the colonization of saline habitats. In European and North African Mediterranean territories, the following taxa occur: *Arthrocnemum macrostachyum* (Moric.) K. Koch; *A. meridionale* Ramírez, Rufo, Sánchez-Mata, and Fuente; *Sarcocornia hispanica* Fuente, Rufo, and Sánchez-Mata; and *S. alpini* (Lag.) Rivas-Martínez, *S. carinata* Fuente, Rufo, Sánchez-Mata, and *S. fruticosa* (L.) A.J. Scott. In contrast, *S. perennis* (Mill) A.J. Scott, *S. pruinosa* Fuente, Rufo, and Sánchez-Mata are limited to the European Atlantic coasts [1–6].

The Chenopodiaceae species are generally characterized by a high content of minerals, polyphenols, and fatty acids, among other compounds of interest. The abundance of inorganic elements (Na^+, K^+, Mg^{2+}, and Ca^{2+}, among others) in the tissues of these plants, together with the wide diversity of bioactive compounds, have been related to their capacity to survive and grow in extreme environments with high salinity and long periods of intense drought [7–10].

Arthrocnemum macrostachyum has recently been used in soil desalination programs [11] due to its capacity to accumulate high concentrations of sodium chloride in its tissues and hence to reduce it in the cultivation medium. El-Naker et al. [12] recorded the presence of a wide range of phytochemical compounds in this genus and identified sixteen that were potentially bioactive, some of which have antioxidant (quercetin, 4-hydroxybenzoic, and caffeic acids), antiviral, antibacterial, and/or anti-tumoral properties (hesperidin, salicylic, chlorogenic, and coumaric acids), including compounds for the treatment of diabetes (rhamnetin).

Several species of *Sarcocornia* have been evaluated as edible plants due to their different nutritional properties, particularly including their antioxidant capacity and lipid composition. Riquelme et al. [13] characterized different phenolic compounds in *Sarcocornia neei* (Lag.) M.A. Alonso and M.B. Crespo, such as kaempferol and quercetin, as well as gallic, ferulic, and coumaric acids, among others. Barreira et al. [8] analyzed the fatty acid profile of *Sarcocornia* spp. collected in the Algarve (Portugal) and reported a predominance of palmitic, linolenic, and linoleic acids. These same authors also detected and quantified greater quantities of these fatty acids in *Arthrocnemum macrostachyum* from the marshes of Praia de Faro in the south of the country. The sustained implementation of the potential of both genera as emerging quality crops began in the 1990s and has continued to the present [14–19].

Our selection of the genera *Sarcocornia* and *Arthrocnemum* was guided by the importance and interest of halophytes in today's agriculture. There are two factors that make halophytes of special interest to be considered: First, their economic potential, considering their productivity in high-salinity and low-water intake environments, and second, their nutritional value in terms of protein, phenolic, and lipid contents, and the great quantity of minerals such as iron, potassium, calcium, and magnesium, as well as other bioactive compounds [8,10]. Samples of the species *S. alpini*, *S. pruinosa*, and *S. perennis* of the genus *Sarcocornia*, which were all from the southwestern Iberian Peninsula (Spain and Portugal), were analyzed. Within the province of Huelva (Spain), the largest number of samples studied originated from a special area of the marshland influenced by the Tinto River. This territory has an abundance of natural heavy metals (especially Cu, Zn, Cr, and Fe) and a slightly acidic pH (6.27–6.35), specifically in the estuarine area that runs from San Juan del Puerto to the river's mouth, together with the Odiel river in the Atlantic Ocean [20,21]. Additionally, in this area, the three species of the genus *Sarcocornia* occupy and dominate a large part of the vegetation of the marshes in an ecological gradient strongly marked by the greater or lesser proximity to the sea, as well as by the dryness of the soil: *S. perennis* occurs in the first vegetation band, almost constantly submerged by the tides; *S. pruinosa* appears in an upper band, occasionally influenced by the tides; and, finally, *S. alpini* dominates in soils that are further away from the tidal influence and are drier [4]. In turn, *A. macrostachyum* also dominates in the southwestern Iberian region, in the driest salt marshes with the highest saline concentration and in an ecological zone farther from the sea [22]. Additionally, samples of *A. macrostachyum* from the interior of the Iberian Peninsula have also been analyzed; these substrates undergo strong summer desiccation with the outcrop of saline efflorescence, in addition to increasing their Ca cation content [23]. It has been proven that the greater or lesser exposure to saline conditions of halophytes in their natural environment has an influence on the content of the bioactive phytochemicals present in them, especially as a protection mechanism against the oxidizing agents produced in these extreme environmental conditions [24]. There is great interest in studying the role that compounds play in the adaptation of halophytes to these environments. Thus, the main objective of this work was to determine the bioactive compounds (phenolic compounds and fatty acids) in various species of *Sarcocornia* (*S. alpini*, *S. pruinosa*, and *S. perennis*) and *Arthrocnemum* (*A. macrostachyum*) in coastal and inland saltmarshes of the Iberian Peninsula.

2. Results

2.1. Total Phenolic Compounds (TPC) and Phenolic Acids

Our values for the total phenolic compounds in the genus *Sarcocornia* expressed as gallic acid equivalent (G.A.E.) were between 3.892 mg G.A.E./g plant dw (dry weight) and 3.231 mg G.A.E./g plant dw (Table 1). The phenolic acids found in the species of *Sarcocornia* and *Arthrocnemum* were benzoic acids (salicylic and veratric) and hydroxycinnamic acids (trans-cinnamic, caffeic, coumaric, and ferulic). All the material from the genus *Sarcocornia* presented trans-cinnamic acid, which is the most frequent and abundant compound (Table S1, Figures 1–3).

Table 1. Data on the sample weight and total phenolic compounds (TPC ± SD (mg G.A.E./g plant dw)) for dry material and humidity. Note: gallic acid equivalent (G.A.E.) and standard deviation (SD). $n = 3$.

ID	Sample	TPC ± SD	Humidity
1	*S. alpini*	3.611 ± 0.107	10.23%
2	*S. alpini*	3.569 ± 0.233	10.59%
3	*S. alpini*	3.778 ± 0.231	11.80%
4	*S. alpini*	3.480 ± 0.164	10.40%
5	*S. alpini*	3.430 ± 0.093	11.30%
6	*S. pruinosa*	3.892 ± 0.203	10.18%
7	*S. pruinosa*	3.879 ± 0.207	10.53%
8	*S. pruinosa*	3.404 ± 0.198	14.78%
9	*S. pruinosa*	3.453 ± 0.064	14.44%
10	*S. pruinosa*	3.299 ± 0.156	13.17%
11	*S. pruinosa*	3.231 ± 0.089	13.65%
12	*S. perennis*	3.407 ± 0.004	10.29%
13	*S. perennis*	3.420 ± 0.139	13.16%
14	*S. perennis*	3.460 ± 0.014	12.35%
15	*A. macrostachyum*	4.680 ± 0.036	10.35%
16	*A. macrostachyum*	4.891 ± 0.060	11.80%
17	*A. macrostachyum*	4.803 ± 0.096	11.36%
18	*A. macrostachyum*	4.220 ± 0.014	10.55%
19	*A. macrostachyum*	4.850 ± 0.012	13.71%
20	*A. macrostachyum*	4.770 ± 0.005	13.64%

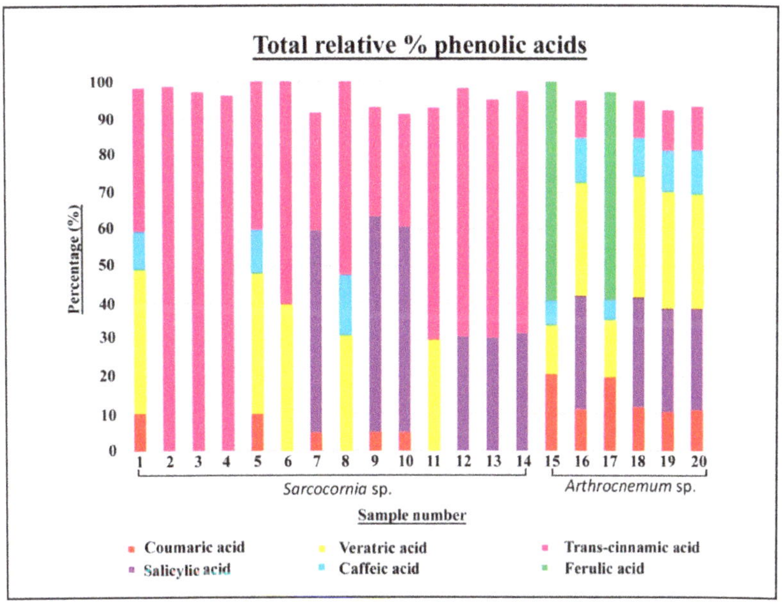

Figure 1. Total relative percentages of the phenolic acids.

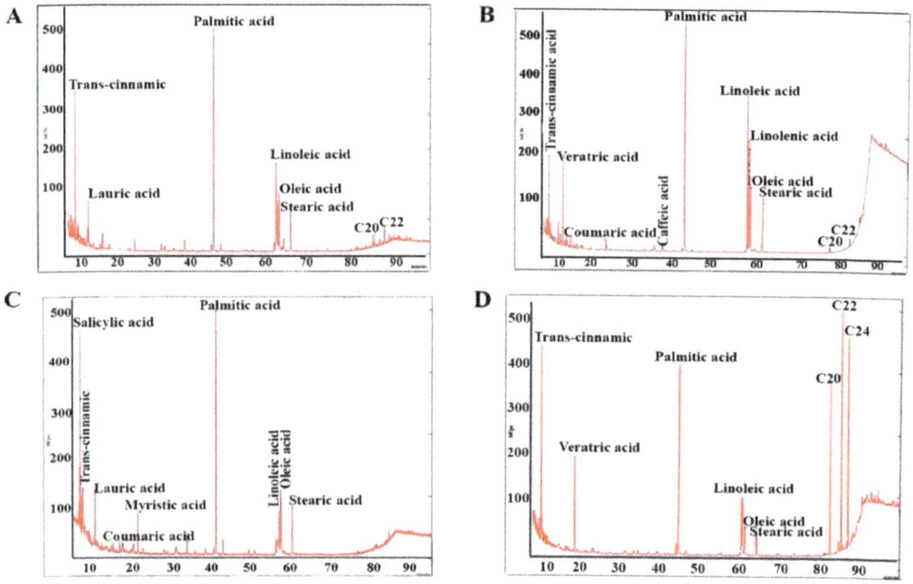

Figure 2. Chromatograms of the samples: phenolic acids and fatty acids in *S. alpini* and *S. pruinosa*. (**A**) *S. alpini* samples 2, 3, and 4. (**B**) *S. alpini* samples 1 and 5. (**C**) *S. pruinosa* samples 7, 9, and 10. (**D**) *S. pruinosa* samples 6 and 8.

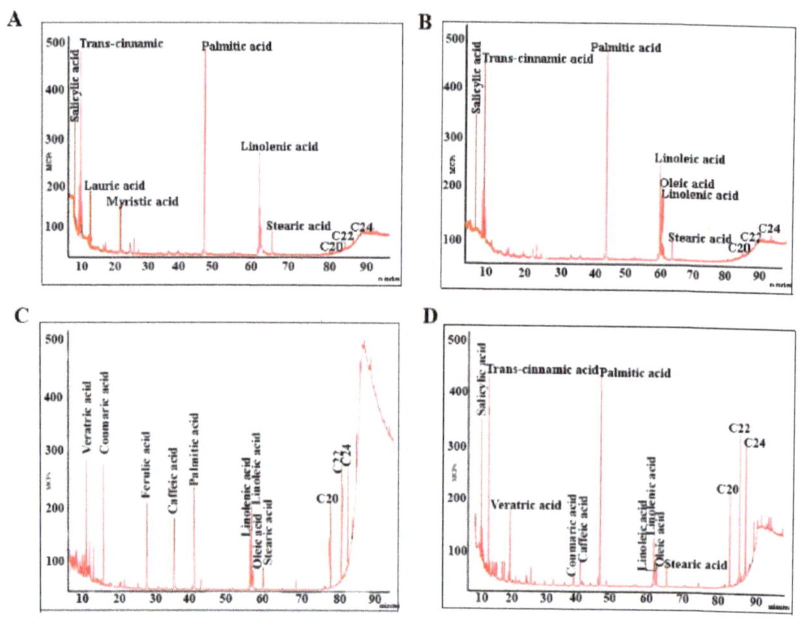

Figure 3. Chromatograms of the samples studied: phenolic acids and fatty acids in *S. perennis* and *A. macrostachyum* (**A**) *S. perennis* sample 12. (**B**) *S. perennis* samples 13 and 14. (**C**) *A. macrostachyum* samples 15 and 17. (**D**) *A. macrostachyum* samples 16 and 18–20.

Sample 1 of *S. alpini* presented veratric acid and trans-cinnamic acid with 39% of relative content, while coumaric and caffeic acid accounted for around 10%. For *S. alpini* material corresponding to samples 2, 3, and 4, only trans-cinnamic acid was detected between 96% and 98%.

In *S. pruinosa*, sample 6 had a content of 60% of trans-cinnamic acid and 39% of veratric acid. Sample 7 contained 60% salicylic acid, 32% trans-cinnamic acid, and a minority of coumaric acid with 4%. Samples 9 and 10 had salicylic acid between 56% and 58%, trans-cinnamic acid with 30%, and, finally, 4% coumaric acid. Sample 11 presented 62% trans-cinnamic acid and 30% veratric acid.

All the material of *S. perennis* presented trans-cinnamic acid between 64% and 67%, and salicylic acid between 30% and 32%.

The total phenolic compounds in *A. macrostachyum* were between 4.891 mg G.A.E./g plant dw and 4.220 mg G.A.E./g plant dw.

Sample 15 contained 59% ferulic acid, which was notable, 20% coumaric acid, 13% veratric acid, and 5% caffeic acid. Sample 17 had a content of 56% ferulic acid, 19% coumaric acid, 15% veratric acid, and 6% caffeic acid.

Samples 16, 18, 19, and 20 had salicylic acid and veratric acid at 30%, in addition to caffeic acid and trans-cinnamic acid, both at 10%. These samples did not contain ferulic acid.

2.2. Flavonoids and Hydroxycinnamic Acids

All the samples of *Sarcocornia* and *Arthrocnemum* studied contained luteolin and this was the only flavonoid present in *S. perennis*. Cyanidin-3-O-arabinoside and luteolin-7-glucoside (Table 2 and Table S3) were found in *S. alpini*, while *S. pruinosa* contained dihydroquercetin and p-Coumaroyl tyrosine. *A. macrostachyum* contained dihydroquercetin and p-Coumaroyl-glucoside. The chemical structures of these compounds are shown in Figure S1.

Table 2. Tentative identification of flavonoids in *Sarcocornia* and *Arthrocnemum*.

Species	Flavonoid Compound	Experimental Mass M-H m/z	MS/MS (m/z)
S. alpini, S. pruinosa, S. perennis, A. macrostachyum	Luteolin	287	285/290
S. alpini	Cyanidin-3-O-arabinoside	419	415/422
	Luteolin-7-glucoside	448	446/450
S. pruinosa and A. macrostachyum	Dihydroquercetin	304	303/310
S. pruinosa	p-Coumaroyl tyrosine	327	322/330
A. macrostachyum	p-Coumaroyl-glucoside	326	322/330

2.3. Fatty Acids

Our results show that the total proportion of saturated fatty acids in the genus *Sarcocornia* represented a mean of 61.5% relative percentage, with lower proportions of monounsaturated fatty acids at 2.7% (Table S2, Figures 2–4). Polyunsaturated fatty acids were at 19.20% relative percentage.

In the genus *Arthrocnemum*, saturated fatty acids represented a mean of 65.2% relative percentage, monounsaturated fatty acids accounted for 7.8% relative percentage, and polyunsaturated fatty acids were 24.1% relative percentage.

Among the saturated fatty acids, the palmitic acid was notable, which was present between 30% and 20% relative percentage in all samples of both *Sarcocornia* and *Arthrocnemum*; this was also the case for stearic acid but with lower percentages between 20% and 10% relative percentage.

Lauric and myristic acids were found only in material from the genus *Sarcocornia*. Lauric acid was notable in *S. pruinosa*, *S. alpini*, and *S. perennis*, with a relative content of between 15% and 8% relative percentage. Myristic acid was detected in *S. alpini* and *S. perennis* with a content of over 10% relative percentage.

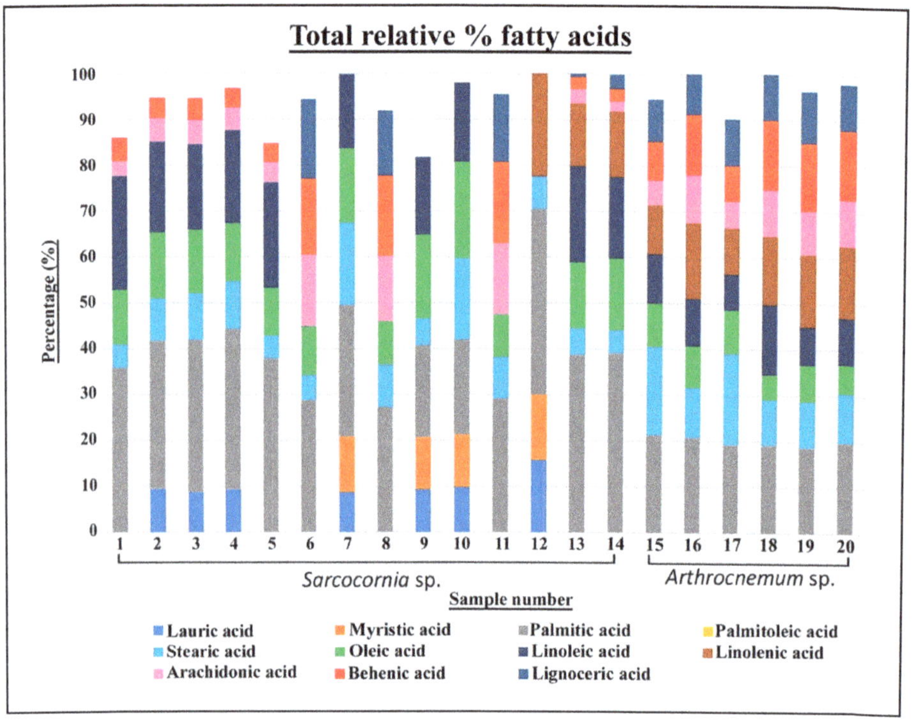

Figure 4. Total relative percentages of the fatty acids.

Among monounsaturated fatty acids, the study material presented oleic acid with values between 20% and 10%, particularly in *S. alpini* and *S. perennis.*

Linoleic acid was the predominant polyunsaturated fatty acid in the genus *Sarcocornia*, with values between 24% and 18% in *S. alpini*, 17% in *S. pruinosa*, and between 21% and 17% in *S. perennis*. This acid decreased to a relative content of 10% in *A. macrostachyum*.

Finally, linolenic acid was present only in *S. perennis* between 22% and 17%, and in *A. macrostachyum* with a relative content of between 16% and 10%.

Arachidonic and behenic acids were found in all the samples of *A. macrostachyum* and *S. alpini*, with values ranging from 15% to 5%, whereas in *S. pruinosa* and *S. perennis*, the content varied between the different samples.

Lignoceric acid was identified in the samples of *S. pruinosa* and *S. perennis* from the estuary of the Tinto River (samples 6, 8, 11, 13, and 14), but was absent in *S. alpini*. This fatty acid was found in all the material of *A. macrostachyum* in proportions of between 11% and 9% relative percentage.

3. Discussion

3.1. Total Phenolic Compounds (TPC)

Our data were within a range of 3.231 to 4.803 mg G.A.E./g plant dw in 20 samples from different populations of *Sarcocornia* and *Arthrocnemum* in the southwest and interior region of the Iberian Peninsula.

Other authors in localities in Portugal have reported values of 20 mg G.A.E./g plant dw for *S. alpini*, of *S. perennis* in populations in Castro Marim, and of 49 mg G.A.E./g plant dw for *A. macrostachyum* collected in Faro, notably both localities in Portugal [8].

This difference can be due to several factors related to the culture conditions of the fresh plant, including the salt stress conditions and environmental changes. Halophytes live in extremely harsh environments with high salinities and UV radiation, and these stressful conditions lead to the production of secondary metabolites such as the phenolic compounds in different concentrations [25]. The mere fact of detecting these total phenolic compounds emphasizes the antioxidant capacity of the halophytes in the study [26–28] and demonstrates that *Sarcocornia* and *Arthrocnemum* have a potential food use.

3.2. Phenolic Acids

Trans-cinnamic acid was significant in the species of the genus *Sarcocornia,* analyzed particularly in *S. alpini* and *S. perennis*, which were collected in the estuary of the Tinto River and in the mouth of the Guadiana River.

Trans-cinnamic acid was very scarce in *Arthrocnemum macrostachyum*, with a proportion of a little over 10%; the abundance of this acid differs between these genera.

Trans-cinnamic acid reduces adipogenesis and lipogenesis, emphasizing its potential for treating obesity [29].

Salicylic acid was the predominant phenolic acid in *S. pruinosa* with over 55%; however, its content ranged between 30% and 25% in *S. perennis*. Salicylic acid accounted for over 60% of the relative content in the material of *A. macrostachyum* from the Tinto River. This acid is involved in regulating plants' response to drought through the genetic expression of the genes PR1 and PR2. The induction of these genes increases the accumulation of salicylic acid as a protection mechanism at times of water stress [30].

Veratric acid was significant in the samples that were not affected by the influence of the tides and occupied drier environments in the salt marshes. This was the case of the populations of *S. alpini* and *S. pruinosa* in the southwest of the Iberian Peninsula, and the populations of *A. macrostachyum* in the interior and southwest. Veratric acid has antibacterial, anti-inflammatory, and anti-hypertensive activities [31].

A slightly greater diversity of phenolic acids has been shown more in *A. macrostachyum* than in species of the genus *Sarcocornia*. There is a clear difference in the content of coumaric acid in these genera within the Iberian Peninsula. Five out of fifteen samples of the genus *Sarcocornia* (samples 1, 5, 7, 9, and 10) were identified as having a relative content between 10% and 4%, while this acid was detected in the six samples analyzed of *Arthrocnemum*, with content ranging between 20% and 10%. This acid has been described in *Salicornia patula* Duval–Jouve on the Iberian Peninsula, where it was determined to be infrequent, as it was found in only two samples of the thirteen evaluated [10].

Caffeic acid is the lowest phenolic acid in the *Sarcocornia* material, as indicated by other authors such as Bertin et al. [32] and Costa et al. [33], with data of 0.402 mg/g in *S. ambigua* from Brazil. This acid thickens the plant cell walls and increases resistance to the ionic toxicity of sodium and heavy metal stress [34], suggesting that the presence of caffeic acid in these halophytes may allow them to adapt to highly saline environments.

Ferulic acid was only identified in *A. macrostachyum* collected in the localities of La Rábida and Belchite, with contents of between 59% and 56%. This phenolic acid has been described in *S. ambigua* [33] and *Salicornia europaea* L. [35]. Deng et al. (2015) observed a positive correlation between the ferulic acid content in the cuticle of *Limonium bicolor* (Bunge) Kuntze and the speed of sodium secretion, suggesting that ferulic acid is directly involved in the secretion of salt through saline glands [36]. These glands have not been described in species of *Arthrocnemum* and *Sarcocornia* [3,37]; however, the detection of ferulic acid in two of our populations of *Arthrocnemum* may point to its implication in certain mechanisms of tolerance to salinity.

Additionally, it has been proven that plants exposed to environments with heavy metals produce a high diversity of secondary metabolites, such as phenolic acids [38]. In our study, the diversity of phenolic acids found seems to correspond more closely with the plant species used (*S. alpini*, *S. pruinosa*, *S. perennis*, and *A. macrostachyum*) rather than with the influence of a medium with a high content of heavy metals, such as the Tinto River. However, the subject really deserves a detailed study in this regard, especially concerning wild plants that grow under the influence of the Tinto River in the province of Huelva.

3.3. Flavonoids and Hydroxycinnamic Acids

All the samples of *Sarcocornia* presented luteolin, which was previously identified in other Salicornioideae such as *S. europaea* [39].

Most flavonoids are present in plants in the form of esters, glucosides, and polymers. The chemical structure of these flavonoids determines their range of intestinal absorption and confers their beneficial uses for halophytes as edible plants. Glycosylation guarantees selective absorption and endows these compounds with prebiotic actions [40]. The species of *Sarcocornia* studied, namely *S. pruinosa* and *S. alpini*, also contained a glycosylated flavonoid with greater molecular weight (cyanidin-3-*O*-arabinoside, luteolin-7-glucoside, and dihydroquercetin). p-Coumaroyl glucose was found in *Arthrocnemum*. The presence of these compounds could be explained by the fact that halophytes increase their antioxidant requirements as a defense against extreme environments, forming macromolecular antioxidants [41]. The detection of these compounds also highlights the value of their use as edible plants.

In addition, apigenin-7-glucoside or rutin were identified in the 20 samples of the analyzed genera *Sarcocornia* and *Arthrocnemum* from material from the Iberian salt marshes; these two antioxidant compounds were identified in the genus *Salicornia* and, in the case of rutin, are associated to its tolerance of salinity [10].

3.4. Fatty Acids

Fatty acid composition affects the ability to tolerate salt stress [42,43]. Ten different fatty acids were found in the samples from the genus *Sarcocornia* and eight in *Arthrocnemum*. These included saturated, monounsaturated, and polyunsaturated fatty acids that provide halophytes an adaptive advantage, as they prevent the oxidative damage caused by the saline stress that is habitual in these environments [24].

Our results show that the saturated fatty acid present in the highest proportion in all species of *Sarcocornia* and *A. macrostachyum* was palmitic acid, which may account for over 90%. Values of 20% were found in other species of *Sarcocornia*, such as *S. ambigua* [44] and in *Arthrocnemum* from Tunisia, with a content of between 19% and 11% [45]. Stearic acid was another important acid that was present in all the halophyte samples studied, with values ranging from 19% to 5%. Other authors have reported similar results between 18% and 12% in *Sarcocornia* from Alcochete in Portugal [46]. Custodio et al. [47] identified 6% stearic acid in *A. macrostachyum*, which was also collected in Faro, Portugal. These bioactive compounds prevent the development of cardiovascular disorders, reduce insulin resistance, and strengthen the immune system [48].

No palmitoleic acid was found in the material from *Sarcocornia* and *Arthrocnemum* in our study. This monounsaturated fatty acid has been identified by other authors in *S. perennis* and *S. alpini* from Portugal, with values of between 21% and 17% [8]. These authors describe a content of between 6% and 4%, while Custodio et al. [47] reported values of between 11% and 4% in *A. macrostachyum* from Portugal.

There was a notable content of polyunsaturated fatty acids, specifically linoleic and linolenic acid. in *S. perennis* and *A. macrostachyum*, with a content of between 22% and 7%, which was much higher than the value described by Barreira et al. [8] in *S. perennis*, namely between 2% and 0.81%. This group of fatty acids have been considered the most important compounds against saline stress and their action has been proposed as an antioxidant [49]. Polyunsaturated fatty acids are bioactive compounds with antifungal activity, in addition to inhibiting carcinogenesis and the progression of atherosclerosis [50].

Long-chain fatty acids, such as arachidonic, behenic, and lignoceric acid, have content values of 17% in the species studied from the genus *Sarcocornia*; these values are higher than those published by Barreira et al. [8], who reported data between 11% and 4% for *S. perennis* and *S. alpini*, for populations from Portugal.

In *A. macrostachyum*, long-chain fatty acids had a relative content of over 15%, notably higher than the values published by Barreira et al. [8] and Custodio et al. [47] for the same species in Portugal.

3.5. Nutritional Importance and Future Implications

Halophytic plants of the Salicornioideae subfamily are known as "sea asparagus", "glasswort", "samphire", and "pickleweed" [51,52]. The plants most consumed as gourmet foods are those annual species of the genus *Salicornia*, especially those named under *S. europaea*, which may include other species given the taxonomic complexity of this group [53]. In fact, the difficulty of distinguishing between these types of plants has led many European markets and restaurants to use these halophytes as a mixture of several species, both annual and perennial [18,51].

Perennial halophytes, such as some species of the genera *Sarcocornia* and *Arthrocnemum* (evolutionarily close to *Salicornia*), have also shown to possess interesting nutritional properties for consumption [8,14,19]. *Sarcocornia* and *Arthrocnemum* produce succulent shoots which can be used for food as green leafy vegetables, as fresh ingredients for salads, and for spicing or substituting salt considering their great sodium amounts [48].

In our study, we analyzed the phenolic compounds and fatty acids of perennial plants of the genera *Sarcocornia* (*S. alpini*, *S. perennis* and *S. pruinosa*) and *Arthrocnemum* (*A. macrostachyum*), reaffirming that they are halophytes that also present properties with authentic nutritional potential for its consumption (potential foods with antioxidant properties, contribution of essential fatty acids for the human diet, etc.). The selective introduction of these underused species in markets and in traditional and healthy cuisine represents a future challenge to be implemented.

4. Materials and Methods

4.1. Materials

The material was collected in the southwest of the Iberian Peninsula in several localities in the Tinto River basin, such as La Rábida, San Juan del Puerto, and the river estuary. Samples were also collected in localities in the mouth of the Guadiana River (Ayamonte and Castro Marim) and in other points of southeast Portugal (Tavira and Santa Luzia). Salt marshes in Madrid and Zaragoza were selected from the interior of the Iberian Peninsula.

Table 3 shows the data for each of the species studied and fresh plants were collected as follows: Upon reception, a portion of fresh plants was stored in airtight plastic bags (anaerogen ™ 3.5 L, Thermo Scientific, Waltham, MA, USA) for one day until its analyses were performed. Then all material were dehydrated in a recirculated air stove (MEMMERT) to 100 °C for six hours for the subsequent analysis of the bioactive compounds: *Sarcocornia alpini* (1–5), *Sarcocornia pruinosa* (6–11), *Sarcocornia perennis* (12–14), and *Arthrocnemum macrostachyum* (15–20).

4.2. Methods

4.2.1. Determination of Humidity

Humidity was determined by drying in an oven (984.25-AOAC, 2005): 5 g of dried samples were weighed in previously dried and tared capsules, and it was placed in a dryer. The samples were placed in a recirculated air stove (MEMMERT) to 100 °C for six hours until the elimination of the water present in the sample (constant weighing).

4.2.2. Preparation of Methanol Extract

Five hundred milligrams of dried plant sample were extracted with a solution of 40 mL of methanol at 25 °C. It was kept in magnetic stirring for 60 min. The extracts were filtered using a Whatman No. 4 filter. The solid residue was recovered and extracted with 40 mlf of methanol. The extracts were filtered again, combined, and evaporated (35 °C under vacuum of the methanolic extracts). Redissolve with methanol to obtain a 30 mg/mL of extract solution, from which different dilutions were made (from 0.03125 mg/mL to 16 mg/mL), was conducted. The extractions were performed in triplicates and were stored at 4 °C until the execution of the analyses.

Table 3. Data on the material: geographic locations, collection date, and MGRS (Military Grid Reference System) of the *S. alpini*, *S. pruinosa*, *S. perennis*, and *A. macrostachyum* samples from the Iberian Peninsula (Spain and Portugal).

ID	Sample	Geographic Location	Collection Date	MGRS Coordinates
1	*S. alpini*	Spain, Huelva, and San Juan del Puerto	15 December 2017	29SPB9230
2	*S. alpini*	Spain, Huelva, and La Rábida	17 July 2018	29SPB8320
3	*S. alpini*	Spain, Huelva, and Ayamonte	18 July 2018	29SPB4122
4	*S. alpini*	Portugal and Esteiro de Carrasqueira	18 July 2018	29SPB3918
5	*S. alpini*	Spain, Huelva, San Juan del Puerto, and saltmarshes of the "Embarcadero de Buitrón"	7 August 2019	29SPB9031
6	*S. pruinosa*	Spain, Huelva, and Tinto river estuary	14 December 2017	29SPB8220
7	*S. pruinosa*	Portugal, Tavira saltmarshes, and Santa Luzia	18 July 2018	29SPB1906
8	*S. pruinosa*	Spain, Huelva, and Punta del Moral	18 July 2018	29SPB4717
9	*S. pruinosa*	Spain, Huelva, and Odiel saltmarshes	7 August 2019	29SPB7926
10	*S. pruinosa*	Spain, Huelva, and La Rábida	17 July 2018	29SPB8320
11	*S. pruinosa*	Spain, Huelva, Odiel River, and "Cabeza Alta"	14 December 2017	29SPB8024
12	*S. perennis*	Portugal, Tavira saltmarshes, and Santa Luzia	18 July 2018	29SPB1806
13	*S. perennis*	Spain, Huelva and Tinto river estuary	14 December 2017	29SPB8220
14	*S. perennis*	Spain, Huelva and Punta del Moral,	18 July 2018	29SPB4717
15	*A. macrostachyum*	Spain, Zaragoza, and Belchite	19 January 2017	30TXL8875
16	*A. macrostachyum*	Spain, Madrid, and Colmenar de Oreja	11 July 2018	31TVK5143
17	*A. macrostachyum*	Spain, Huelva, and La Rábida	15 December 2017	29SPB8320
18	*A. macrostachyum*	Portugal, Marismas de Tavira, and Santa Luzia	18 July 2018	29SPB1906
19	*A. macrostachyum*	Spain, Huelva, San Juan del Puerto, and saltmarshes of the "Embarcadero de Buitrón"	7 August 2019	29SPB9131
20	*A. macrostachyum*	Spain, Huelva, and Ayamonte	18 July 2018	29SPB4218

4.2.3. Total Phenolic Compounds (TPC)

The total phenolic content was determined by the Folin–Ciocalteu method [54] using gallic acid as the recommended standard [55]. An 0.5 mL of aliquot of methanolic extract was taken from the extracts obtained previously and 2.5 mL of the Folin–Ciocalteu reagent was added and left to react for 3 min. Then, 2 mL of Na_2CO_3 solution was added and mixed in a Heidolph shaker (Berlin, Germany). The solution was incubated at a temperature of 40 °C and stored in the dark for 1 h. The absorbance was measured at 765 nm with a spectrophotometer and the results were expressed as gallic acid equivalents (G.A.E.).

4.2.4. Gas Chromatography Coupled with Mass Spectrometry (GC-MS) for Phenolic and Fatty Acid Analysis

Chromatographic separation was performed as follows: Methanol extracts were brought to dryness in a Rotavapor Fischer rotary evaporator (USA) and later in the Telstar lyophilizer (Barcelona, Spain). The amount of the total sample obtained was weighed. The samples were then subjected to derivatization with a 0.2 N methanolic solution of m-trifluoromethylphenyl trimethylammonium hydroxide Meth Prep II (Fisher, Loughborough, UK). This one-step reagent simplifies the transesterification of triglycerides to methyl esters. In total, 5 μL was injected into GC/MS Agilent 6120 (Santa Clara, CA, USA). All standards were from Sigma Aldrich (Sant Louis, MI, USA) at ≥95.0% (HPLC).

The chromatography-mass spectrometry was carried out with the Interdepartmental Research Service at the Universidad Autónoma de Madrid (UAM).

4.2.5. High-Performance Liquid Chromatography-Electrospray Ionization Mass Spectrometry (HPLC-MS/ESI) for Flavonoid and Hydroxycinnamic Acid Analysis

Flavonoids were determined using a HPLC-MS/ESI Agilent 1100 (Santa Clara, CA, USA) in a C20 column ACE 3 C18 PFP, 150 mm × 4.6 mm, which was maintained at 35 °C. The solvent system used was a gradient of water (solvent A) and formic acid 0.1% (solvent A), and the acetonitrile and formic acid 0.1% (solvent B) as follows. For solvent A: 0 min, 96% of solvent A; 10 min, 90% of solvent A; 20 min, 80% of solvent A; 35 min, 60% of solvent A; 40 min, 40% of solvent A; 45 min, 40% of solvent A; 55 min, 96% of solvent A; and 60 min, 96% of solvent A. For solvent B: 0 min, 4% of solvent B; 10 min, 10% of solvent B; 20 min, 20% of solvent B; 35 min, 40% of solvent B; 40 min, 60% of solvent B; 45 min, 60% of solvent B; 55 min, 4% of solvent B; and 60 min, 4% of solvent B. The flow rate was 0.5 mL/min and runs were monitored with an ESI detector set at 280 nm (phenolic acids) and 360 nm (flavonols) for a total chromatogram time of 50 min. The fragmenter worked with 100 V. The drying gas flow was 10 L/min. The nebulizer pressure was 40 psig. The drying gas temperature was 350 °C. The vaporizer temperature was 250 °C. The capillary voltage was 4000 V. The charging voltage was 2000 V. An injection volume of 10 μL was taken from 1.2 mg/2 mL. This technique was used to identity the flavonoids in the extract according to their protonation [M+H] and to calculate the relative retention time of each peak in the chromatograms obtained by HPLC, with a mass range from 700 to 600 umas.

5. Conclusions

The bioactive compounds (phenolic compounds and fatty acids) present in *S. alpini*, *S. pruinosa*, *S. perennis*, and *A. macrostachyum* from different territories of Spain and Portugal were described. Samples of the genus *Sarcocornia* highlighted the presence of veratric acid material from dryer environments. A slightly greater diversity of phenolic acids was shown in *A. macrostachyum* than in species of the genus *Sarcocornia*. Ferulic acid was also detected in two of the samples from this genus but was not present in the genus *Sarcocornia*. The composition of the flavonoids detected in these species showed glycosylated structures that conferred prebiotic properties of these halophytes. The material from *S. alpini*, *S. pruinosa*, and *A. macrostachyum* contained macromolecular antioxidants, namely cyanidin-3-*O*-arabinoside, luteolin-7-glucoside, dihydroquercetin, and p-Coumaroyl glucoside, thus increasing their antioxidant requirements as a defense against extreme environments. The lipid profile revealed palmitic, linoleic, and oleic acids as the main fatty acids in the genus *Sarcocornia*, while the palmitic, linolenic, and stearic acid content was particularly notable in the genus *Arthrocnemum*. The presence of these compounds in different halophytes confirms their value for survival in conditions of extreme salinity and drought, and also adds to their value for consumption.

Supplementary Materials: The following are available online at https://www.mdpi.com/article/10.3390/plants10102218/s1, Figure S1: Tentative flavonoid compounds identified in *Sarcocornia* and *Arthrocnemum* extracts. All structures are based on Phenol Explorer; Table S1: Relative content of the phenolic compounds (%) in *Sarcocornia* and *Arthrocnemum* samples by GC-MS. Legend: nd: not detected. Numbers 1–20 refer to the materials shown in Table 3; and Table S2: Relative content of the fatty acids (%) in *Sarcocornia* and *Arthrocnemum* by GC-MS. Legend: lauric acid (C12:0); myristic acid (C14:0); palmitic acid (C16:0); stearic acid (C18:0); oleic acid (C18:1); linoleic acid (C18:2); linolenic acid (C18:3); arachidonic acid (C20:0); behenic acid (C22:0); lignoceric acid (C24:0); SFAs (saturated fatty acids); MUFAs (monounsaturated fatty acids); and PUFAs (polyunsaturated fatty acids). nd: not detected. Numbers 1–20 refer to the materials shown in Table 3.

Author Contributions: V.d.l.F.G.: study conception and design. I.S.-G.: data collection, analysis, and interpretation of the results. E.R.C.: preparation of the draft manuscript. All authors have read and agreed to the published version of the manuscript.

Funding: This research study was funded by a grant from the Ministerio de Ciencia e Innovación, CGL2015-66242-R.

Institutional Review Board Statement: Not applicable.

Informed Consent Statement: Not applicable.

Data Availability Statement: The data are contained within the article or Supplementary Material.

Acknowledgments: We would like to thank the staff of the Interdepartmental Research Department (SiDi-UAM), especially Rosa Sedano Pérez.

Conflicts of Interest: The authors declare no conflict of interest.

References

1. De la Fuente, V.; Oggerin, M.; Rufo, L.; Rodríguez, N.; Ortuñez, E.; Sánchez-Mata, D.; Amils, R. A micromorphological and phylogenetic study of *Sarcocornia* AJ Scott (Chenopodiaceae) on the Iberian Peninsula. *Plant Biosyst.* **2013**, *147*, 158–173. [CrossRef]
2. De la Fuente, V.; Rufo, L.; Rodríguez, N.; Sánchez-Mata, D.; Franco, A.; Amils, R. A study of *Sarcocornia* AJ Scott (Chenopodiaceae) from Western Mediterranean Europe. *Plant Biosyst.* **2016**, *150*, 343–356. [CrossRef]
3. De la Fuente, V.; Sánchez-Gavilán, I.; Ramírez, E.; Rufo, L.; Sánchez-Mata, D. Morphological Variability of Halophytes: Salicornioideae on Iberian Peninsula. In *Handbook of Halophytes*; Grigore, M.N., Ed.; Springer: Cham, Switzerland, 2021. [CrossRef]
4. Rufo, L.; De La Fuente, V.; Sánchez-Mata, D. *Sarcocornia* plant communities of the Iberian Peninsula and the Balearic Islands. *Phytocoenologia* **2016**, *46*, 383–396. [CrossRef]
5. Rufo, L.; Iglesias-López, M.T.; De La Fuente, V. The endemic halophyte *Sarcocornia carinata* Fuente, Rufo & Sánchez-Mata (Chenopodiaceae) in relation to environmental variables: Elemental composition and biominerals. *Plant Soil* **2021**, *460*, 189–209.
6. Ramírez, E.; Rufo, L.; Sánchez-Mata, D.; De La Fuente, V. *Arthrocaulon meridionalis* (Chenopodiaceae), a new species of Mediterranean flora. *Mediterr. Bot.* **2019**, *40*, 33–41. [CrossRef]
7. Alhdad, G.M.; Seal, C.E.; Al-Azzawi, M.J.; Flowers, T.J. The effect of combined salinity and waterlogging on the halophyte *Suaeda maritima*: The role of antioxidants. *Environ. Experiment. Bot.* **2013**, *87*, 120–125. [CrossRef]
8. Barreira, L.; Resek, E.; Rodrigues, M.J.; Rocha, M.I.; Pereira, H.; Bandarra, N.; da Silva, M.M.; Varela, J.; Custódio, L. Halophytes: Gourmet food with nutritional health benefits? *J. Food Comp. Analysis* **2017**, *59*, 35–42. [CrossRef]
9. De la Fuente, V.; Rufo, L.; Sánchez-Gavilán, I.; Ramírez, E.; Rodríguez, N.; Amils, R. Plant tissues and embryos biominerals in *Sarcocornia pruinosa*, a halophyte from the Río Tinto salt marshes. *Minerals* **2018**, *8*, 505. [CrossRef]
10. Sánchez-Gavilán, I.; Ramírez, E.; de la Fuente, V. Bioactive Compounds in *Salicornia patula* Duval-Jouve: A Mediterranean Edible Euhalophyte. *Foods* **2021**, *10*, 410. [CrossRef]
11. Barcia-Piedras, J.M.; Perez-Romero, J.A.; Mateos-Naranjo, E.; Camacho, M.; Redondo-Gómez, S. Effect of prior salt experience on desalination capacity of the halophyte *Arthrocnemum macrostachyum*. *Desalination* **2019**, *463*, 50–54. [CrossRef]
12. El-Naker, N.A.; Yousef, A.F.; Yousef, L.F. A review of *Arthrocnemum* (*Arthrocaulon*) *macrostachyum* chemical content and bioactivity. *Phytochem. Rev.* **2020**, *19*, 1427–1448. [CrossRef]
13. Riquelme, J.; Olaeta, J.A.; Galvez, L.; Undurraga, P.; Fuentealba, C.; Osses, A.; Orellana, J.; Gallardo, J.; Pedreschi, R. Nutritional and functional characterization of wild and cultivated *Sarcocornia neii* grown in Chile. *Cien. Investig. Agr.* **2016**, *43*, 283–293. [CrossRef]
14. Ventura, Y.; Wuddineh, W.A.; Myrzabayeva, M.; Alikulov, Z.; Khozin-Goldberg, I.; Shpigel, M.; Sagi, M. Effect of seawater concentration on the productivity and nutritional value of annual *Salicornia* and perennial *Sarcocornia* halophytes as leafy vegetable crops. *Scientia Horticult.* **2011**, *128*, 189–196. [CrossRef]
15. Ventura, Y.; Sagi, M. Halophyte crop cultivation: The case for *Salicornia* and *Sarcocornia*. *Environ. Experiment. Bot.* **2013**, *92*, 144–153. [CrossRef]
16. Singh, D.; Buhmann, A.K.; Flowers, T.J.; Seal, C.E.; Papenbrock, J. *Salicornia* as a crop plant in temperate regions: Selection of genetically characterized ecotypes and optimization of their cultivation conditions. *AoB Plants* **2014**, *6*, plu071. [CrossRef]
17. Paraskevopoulou, A.; Mitsios, I.; Fragkakis, I.; Nektarios, P.; Ntoulas, N.; Londra, P.; Papafotiou, M. The growth of *Arthrocnemum macrostachyum* and *Halimione portulacoides* in an extensive green roof system under two watering regimes. *Agricult. Agr. Sci. Proc.* **2015**, *4*, 242–249. [CrossRef]
18. Urbano, M.; Tomaselli, V.; Bisignano, V.; Veronico, G.; Hammer, K.; Laghetti, G. *Salicornia patula* Duval- Jouve: From gathering of wild plants to some attempts of cultivation in Apulia region (southern Italy). *Genet. Resour. Crop Evol.* **2017**, *64*, 1465–1472. [CrossRef]
19. Loconsole, D.; Cristiano, G.; De Lucia, B. Glassworts: From Wild Salt Marsh Species to Sustainable Edible Crops. *Agriculture* **2019**, *9*, 14. [CrossRef]
20. Rufo, L.; Rodríguez, N.; Amils, R.; de la Fuente, V.; Jiménez-Ballesta, R. Surface geochemistry of soils associated to the Tinto River (Huelva, Spain). *Sci. Total Environ.* **2007**, *378*, 223–227. [CrossRef]
21. Rufo, L.; de la Fuente, V. Chemical and mineralogical characterization of the soils of the main plant communities of the "Río Tinto" Basin. *Schironia* **2010**, *9*, 5–13.
22. Ramírez, E.; Rufo, L.; Sánchez-Mata, D.; Sánchez-Gavilán, I.; de la Fuente, V. Arthrocnemum macrostachyum plant communities in the Iberian Peninsula, Balearic and Canary Islands (Spain and Portugal). In *Tools for Landscape-Scale Geobotany and Conservation*; Springer: Cham, Switzerland, 2021; pp. 231–245.

23. Sánchez-Gavilán, I.; Rufo, L.; Rodriguez, N.; de la Fuente, V. On the elemental composition of the Mediterranean euhalophyte *Salicornia patula* Duval-Jouve (Chenopodiaceae) from saline habitats in Spain (Huelva, Toledo and Zamora). *Environ. Sci. Pollut. Res.* **2021**, *28*, 2719–2727. [CrossRef]
24. Nikalje, G.C.; Bhaskar, S.D.; Yadav, K.; Penna, S. Halophytes: Prospective Plants for Future. In *Ecophysiology, Abiotic Stress Responses and Utilization of Halophytes*; Hasanuzzaman, M., Nahar, K., Öztürk, M., Eds.; Springer: Singapore, 2019. [CrossRef]
25. Al Jitan, S.; Alkhoori, S.A.; Yousef, L.F. Phenolic acids from plants: Extraction and application to human health. *Stud. Nat. Prod. Chem.* **2018**, *58*, 389–417.
26. Castañeda-Loaiza, V.; Oliveira, M.; Santos, T.; Schüler, L.; Lima, A.R.; Gama, F.; Salazar, M.; Neng, N.R.; Varela, J.; Barreira, L. Wild vs cultivated halophytes: Nutritional and functional differences. *Food Chem.* **2020**, *333*, 127536. [CrossRef] [PubMed]
27. Barroca, M.J.; Guiné, R.P.; Amado, A.M.; Ressurreição, S.; da Silva, A.M.; Marques, M.P.M.; de Carvalho, L.B. The drying process of *Sarcocornia perennis*: Impact on nutritional and physico-chemical properties. *J. Food Sci. Tech.* **2020**, *57*, 4443–4458. [CrossRef]
28. Oliveira-Alves, S.C.; Andrade, F.; Prazeres, I.; Silva, A.B.; Capelo, J.; Duarte, B.; Caçador, I.; Coelho, J.; Serra, A.T.; Bronze, M.R. Impact of Drying Processes on the Nutritional Composition, Volatile Profile, Phytochemical Content and Bioactivity of *Salicornia ramosissima*. J. Woods. *Antioxidants* **2021**, *10*, 1312. [CrossRef] [PubMed]
29. Wang, Z.; Ge, S.; Li, S.; Lin, H.; Lin, S. Anti-obesity effect of transcinnamic acid on HepG2 cells and HFD-fed mice. *Food Chem. Toxicol.* **2020**, *137*, 111–148. [CrossRef] [PubMed]
30. Miura, K.; Tada, Y. Regulation of water, salinity, and cold stress responses by salicylic acid. *Front. Plant Sci.* **2014**, *5*, 4. [CrossRef]
31. Patra, A.K. An overview of antimicrobial properties of different classes of phytochemicals. In *Dietary Phytochemicals and Microbes*; Springer: Dordrecht, The Netherlands, 2012; pp. 1–32.
32. Bertin, R.L.; Gonzaga, L.V.; Borges, G.D.S.C.; Azevedo, M.S.; Maltez, H.F.; Heller, M.; Fett, R. Nutrient composition and, identification/quantification of major phenolic compounds in *Sarcocornia ambigua* (Amaranthaceae) using HPLC–ESI-MS/MS. *Food Res. Int.* **2013**, *55*, 404–411. [CrossRef]
33. Costa, C.S.B.; Chaves, F.C.; Rombaldi, C.V.; Souza, C.R. Bioactive compounds and antioxidant activity of three biotypes of the sea asparagus *Sarcocornia ambigua* (Michx.) MA Alonso & MB Crespo: A halophytic crop for cultivation with shrimp farm effluent. *S. Afr. J. Bot.* **2018**, *117*, 95–100.
34. Riaz, U.; Kharal, M.A.; Murtaza, G.; uz Zaman, Q.; Javaid, S.; Malik, H.A.; Aziz, H.; Abbas, Z. Prospective roles and mechanisms of caffeic acid in counter plant stress: A mini review. *Pakistan J. Agricult. Res.* **2019**, *32*, 8. [CrossRef]
35. Panth, N.; Park, S.-H.; Kim, H.J.; Kim, D.H.; Oak, M.H. Protective Effect of *Salicornia europaea* Extracts on High Salt Intake-Induced Vascular Dysfunction and Hypertension. *Int. J. Mol. Sci.* **2016**, *17*, 1176. [CrossRef]
36. Deng, Y.Q.; Feng, Z.T.; Yuan, F.; Guo, J.R.; Suo, S.S.; Wang, B.S. Identification and functional analysis of the autofluorescent substance in *Limonium bicolor* salt glands. *Plant Physiol.* **2015**, *97*, 20–27. [CrossRef]
37. Grigore, M.N.; Toma, C. Anatomical adaptations of halophytes. In *A Review of Classic Literature and Recent Findings*; Springer International Publishing: Cham, Switzerland, 2017; p. 338.
38. Asgari Lajayer, B.; Ghorbanpour, M.; Nikabadi, S. Heavy metals in contaminated environment: Destiny of secondary metabolite biosynthesis, oxidative status and phytoextraction in medicinal plants. *Ecotoxicol. Environ. Saf.* **2017**, *145*, 377–390. [CrossRef] [PubMed]
39. Grigore, M.N.; Oprica, L. Halophytes as possible source of antioxidant compounds, in a scenario based on threatened agriculture and food crisis. *Iran. J. Public Health* **2015**, *44*, 11–53.
40. D'Archivio, M.; Filesi, C.; Varì, R.; Scazzocchio, B.; Masella, R. Bioavailability of the Polyphenols: Status and Controversias. *Int. J. Mol. Sci.* **2010**, *11*, 1321–1342. [CrossRef] [PubMed]
41. Pérez-Jiménez, J.; Saura-Calixto, F. Macromolecular antioxidants or non-extractable polyphenols in fruit and vegetables: Intake in four European countries. *Food Res. Int.* **2015**, *74*, 315–323. [CrossRef]
42. Buhmann, A.; Papenbrock, J. An economic point of view of secondary compounds in halophytes. *Funct. Plant Biol.* **2013**, *40*, 952–967. [CrossRef] [PubMed]
43. Zhang, J.; Yang, D.; Li, M.; Shi, L. Metabolic Profiles Reveal Changes in Wild and Cultivated Soybean Seedling Leaves under Salt Stress. *PLoS ONE* **2016**, *11*, 7. [CrossRef] [PubMed]
44. D'oca, M.G.; Morón-Villarreyes, J.A.; Lemões, J.S.; Costa, C.S. Fatty acids composition in seeds of the South American glasswort *Sarcocornia ambigua*. *Anais Acad. Brasil. Ciên.* **2012**, *84*, 865–870. [CrossRef] [PubMed]
45. Maatallah Zaier, M.; Ciudad-Mulero, M.; Cámara, M.; Pereira, C.; Ferreira, I.; Achour, L.; Kacem, A. Revalorization of Tunisian wild Amaranthaceae halophytes: Nutritional composition variation in two different phenotypes stages. *J. Food Comp. Anal.* **2020**, *89*, 103463. [CrossRef]
46. Duarte, B.; Caçador, I.; Matos, A.R. Lipid landscape remodelling in *Sarcocornia fruticosa* green and red physiotypes. *Plant Physiol. Biochem.* **2020**, *157*, 128–137. [CrossRef]
47. Custódio, L.; Ferreira, A.C.; Pereira, H.; Silvestre, L.; Vizetto-Duarte, C.; Barreira, L.; Rauter, A.P.; Alberício, F.; Varela, J. The marine halophytes *Carpobrotus edulis* L. and *Arthrocnemum macrostachyum* L. are potential sources of nutritionally important PUFAs and metabolites with antioxidant, metal chelating and anticholinesterase inhibitory activities. *Bot. Mar.* **2012**, *55*, 281–288. [CrossRef]

48. Cámara, M.; Fernández-Ruiz, V.; Ruiz-Rodríguez, B.M. Wild Edible Plants as Sources of Carotenoids, Fibre, Phenolics and Other Non-Nutrient Bioactive Compounds. In *Mediterranean Wild Edible Plants. Ethnobotany and Food Composition Tables*; Sánchez-Mata, M.C., Tardío, J., Eds.; Springer: Cham, Switzerland, 2016; pp. 187–205.
49. Surówka, E.; Latowski, D.; Libik-Konieczny, M.; Miszalski, Z. 11 ROS Signalling, and Antioxidant Defence Network in Halophytes. In *Halophytes and Climate Change: Adaptive Mechanisms and Potential Uses*; CABI: Boston, MA, USA, 2019; p. 179.
50. Vizetto-Duarte, C.; Figueiredo, F.; Rodrigues, M.J.; Polo, C.; Rešek, E.; Custódio, L. Sustainable valorization of halophytes from the mediterranean area: A comprehensive evaluation of their fatty acid profile and implications for human and animal nutrition. *Sustainability* **2019**, *11*, 2197. [CrossRef]
51. Antunes, M.D.; Gago, C.; Guerreiro, A.; Sousa, A.R.; Julião, M.; Miguel, M.G.; Faleiro, M.L.; Panagopoulos, T. Nutritional Characterization and Storage Ability of *Salicornia ramosissima* and *Sarcocornia perennis* for Fresh Vegetable Salads. *Horticulturae* **2021**, *7*, 6. [CrossRef]
52. Patel, S. *Salicornia*: Evaluating the halophytic extremophile as a food and a pharmaceutical candidate. *3 Biotech* **2016**, *6*, 104. [CrossRef] [PubMed]
53. Kadereit, G.; Ball, P.; Beer, S.; Mucina, L.; Sokoloff, D.; Teege, P.; Yaparak, A.; Freitag, H. A taxonomic nightmare comes true: Phylogeny and biogeography of glassworts (*Salicornia*, L., Chenopodiaceae). *Taxon* **2007**, *56*, 1143–1170. [CrossRef]
54. Slinkard, K.; Singleton, V.L. Total phenol analysis: Automation and comparison with manual methods. *Am. J. Enol. Vitic.* **1977**, *28*, 49–55.
55. Andrés-Lacueva, C.; Medina-Remón, A.; Llorach, R.; Urpi-Sarda, M.; Khan, N.; Chiva-Blanch, G.; Zamora-Ros, R.; Rotches-Ribalta, M.; Lamuela-Raventos, R.M. Phenolic compounds: Chemistry and occurrence in fruits and vegetables. In *Fruit and Vegetable Phytochemicals: Chemistry, Nutritional Value and Stability*; De la Rosa, L., Alvarez-Parrilla, E., Gonzalez-Aguilar, G.A.E., Eds.; John Wiley and Sons: Cambridge, UK, 2010; pp. 53–88.

plants

Article

Heavy Metal Pre-Conditioning History Modulates *Spartina patens* Physiological Tolerance along a Salinity Gradient

João Carreiras [1], Jesús Alberto Pérez-Romero [2], Enrique Mateos-Naranjo [2], Susana Redondo-Gómez [2], Ana Rita Matos [3,4], Isabel Caçador [1,4] and Bernardo Duarte [1,4,*]

1 MARE—Marine and Environmental Sciences Centre, Faculty of Sciences of the University of Lisbon, Campo Grande, 1749-016 Lisbon, Portugal; jgcarreiras@fc.ul.pt
2 Departamento de Biología Vegetal y Ecología, Facultad de Biología, Universidad de Sevilla, Av. Reina Mercedes s/n, 41012 Sevilla, Spain; jesusalperezromero@gmail.com (J.A.P.-R.); emana@us.es (E.M.-N.); susana@us.es (S.R.-G.)
3 Plant Functional Genomics Group, BioISI—Biosystems and Integrative Sciences Institute, Departamento de Biologia Vegetal, Faculdade de Ciências da Universidade de Lisboa, Campo Grande, 1749-016 Lisboa, Portugal; armatos@fc.ul.pt
4 Departamento de Biologia Vegetal, Faculdade de Ciências da Universidade de Lisboa, Campo Grande, 1749-016 Lisbon, Portugal; micacador@fc.ul.pt
* Correspondence: baduarte@fc.ul.pt

Abstract: Land salinization, resulting from the ongoing climate change phenomena, is having an increasing impact on coastal ecosystems like salt marshes. Although halophyte species can live and thrive in high salinities, they experience differences in their salt tolerance range, being this a determining factor in the plant distribution and frequency throughout marshes. Furthermore, intraspecific variation to NaCl response is observed in high-ranging halophyte species at a population level. The present study aims to determine if the environmental history, namely heavy metal pre-conditioning, can have a meaningful influence on salinity tolerance mechanisms of *Spartina patens*, a highly disperse grass invader in the Mediterranean marshes. For this purpose, individuals from pristine and heavy metal contaminated marsh populations were exposed to a high-ranging salinity gradient, and their intraspecific biophysical and biochemical feedbacks were analyzed. When comparing the tolerance mechanisms of both populations, *S. patens* from the contaminated marsh appeared to be more resilient and tolerant to salt stress, this was particularly present at the high salinities. Consequently, as the salinity increases in the environment, the heavy metal contaminated marsh may experience a more resilient and better adapted *S. patens* community. Therefore, the heavy metal pre-conditioning of salt mash populations appears to be able to create intraspecific physiological variations at the population level that can have a great influence on marsh plant distribution outcome.

Keywords: halophytes; osmotic stress; pre-conditioning; intraspecific variability

Citation: Carreiras, J.; Pérez-Romero, J.A.; Mateos-Naranjo, E.; Redondo-Gómez, S.; Matos, A.R.; Caçador, I.; Duarte, B. Heavy Metal Pre-Conditioning History Modulates *Spartina patens* Physiological Tolerance along a Salinity Gradient. *Plants* **2021**, *10*, 2072. https://doi.org/10.3390/plants10102072

Academic Editors: Oscar Vicente, Juan Barceló and Marius-Nicusor Grigore

Received: 30 August 2021
Accepted: 28 September 2021
Published: 30 September 2021

1. Introduction

According to the analysis of the data gathered, through this and the last century, the Intergovernmental Panel on Climate Change (IPCC) report shows a worldwide intensification of abiotic stresses with alarming environmental and economic implications, notably the increase and intensity of extreme climate events, droughts, floods, sea-level rise, water, and land salinity variations and others [1]. Earth can be considered a salt planet since approximately, 97.5% of all planet's water content is saltwater, occupying roughly 70% of the surface encompassed in oceans, lakes, and groundwater [2], furthermore, it has been estimated that high soil salinity is affecting 20% of total Earth's land surface and 33% of agricultural irrigated lands [3,4].

In coastal regions, especially in the high populated low-elevation coastal lands and estuaries, climate change will likely increase, at an elevated rate, the soil, and water salinity, mostly due to predicted storm surges, tidal surges, and sea-level rise causing an onward saltwater land inundation. Therefore, it can be presumed that soil and water salinity-induced stress is and will be one of the major plant abiotic stresses. Usually, salt stress in plants is a powerful limiting production factor, upsetting every major crop development and productivity [5]. Most of the crop plants when exposed to NaCl concentrations from 40 mM to 200 mM become severely damage or die, plants exposed to elevated salt concentrations result in several complex biochemical, physiological, and morphological damages, such as nutrient uptake and assimilation [6–8]. On the other hand, and contrary to 99% of all the plant species, halophytic vegetation species can not only survive but be highly productive in saline environments. Halophytes are, by definition, plants that can live normally and complete their life cycle under a salt concentration of at least 200 mM, with most plants exhibiting tolerance to a remarkable amplitude of NaCl concentration [9]. However, it is known that different halophytes species have unlike responses to the same salinity, ranging from species having optimal performance in salt-free environments to high NaCl concentrations such as 400 mM [9,10]. Species salinity tolerance responses variations are relevant when taking into consideration the latent alterations to salinity in the environment, which will most likely change salt-tolerant plant habitat availability and distribution, within an ecosystem. This is evident in most halophytes inhabiting salt marshes, where species distribution is organized across salinity gradients and micro-habitats salinity variations associated with marsh topography and morphology, where plants are arranged according to their salinity tolerance [11,12]. Nonetheless, intraspecies phenological and physiological variation phenomenon can occur to a greater degree in highly tolerant species that are capable to adapt to environments that largely differ in their abiotic conditions [13–15]. When intraspecific NaCl response is taken into consideration it may show a different response to the same NaCl concentration, therefore it is important to understand coastal ecosystem modification and evolution once exposed to salinity changes.

Tagus estuary wetland is considered one of the more important in Europe and encompasses the most extensive and continuous salt marsh area in Portugal, presenting a great concentration of organic matter and biological productivity [12]. Salt marshes located within this estuary share most of the colonized halophyte species, such as the halophyte *Spartina patens* (Aiton) Muhl, a highly tolerant, invasive salt-excreting grass that is now spreading across Mediterranean marshes [16]. Moreover, neighboring marshes within the Tagus estuary, although being under mostly similar abiotic conditions, like salinity and temperature, can, due to anthropogenic actions, display a significant difference in their soil chemistry, when comparing marshes located within natural reserves to industrially contaminated marshes, notably caused heavy metals pollution [17]. These aspects make this species a suitable model to understand the effects of metal pollution pre-conditioning on tolerance range to salinity variations and ascertain to what extent different populations could otherwise respond to future changes in soil and water salinity. Additionally, several studies have suggested that intraspecific salt tolerance variations can occur in different *S. patens* populations [18,19], as well as heavy metal pre-conditioning can have a significant impact on this plant abiotic tolerance mechanisms [20].

The present work intends to determine heavy metal cross-tolerance thru pre-conditioning to salinity stresses in *S. patens*. Employing imposing salinity treatments on two populations, one from a heavy metal contaminated marsh and the other from a pristine one, it was possible to evaluate significant intraspecific variability in the physiological salt tolerance mechanisms. Given the ongoing climate change, it is relevant knowledge of the differently salt marshes species potential to adapt and respond, as well as the perception of a more complex reaction directly related to future salt-induced habitat modifications, concerning plant distribution and frequency.

2. Results

2.1. Photochemical Processes

When exposed to a salinity gradient, *S. patens* showed substantial variances in terms of their photochemical responses at a population level. The relative electron transport rates (rETR; Figure 1a) variation was shown to be significantly different, between populations, at 800 mM NaCl, higher values in the pristine marsh population. The photosynthetic efficiency (α; Figure 1b), measured within populations, displayed stability along the applied salinity gradient, with only individuals from the heavy metal contaminated marsh showing a significant difference at 400 mM NaCl. Nevertheless, significant differences were found between populations, contaminated marsh individuals displayed higher photosynthetic efficiency at 400 mM NaCl however at 800 mM NaCl the opposite was found. Maximum electron transport rates (ETR_{max}; Figure 1c) measurements showed a significant variation in 800 mM NaCl conditions between the populations, lower in the contaminated site individuals.

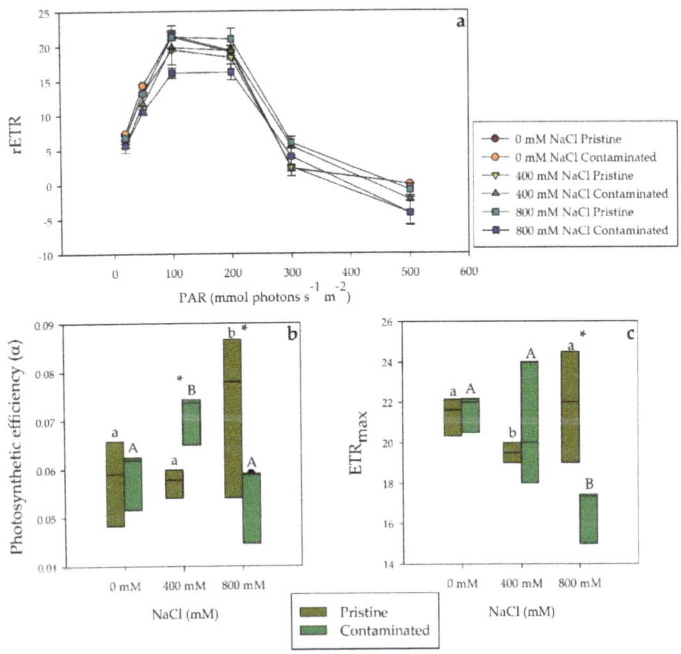

Figure 1. (**a**) Relative electron transport rates (rETR), (**b**) photosynthetic efficiency (α), and (**c**) maximum electron transport rate (ETR_{max}) in *S. patens* dark-adapted leaves from pristine and heavy metal contaminated sites (average $\pm$ standard error, N = 5), along with the tested NaCl concentrations. Letters indicate significant differences between treatments at $p < 0.05$; asterisks mark significant differences between populations at $p < 0.05$.

Energetic fluxes per leaf cross-section of the salt-treated chloroplasts showed a decrease in both populations under saline conditions in absorbed energy flux along a salinity gradient (ABS/CS; Figure 2a), significant in the 400 mM NaCl treated samples from the contaminated marsh. In trapped energy flux (TR/CS; Figure 2b) a similar significant reduction was observed in both populations although this decrease was not significant in the population from the heavy metal contaminated location at 800 mM NaCl, possibly due to the comparatively lower values exhibited in the 0 mM NaCl exposed individuals. Electron transport energy flux (ET/CS; Figure 2c) displayed a significant reduction at 400 and 800 mM NaCl, being this reduction more acute in *S. patens* from the pristine site.

Dissipation energy flux (DI/CS; Figure 2d) showed unlike and significant responses to salinity stress between salt marsh populations. Plants from the pristine site when exposed to 400 mM NaCl showed increase energy dissipation whilst the contaminated site individuals exposed exhibited a reduction in dissipation. Finally, oxidized reaction centers (RC/CS; Figure 2e) significantly decreased with salinity concentration to a similar degree in both sampling populations.

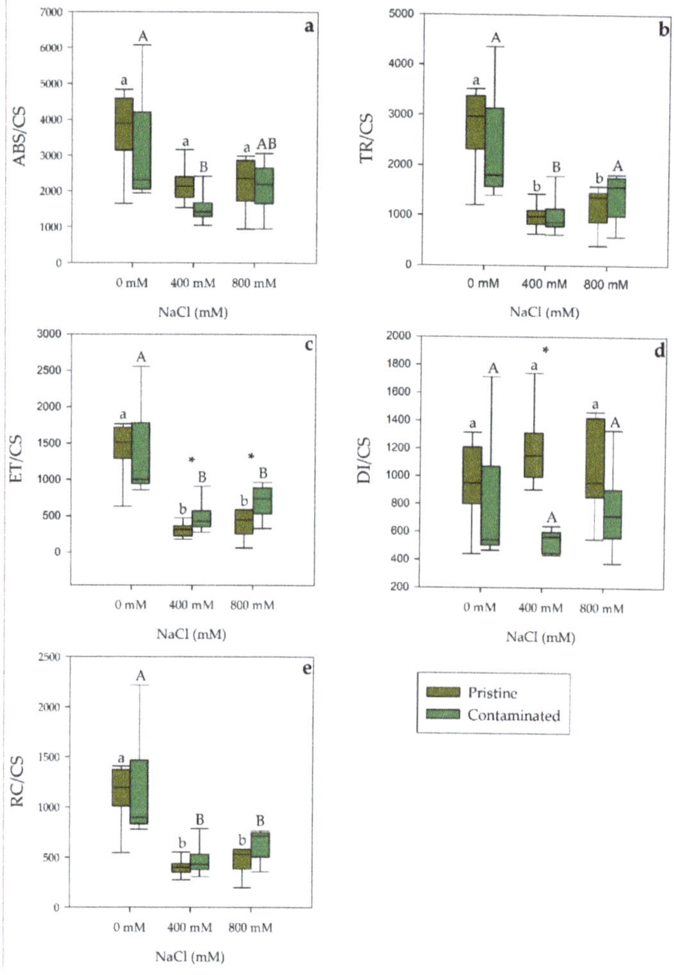

Figure 2. Phenomenological energetic parameters, (**a**) absorbed energy flux (ABS/CS), (**b**) trapped energy flux (TR/CS), (**c**) electron transport energy flux (ET/CS), (**d**) dissipation energy flux (DI/CS), and (**e**) oxidized reaction centers (RC/CS) on a cross-section basis, in *S. patens* dark-adapted leaves from pristine and heavy metal contaminated sites (average ± standard error, N = 5), along with the tested NaCl concentrations. Letters indicate significant differences between treatments at $p < 0.05$; asterisks mark significant differences between populations at $p < 0.05$.

Considering the total number of electrons transferred into the electron transport chain (N; Figure 3a), contaminated marsh individuals, when salt exposed, exhibited no significant changes while pristine site plants showed an increasing trend, displaying a significant increase at 400 mM NaCl and a highly significant increase at 800 mM NaCl, in relation to the contaminated site population as well at both salinities. Regarding the net rate of

PS II reaction centers closure (M_0; Figure 3b), a significantly higher value was evident in the 400 mM salt treatments of both populations, nonetheless contaminated site samples showed significantly lower M_0 at 0 and 400 mM NaCl. Electron movement efficiency from the reduced intersystem electron acceptors to the PS I end electron movement (δR_0; Figure 3c) showed an increase through salinity treatments in *S. patens* from the pristine mash, significant at 400 mM NaCl. The oxidized quinone pool size (Figure 3d) showed a similar pattern when comparing populations, the only difference was exhibited at 800 mM NaCl, a significantly lower size in the contaminated marsh population. Considering the grouping probability (P_G; Figure 3e), a significantly higher PS II antennae connectivity was exhibited at 800 mM NaCl in both site samples, while at 400 mM NaCl *S. patens* from the contaminated marsh, showed a significantly higher value within and between populations.

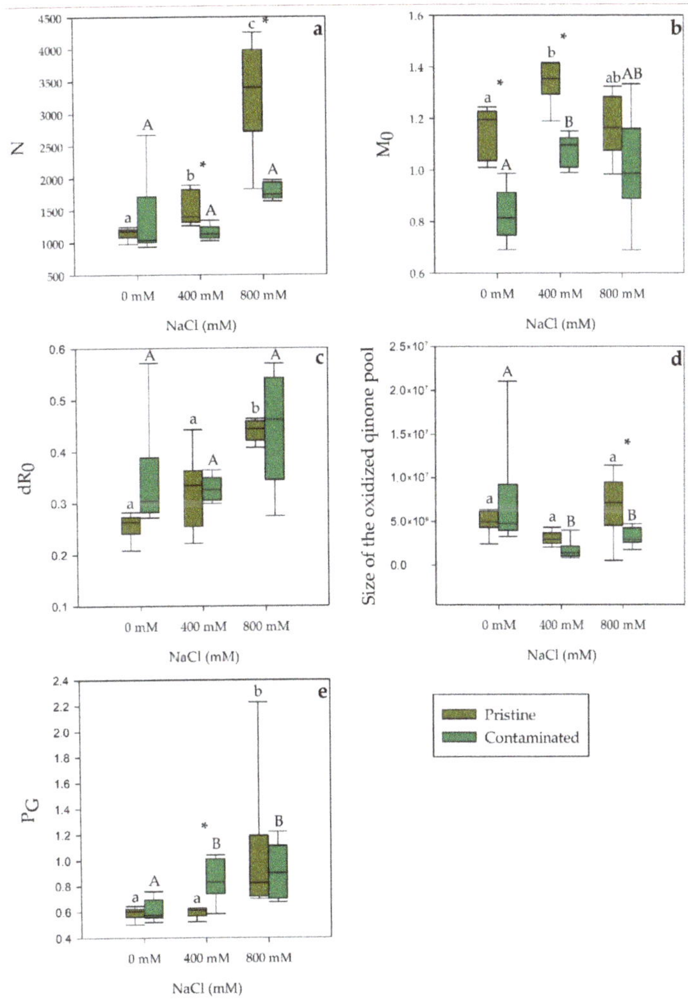

Figure 3. OJIP derived parameters, (**a**) the total number of electrons transferred into the electron transport chain (N), (**b**) the net rate of PS II reaction centers closure (M_0), (**c**) PS I efficiency in reducing its electron acceptors (δR_0), (**d**) size of the oxidized quinone pool, and (**e**) grouping probability (P_G) in *S. patens* dark-adapted leaves from pristine and heavy metal contaminated sites (average ± standard error, N = 5), along with the tested NaCl concentrations. Letters indicate significant differences between treatments at $p < 0.05$; asterisks mark significant differences between populations at $p < 0.05$.

2.2. Photosynthetic Pigments Profile

Regarding leaf pigments concentration, we found significant differences between salinity treatments and populations (Figure 4). Thus, total chlorophyll concentration (chl *a* and chl *b*) was significantly higher in the salt-treated pristine marsh individuals compared with their contaminated marsh counterparts (Figure 4a). In addition, in the pristine marsh *S. patens*, higher pigment concentrations were also found, with significance, in auroxanthin in all treatments, in lutein, neoxanthin, and violaxanthin at 800 mM NaCl and β-carotene and zeaxanthin when exposed to 400 and 800 mM NaCl concentrations (Figure 4b). The total carotenoid to total chlorophyll ratio (Figure 5b) displayed a similar pattern, with no significant differences, between populations. Contaminated site samples exhibited a significantly higher Chlorophyll *a*/*b* ratio than the pristine marsh population at 800 mM NaCl (Figure 5a). Furthermore, at 800 mM NaCl, the contaminated marsh *S. patens* showed a significantly lower chlorophyll degradation index (CDI, Figure 5c) and de-epoxidation state (DES, Figure 5d).

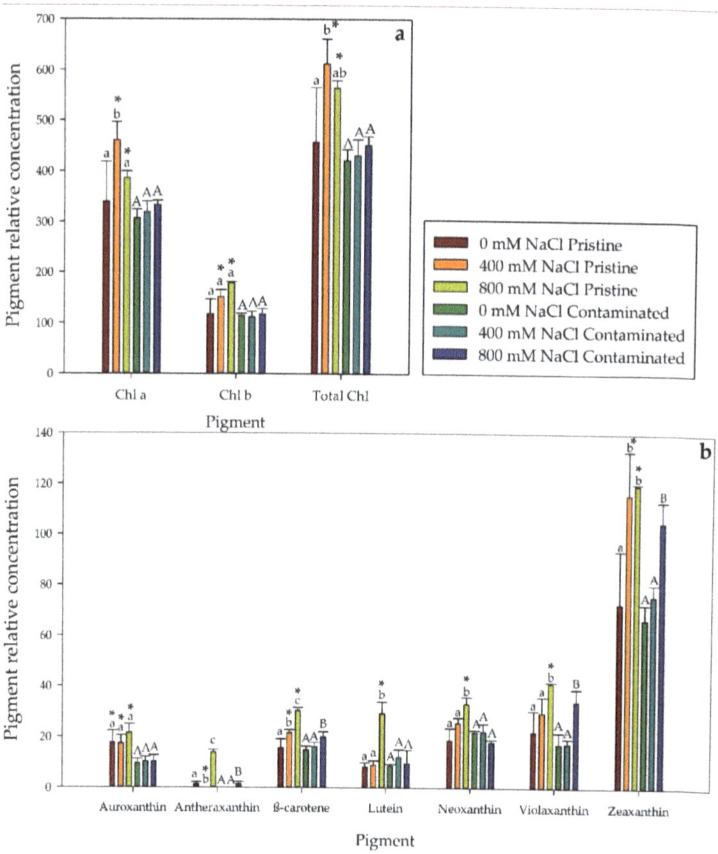

Figure 4. Pigment relative concentration. (**a**) Leaf chlorophyll *a* (Chl *a*), chlorophyll *b* (Chl *b*), total chlorophyll (Total Chl), (**b**) auroxanthin, antheraxanthin, β-carotene, lutein, neoxanthin, violaxanthin and zeaxanthin concentration (μg g^{-1} FW) in *S. patens* individuals from pristine and heavy metal contaminated sites (average ± standard error, N = 5), along with the tested NaCl concentrations. Letters indicate significant differences between treatments at $p < 0.05$; asterisks mark significant differences between populations at $p < 0.05$.

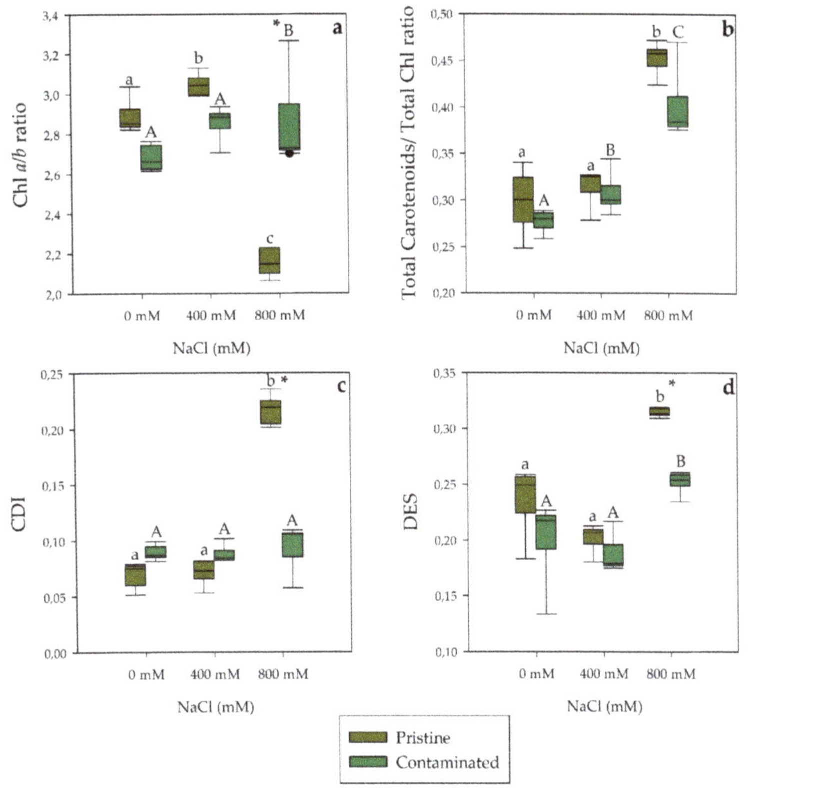

Figure 5. Leaves pigment ratios, (**a**) chlorophyll *a*/*b* ratio (Chl *a/b* ratio), (**b**) total carotenoid to total chlorophyll ratio, (**c**) chlorophyll degradation index (CDI), and (**d**) de-epoxidation state (DES) in *S. patens* individuals from pristine and heavy metal contaminated sites (average ± standard error, N = 5), along with the tested NaCl concentrations. Letters indicate significant differences between treatments at $p < 0.05$; asterisks mark significant differences between populations at $p < 0.05$.

2.3. *Antioxidant Enzymatic Activities*

Catalase activity presented a highly significant increase in salt treatments, being these values considerably higher in plants from the pristine marsh (Figure 6a). Contrarily ascorbate peroxidase activity and superoxide dismutase activity did not show any significant variations between both tested populations and salinity concentrations (Figure 6b,d). Guaiacol peroxidase activity showed a significant activity decrease through NaCl concentration gradient in individuals from the contaminated marsh, while its values did not vary with salinity concentration in those collected in the pristine site (Figure 6c). Finally, regarding the total protein content of the leaves, a decreasing tendency was observed in the pristine site samples, significant at 800 mM NaCl within and between population groups (Figure 6e).

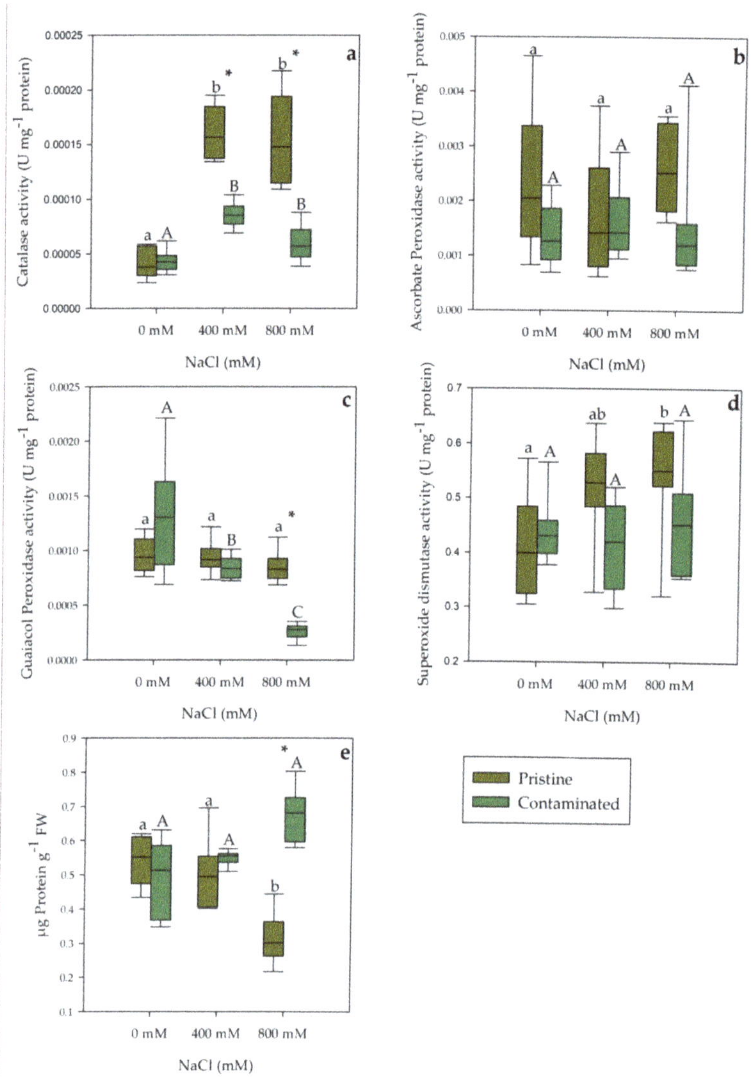

Figure 6. (**a**) Catalase, (**b**) ascorbate peroxidase, (**c**) guaiacol peroxidase, and (**d**) superoxide dismutase activities (U mg^{-1} protein) and (**e**) total protein content (μg Protein g^{-1} FW) in the leaves of *S. patens* from pristine and heavy metal contaminated sites (average ± standard error, N = 5), along with the tested NaCl concentrations. Letters indicate significant differences between treatments at $p < 0.05$; asterisks mark significant differences between populations at $p < 0.05$.

2.4. Fatty Acid Composition

Regarding fatty acid leaf content under salinity exposure (Table 1), the most abundant fatty acids found in the tested groups were palmitic acid (C16:0), linoleic acid (C18:2), and linolenic acid (C18:3). *Spartina patens* individuals from the pristine marsh when exposed to salinity showed an increase in palmitic acid, while the individuals from the contaminated site showed a decrease, displaying a significantly different trend between populations. An opposite trend, between the *S. patens* populations, was also found in the stearic acid (C18:0) concentration, increasing and decreasing through salinity treatments in the individuals from the contaminated and pristine marsh respectively. Trans-delta 3-hexadecenoic acid

(C16:1t) displayed a significantly higher percentage in the pristine marsh group in 0 mM and 800 mM NaCl. Both populations displayed a decrease in linolenic acid content as a result of NaCl treatments. Considering the fatty acid saturation classes in salt-treated leaves, both marsh populations displayed similar trends. However, saturated fatty acid (SFA) at 800 mM NaCl was found to be significantly higher in the contaminated site samples (Figure 7). The total fatty acid content of *S. patens* presents highly significant increases in both population treatments (Figure 8a). Contaminated site individuals, when exposed to increasing salinities, displayed an increasing trend in the C18:2/C18:3 ratio (Figure 8b), as well as an inverse trend in the double bond index (DBI; Figure 8c). In contrast, in NaCl treated plants from the pristine site no significant changes were observed.

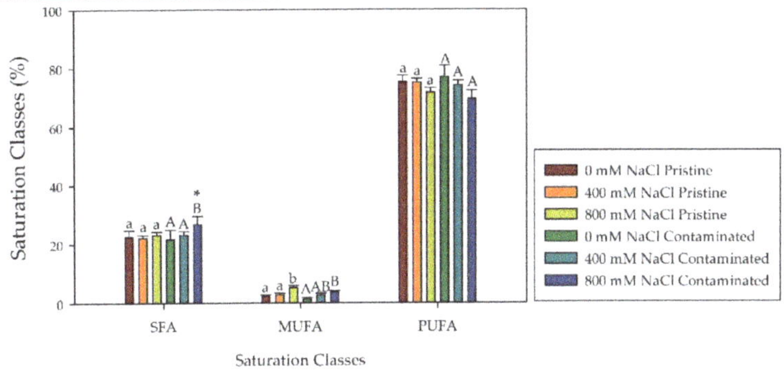

Figure 7. Saturated fatty acid (SFA), monounsaturated fatty acid (MUFA) and polyunsaturated fatty acid (PUFA) relative concentration (%, average ± standard error, N = 5) in *S. patens* leaves from pristine and heavy metal contaminated sites, along with the tested NaCl concentrations. Letters indicate significant differences between treatments at $p < 0.05$; asterisks mark significant differences between populations at $p < 0.05$.

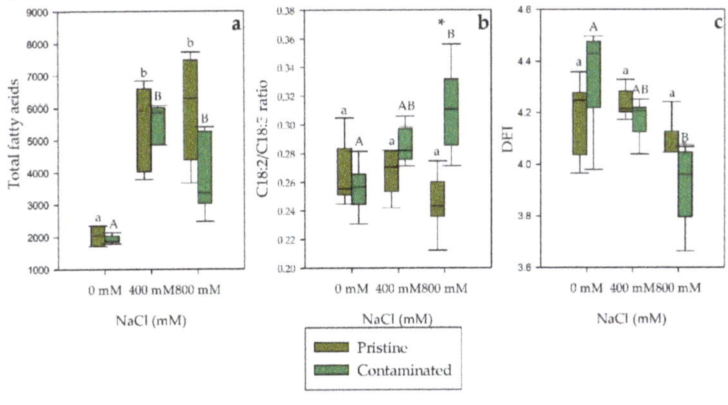

Figure 8. (**a**) Total fatty acid content (μg.g^{-1} FW), (**b**) linoleic acid to linolenic acid ratio (C18:2/C18:3 ratio), and (**c**) double-bound index (DBI) in *S. patens* leaves from pristine and heavy metal contaminated sites (average ± standard error, N = 5), along with the tested NaCl concentrations. Letters indicate significant differences between treatments at $p < 0.05$; asterisks mark significant differences between populations at $p < 0.05$.

Table 1. Fatty acid relative content (%, average ± standard error, N = 5) namely, pentadecanoic acid (C15:0), palmitic acid (C16:0), trans-delta 3-hexadecenoic acid (C16:1t), hexadecatrienoic acid (C16:3), stearic acid (C18:0), oleic acid (C18:1), linoleic acid (C18:2) and linolenic acid (C18:3) in *S. patens* leaves from pristine and heavy metal contaminated sites, along with the tested NaCl concentrations. Letters indicate significant differences between treatments at $p < 0.05$; asterisks mark significant differences between populations at $p < 0.05$.

Salinity (mM)		15:0	16:0		16:1t		16:3	18:0	18:1	18:2	18:3
0	Pristine	4.16 ± 0.42 ^{a}	12.12 ± 0.72 ^{a}	*	1.34 ± 0.37 ^{a}	*	1.13 ± 0.44 ^{a}	6.26 ± 1.68 ^{a}	0.97 ± 0.38 ^{a}	15.55 ± 0.66 ^{a}	58.46 ± 3.00 ^{a}
	Contaminated	2.82 ± 1.18 ^{A}	15.34 ± 1.78 ^{A}		0.42 ± 0.24 ^{A}		3.46 ± 1.57 ^{A}	0.94 ± 0.38 ^{A}	0.00 ± 0.00 ^{A}	15.64 ± 0.61 ^{A}	61.37 ± 3.65 ^{A}
400	Pristine	4.07 ± 0.85 ^{a}	12.78 ± 0.59 ^{a}	*	1.02 ± 0.41 ^{a}		0.00 ± 0.00 ^{a}	5.35 ± 0.87 ^{a}	1.74 ± 0.27 ^{a}	15.79 ± 0.96 ^{a}	59.21 ± 1.01 ^{a}
	Contaminated	3.20 ± 0.47 ^{A}	14.92 ± 1.03 ^{A}		0.96 ± 0.21 ^{A}		5.08 ± 0.98 ^{A}	1.71 ± 0.50 ^{A}	0.00 ± 0.00 ^{A}	16.49 ± 0.71 ^{A}	57.65 ± 1.41 ^{A}
800	Pristine	4.66 ± 1.20 ^{a}	16.03 ± 1.28 ^{b}	*	3.84 ± 0.86 ^{b}	*	2.47 ± 0.47 ^{a}	1.36 ± 0.18 ^{a}	0.23 ± 0.29 ^{a}	14.06 ± 1.25 ^{a}	57.25 ± 1.34 ^{a}
	Contaminated	4.01 ± 0.39 ^{B}	13.77 ± 0.93 ^{A}		1.59 ± 0.37 ^{A}		9.00 ± 3.05 ^{A}	2.17 ± 0.74 ^{A}	0.00 ± 0.00 ^{A}	16.41 ± 1.31 ^{A}	53.05 ± 2.87 ^{A}

2.5. Multivariate Classification

Gathering all the photochemical data (full Kautsky induction curve dataset) into a unifying canonical analysis of principal coordinates (CAP) the abovementioned differences and traits are highlighted in an integrative form. Moreover, the cross-validation step of this canonical analysis presented a highly elevated classification efficiency of more than 95% for allocation within groups, reinforcing the statical differences observed at the individual level of each of the photochemical traits as efficient descriptors of the populations' behavior along the tested salinity gradient (Figure 9a). The pristine marsh individuals were grouped and identified, sharing similar photochemical traits, while the individuals from the contaminated site evidence a clear separation under the exposure to different salinity values. A similar approach was performed regarding the leaf fatty acid profile, with the CAP projection based on these traits producing a different grouping profile. Intermediate salinity exposed samples from both populations shoed similar fatty acid profiles being grouped in the center of the projection alongside the samples from individuals collected at the contaminated site exposed to 0 mM NaCl (Figure 9b). Samples from the pristine site exposed to the lowest and highest salinity treatment tested were grouped differentially from the remaining samples. Worth noticing that the CAP analysis based on the fatty acid analysis showed a lower classification efficiency (approximately 70%). Both these CAP analyses show to highlight the different impacts of the tested salinity treatments in the photochemical and fatty acid metabolism, and the different feedbacks from each of the *S. patens* populations.

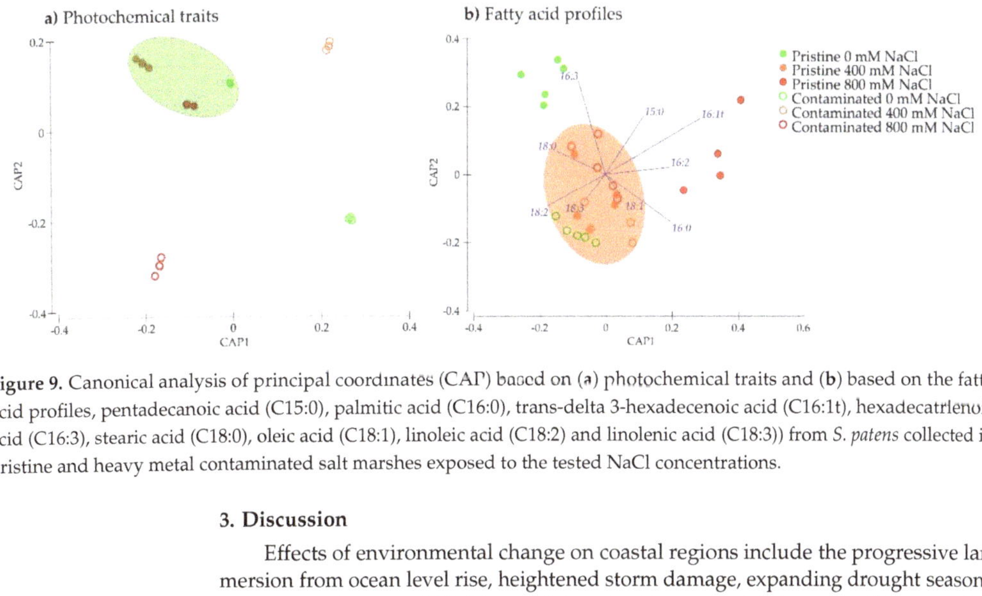

Figure 9. Canonical analysis of principal coordinates (CAP) based on (**a**) photochemical traits and (**b**) based on the fatty acid profiles, pentadecanoic acid (C15:0), palmitic acid (C16:0), trans-delta 3-hexadecenoic acid (C16:1t), hexadecatrienoic acid (C16:3), stearic acid (C18:0), oleic acid (C18:1), linoleic acid (C18:2) and linolenic acid (C18:3)) from *S. patens* collected in pristine and heavy metal contaminated salt marshes exposed to the tested NaCl concentrations.

3. Discussion

Effects of environmental change on coastal regions include the progressive land immersion from ocean level rise, heightened storm damage, expanding drought seasons, and temperature increase [21]. These abiotic alterations reveal to have significant implications on the environmental salinity gradient, with recent studies forecasting disturbing impacts in salinity concerning waterfront areas [22–24]. Therefore, salt marshes ecosystems will be largely affected, especially when considering the salinity concentration to be one of the major constraints of species frequency, distribution, and zonation along with the marsh profile [25,26]. However, complex and significant interspecific variations in the salinity responses of halophytes due to pre-conditioned histories can be a factor in the adaptation of neighboring marsh populations [20].

Photochemical analysis of *S. patens*, when exposed to salt treatments, confirmed that this species has a high degree of tolerance to salinity even at high NaCl concentrations as shown in previous studies [27,28]. Nevertheless, noticeable differences were shown between NaCl treatments as well as between pristine and contaminated marsh populations. Considering the electron transport chain and related parameters in salt-treated *S. patens*, it was observed a significant intraspecific difference at 400 and 800 mM NaCl. The individuals from the contaminated marsh, at 400 mM NaCl, displayed a significantly higher photosynthetic light efficiency, coupled with relatively low dissipation energy (DI/CS), which suggests a high electronic transport chain efficiency [20]. On the other hand, at 800 mM NaCl, contaminated marsh samples show a significantly lower maximum ETR and photosynthetic efficiency suggesting an inferior electron transport chain proficiency at higher salinity concentration when compared to the heavy metal contaminated individuals. According to the data acquired, in the salt exposed groups, there is a linear relationship between the variation found in the oxidized quinone pool size and the electron transport energy flux (ET/CS), the absorbed energy flux (ABS/CS), the trapped energy flux (TR/CS), as well as the available reaction centers (RC/CS), nonetheless, the populations showed significant differences among these parameters. Even though the contaminated site *S. patens*, at 400 and 800 mM NaCl, when compared to the pristine site samples, showed a more significant decrease in quinone pool size, the reduction in the electron transport energy flux was less significant, this can be explained by a better PS II efficiency associated with the lower dissipation energy flux found in the individuals from the contaminated marsh, especially at 400 mM NaCl was the intraspecific differences were found to be more significant [29]. Although the size of the oxidized quinone pool, in the individuals from the pristine location, displayed no significant changes when subjected to salinity, the number of quinone turnovers increased, showing lower quinone pool reduction rates [30]. In contrast, the individuals from the contaminated marsh exhibited a significant reduction of the quinone pool size and no significant changes in its turnover time, indicating tolerance mechanisms that allowed the maintenance of electronic flow rate from the reduced quinone pool to the electron transport chain [10].

The Xanthophyll cycle is a well-described mechanism of energy dissipation, commonly observed in halophytic plants [10,30,31]. To reduce energy overload within light-harvesting complexes (LHCs), the de-epoxidation of the violaxanthin pool towards the zeaxanthin is normally activated [32]. *Spartina patens* when exposed to salinity, only in 800 mM NaCl treatments in both population samples, showed a highly significant increase in de-epoxidation, reflection of a higher activity of the xanthophyll cycle attempting to scatter the excessive redox potential amassed inside the stroma. The activity shift to photoprotection in the higher salinities was also clear due to the significant rise of the total carotenoid to total chlorophyll ratio. As a possible countermeasure against reactive oxygen species [31] significant increases were observed in β-carotene and lutein, antioxidant acting, pigment concentrations in the salt exposed populations, in particular, this phenomenon was found with more significance at the higher salt concentration in the individuals from the pristine marsh. A highly significant increase in β-carotene was present in the contaminated site samples at 800 mM NaCl and both salinity treatment in the plants from the pristine location, as well as a highly significant increase in lutein at 800 mM NaCl. This may indicate a better ROS savaging capability by *S. patens* from the contaminated marsh.

The interaction of high NaCl concentrations with the cell organelles leads to increased production of ROS resulting in potentially harmful physiological reactions within the plant cells, affecting among others, proteins production and metabolism [33,34]. Halophytes built up a highly proficient system of enzymatic rapid responses toward salinity changes, immediately activated when the environmental conditions shift aside from the saline comfort zone [35]. When assessing the oxidative stress biomarkers in *S. patens*, discrepancies in the responses to the salinity stress between populations are clear. Contrary to what was found in the contaminated marsh plants, the pristine site individuals displayed an increase of antioxidant enzyme activities, revealed by the increase of superoxide dismutase and catalase activity when salt treated, significant at 800 mM NaCl [36]. This, coupled with a decrease in total protein content found in this same group, suggests that, when exposed to 800 mM NaCl, *S. patens* from the pristine marsh, when comparing to the contaminated marsh population, as a higher ROS production, as well as comparatively inferior scavenging mechanism of the ROS species [37,38].

The fatty acid profiles of the salt exposed halophytes presented similar responses amongst populations, but some differences are noteworthy. The linoleic (C18:2) and linolenic (C18:3) ratios, considered a salt stress indication, when under stress conditions the ratio shifts towards linolenic, since it is a membrane restructuring with lower amounts of polyunsaturated acids, thus inferior C18:3 concentration in the leaf is considered an adaptation to salt exposure [39]. The C18:2/ C18:3 ratio increased exclusively in the salinity treated *S. patens* from the contaminated site, therefore it can be suggested that these individuals are less stressed than those from the pristine location. Furthermore, C18:3 can also act as a direct non-enzymatic reactive oxygen species scavenger [40], which complies with, comparatively, lower ROS consequences found previously in the *S. patens* from the contaminated marsh. Furthermore, the population from the contaminated marsh displayed a highly significant rise in oleic acid (C18:1), known for improving the stabilization of light-harvesting complexes [41], seen by the positive significant correlation between LHC stress indicator chl *a*/ chl *b* ration and the C18:1 fatty acid significant correlation ($r^2 = 0.921$; $p < 0.05$). On the other hand, in the individuals from the pristine site, the correlation between these two variables is quite low ($r^2 = 0.161$; $p < 0.05$), indicating that this mechanism only occurs in the plants from the contaminated marsh. Tras-delta-3 hexadecenoic acid (C16:1t), exclusive to plastids [42] and the only strictly light-dependent fatty acid, enables the correct organization of light-harvesting antennae complexes [30,43–45]. When comparing the individuals subjected to 0 mM NaCl from both populations, a significant increasing trend was found in the C16:1t concentration of the individuals from the contaminated marsh, concomitant with the, previously determined, lower energy dissipation and reduced reaction centers turnover and closure rates found in the plants from the contaminated marsh, comparatively to those found in the individuals from the pristine marsh, proposing a better LHC organization and heath in the heavy metal affected *S. patens* when exposed to salt stress.

The overall physiological shift was observed in the CAP analysis where it was compared the physiological and photochemical variations of the individuals under the different NaCl treatments. The cross-validation provided an efficient approach to classify and assess the changes and effects in both populations [46]. When observing the multivariate analysis, NaCl treated *S. patens* from the pristine marsh showed a clear grouping at the photochemical changes, however, when using the fatty acids profile as the basis the grouping was seen in the contaminated marsh populations. This distinct classification efficiently displays *S. patens* intraspecific variation. The higher degree of efficiency in the classification of the samples observed in the photochemical traits-based CAP analysis indicates that not only this metabolism is more affected (thus producing more pronounced differences between sample groups) but also that has a higher ability to be used as biomarkers in similar studies comparing not only salinity treatments but also plant populations. Although fatty acid profiles are known to be sensitive to osmotic stress in this particular species as well as in other halophyte species when comparing the same species along a salinity gradient [28,47,48], this canonical approach loses sensitivity when comparing populations of the same species exposed to the same salinity treatments, pointing out to a prevalent role of the salinity treatment over the population origin, in this case, thus leading to less efficient fatty acid-based canonical analysis.

4. Material and Methods

4.1. Sampling Sites and Plant Material Collection

Sampling was carried out on the Tagus estuary, located in the western coast of Portugal, one of the larger estuaries in occidental Europe with an area of approximately 320 km^2 (38°44′ N, 9°03′ W; Figure 10). The estuary involves a watershed superior to 80,000 km^2 in

Spain and Portugal territories, being the second most significant hydrological basin in the whole Iberian Peninsula.

Spartina patens sampling was done during low tide in the southern part of the Tagus estuary in September 2017, on the same day and tidal period at two sampling sites: Alcochete salt marsh (38°45′ N, 8°56′ W) situated within the Tagus Estuary Natural Reserve and Rosário salt marsh (38°40′ N, 9°01′ W) in the vicinity of a former industrial area (Figure 10). Whole plants were excavated from the sediment and intact individuals were transported individually to the laboratory (in refrigerated bags and quickly transported (less than an hour) to the laboratory. Due to the high proximity between both sampling sites, the plant phenological cycle is not different, with very similar plants in terms of morphology and biomass between both sites (data not shown). The geographical location of both marshes prompts a differential metal contamination exposure from anthropogenic origins [28,31]. This is reflected in the bioavailable metal concentrations found in both marshes, with Alcochete sediments showing non-detectable bioavailable Cd concentrations, 0.023 ppm of Cu, 0.001 ppm of Ni, 0.022 ppm of Pb, and 0.052 ppm of Zn on average [31]. On the other hand, Rosário salt marsh sediments exhibited much higher bioavailable metals average concentration values, presenting 0.001 ppm of Cd, 0.034 ppm of Cu, 0.003 ppm of Ni, 0.116 ppm of Pb, and 0.233 ppm of Zn [20,33]. Considering these values, Alcochete marsh was classified as pristine and Rosário marsh as heavy metal contaminated.

At the laboratory, plant samples were gently washed to remove dust and sediments. *Spartina patens* intact tussocks were set in pots (N = 5) filled with perlite and irrigated with 1/4 Hoagland nutrient solution [49]. For experimental proposes, individuals were chosen to have all experimental units with individuals presenting similar height and apparent biomass (data not shown). Plants were placed in a phytoclimatic chamber programmed to simulate a natural light environment using a sinusoidal function (maximum PAR 300 µmol photons m^2 s^{-1}, 16/8 h day/night rhythm, 20/18 °C day/night temperature amplitude, relative humidity, 50 ± 2%), and kept under these conditions for 2 months to acclimate to the new growth conditions.

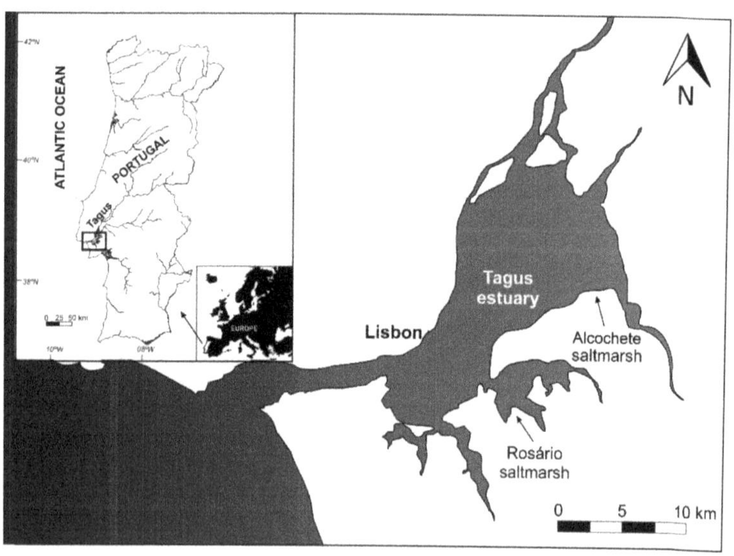

Figure 10. Tagus Estuary map with Alcochete (pristine) and Rosário (heavy metal contaminated) salt marshes sampling stations marked [50].

4.2. Experimental Setup

After the abovementioned adaptation period, *S. patens* individuals from both sites (pristine and contaminated) were separated into 3 groups with 5 replicate individuals (pots).

The sample groups were placed in a phytoclimatic chamber programmed to simulate a natural light environment using a sin function (maximum PAR 300 µmol photons m^2 s^{-1}, 16/8 h day/night rhythm, 20/18 °C day/night temperature amplitude, relative humidity, 50 ± 2%) and the Hoagland nutrient replaced, in two sample groups, with salinity treatment solution of 1/4 Hoagland solution supplemented with NaCl to attain the desired target salinities (400 and 800 mM). Exposure trials lasted for 7 days after which chlorophyll fluorescence measurements were made and consecutively, plants were harvested. Leaf samples for biochemical measurements were immediately flash-frozen in liquid-N_2 and stored at −80 °C until analysis.

4.3. Pulse Amplitude Modulated (PAM) Fluorometry

Modulated chlorophyll fluorescence measurements were made in attached leaves with a FluoroPen FP100 PAM (Photo System Instruments, Czech Republic). All the measurements in the dark-adapted state were made after the darkening of the leaves for at least 30 min. Rapid light curves (RLC) measurements, in dark-adapted leaves, were attained using the preprogrammed LC1 protocol of the FluorPen, consisting of a sequence of pulses from 0 to 500 µmol m^{-2} s^{-1}. Each ΦPS II measurement was used to calculate the electron transport rate (ETR) through photosystem II using the following equation: ETR = ΦPS II × PAR × 0.5, where PAR is the actinic photosynthetically active radiation generated by the FluoroPen and 0.5 assumes that the photons absorbed are equally partitioned between PS II and PSI [51]. Without knowledge of the actual amount of light being absorbed, fluorescence measurements can only be used as an approximation for electron transport [52–54]. Rapid light curves (RLC) were generated from the calculated ETRs versus irradiance applied plot and fitted to a double exponential decay function to quantify the characteristic parameters, alpha and ETR_{max} [55,56]. The OJIP transient (or Kautsky curves) depicts the rate of reduction kinetics of various components of PS II. This is obtained when a dark-adapted leaf is illuminated with the saturating light intensity of 3500 µmol m^{-2} s^{-1} then it exhibits a polyphasic rise in fluorescence (OJIP): level O represents all the open reaction centers at the onset of illumination with no reduction of Q_A (fluorescence intensity lasts for 10 ms); O to J transient indicates the net photochemical reduction of Q_A (the stable primary electron acceptor of PS II) to Q_A^- (lasts for 2 ms); the J to I transition is due to all reduced states of closed RCs such as Q_A^- Q_B^-, Q_A Q_B^{2-} and Q_A^- Q_B H_2 (lasts for 2–30 ms); P-step coincides with a maximum concentration of Q_A^- Q_B^2 with plastoquinol pool maximally reduced and also reflects a balance between the light incident at the PS II side and the rate of utilization of the chemical (potential) energy and the rate of heat dissipation [57]. Table 2 summarizes all the parameters that could be calculated from the fluorometric analysis.

Table 2. Summary of fluorometric analysis parameters and their description.

JIP-Test	
Rapid Light Curves (RLCs)	
rETR	Relative electron transport rate at each light intensity (rETR = QY × PAR × 0.5).
ETR_{max}	Maximum ETR after which photo-inhibition can be observed.
α	Photosynthetic efficiency, obtained from the initial slope of the RLC.
Area	Corresponds to the oxidized quinone pool size available for reduction and is a function of the area above the Kautsky plot.
N	Reaction center turnover rate.
S_M	Corresponds to the energy needed to close all reaction centers.
M_0	The net rate of PS II RC closure.
δR_0	PS I efficiency in reducing its electron acceptors.
P_G	Grouping probability, directly related to PS II antennae connectivity.
ABS/CS	Absorbed energy flux per cross-section.
TR/CS	Trapped energy flux per cross-section
ET/CS	Electron transport energy flux per cross-section.
DI/CS	Dissipated energy flux per cross-section.
RC/CS	The number of available reaction centers per cross-section.

4.4. Pigment Profiling

Ground freeze-dried leaf samples were extracted with 100% acetone added and subjected to an ultra-sound bath for 1 min to ensure complete disaggregation of the leaf material. Extraction occurred in the dark for 24 h at −20 °C, after which the samples were centrifuged at 4000× *g* at 4 °C for 15 min. Supernatants were scanned from 350 nm to 750 nm in 1 nm steps, using a dual-beam spectrophotometer (Shimadzu UV/VIS UV1601 Spectrophotometer). Finally, the detected pigment sample absorption spectra were analyzed and quantified employing Gauss-Peak Spectra (GPS) method [58]. The sample spectrum was analyzed, through the GPS fitting library, using SigmaPlot Software. This method is based on the sample spectrum fitting, by a linear combination, to the Gauss-peak spectra, that describes each pigment in the detected spectrum, identifying the samples pigment profile, chlorophyll *a*, chlorophyll *b*, auroxanthin, antheraxanthin, β-carotene, lutein, violaxanthin, and zeaxanthin.

For a better evaluation of the light-harvesting and photoprotection mechanisms, the De-Epoxidation State (DES) was calculated as:

$$\text{DES} = \frac{([\text{Antheraxanthin}] + [\text{Zeaxanthin}])}{([\text{Violaxanthin}] + [\text{Antheraxanthin}] + [\text{Zeaxanthin}])} \quad (1)$$

4.5. Leaf Fatty Acid Composition

Leaf fatty acid analyses were performed by direct trans-esterification of leaf samples as previously described [20,59,60]. Fatty acid methyl esters (FAME) were prepared in glass tubes containing the internal standard heptadecanoate (C17:0), methanol, and sulphuric acid, at 70 °C for one hour. After cooling down the FAME were extracted by adding petroleum and water, vortexed, centrifuged at 4000× *g* for 5 min. The upper layer was dried under a nitrogen stream in a water bath set to 37 °C. After evaporation, 50 µL of hexane was added to the residue and one µL of the solution separated in a gas chromatograph (Varian 3900, Palo Alto, CA, USA) equipped with a hydrogen flame-ionization detector using a fused silica 0.25 mm i.d. × 50 m capillary column (WCOT Fused Silica, CP-Sil 88 for FAME; Varian). The double-bound index (DBI) was calculated using the equation:

$$\text{DBI} = \frac{2\times\ ((16:1t+18:1)+\ 2\times 18:2+3\times\ (18:3+16:3))}{100} \quad (2)$$

4.6. Oxidative Stress Biomarkers

For enzyme extractions of *S. patens* leaf samples were retrieved from −80°C storage and extractions were performed according to Tiryakioglu et al. [61], at 4°C. Frozen leaves were homogenized in 50 mM sodium phosphate buffer (pH 7.6) supplemented with 0.1 mM Na-EDTA in a ceramic mortar with a proportion of 500 mg (FW) to 8 mL respectively. The homogenate was centrifuged at 8890× *g* for 20 minutes at 4 °C, and the supernatant was transferred to a test tube and used for the antioxidant enzyme analyses.

The enzyme activity measurements of catalase (CAT, EC.1.11.1.6.), Ascorbate peroxidase (APx, E.C. 1.11.1.11), Guaiacol peroxidase (GPX, E.C. 1.11.1.7), and Superoxide dismutase (SOD, E.C. 1.15.1.1) were performed in a dual-beam spectrophotometer (Shimadzu UV/VIS UV1601 Spectrophotometer) using quartz cuvettes. Catalase activity assays were performed according to the method of Teranishi et al. [62], by monitoring the H_2O_2 consumption and consequent decrease in absorbance at 240 nm (molar extinction coefficient of 39.4 mM^{-1} cm^{-1}). Ascorbate peroxidase was measured according to Tiryakioglu et al. [61], by observing the ascorbate oxidation and consequent absorbance reduction at 290 nm (molar extinction coefficient of 2.8 mM^{-1} cm^{-1}). Guaiacol peroxidase measurement was performed according to Bergmeyer et al. [63], by monitoring guaiacol oxidation products formation and its increase in absorbance during 60 seconds at 470 nm (molar extinction coefficient of 26.6 mM^{-1} cm^{-1}). Superoxide dismutase total activity was assayed according to the method of Marklund and Marklund [64], by measuring the oxidation rate of pyrogallol monitored at 325 nm. The autoxidation of pyrogallol was read without enzymatic extract during the same period and time interval for comparison enabling. Protein quantification was determined using the Bradford method [65].

Membrane lipid peroxidation quantification was performed in *S. patens* leaf samples according to Heath & Packer [66]. First, leaf samples were homogenized in a freshly prepared Thiobarbituric acid (TBA) solution (0.5% (w/v) TBA in 20% (w/v) Trichloroacetic acid), in a proportion of 100 mg FW to 2 mL of solution. The homogenate was incubated for 30 min at 95 °C, cooled on ice to stop the reaction, and centrifuged at 4000× g for 5 min at 4 °C. The absorbance was read at 532 nm and 600 nm in a Shimadzu UV-1601 spectrophotometer. Malondialdehyde (MDA) concentration was calculated using the molar extinction coefficient, 155 mM^{-1} cm^{-1} when applying the following equation:

$$A_{532\ nm} - A_{600nm} = [\mathrm{MDA}]mM \times \varepsilon\mathrm{MDA} \quad (3)$$

4.7. Statistical Analysis

Statistical analysis of the data derived from the previous analysis was made based on non-parametric tests, due to a lack of normality and homogeneity. The resultant effects of warming treatments in the different populations and salinity treatments were compared by performing Kruskal–Wallis test using Statistica Software (Statasoft, Tulsa, OK, USA). Significant and highly significant values were assumed when the probability value (p-value) was smaller than 0.05 and 0.01 respectively. Multivariate analysis was also conducted using Primer 6 software [67]. A Canonical Analysis of Principal Components (CAP) was also performed using the physiological traits as inputs, to test the efficiency of the variables in describing the populations' behavior under altered thermal environments, but also to analyze this behavior, producing a statistically tested canonical plot. To evaluate the changes in photochemical and fatty acid metabolism as a whole, a multivariate approach was applied [46]. Canonical analysis of principle (CAP) coordinates, using Euclidean distances, were used to visualize differences in multivariate space regarding studied photochemical variables and fatty acid relative composition, as well as to determine the allocation efficiency to different treatment groups. This multivariate approach is insensitive to heterogeneous data and frequently used to compare different sample groups using the intrinsic characteristics of each group (metabolic characteristics) [30,46,68]. Multivariate statistical analyses were conducted using Primer 6 software [67].

5. Conclusions

This study provides new insights on the relationship between environmental history and tolerance variation of *Spartina patens* to salinity. Biophysical and biochemical intraspecific data variation suggests that heavy metal pre-conditioning has a considerable and significant influence on the salinity tolerance mechanisms and salinity resistance of these plants. When comparing marshes, individuals from the pristine site appear to withstand the harshest photochemical consequences as seen by the decrease of the chlorophyll a/b ratio, through salt concentrations, opposite to the increasing tendency found in the pre-conditioned *S. patens*. These responses were correlated with the highly significant increase in oleic acid found only in *S. patens* from the contaminated marsh, indicating that these plants have an effective light-harvesting complexes stabilization mechanism. Moreover, individual from the pristine marsh exhibited impairments in the LHC mechanisms, coupled with the comparatively deficient energy dissipation mechanisms at high salinities, seems to lead to higher ROS generation and as a consequence of higher plant damage degree. Therefore, it could be concluded that, as salinity increases, the heavy metal contaminated marsh (i.e., Rosario) may generate a more aggressive *S. patens* invasion and spreading, and consequently a more negative ecological effect in the marsh biodiversity especially at high salinities (800 mM NaCl) where the fitness variation between populations is more significant. Therefore, pre-conditioning history seems to potentially be a key factor in the understanding of intraspecies response to future constraints and, subsequently, essential when considering ecological evolution to climate change realities.

Author Contributions: Conceptualization, J.C. and B.D.; methodology, J.C., A.R.M. and B.D; software, J.C. and B.D.; validation, E.M.-N., S.R.-G. and I.C.; formal analysis, J.C.; investigation, J.C. and J.A.P.-R.; resources, A.R.M. and B.D.; data curation, J.C. and B.D.; writing—original draft preparation, J.C.; writing—review and editing, E.M.-N., S.R.-G., A.R.M., I.C. and B.D.; visualization, supervision, project administration and funding acquisition, I.C. and B.D. All authors have read and agreed to the published version of the manuscript.

Funding: This research was funded by Fundação para a Ciência e a Tecnologia (FCT), via the project grant UIDB/04292/2020, UID/MULTI/04046/2019 and PTDC/CTA-AMB/30056/2017. Work was also funded by MAR 2020 program via the Project RESTAURA2020 (16-01-04-FMP-0014). B. Duarte was supported by investigation contracts (CEECIND/00511/2017).

Conflicts of Interest: The authors declare no conflict of interest.

References

1. Barros, V.R.; Field, C.B.; Dokken, D.J.; Mastrandrea, M.D.; Mach, K.J.; Bilir, T.E.; Chatterjee, M.; Ebi, K.L.; Estrada, Y.O.; Genova, R.C.; et al. *Climate Change 2014 Impacts, Adaptation and Vulnerability Part B: Regional Aspects: Working Group Ii Contribution to the Fifth Assessment Report of the Intergovernmental Panel on Climate Change*; Barros, V.R., Field, C.B., Dokken, D.J., Mastrandrea, M.D., Mach, K.J., Eds.; Cambridge University Press: Cambridge, UK, 2014. [CrossRef]
2. Shiklomanov, I.A. Appraisal and assessment of world water resources. *Water Int.* **2000**, *25*. [CrossRef]
3. Glick, B.R.; Cheng, Z.; Czarny, J.; Duan, J. Promotion of Plant Growth by ACC Deaminase-Producing Soil Bacteria. *Eur. J. Plant Pathol.* **2007**. [CrossRef]
4. Shrivastava, P.; Kumar, R. Soil Salinity: A Serious Environmental Issue and Plant Growth Promoting Bacteria as One of the Tools for Its Alleviation. *Saudi J. Biol. Sci.* **2015**. [CrossRef]
5. Bacilio, M.; Rodriguez, H.; Moreno, M.; Hernandez, J.P.; Bashan, Y. Mitigation of Salt Stress in Wheat Seedlings by a Gfp-Tagged Azospirillum Lipoferum. *Biol. Fertil. Soils* **2004**, *40*. [CrossRef]
6. Munns, R.; Tester, M. Mechanisms of Salinity Tolerance. *Annu. Rev. Plant Biol.* **2008**. [CrossRef] [PubMed]
7. Akbarimoghaddam, H.; Galavi, M.; Ghanbari, A.; Panjehkeh, N. Salinity Effects on Seed Germination and Seedling Growth of Bread Wheat Cultivars. *Trakia J. Sci.* **2011**, *9*, 43–50.
8. Jamil, A.; Riaz, S.; Ashraf, M.; Foolad, M.R. Gene Expression Profiling of Plants under Salt Stress. *Crit. Rev. Plant Sci.* **2011**. [CrossRef]
9. Flowers, T.J.; Colmer, T.D. Salinity Tolerance in Halophytes. *New Phytol.* **2008**. [CrossRef] [PubMed]
10. Duarte, B.; Santos, D.; Marques, J.C.; Caçador, I. Biophysical Probing of Spartina Maritima Photo-System II Changes during Prolonged Tidal Submersion Periods. *Plant Physiol. Biochem.* **2014**, *77*. [CrossRef]
11. Adam, P. Saltmarshes in a Time of Change. *Environ. Conserv.* **2002**, *29*. [CrossRef]
12. Caçador, I.; Tibério, S.; Cabral, H.N. Species zonation in corroios salt marsh in the tagus estuary (Portugal) and its dynamics in the past fifty years. *Hydrobiologia* **2007**, *587*, 205–211. [CrossRef]
13. Jump, A.S.; Peñuelas, J. Running to Stand Still: Adaptation and the Response of Plants to Rapid Climate Change. *Ecol. Lett.* **2005**, *8*, 1010–1020. [CrossRef]
14. Gratani, L.; Crescente, M.F. Phenology and Leaf Adaptive Strategies of Mediterranean Maquis Plants. *Ecol. Mediterr.* **1997**, *23*, 11–19. [CrossRef]
15. Valladares, F.; Gianoli, E.; Gómez, J.M. Ecological Limits to Plant Phenotypic Plasticity. *New Phytol.* **2007**, *176*, 749–763. [CrossRef] [PubMed]
16. Duarte, B.; Baeta, A.; Rousseau-Gueutin, M.; Ainouche, M.; Marques, J.C.; Caçador, I. A Tale of Two Spartinas: Climatic, Photobiological and Isotopic Insights on the Fitness of Non-Indigenous versus Native Species. *Estuar. Coast. Shelf Sci.* **2015**, *167*, 178–190. [CrossRef]
17. Caçador, I.; Vale, C.; Catarino, F. The Influence of Plants on Concentration and Fractionation of Zn, Pb and Cu in Salt Marsh Sediments (Tagus Estuary, Portugal). *J. Aquat. Ecosyst. Stress Recov.* **1996**, *5*. [CrossRef]
18. Flowers, T.J. Improving crop salt tolerance. *J. Exp. Bot.* **2004**, *55*. [CrossRef]
19. Hester, M.W.; Mendelssohn, I.A.; Mckee, K.L. Intraspecific Variation in Salt Tolerance and Morphology in the Coastal Grass Spartina Patens (Poaceae). *Am. J. Bot.* **1996**, *83*. [CrossRef]
20. Carreiras, J.; Alberto Pérez-Romero, J.; Mateos-Naranjo, E.; Redondo-Gómez, S.; Rita Matos, A.; Caçador, I.; Duarte, B. The Effect of Heavy Metal Contamination Pre-Conditioning in the Heat Stress Tolerance of Native and Invasive Mediterranean Halophytes. *Ecol. Indic.* **2020**, *111*. [CrossRef]
21. Lee, J.-Y.; Marotzke, J.; Nala, G.; Cao, L.; Corti, S.; Dunne, J.P.; Engelbrecht, F.; Fisher, E.; Fyfe, J.C.; Jones, C.; et al. Climate Change 2021: The Physical Science Basis. Contribution of Working Group I to the Sixth Assessment Report of the Intergovernmental Panel on Climate Change. *Cambridge Univ. Press* **2021**, *No. In Press*, 3949.
22. Rodrigues, M.; Fortunato, A.B. Assessment of a Three-Dimensional Baroclinic Circulation Model of the Tagus Estuary (Portugal). *AIMS Environ. Sci.* **2017**, *4*, 763–787. [CrossRef]

23. Dasgupta, S.; Hossain, M.M.; Huq, M.; Wheeler, D. Climate Change and Soil Salinity: The Case of Coastal Bangladesh. *Ambio* **2015**, *44*. [CrossRef]
24. Wheeler, D. Quantifying Vulnerability to Climate Change: Implications for Adaptation Assistance. *SSRN Electron. J.* **2012**. [CrossRef]
25. Duarte, B.; Sleimi, N.; Cagador, I. Biophysical and Biochemical Constraints Imposed by Salt Stress: Learning from Halophytes. *Front. Plant Sci.* **2014**, *5*. [CrossRef] [PubMed]
26. Silvestri, S.; Defina, A.; Marani, M. Tidal Regime, Salinity and Salt Marsh Plant Zonation. *Estuar. Coast. Shelf Sci.* **2005**, *62*. [CrossRef]
27. Duarte, B.; Santos, D.; Marques, J.C.; Caçador, I. Ecophysiological Constraints of Two Invasive Plant Species under a Saline Gradient: Halophytes versus Glycophytes. *Estuar. Coast. Shelf Sci.* **2015**, *167*, 154–165. [CrossRef]
28. Duarte, B.; Matos, A.R.; Marques, J.C.; Caçador, I. Leaf Fatty Acid Remodeling in the Salt-Excreting Halophytic Grass Spartina Patens along a Salinity Gradient. *Plant Physiol. Biochem.* **2018**, *124*, 112–116. [CrossRef] [PubMed]
29. Kalaji, H.M.; Govindjee; Bosa, K.; Kościelniak, J.; Zuk-Gołaszewska, K. Effects of Salt Stress on Photosystem II Efficiency and CO2 Assimilation of Two Syrian Barley Landraces. *Environ. Exp. Bot.* **2011**, *73*. [CrossRef]
30. Duarte, B.; Cabrita, M.T.; Gameiro, C.; Matos, A.R.; Godinho, R.; Marques, J.C.; Caçador, I. Disentangling the Photochemical Salinity Tolerance in Aster Tripolium, L.: Connecting Biophysical Traits with Changes in Fatty Acid Composition. *Plant Biol.* **2017**, *19*. [CrossRef] [PubMed]
31. Duarte, B.; Santos, D.; Marques, J.C.; Caçador, I. Ecophysiological Adaptations of Two Halophytes to Salt Stress: Photosynthesis, PS II Photochemistry and Anti-Oxidant Feedback-Implications for Resilience in Climate Change. *Plant Physiol. Biochem.* **2013**, *67*. [CrossRef] [PubMed]
32. Wentworth, M.; Ruban, A.V.; Horton, P. Chlorophyll Fluorescence Quenching in Isolated Light Harvesting Complexes Induced by Zeaxanthin. *FEBS Lett.* **2000**, *471*. [CrossRef]
33. Duarte, B.; Santos, D.; Caçador, I. Halophyte anti-oxidant feedback seasonality in two salt marshes with different degrees of metal contamination: Search for an efficient biomarker. *Funct. Plant Biol.* **2013**, *40*. [CrossRef]
34. Kaya, C.; Ashraf, M.; Alyemeni, M.N.; Ahmad, P. The Role of Endogenous Nitric Oxide in Salicylic Acid-Induced up-Regulation of Ascorbate-Glutathione Cycle Involved in Salinity Tolerance of Pepper (*Capsicum Annuum* L.) Plants. *Plant Physiol. Biochem.* **2020**, *147*, 10–20. [CrossRef]
35. Zhou, H.; Lin, H.; Chen, S.; Becker, K.; Yang, Y.; Zhao, J.; Kudla, J.; Schumaker, K.S.; Guo, Y. Inhibition of the Arabidopsis Salt Overly Sensitive Pathway by 14-3-3 Proteins. *Plant Cell* **2014**, *26*. [CrossRef] [PubMed]
36. Sheteiwy, M.S.; Shao, H.; Qi, W.; Hamoud, Y.A.; Shaghaleh, H.; Khan, N.U.; Yang, R.; Tang, B. GABA-Alleviated Oxidative Injury Induced by Salinity, Osmotic Stress and Their Combination by Regulating Cellular and Molecular Signals in Rice. *Int. J. Mol. Sci.* **2019**, *20*, 5709. [CrossRef] [PubMed]
37. Ahanger, M.A.; Aziz, U.; Alsahli, A.A.; Alyemeni, M.N.; Ahmad, P. Influence of Exogenous Salicylic Acid and Nitric Oxide on Growth, Photosynthesis and Ascorbate- Glutathione Cycle in Salt Stressed Vigna Angularis. *Biomolecules* **2020**, *10*, 42. [CrossRef] [PubMed]
38. Khan, I.; Raza, M.A.; Awan, S.A.; Shah, G.A.; Rizwan, M.; Ali, B.; Tariq, R.; Hassan, M.J.; Alyemeni, M.N.; Brestic, M.; et al. Amelioration of Salt Induced Toxicity in Pearl Millet by Seed Priming with Silver Nanoparticles (AgNPs): The Oxidative Damage, Antioxidant Enzymes and Ions Uptake Are Major Determinants of Salt Tolerant Capacity. *Plant Physiol. Biochem.* **2020**, *156*, 221–232. [CrossRef] [PubMed]
39. Ben Hamed, K.; Ben Youssef, N.; Ranieri, A.; Zarrouk, M.; Abdelly, C. Changes in Content and Fatty Acid Profiles of Total Lipids and Sulfolipids in the Halophyte Crithmum Maritimum under Salt Stress. *J. Plant Physiol.* **2005**, *162*. [CrossRef]
40. Mène-Saffrané, L.; Dubugnon, L.; Chételat, A.; Stolz, S.; Gouhier-Darimont, C.; Farmer, E.E. Nonenzymatic Oxidation of Trienoic Fatty Acids Contributes to Reactive Oxygen Species Management in Arabidopsis. *J. Biol. Chem.* **2009**, *284*. [CrossRef]
41. Siefermann-Harms, D.; Angerhofer, A. Evidence for an O2-Barrier in the Light-Harvesting Chlorophyll-a/b-Protein Complex LHC II. *Photosynth. Res.* **1998**, *55*. [CrossRef]
42. Trémolières, A.; Siegenthaler, P.-A. Reconstitution of photosynthetic structures and activities with lipids. *Lipids Photosynth. Struct. Funct. Genet.* **1998**, *6*. [CrossRef]
43. Hobe, S.; Prytulla, S.; Kühlbrandt, W.; Paulsen, H. Trimerization and Crystallization of Reconstituted Light-Harvesting Chlorophyll a/b Complex. *EMBO J.* **1994**, *13*. [CrossRef]
44. Hobe, S.; Förster, R.; Klingler, J.; Paulsen, H. N-Proximal Sequence Motif in Light-Harvesting Chlorophyll a/b-Binding Protein Is Essential for the Trimerization of Light-Harvesting Chlorophyll Alb Complex. *Biochemistry* **1995**, *34*. [CrossRef] [PubMed]
45. Nußberger, S.; Dörr, K.; Wang, D.N.; Kühlbrandt, W. Lipid-Protein Interactions in Crystals of Plant Light-Harvesting Complex. *J. Mol. Biol.* **1993**, *234*. [CrossRef] [PubMed]
46. Duarte, B.; Prata, D.; Matos, A.R.; Cabrita, M.T.; Caçador, I.; Marques, J.C.; Cabral, H.N.; Reis-Santos, P.; Fonseca, V.F. Ecotoxicity of the Lipid-Lowering Drug Bezafibrate on the Bioenergetics and Lipid Metabolism of the Diatom Phaeodactylum Tricornutum. *Sci. Total Environ.* **2019**, *650*. [CrossRef]
47. Duarte, B.; Caçador, I.; Matos, A.R. Lipid Landscape Remodelling in Sarcocornia Fruticosa Green and Red Physiotypes. *Plant Physiol. Biochem.* **2020**, *157*. [CrossRef]

48. Duarte, B.; Matos, A.R.; Caçador, I. Photobiological and Lipidic Responses Reveal the Drought Tolerance of Aster Tripolium Cultivated under Severe and Moderate Drought: Perspectives for Arid Agriculture in the Mediterranean. *Plant Physiol. Biochem.* **2020**, *154*. [CrossRef]
49. Hoagland, D.R.; Arnon, D.I. Preparing the Nutrient Solution. *Water-Cult. Method Grow. Plants Soil* **1950**, *347*, 2–33.
50. Pedro, S.; Duarte, B.; Raposo de Almeida, P.; Caçador, I. Metal Speciation in Salt Marsh Sediments: Influence of Halophyte Vegetation in Salt Marshes with Different Morphology. *Estuar. Coast. Shelf Sci.* **2015**, *167*. [CrossRef]
51. Genty, B.; Briantais, J.M.; Baker, N.R. The Relationship between the Quantum Yield of Photosynthetic Electron Transport and Quenching of Chlorophyll Fluorescence. *Biochim. Biophys. Acta-Gen. Subj.* **1989**, *990*. [CrossRef]
52. Beer, S.; Björk, M. Measuring Rates of Photosynthesis of Two Tropical Seagrasses by Pulse Amplitude Modulated (PAM) Fluorometry. *Aquat. Bot.* **2000**, *66*. [CrossRef]
53. Beer, S.; Vilenkin, B.; Weil, A.; Veste, M.; Susel, L.; Eshel, A. Measuring Photosynthetic Rates in Seagrasses by Pulse Amplitude Modulated (PAM) Fluorometry. *Mar. Ecol. Prog. Ser.* **1998**, *174*. [CrossRef]
54. Runcie, J.W.; Durako, M.J. Among-Shoot Variability and Leaf-Specific Absorptance Characteristics Affect Diel Estimates of in Situ Electron Transport of Posidonia Australis. *Aquat. Bot.* **2004**, *80*. [CrossRef]
55. Marshall, H.L.; Geider, R.J.; Flynn, K.J. A Mechanistic Model of Photoinhibition. *New Phytol.* **2000**, *145*. [CrossRef]
56. Platt, T.; Gallegos, C.L.; Harrison, W.G. Photoinhibition of photosynthesis in natural assemblages of marine phytoplankton. *J. Mar. Res.* **1980**, *38*, 687–701.
57. Zhu, X.G.; Govindjee; Baker, N.R.; DeSturler, E.; Ort, D.R.; Long, S.P. Chlorophyll a Fluorescence Induction Kinetics in Leaves Predicted from a Model Describing Each Discrete Step of Excitation Energy and Electron Transfer Associated with Photosystem II. *Planta* **2005**, *223*. [CrossRef] [PubMed]
58. Küpper, H.; Seibert, S.; Parameswaran, A. Fast, Sensitive and Inexpensive Alternative to Analytical Pigment HPLC: Quantification of Chlorophylls and Carotenoids in Crude Extracts by Fitting with Gauss Peak Spectra. *Anal. Chem.* **2007**, *79*. [CrossRef] [PubMed]
59. Duarte, B.; Pedro, S.; Marques, J.C.; Adão, H.; Caçador, I. Zostera Noltii Development Probing Using Chlorophyll a Transient Analysis (JIP-Test) under Field Conditions: Integrating Physiological Insights into a Photochemical Stress Index. *Ecol. Indic.* **2017**, *76*, 219–229. [CrossRef]
60. Esquível, M.G.; Matos, A.R.; Marques Silva, J. Rubisco Mutants of Chlamydomonas Reinhardtii Display Divergent Photosynthetic Parameters and Lipid Allocation. *Appl. Microbiol. Biotechnol.* **2017**, *101*. [CrossRef] [PubMed]
61. Tiryakioglu, M.; Eker, S.; Ozkutlu, F.; Husted, S.; Cakmak, I. Antioxidant Defense System and Cadmium Uptake in Barley Genotypes Differing in Cadmium Tolerance. *J. Trace Elem. Med. Biol.* **2006**, *20*. [CrossRef]
62. Teranishi, Y.; Tanaka, A.; Osumi, M.; Fukui, S. Catalase Activities of Hydrocarbon-Utilizing Candida Yeasts. *Agric. Biol. Chem.* **1974**, *38*. [CrossRef]
63. Bergmeyer, H.U.; Gawehn, K.; Grassl, M. Lactatedehydrogenase, UV-assay with pyruvate and NADH. *Methods Enzym. Anal.* **1974**, *2*, 574–579. [CrossRef]
64. Marklund, S.; Marklund, G. Involvement of the Superoxide Anion Radical in the Autoxidation of Pyrogallol and a Convenient Assay for Superoxide Dismutase. *Eur. J. Biochem.* **1974**, *47*. [CrossRef]
65. Bradford, M.M. A Rapid and Sensitive Method for the Quantitation of Microgram Quantities of Protein Utilizing the Principle of Protein-Dye Binding. *Anal. Biochem.* **1976**, *72*, 248–254. [CrossRef]
66. Heath, R.L.; Packer, L. Photoperoxidation in Isolated Chloroplasts. I. Kinetics and Stoichiometry of Fatty Acid Peroxidation. *Arch. Biochem. Biophys.* **1968**, *125*. [CrossRef]
67. Clarke, K.R.; Gorley, R.N. *PRIMER v6: User Manual/Tutorial.*, 1st ed.; PRIMER-E: Plymouth, UK, 2006; 192p.
68. Cabrita, M.T.; Duarte, B.; Gameiro, C.; Godinho, R.M.; Caçador, I. Photochemical Features and Trace Element Substituted Chlorophylls as Early Detection Biomarkers of Metal Exposure in the Model Diatom Phaeodactylum Tricornutum. *Ecol. Indic.* **2018**, *95*. [CrossRef]

plants

Article

Plant Growth Regulators Application Enhance Tolerance to Salinity and Benefit the Halophyte *Plantago coronopus* in Saline Agriculture

Milagros Bueno [1,*] and María del Pilar Cordovilla [1,2]

1 Plant Physiology Laboratory, Department Animal Biology, Plant Biology and Ecology, Faculty of Experimental Science, University of Jaén, Paraje Las Lagunillas, E-23071 Jaén, Spain; mpilar@ujaen.es
2 Center for Advances Studies in Olive Grove and Olive Oils, Faculty of Experimental Science, University of Jaén, Paraje Las Lagunillas, E-23071 Jaén, Spain
* Correspondence: mbueno@ujaen.es; Tel.: +34-953-212-140

Abstract: Climate change, soil salinisation and desertification, intensive agriculture and the poor quality of irrigation water all create serious problems for the agriculture that supplies the world with food. Halophyte cultivation could constitute an alternative to glycophytic cultures and help resolve these issues. *Plantago coronopus* can be used in biosaline agriculture as it tolerates salt concentrations of 100 mM NaCl. To increase the salt tolerance of this plant, plant growth regulators such as polyamine spermidine, salicylic acid, gibberellins, cytokinins, and auxins were added in a hydroponic culture before the irrigation of NaCl (200 mM). In 45-day-old plants, dry weight, water content, osmolyte (sorbitol), antioxidants (phenols, flavonoids), polyamines (putrescine, spermidine, spermine (free, bound, and conjugated forms)) and ethylene were determined. In non-saline conditions, all plant regulators improved growth while in plants treated with salt, spermidine application was the most effective in improving growth, osmolyte accumulation (43%) and an increase of antioxidants (24%) in *P. coronopus*. The pretreatments that increase the sorbitol content, endogenous amines (bound spermine fraction), phenols and flavonoids may be the most effective in protecting to *P. coronopus* against stress and, therefore, could contribute to improving the tolerance to salinity and increase nutritional quality of *P. coronopus*.

Keywords: antioxidants; climate change; growth; osmolyte accumulation; phytohormones; polyamines; salicylic acid; biosaline agriculture; salt tolerance

Citation: Bueno, M.; Cordovilla, M.d.P. Plant Growth Regulators Application Enhance Tolerance to Salinity and Benefit the Halophyte *Plantago coronopus* in Saline Agriculture. *Plants* **2021**, *10*, 1872. https://doi.org/10.3390/plants10091872

Academic Editors: Juan Barceló, Oscar Vicente and Marius-Nicusor Grigore

Received: 6 August 2021
Accepted: 7 September 2021
Published: 10 September 2021

1. Introduction

The increase in the world's population, intensive agriculture, poor quality irrigation water, the decrease in the amount of arable land, desertification, soil salinization, and climate change are all factors that have provoked a decrease in crop quality and yields; therefore, application of innovative techniques could improve crop performance [1–3]. Glycophytes are normally used in agriculture, but in a saline environment, they are subjected to osmotic stress and ionic toxicity, factors that negatively affect germination, growth, and crop yield; thus, identifying alternative salt-tolerant crops that can facilitate ecological rehabilitation and restoration and biosaline agriculture should be a priority research area in current agriculture (http://www.sussex.ac.uk/affiliates/halophytes, accessed on 14 June 2021) [4]. Plants halophytes thrive in saline habitats, and can survive in extreme conditions (arid inlands, subtropical habitats, and temperate zones) [5,6]; in addition, these plants possess a series of strategies at anatomical, morphological, physiological, biochemical, and genetic level that allow them to survive to different habitats [7,8]. These strategies are wide-ranging and include phenotypic plasticity, dilution or salt excretion (succulence, salt glands, bladder hairs), decreased transpiration, stomatic and CO_2 resistance control, water-use efficiency, C3-C4-CAM pathway, high K^+/Na^+ compartmentalization (through the Na^+/H^+

antiporter of tonoplast and plasma membrane), osmolyte accumulation (polysaccharides, amino acids, polyols), antioxidant systems activation (for protection of photosynthetic apparatus, biomembranes and nucleic acids), the modulation of plant growth regulators, and the expression of certain gene (up-regulating osmolytes and antioxidants) that allows them to survive in a wide variety of environmental conditions [7–11]. On the other hand, halophytes can be used directly as a possible alternative to glycophytes, biofuel-producing crops, fodder and animal feeds, oilseeds and proteins crops, medicinal plants, and in phytoremediation [12–19]. Biosaline agriculture has three main advantages: the recovery of saline and degraded soils, its ability to use wastewater from agriculture, and the increase in the production of metabolites with better nutritional quality [12–14,16–20].

In general, plant growth regulators (PGRs) are used to improve crop production and increase to abiotic stress tolerance in glycophytes [21,22]. In saline conditions, PGRs could improve halophytes tolerance for a better crop production. Nevertheless, little is known about PGRs in halophytes and their responses to abiotic stress [23]. These compounds modulate different stages from seed germination to fruit development, ripening and senescence. They also are related to abiotic stress tolerance, and regulate the root: shoot ratio, control stomatal resistance, regulate antioxidant enzymes, delay leaf senescence and act as signal molecules [21]. Auxins regulate cell elongation, vascular tissue development and apical dominance [24]. Cytokinins control cell division, chloroplast biogenesis, leaf senescence, shoot differentiation, anthocyanin production and photomorphogenic development [25]. Gibberellic acid induces seed germination, leaf and stem elongation, favours flowering and fruit development [26,27]. Polyamine application [putrescine (Put), spermidine (Spd), and spermine (Spm)] in agricultural crops serve to protect plants against stress, modulating the homeostasis of reactive oxygen species (ROS), regulating antioxidant systems, cation transport across plant membrane, osmoregulation, and directly or indirectly regulate gene expression [28,29]. Finally, salicylic acid treatment favours the accumulation of osmolytes, alleviates photosynthesis and enhance the upregulation of antioxidant systems in some species [30,31]. We focused our study of PGRs irrigation on the cultivation of *Plantago coronopus*, a halophyte native to the Mediterranean region (South Spain) [32].

Plantago coronopus L. (Family Plantaginaceae) inhabits marine cliffs, marshes, and endorheic basins at altitudes up to 800 m (a.s.l.). This halophyte is annual or biennial, with leaves with central veins arranged in basal rosettes measuring 2–20 cm length. Its flowers are produced in spikes and appear in April–October; its seeds are small and brown. It is typically found in saltmarshes in SE Spain [33]. This plant has photosynthesis pathway C3, osmolytes (sorbitol and proline) [34], and antioxidants (phenols and polyamines) [32]. Its mechanisms of tolerance to salinity have been investigated by several authors [32,34–36]. Transport of toxic ions (Na^+ and Cl^-) to aerial part, and their accumulation in vacuole, in addition to osmotic adjustment in its cytoplasm due to high concentrations of osmolytes allow develop succulence and therefore tolerate a certain degree of salinity [34]. On the other hand, this halophyte is used in biosaline agriculture as its edible leaves are greatly appreciated in salads due to their mild salty taste, crunchy texture, and excellent nutritional value [high content of phenols, amino acids (phenylalanine, tyrosine) and minerals (potassium, calcium, magnesium, sodium, etc.). *Plantago coronopus* showed a higher chlorophyll and flavonoids contents when it was grown in a Se enriched medium. These microgreens showed better nutraceutical value. On the other hand, these herbs grown in the open air presented a better development that in greenhouses, demonstrating the potential of this halophyte in saline agriculture [20,35,37–41].

The following PGRs were added to a hydroponic culture of *P. coronopus*: auxins (indole-acetic acid), cytokinins (Kinetin), gibberellic acid (GA_3), polyamine (spermidine) and salicylic acid before NaCl (200 mM) application. After 21 days of growth in the absence or presence of salt, dry weight, water content, sorbitol, phenols, flavonoids, endogenous polyamines [putrescine, spermidine, spermine (free, conjugated and bound)], and ethylene were determined. We wanted to identify which PGRs produced the best results to investigate: (1) ways to improve its tolerance to salinity, (2) boost its growth, and (3) increase the nutritional quality of this species. The results could provide technical guidance for increasing the cultivation of this halophyte and the benefits that it provides.

2. Results

2.1. Effect of Plant Growth Regulators (PGRs) Application on Growth of P. coronopus

Previous works by our research group showed that *P. coronopus* seeds collected from Brujuelo saltmarsh in Jaén (Spain) and cultivated hydroponically showed similar dry weight at 0 and 100 mM NaCl and a decrease at 200 mM NaCl (whole plant) [32]. We decided to choose 0 mM and 200 mM NaCl for the cultivation of this halophyte. Growth parameters such as dry weight and water content, at 45 days old, are shown in Table 1. In non-saline conditions, a positive effect on stem + leaves dry weight (SLDW) and root dry weight (RDW) was observed; being pretreatments Spd and SA ($p \leq 0.05$) whose having the highest values, above all in RDW (A in Table 1). In the case of Spd, the increases in SLDW and RDW were 47% and 86%, respectively. In water content Spd and SA also had the highest values of all studied pretreatments, especially in roots (increase of 9% compared to control). In saline pretreatments (B in Table 1), Kinetin + salt and Spd + salt obtained the best results for SLDW, while IAA + salt and Spd + salt had the best values for RDW. In the pretreatment Spd + salt the increases were 174% and 197% for SLDW and RDW, respectively, compared to the controls (salt). Growth with the treatments Kinetin + salt and Spd + salt are shown in Figure 1.

Table 1. (**A**) Effect of PGRs (plant growth regulators) application in salt-free pretreatment, (**B**) effect of PGRs application under saline conditions (200 mM NaCl) in *Plantago coronopus*, at 45 days of culture on SLDW (stem + leaves dry weight), RDW (root dry weight), SLWC (stem + leaves water content), and roots water content (RWC). Means ± SE (n = 16). Different letters within the same row represent significant differences between treatments, according to Tukey's test ($p \leq 0.05$).

A. PGRs Application without Salt	**SLDW (g/plant)**	**RDW (g/plant)**	**SLWC (%)**	**RWC (%)**
Control (no PGR)	0.134 ± 0.0038^{c}	0.0198 ± 0.0014^{b}	94.16 ± 0.47^{a}	85.07 ± 0.24^{c}
IAA	0.171 ± 0.0056^{ab}	0.0321 ± 0.0037^{a}	95.45 ± 0.52^{a}	92.05 ± 0.28^{ab}
Kinetin	0.179 ± 0.0062^{a}	0.0214 ± 0.0016^{b}	95.67 ± 0.55^{a}	90.63 ± 0.63^{ab}
GA_3	0.147 ± 0.0052^{bc}	0.0199 ± 0.0018^{b}	95.51 ± 0.46^{a}	89.80 ± 0.47^{b}
Spd	0.197 ± 0.0089^{a}	0.0369 ± 0.0021^{a}	95.91 ± 0.56^{a}	92.88 ± 0.66^{a}
SA	0.186 ± 0.0063^{a}	0.0341 ± 0.0012^{a}	95.74 ± 0.39^{a}	92.53 ± 0.45^{a}
B. PGRs Application with Salt	**SLDW (g/plant)**	**RDW (g/plant)**	**SLWC (%)**	**RWC (%)**
Control (salt)	0.080 ± 0.0110^{d}	0.0101 ± 0.0005^{d}	91.54 ± 0.35^{b}	82.63 ± 0.33^{c}
IAA + salt	0.132 ± 0.0078^{bc}	0.0167 ± 0.0005^{b}	92.50 ± 0.38^{b}	85.54 ± 0.38^{ab}
Kinetin + salt	0.161 ± 0.0064^{b}	0.0128 ± 0.0004^{c}	93.18 ± 0.42^{b}	83.55 ± 0.90^{bc}
GA_3 + salt	0.085 ± 0.0063^{d}	0.0117 ± 0.0006^{cd}	92.11 ± 0.52^{b}	83.09 ± 0.56^{bc}
Spd + salt	0.219 ± 0.0100^{a}	0.0300 ± 0.0058^{a}	95.72 ± 0.45^{a}	87.55 ± 0.40^{a}
SA + salt	0.092 ± 0.0090^{cd}	0.0094 ± 0.0008^{d}	92.50 ± 0.44^{b}	82.43 ± 0.70^{c}

2.2. Effect of PGRs Application on Sorbitol Content

It is well known that soluble carbohydrates (sorbitol) are plentiful in the family *Plantaginaceae*. For this reason, in the leaves of *P. coronopus* this osmolyte was analyzed at 45 days of culture. In pretreatments without salt (Figure 2A) no significant differences were found between PGR pretreatments compared to the control (without PGRs). However, in saline conditions the Spd + salt and Kinetin + salt had higher values ($p \leq 0.05$) of osmolyte accumulation (Figure 2B), (increase 0.43-fold and 0.33-fold) respectively, compared to untreated plants (without PGRs + salt). It should also be noted that in all pretreatments under both saline and non-saline conditions, sorbitol concentrations were high even in the control treatments (no PGRs) and (salt).

Figure 1. *Plantago coronopus* cultivated with Kinetin + salt compared to control (**A**), and *P. coronopus* cultivated with Spd + salt compared to control (**B**).

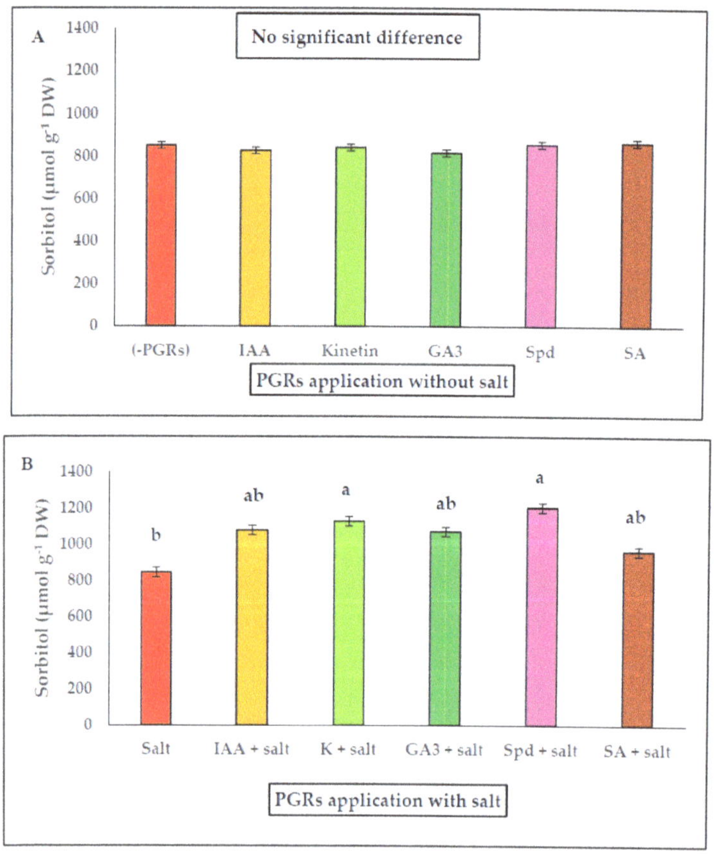

Figure 2. (**A**) Sorbitol content in *P. coronopus* leaves with the following PGRs: IAA, Kinetin, GA_3, Spd, and SA compared to control (-PGRs). (**B**) Sorbitol content with PGRs in saline conditions: IAA + salt, Kinetin + salt, GA_3 + salt, Spd + salt, and SA + salt, compared with salt (200 mM NaCl). Means ± SE (n = 3). Different letters above bars represent significant difference between treatments ($p \leq 0.05$).

2.3. Effect of PGRs Application on the Total Amount of Phenols and Flavonoids in Saline Conditions

We studied the effect of PGRs application with salt on the antioxidant content (measured as total phenols and flavonoids) in the leaves of *P. coronopus* at 45 days of culture to observe whether any pretreatment PGRs increased phenols and flavonoids content. The results shown in Table 2 indicate that treatments with Kinetin + salt and Spd + salt significantly ($p \leq 0.01$) increased the content of phenols and flavonoids by 24% compared to untreated plants (only salt). The values of phenols and flavonoids under non-saline conditions did not have relevant results or show any significant differences between pretreatments (data not shown).

Table 2. Effect of PGRs application with salt (200 mM NaCl) in *P. coronopus* leaves, at 45 days old, on total phenols and flavonoids. The values represent means ± SE (n = 3). The total phenols was expressed in mg gallic acid (GAE) per gr dry weight, and total flavonoids was expressed in mg of catechin (CE) per gr dry weight. Different letters within the same row represent significant difference among treatments, according to Tukey's test ($p \leq 0.01$).

PGRs Application with Salt	Total Phenols (mg GAE g^{-1} DW)	Total Flavonoids (mg CE g^{-1} DW)
Control (Salt)	4.5 ± 0.11 ^{c}	3.1 ± 0.11 ^{b}
IAA + salt	4.9 ± 0.12 ^{bc}	3.5 ± 0.12 ^{ab}
Kinetin + salt	5.3 ± 0.10 ^{ab}	3.7 ± 0.23 ^{ab}
GA_3 + salt	4.6 ± 0.17 ^{bc}	2.9 ± 0.23 ^{b}
Spd + salt	5.9 ± 0.21 ^{a}	4.0 ± 0.14 ^{a}
SA + salt	5.0 ± 0.20 ^{bc}	3.2 ± 0.20 ^{b}

2.4. Effect of PGRs Application on Endogenous Free, Bound and Conjugated Polyamines and Ethylene

In general, the pretreatments Spd without salt, and Spd with salt gave the greatest growth results in *P. coronopus*. Therefore, we considered it necessary to analyze PGRs application on the endogenous PA content (free, bound and conjugated) in the absence or presence of salt. The data are shown in Figures 3–5. In salt-free PGRs pretreatment, endogenous Put, Spd, and Spm (free, bound, and conjugated forms) increased compared to the control (-PGRs); the pretreatments with Kinetin, Spd, and SA had the highest values for endogenous Put (Figure 3a), endogenous Spd (Figure 4a) and endogenous Spm (Figure 5a), which corresponded to a greater increase in DW and WC for *P. coronopus*. This increase mainly occurs in bound and free PA fractions. However, under saline conditions, PA levels are modulated by salt. We detected a decreased of endogenous Put (free, bound and conjugated) in pretreatments Kinetin + salt and Spd + salt (Figure 3b) compared to values in Figure 3a; nevertheless, no significant difference was observed in pretreatments with salt due to the low amount of Put detected. However, endogenous Spm did increase in free and, above all, bound forms (Figure 5b), and these values being always higher than observed in saline-free pretreatments. The most significant increase was observed for pretreatment Spd with salt: where endogenous Spm increased two-fold (free form), 2.7-fold (bound form) and 2-fold (conjugated form) compared to the control salt (Figure 5b). Therefore, pretreatments Spd + salt and Kinetin + salt decreased endogenous Put (free, bound and conjugated) and increased endogenous Spm content (above all endogenous Spm bound). In pretreatment Spd + salt the increase of endogenous Spm (bound fraction) (Figure 5b) was higher by 5.3-fold than endogenous Spm (bound fraction) in pretreatment Spd (Figure 5a).

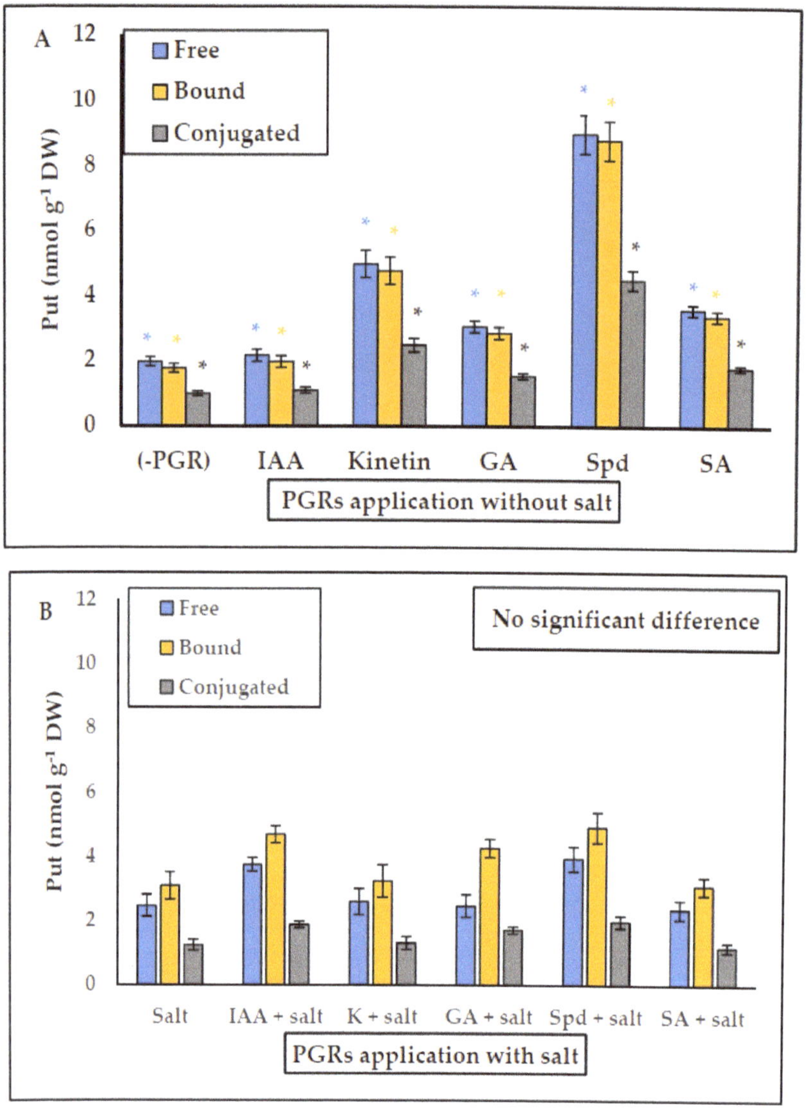

Figure 3. (**A**) Effect of PGRs application without salt, and (**B**) PGRs with salt (200 mM NaCl), in *Plantago coronopus* leaves at 45-day-old, on Putrescine (Free, Bound and Conjugated) content. Means ± SE (n = 3). The asterisk above the column represents significant difference between treatments of free Put, significant difference between treatments of bound Put and significant difference between treatments of conjugated Put, according to Tukey's test ($p \leq 0.01$).

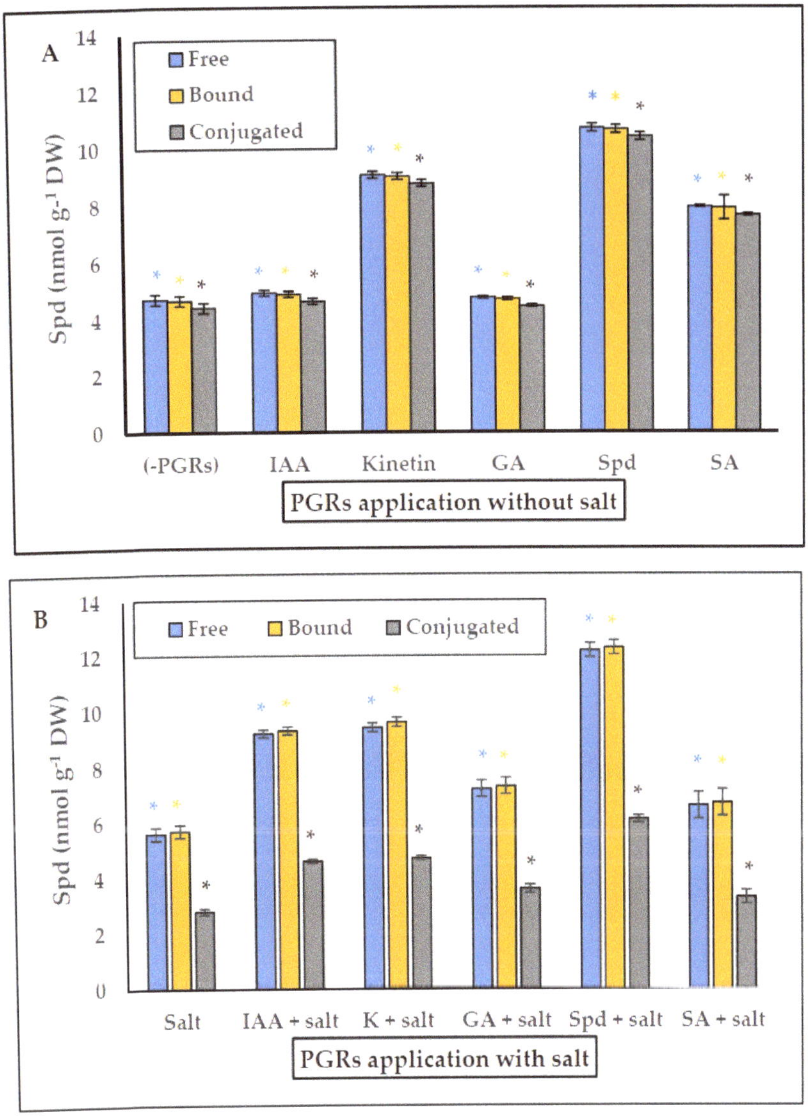

Figure 4. (**A**) Effect of PGRs application without salt, and (**B**) PGRs with salt (200 mM NaCl), in *Plantago coronopus* leaves at 45-day-old, on Spermidine (Free, Bound and Conjugated) content. Means ± SE ($n = 3$). The asterisk above the column represents significant difference between treatment of free Spd, significant difference between treatments of bound Spd and significant difference between treatments of conjugated Spd, according to Tukey's test ($p \leq 0.01$).

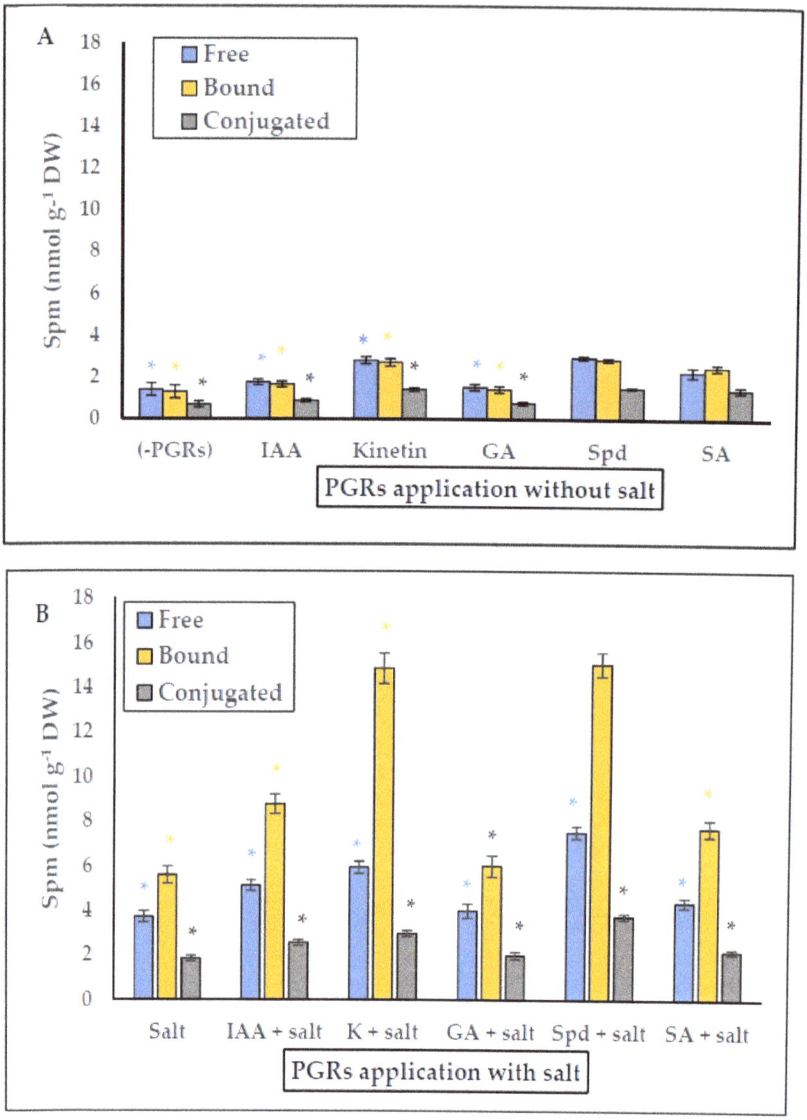

Figure 5. (**A**) Effect of PGRs application without salt, and (**B**) PGRs with salt (200 mM NaCl), in *Plantago coronopus* leaves at 45-day-old, on Spermine (Free, Bound and Conjugated) content. Means ± SE (n = 3). The asterisk above the column represents significant difference between treatments of free Spm, significant difference between treatments of bound Spm and significant difference between treatments of conjugated Spm, according to Tukey's test ($p \leq 0.01$).

Table 3 shows total PAs (Put (free, conjugated and bound forms) + Spd (free, conjugated and bound forms) + Spm (free, conjugated and bound forms)) under saline and non-saline conditions. All pretreatments increased PA content, especially under saline conditions, the highest values being Spd − salt and Spd + salt. Ethylene production in the leaves of this halophyte were compared to the total PAs. The results indicated a decrease in ethylene production that may contribute to increase PA content due to sharing a common synthesis pathway that we explain in discussion.

Table 3. (**A**) Effect of PGRs application without salt, and (**B**) PGRs with salt (200 mM NaCl) in *P. coronopus* leaves on total PAs [Put (Free, Bound and Conjugated) + Spd (Free, Bound and Conjugated) + Spm (Free, Bound and Conjugated)] and ethylene. Means ± SE ($n = 3$). Different letters within the same row represent significant differences among treatments, according to Tukey's test ($p \leq 0.01$).

A. PGRs Application without Salt	**Total PAs (nmol g^{-1} DW)**	**Ethylene (nL g^{-1} FW h^{-1})**
Control (no PGRs)	21.87 ± 0.68 d	10.30 ± 0.64 a
IAA	24.54 ± 0.75 d	8.35 ± 0.61 ab
Kinetin	46.00 ± 1.81 b	5.62 ± 0.71 b
GA_3	25.01 ± 0.88 d	7.68 ± 0.67 ab
Spd	61.29 ± 1.45 a	6.35 ± 0.53 b
SA	36.50 ± 0.69 c	5.81 ± 0.65 b
B. PGRs Application with Salt	**Total PAs (nmol g^{-1} DW)**	**Ethylene (nL g^{-1} FW h^{-1})**
Control (salt)	32.11 ± 0.58 c	7.61 ± 0.52 a
IAA + salt	49.94 ± 0.42 b	3.30 ± 0.68 b
Kinetin + salt	54.70 ± 2.42 b	3.77 ± 0.47 b
GA_3 + salt	36.76 ± 1.16 c	5.23 ± 0.59 ab
Spd + salt	67.79 ± 2.54 a	3.74 ± 0.64 b
SA + salt	37.50 ± 1.94 c	5.76 ± 0.53 ab

The correlation between Spm and sorbitol (r = 0.8465; $p \leq 0.01$), the total PAs and Spd (r = 0.9193; $p \leq 0.01$), the total PAs and Spm (r = 0.7184; $p \leq 0.01$) were always positive. However, the negative correlation between ethylene (C_2H_2) and Spm (r = −0.732; $p \leq 0.01$) and ethylene and total PAs (r = −0.723; $p \leq 0.01$) indicated that these metabolites (sorbitol. Spd and Spm) are necessary, especially under saline conditions, for enhancing salt tolerance and mitigating the adverse effect of stress (Table 4).

Table 4. Simple correlation coefficient (Pearson method) among all parameters studied in saline and non-saline conditions in *P. coronopus* ($p \leq 0.05$ *; $p \leq 0.01$ **).

	SLDW	**RDW**	**SLWC**	**RWC**	**SOR**	**PUT**	**SPD**	**SPM**	**Total PAs**	**C_2H_2**
SLDW	1									
RDW	0.7522 **	1								
SLWC	0.7629 **	0.7559 **	1							
RWC	0.7049 **	0.8769 **	0.7941 **	1						
SOR	0.0723	−0.2711	−0.2874	−0.440 **	1					
PUT	0.4624 **	0.4943 **	0.3212	0.4720 **	−0.0267	1				
SPD	0.6824 **	0.4669 **	0.3550 *	0.3892 *	0.3662 *	0.7729 **	1			
SPM	0.1200 *	−0.3134	−0.3501 *	−0.470 **	0.8465 **	−0.0308	0.4360 **	1		
Total PAs	0.5148 **	0.2064	0.0816 *	0.0840	0.5897 **	0.6838 **	0.9193 **	0.7184 **	1	
C_2H_2	−0.2150	0.1044	0.2573	0.1496	−0.659 **	-0.2616	−0.588 **	−0.732 **	−0.723 **	1

Parameters studied: SLDW (stem + leaf dry weight); RDW (root dry weight); SLWC (stem + leaf water content); RWC (root water content); SOR (sorbitol); PUT (Free + Bound + Conjugated); SPD (Free + Bound + Conjugated); SPM (Free + Bound + Conjugated); Total PAs (Total PUT + Total SPD + Total SPM); C_2H_2 (ethylene production).

3. Discussion

The benefits that PGRs application have on growth and abiotic stress are well known [42–44]. Under saline conditions, PGRs alleviate the adverse effects of salt on morphological, physiological, biochemical characteristics, and on crop yields and quality [10,29,45].

Previous studies showed a fall in dry weight at 200 mM NaCl in *P. coronopus* [32]. This species is in fact less salt-tolerant that other halophytes such as *Frankenia pulverulenta* and

Atriplex prostrata that also grow in the Brujuelo saltmarsh (Jaén, Spain) [32]. In salt-free pretreatment PGR, the dry weight and water content were increased in *P. coronopus*, especially under pretreatments Spd, SA, and Kinetin in aerial parts and Spd, SA, and IAA in roots (A in Table 1). More specifically, PAs such as Spd are aliphatic biogenic amines. These amines serve as an N reserve for the plant, N:C ratio regulate, favour synthesis of pigments photosynthetic, acid nucleic and proteins. Spermidine applications elevate levels of endogenous PA, but the enzymes involved in its biosynthesis can be increased without altering PA degrading enzymes, such as occurs in zoysia grass subjected to saline stress [46], and therefore these triamine could improve photosynthetic activity and protein synthesis favouring the growth of *P. coronopus* [28,47,48]. On the other hand, SA is a phenolic secondary metabolite, although is more related to abiotic stress tolerance and defensive responses against pathogens, application SA can have beneficious effect on cell and vegetative growth, photosynthesis, and flowering in this halophyte [49]. Regarding auxins and CKs stimulating elongation, cell division, formation of roots, leaves elongation, chloroplast differentiation and photosynthesis, however, a partial effect on growth (IAA stimulated roots and Kinetin stimulated the aerial part) was observed in *P. coronopus*. Crosstalk interaction with other phytohormones, as well as signaling network are very complex. On the other hand, CKs and auxins can have antagonistic effect at low to medium concentration, and only at higher concentration they have adjunctive effect [24,50]. With respect to gibberellins, little effect has in this halophyte, so the effect of each treatment may be genotype-dependent [48]. In saline conditions, the pretreatment Spd with salt was the most effective both in terms of dry weight and water content (B in Table 1). At cellular level, PAs can act as a compatible solute, as scavengers of free radicals, regulate plant membrane transport and act as a signal molecule during stress response [28,51–53]. In plant growth, PAs can offer specific protection to the photosynthetic apparatus (structural organization and functional activity of thylakoids), stabilization of biomembranes, and homeostasis redox [54]. A positive effect on photosynthetic activity and uptake of water seems to occur in *P. coronopus* when Spd was applied (Figure 1). Few studies have ever examined PAs in other halophytes. In crops with high nutritional values such as quinoa (*Chenopodium quinoa*), PAs (especially, an increase in Spd and Spm under saline conditions) may be useful markers of salt-tolerant genotypes [55,56] and may exert a protective effect improving growth on *Cymodocea nodosa* [57] and *Solanum chilense* [58]. Specifically, exogenous application of Spd in *C. nodosa* improving chlorophyll fluorescence levels under different saline treatments, maintaining the photosynthetic apparatus functional, under long-term hypo-osmotic stress [57]. Nevertheless, the positive effect of PAs may vary depending on the type of biotic and abiotic stress, plant species, time of exposure and physiological status of the tissues/organs [59,60], and therefore the effect of pretreatments must be studied in each halophyte.

Halophytes (dicotyledonous) accumulate inorganic ions (mainly Na^+, Cl^-) in their aerial parts and excrete excess salt through saline glands, bladder hairs or by developing succulence in their leaves [61]. For this reason, we focused our studies on *P. coronopus* leaves. In previous studies, we detected a high concentration of ions (Na^+, Cl^-) related to a certain degree of succulence [32]. Al-Hassan et al. [34] concluded that family *Plantaginaceae* have a "constitutive mechanism" of tolerance in which the transport of Na^+ and Cl^- ions (inorganic osmolytes) to the leaves and compartmentalization in the vacuole, contribute to cellular osmotic balance, and increase antioxidant metabolism under saline stress [34,36]. Polyamines are related to ionic transport through at membrane thylakoid, tonoplast, and plasma membranes [52,62,63]. Pottosin and Shabala showed that exogenous PAs application (0.1–1 mM) activated Ca^{2+} efflux, net H^+ fluxes, and activated H^+-ATPase pump under stress, but all these experiments were realized in the roots of glycophyte seedlings [52,62,63]. There are no studies on the application of PAs in halophytes on membrane ion channels. Nevertheless, irrigation for 10 days with PGRs (in saline and non-saline conditions) did not modify significantly ionic content and the "pre-adaptation" to stress proposed by Al-Hassan et al. [34] in *P. coronopus* (therefore ion data were not included). On the other hand, the osmoprotective compounds (proline, glycine-betaine, sugar, and polyols) favor water uptake, act as chaperons to molecular stabilized proteins and membranes, scavenge ROS, and/or protect antioxidant enzymes [64,65]. The family Plantaginaceae preferably accumulates sugars and polyols, sorbitol being the most abundant soluble carbohydrate in all *Plantago* species [66]. Sorbitol accumulation and synthesis is carried out above all under anaerobic conditions such as those present in saltmarshes, with confers on the competitive advantages in the environments in which this halophyte normally grows (e.g., saltmarshes in Jaén, Spain) [32,34]. In *P. coronopus*, sorbitol was found in high concentrations in both saline and non-saline PGR pretreatments (Figure 2), although in pretreatments with salt, the Spd had higher values according to higher increase in dry mass and water content. The osmolyte content is probably modulated by PGRs in saline conditions. Pretreatment PGRs stimulate growth probably because they increase photosynthetic activity (above all pretreated with Kinetin and Spd) and increase the sugar content; of these sugars sorbitol plays the role of osmolyte in *Plantago* species growing in adverse environmental conditions. Sorbitol acts to maintain osmotic homeostasis, scavenging ROS, can regulate the osmotic balance, and sequester Na^+ in the vacuole or apoplast alleviating the toxic effect of saline stress on *P. coronopus* [34]. On the other hand, CKs and PAs mutually regulate different physiological and biochemical processes with strong correlations between CK and PA levels, and act as inter- and intracellular messengers regulating abiotic stress [67].

The selection of productive, fast-growing halophytes with high saline tolerance that give high yields is of vital importance if agriculture is to be successful. *Plantago coronopus* is a source of valuable secondary metabolites of great economic value [37]. Antioxidants such as phenols and flavonoids are an essential part of the human diet and so we used different PGR pretreatments to analyze these two metabolites under saline conditions (Table 2). Previous studies have demonstrated an increase in total phenols as NaCl application increases [32]. These bioactive molecules eliminate large amounts of ROS and protect the cell against oxidative stress on its lipids, proteins, and DNA, in addition act as hydrogen donors, single oxygen quenchers and reducing agents [32,65,68]. The experiments by Boestfleisch et al. [20] have shown that it is possible to manipulate a plant's antioxidant capacity by modifying the saline growth environment, and the development stage. Our results indicate that mixing Spd with salt significantly improved the content of phenol and flavonoids when compared to untreated plants (only salt). Wild edible plants tend to have higher micronutrient contents and secondary metabolites than those of domestication varieties, therefore *P. coronopus* cultivation irrigated with Spd with salt can increase metabolite contents and constitute a good a source of sugar, minerals, vitamins, and antioxidants, might provide health benefits, and could be used as a new gastronomic food [69–71].

The best treatment under both non-saline and saline conditions was the PA Spermidine. Thus, we decided to analyze the endogenous PA content in this halophyte. In the biosynthetic pathway precursors of diamine Put are ornithine and arginine, while the triamine Spd and tetramine Spm are produced by addition of aminopropyl groups from S-adenosyl methionine (SAM) that are sequentially incorporated to Put and Spd by enzymatic reactions catalyzed, respectively, by Spd synthase and Spm synthase. The SAM is decarboxylated by SAMDC (S-adenosyl-methionine decarboxylase) [28,47,53]. Currently, little is known about the endogenous content of PAs in halophytes. Only thirteen halophytes have been studied and PAs have been associated with saline excretion, ionic balance, osmoregulation, protective role on photosynthetic apparatus and biomembranes, high photochemical efficiency in photosystem II, and an increased antioxidant defence system [32,72]. The low levels of free PAs (Put, Spd and Spm) detected under saline conditions in *P. coronopus* [32] made it interesting to study the interconversion between different PA forms under treatment with PGRs. Polyamines can exist in free soluble forms, conjugated to hydroxycinnamic acids (small molecules), or bound to macromolecules such as DNA, lipids, and proteins [51]. In the vegetative stage, salt modulated PA levels, decreased Put content, and increased free and bound forms of Spd and Spm, with values that were always higher than under non-saline conditions. It is interesting underline the drastic increase in bound > free > conjugated Spm forms compared to endogenous Spm (free, bound, and conjugated) in non-saline conditions in pretreatment Spd + salt and pretreatment Spd (Figure 5A,B). We hypothesize that bound forms (above all in Spd pretreatments) can be related to the protection of endogenous cellular structures (mainly biomembranes and photosynthetic apparatus), such as occur in the halophyte *Inula crithmoides* [54]. In addition, Spd treatment increases it endogenous content and enhance also endogenous Spm. More specifically, bound Spd and Spm forms were detected in PSII and LHCII (light-harvesting antenna complex) [73,74]. There are no studies on halophytes, but the exogenous Spd application in some glycophytes showed stabilization of PSII, improving photosynthetic performance and the antioxidant system in chloroplasts under saline conditions [74–76]. We consider that similar effects can occur in *P. coronopus* when Spd is applied. At the level of transgenic plants, the Spd synthase gene (*EsSPDS1*) (an enzyme that synthesizes Spd and increase the content of endogenous Spd and Spm) was cloned and characterized in the obligate halophyte *Eutrema salsugineum* and inserted into transgenic tobacco plant subjected to water and salinity stress. The results showed lower malondialdehyde (MDA, oxidative stress indicator) levels, less ion leakage and ROS levels, which indicates better protection in biomembranes, higher water content and more antioxidant enzymes than in non-transformed plants [77]. Clearly, Spd application improves stress tolerance, probably by protecting membranes and photosynthetic apparatus, and decreasing ROS, which could explain our results regarding the enhanced salinity tolerance in *P. coronopus*.

Ethylene and PAs have a common precursor, SAM. The increase of total PAs was accompanied by a decrease in ethylene production under different PGR treatments, which could contribute to PA accumulation (Table 3), thereby indicating a certain competition between PAs and ethylene for SAM, the common precursor. Therefore, SAM can be derivative to the formation of PAs, above all, during salt stress [78,79]. The correlation coefficient between studied parameters (Table 4) confirms our results. Finally, studies in transgenic *Arabidopsis* plants (with overexpression of *SAMDC* and, therefore, with high levels of Spd and Spm) under abiotic stress revealed better growth, maintaining higher photosynthetic activity, higher Fv/Fm and an increase in the P^{I}_{ABS} (Performance Index Based on Absorption). The enhancement in P^{I}_{ABS} caused a higher efficiency of quantum yield and specific energy fluxes of PSII, and also higher activities of antioxidant enzymes were found in the transformed plant [80].

4. Materials and Methods

4.1. Plant Material and Growth Conditions

Seeds of *P. coronopus* were randomly collected in September 2016 from Brujuelo saltmarsh (GPS location: 37°52′46″ N, 3°40′11″ W) (province of Jaén, Spain). Seeds were kept dry at 4 °C before being washed with sterile distilled water and sown in Petri dishes at 25 ± 1 °C and a photoperiod of 16 h [32]. After 10 days, the most uniform seedlings were transferred to 1.5 L pots with vermiculite as a substrate. Four seedlings per pot were sown and hydroponically cultivated using Hoagland nutrient solution 50% pH 6.5 ± 0.1 [81]. Plants were watered every two days with Hoagland nutrient solution. The environmental conditions in the growth chamber were the followings: photosynthetic photon flux density (PPFD) 500 μmol photon m^{-2} s^{-1}, 400–700 nm, provided by Sylvania Inc., Danvers, MA, USA, lamps, photoperiod 16h/8h in a day/night cycle, temperature (day) 25 °C ± 1 °C and (night) 16 °C ± 1 °C, and relative humidity of 55–75%.

4.2. Experimental Design and Treatments with PGRs and NaCl

In a growth chamber, plants were acclimated (in hydroponic conditions) for two weeks. Subsequently, these plants were treated for 10 days with different growth regulators applied to the nutrient solution when watered. The growth regulators used were the following: auxin: IAA; cytokinin: kinetin; gibberellins: GA_3; polyamine: Spd; and salicylic acid: SA.

Six treatments were established with six pots for each treatment PGRs

1. No PGRs
2. IAA (1 μM)
3. Kinetin (1 μM)
4. GA_3 (1 μM)
5. Spd (0.5 mM)
6. SA (0.5 mM)

Six treatments were established with six pots for each treatment PGRs + salt

7. NaCl (200 mM)
8. IAA (1 μM) + NaCl (200 mM)
9. Kinetin (1 μM) + NaCl (200 mM)
10. GA_3 (1 μM) + NaCl (200 mM)
11. Spd (0.5 mM) + NaCl (200 mM)
12. SA (0.5 mM) + NaCl (200 mM)

Subsequently, these pots were irrigated with two concentrations of NaCl: 0 mM (treatment 1–6) and 200 mM (treatment 7–12). NaCl levels were selected according to previous experiments realized by us [32]. To avoid osmotic shock, NaCl were increased progressively until the final required concentration was reached [32]; After 21 days in saline or non-saline conditions, plants were harvested for further analysis. Plants were 45 days old when harvested (14 days acclimation in pots, then 10 days of pretreatment with PGRs, and finally 21 days under saline or non-saline conditions). Flowering in this halophyte began approximately at 40–45 days old.

4.3. Growth Parameters

The following parameters were determined: fresh weight (FW) (roots, stems, and leaves), dry weight (DW) (leaves + stems and roots), and water content (WC) (leaves + stems and roots). To obtain DW, plants were placed in a forced-air oven at 70v °C for 72–96 h until a constant weight was obtained. This material was used to determine sorbitol, phenols, flavonoids, and endogenous PAs (free, conjugated and bound). Water content was calculated following the formula: WC (%) = (FW − DW/FW) × 100, where SL = stem and leaves, and R = Roots [82]. In fresh material (leaves), the ethylene production was determined.

4.4. Sorbitol Quantification

Sorbitol (Sor) was analyzed following Hassan et al. [34] for *P. coronopus* leaves (mature plants). For 10 min, dry leaves (45-day-old) were boiled in milliQ water and subsequently filtered with filters (0.22 μm). Afterwards, all samples (grown in absence and presence of salt) were injecting (20 μL) in a Waters 717 autosampler into a Prontosil 120-3-amino column (4.6 × 125 mm; 3 μm particle size). The conditions of isocratic flux were: (1 mL/min) of 85% acetonitrile for 25 min in each run. Sor integration peaks were obtained in the Waters Empower software and the quantification was realized compared with the standard calibration curve. A Waters 1525 HPLC (high-performance liquid chromatography) coupled with a 2424 evaporative light scattering (ELS) detector (Markham, ON, Canada) were used to determinate Sor content. The source parameters of ELSD were gain 75, data rate 1 point per second, nebulizer heating 60%, drift tube 50 °C, and gas pressure 2.8 Kg/cm^2. All experiments were conducted at room temperature.

4.5. Determination of Total Phenols and Flavonoids

The method of Boestfleisch et al. [20] was followed. Leaves dry were incubated (10 min) in methanol (80%) with continuous shaking. Subsequently, samples obtained in saline conditions were centrifugation for 5 min at 15,000× *g* and the supernatant was collected.

The quantification of total phenols was performed following the protocols by Dudonné et al. [83]. One hundred μL of water was pipetted into small tubes. Then were added: blank (80% methanol), or gallic acid standard (5–250 μg mL^{-1}) or 10 μL of methanolic extract. The reaction is completed with Folin–Ciocalteu reagent (10 μL). After waiting 8 min sodium carbonate (7%) (100 μL) was added. In the dark and room temperature, tubes were incubated for approximately 90–100 min. Total phenols were calculated using a standard curve. The samples at wavelength of 765 nm were measured in a spectrophotometer VARIAN Cary 4000 UV-VIS (Santa Clara, CA, USA).

The quantification of the total flavonoids was performed following Dewanto et al. [84]. In this case, in each tube was added 150 μL of water. Then we added blank (80% methanol), or catechin hydrate standard (0–400 μg mL^{-1}), or 25 μL of methanolic extract, and $NaNO_3$ (3.75%) (10 μL). After waiting 6 min, the reaction was completed with $AlCl_3$ (10%) (15 μL). After 5 min of incubation, NaOH (1 M) (50 μL) was added. Total flavonoids were calculated from a standard curve. The samples and the curve standard at wavelength 510 nm. were measured in a spectrophotometer VARIAN Cary 4000 UVA-VIS (Santa Clara, CA, USA).

4.6. Analysis of Free, Bound and Conjugated Polyamines

For PAs extraction the method followed by Ghabriche et al. [54] was used with minor modifications. Dry leaf samples (in saline and non-saline conditions) were ground in a mortar and homogenized with HCl (1 M) (*v*/*v*), then centrifuged at 23,000× *g* at 4 °C for 20 min. The supernatant was used to determine free polyamines by dansylation method [85]. The samples were resuspended in methanol (1 mL) and then centrifuged at 13,000× *g* for 15 min. Later, these samples needed to be filtered using microfilters (Chromafil PES-45/15, 0.45 μm; Macherey-Nagel). Twenty μL were injected into a Bio-Rad HPLC system (Hercules, CA, USA) equipped with a Nucleosil 100-5 C18MN 250/04 column (particle size: 5 μm, 4.6 × 250 mm^2). The conditions of HPLC to quantify the integration peaks were the following: a methanol/water stepped gradient program changing from 60% to 100% methanol over 25 min, flow rate 1 mL min^{-1}, and temperature of 35 °C. A Shimadzu RF-10Axl fluorimeter detector (excitation wavelength 320 nm and emission wavelength 510 nm) was used to determine dansylated free polyamines.

Bound forms (covalently bound to macromolecules such as proteins) and conjugated forms (covalently bound with small molecules such as hydroxycinnamic acids) were also analyzed. We added 200 μL of HCL (12 N) to the same amount of supernatant (200 μL) and transferred to dark tightly capped glass tubes. These tubes were placed in a heater and heated at 110 °C for 24 h to realize sample hydrolysis. After HCl was evaporated, the residue was resuspended in 200 μL of perchloric acid (10%) and used for dansylation. The pellet was used to extract bound PAs. This was dissolved in 5 mL of NaOH (1N). The mixture was centrifuged at 23,000× *g* at 4 °C for 20 min., and the supernatant was hydrolysed and dansylated under the same conditions described above. Dansylated free PAs (supernatant), dansylated conjugated PAs (supernatant hydrolysed) and dansylated bound PAs (pellet hydrolysed) were injected (20 μL) in the HPLC, in addition to PA standards (Put, Spd, Spm from Sigma, San Francisco, CA, USA), for quantification.

4.7. Ethylene Production

The method of Bueno et al. [86] was followed: fresh leaves collected of *P. coronopus* (45-day-old) were immediately transferred into a 5 mL flask (containing at the bottom filter paper and 50 μL of distilled water). All flasks were sealed with a silicone-rubber stopper (to prevent gas leakage). Flasks were incubated on a stove for 1 h incubation period, at 30 °C in darkness. Later, a 1 mL gas sample was injected into a HP 5890 series II, Hewlett Packard (Palo Alto, CA, USA) gas chromatograph fitted with a flame ionization detector and a 2 m × 4 mm stainless-steel column packed with 50–80 mesh Poropack-R. The conditions of chromatograph were: N_2, H_2 and synthetic air flow rates 50, 86, and 400 mL min^{-1}, respectively. To analyze and quantify ethylene production, peaks integration was compared with the retention time of ethylene (C_2H_4) standard, (purity 99.9%).

4.8. Data Analysis

A randomized block design was used in our experiments. Data are presented as mean ± standard error (SE). A Statgraphics Centurion v. 17 (University of Jaén) was used to perform analyses of variance (ANOVA). Significant differences between means were determined using Tukey's multiple range test ($p \leq 0.05$ and $p \leq 0.01$). All parameters in the absence or presence of salt were compared using Pearson's correlation coefficients.

5. Conclusions

Halophyte cultivation as a part of biosaline agriculture could help improve productivity and crop quality and be used to restore saline and degraded land. In *P. coronopus* cultivation, exogenous Spd application (0.5 mM) to the nutritive solution can improve growth and increase salt-stress tolerance, as well as increasing the osmolyte (sorbitol) and antioxidant compounds (phenols and flavonoids) under saline conditions. The increase in the endogenous PA pool, especially Spd and Spm (bound forms), is probably related to the protection of subcellular structures, the maintenance of photosynthetic activity, osmotic adjustment, ionic homeostasis and the improvement of antioxidant activity. In addition, the increase in Spd levels showed a negative correlation with ethylene, indicating than the decrease in ethylene also can contribute to PA accumulation. Auxins, CKs, GAs and SA pretreatments stimulated growth under non-saline conditions, but these PGRs were unable to mitigate the adverse effects of stress. Therefore, Spd application is the best pretreatment for *P. coronopus* cultivation and can contribute to improving the tolerance to salinity and nutritional quality of this halophyte, although it will be necessary to research in each halophyte which is the better treatment to apply.

Author Contributions: Conceptualization, M.d.P.C. and M.B.; methodology, M.d.P.C. and M.B.; software, M.d.P.C. and M.B.; validation, M.d.P.C. and M.B.; formal analysis, M.d.P.C. and M.B.; investigation, M.d.P.C. and M.B.; resources, M.B.; data curation, M.d.P.C. and M.B.; writing—original draft preparation, M.B.; writing—review and editing, M.B.; visualization, M.d.P.C. and M.B.; supervision, M.d.P.C. and M.B.; project administration, M.d.P.C.; funding acquisition, M.d.P.C. All authors have read and agreed to the published version of the manuscript.

Funding: This research was funded by Spanish Ministry of Science and Innovation, grant number CGL2006-08830, The involvement of growth regulators in plant response to salinity.

Acknowledgments: We want to thank the Technical Research Services of the University of Jaén for their invaluable help in the quantification of metabolites (sorbitol, polyamines, and ethylene). Our thanks to the designer Sonia Reyes for her valuable help in preparing high-quality tables, graphics, and photographs.

Conflicts of Interest: The authors declare no conflict of interest.

References

1. Thornton, P.K.; Ericksen, P.J.; Herrero, M.; Challinor, A.J. Climate variability and vulnerability to climate change: A review. *Glob. Chang. Biol.* **2014**, *20*, 3313–3328. [CrossRef]
2. Leisner, C.P. Review: Climate change impacts on food security-focus on perennial cropping systems and nutritional value. *Plant Sci.* **2020**, *293*, 110412. [CrossRef]
3. Shahzad, A.; Ullah, S.; Dar, A.A.; Sardar, M.F.; Mehmood, T.; Tufail, M.A.; Shakoor, A.; Haris, M. Nexus on climate change: Agriculture and possible solution to cope future climate change stress. *Environ. Sci. Pollut. Res. Int.* **2021**, *28*, 14211–14232. [CrossRef]
4. Santos, J.; Al-Azzawi, M.; Aronson, J.; Flowers, T.J. eHALOP a database of salt-tolerant plants: Helping put halophytes to work. *Plant Cell Physiol.* **2016**, *57*, e10. [CrossRef]
5. Grigore, M.N. Definition and classification of halophytes as an ecological group of plants. In *Handbook of Halophytes: From Molecules to Ecosystems towards Biosaline Agriculture*, 1st ed.; Grigore, M.N., Ed.; Springer: Cham, Switzerland, 2021; pp. 1–48. [CrossRef]
6. Gul, B.; Ansari, R.; Flowers, T.J.; Khan, M.A. Germination strategies of halophyte seeds under salinity. *Environ. Exp. Bot.* **2013**, *92*, 4–18. [CrossRef]
7. Kumari, A.; Das, P.; Parida, A.K.; Agarwal, P.K. Proteomics, metabolomics, and ionomics perspectives of salinity tolerance in halophytes. *Front. Plant Sci.* **2015**, *6*, 537. [CrossRef]

8. Bueno, M.; Cordovilla, M.P. Ecophysiology and uses of halophytes in diverse habitats. In *Handbook of Halophytes: From Molecules to Ecosystems towards Biosaline Agriculture*, 1st ed.; Grigore, M.N., Ed.; Springer: Cham, Switzerland, 2021; pp. 1–25. [CrossRef]
9. Grigore, M.N.; Toma, C. Integrative anatomy of halophytes from Mediterranean climate. In *Handbook of Halophytes: From Molecules to Ecosystems towards Biosaline Agriculture*, 1st ed.; Grigore, M.N., Ed.; Springer: Cham, Switzerland, 2021; pp. 1–35. [CrossRef]
10. Shaid, M.A.; Sarkhosh, A.; Khan, N.; Balal, R.M.; Ali, S.; Rossi, L.; Gómez, C.; Mattson, N.; Nasim, W.; García-Sánchez, F. Insights into physiological and biochemical impacts of salt stress on plant growth and development. *Agronomy* **2020**, *10*, 938. [CrossRef]
11. Borsai, O.; Al Hassan, M.; Negrusier, C.; Raigón, M.D.; Boscaiu, M.; Sestraş, R.E.; Vicente, O. Responses to salt stress in *Portulaca*: Insight into its tolerance mechanism. *Plants* **2020**, *9*, 1660. [CrossRef]
12. Panta, S.; Flowers, T.; Lane, P.; Doyle, R.; Haros, G.; Shabala, S. Halophyte agriculture: Success stories. *Environ. Exp. Bot.* **2014**, *107*, 71–83. [CrossRef]
13. Atzori, G. The potential of edible halophytes as new crops in saline agriculture. The ice plant (*Mesembryanthemum crystallinum* L.) case study. In *Future of Sustainable Agriculture in Saline Environments*, 1st ed.; CRC Press: Boca Raton, NY, USA, 2021; pp. 443–460. [CrossRef]
14. Nikalje, G.C.; Srivastava, A.K.; Pandey, G.K.; Suprasanna, P. Halophytes in biosaline agriculture: Mechanism, utilization, and value addition. *Land Degrad. Dev.* **2018**, *29*, 1081–1095. [CrossRef]
15. Christiansen, A.H.C.; Lyra, D.A.; Jørgensen, H. Increasing the value of *Salicornia bigelovii* green biomass grown in a desert environment through biorefining. *Ind. Crops Prod.* **2021**, *160*, 113105. [CrossRef]
16. Duarte, B.; Ca ador, I. Iberian halophytes as agroecological solutions for degraded lands and biosaline agriculture. *Sustainability* **2021**, *13*, 1005. [CrossRef]
17. Shaygan, M.; Mulligan, D.; Baumgartl, T. The potential of three halophytes (*Tecticornia pergranulata, Sclerolaena longicuspis,* and *Frankenia serpyllifolia*) for the rehabilitation of brine-affected soils. *Land Degrad. Dev.* **2018**, *29*, 2002–2014. [CrossRef]
18. Hasanuzzaman, M.; Nahar, K.; Alam, M.M.; Bhowmik, P.C.; Hossain, M.A.; Rahman, M.M.; Vara Prasad, M.N.; Ozturk, M.; Fujita, M. Potential use of halophytes to remediate saline soils. *BioMed Res. Int.* **2014**, *2014*, 589341. [CrossRef] [PubMed]
19. Jesus, J.M.; Danko, A.S.; Fiúza, A.; Borges, M.T. Phytoremediation of salt-affected soils: A review of processes, applicability, and the impact of climate change. *Environ. Sci. Poll. Res.* **2015**, *22*, 6511–6525. [CrossRef]
20. Boestfleisch, C.; Wagenseil, N.B.; Buhmann, A.K.; Seal, C.E.; Wade, E.M.; Muscolo, A.; Papenbrock, J. Manipulating the antioxidant capacity of halophytes to increase their cultural and economic value through saline cultivation. *AoB Plants* **2014**, *6*, plu046. [CrossRef]
21. Zheng, L.; Ma, H.; Jiao, Q.; Ma, C.; Wang, P. Phytohormones: Important participators in plant salt tolerance. *Intl. J. Agric. Biol.* **2020**, *24*, 319–332. [CrossRef]
22. Ashfaq, M.; Khan, S. Role of phytohormones in improving the yield of oilseed crops. In *Oils Seed Crops: Yield and Adaptation under Environmental Stress*, 1st ed.; Ahmad, P., Ed.; John Wiley & Sons Ltd.: Hoboken, NJ, USA, 2017; Volume 9, pp. 165–183. [CrossRef]
23. Llanes, A.; Reginato, M.; Devinar, G.; Luna, V. What is known about phytohormones in halophytes? A review. *Biologia* **2018**, *73*, 727–742. [CrossRef]
24. Szepesi, A. Halotropism: Phytohormonal aspects and potential applications. *Front. Plant Sci.* **2020**, *11*, 571025. [CrossRef] [PubMed]
25. Berková, V.; Kameniarová, M.; Ondrisková, V.; Berka, M.; Menšiková, S.; Kopecká, R.; Luklová, M.; Novák, J.; Spichal, L.; Rashotte, A.M.; et al. *Arabidopsis* response to inhibitor of cytokinin degradation INCYDE: Modulations of cytokinin signaling and plant proteome. *Plants* **2020**, *9*, 1563. [CrossRef]
26. Fahad, S.; Hussain, S.; Matloob, A.; Khan, F.A.; Khaliq, A.; Saud, S.; Hassan, S.; Shan, D.; Khan, F.; Ullah, N.; et al. Phytohormones and plant responses to salinity stress: A review. *Plant Growth Regul.* **2015**, *75*, 391–404. [CrossRef]
27. Marković, M.; Momcilov, T.; Uzelac, B.; Cingel, A.; Milošević, S.; Jevremović, S.; Subotić, A. Breaking the dormancy of snake's head fritillary (*Fritillary meleagris* L.) in vitro bulbs-Part 1: Effect of GA_3, GA inhibitors and temperature on fresh weight, sprouting and sugar content. *Plants* **2020**, *9*, 1449. [CrossRef]
28. Liu, J.H.; Wang, W.; Wu, H.; Gong, X.; Moriguchi, T. Polyamines function in stress tolerance: From synthesis to regulation. *Front. Plant Sci.* **2015**, *6*, 827. [CrossRef] [PubMed]
29. Alcázar, R.; Bueno, M.; Tiburcio, A.F. Polyamines: Small amines with large effects on plant abiotic stress tolerance. *Cells* **2020**, *9*, 2373. [CrossRef] [PubMed]
30. Azad, N.; Rezayian, M.; Hassanpour, H.; Nikman, V.; Ebrahimzadeh, H. Physiological mechanisms of salicylic acid in *Mentha pulegium* L. under salinity and drought stress. *Braz. J. Bot.* **2021**, *44*, 359–369. [CrossRef]
31. Jayakannan, M.; Bose, J.; Babourina, O.; Rengel, Z.; Shabala, S. Salicylic acid in plant salinity stress signalling and tolerance. *Plant Growth Regul.* **2015**, *76*, 25–40. [CrossRef]
32. Bueno, M.; Lendínez, M.L.; Calero, J.; Cordovilla, M.P. Salinity responses of three halophytes from inland saltmarshes of Jaén (southern Spain). *Flora* **2020**, *266*, 151589. [CrossRef]
33. Consejería del Medio Ambiente. *Flora Vascular de Andalucía Oriental*; Junta de Andalucía: Sevilla, Spain, 2009.
34. Al Hassan, M.; Pacurar, A.; López-Gresa, M.P.; Donat-Torres, M.P.; Llinares, J.V.; Boscaiu, M.; Vicente, O. Effects of salt stress on three ecologically distinct *Plantago* species. *PLoS ONE* **2016**, *11*, e0160236. [CrossRef]
35. Koyro, H.W. Effect of salinity on growth, photosynthesis, water relations and solute compositions of the potential cash crop halophyte *Plantago coronopus* (L.). *Environ. Exp. Bot.* **2006**, *56*, 136–146. [CrossRef]

36. Ltaeif, H.B.; Sakhraoui, A.; González-Orenga, S.; Faz, A.L.; Boscaiu, M.; Vicente, O.; Rouz, S. Responses to salinity in four *Plantago* species from Tunisia. *Plants* **2021**, *10*, 1392. [CrossRef] [PubMed]
37. Jdey, A.; Falleh, H.; Ben Jannet, S.; Mkadmini Hammi, K.; Dauvergne, X.; Ksouri, R.; Magné, C. Phytochemical investigation and antioxidant, antibacterial and anti-tyrosinase performances for six medicinal halophytes. *South Afr. J. Bot.* **2017**, *112*, 508–514. [CrossRef]
38. Petropoulos, S.A.; Karkanis, A.; Martins, N.; Ferreira, I.C. Edible halophytes of the Mediterranean basin: Potential candidates for novel food products. *Trends Food Sci. Technol.* **2018**, *74*, 69–84. [CrossRef]
39. Ceccanti, C.; Brizzi, A.; Landi, M.; Incrocci, L.; Pardossi, A.; Guidi, L. Evaluation of major minerals and trace elements in wild and domesticated edible herbs traditionally used in the Mediterranean area. *Biol. Trace Elem. Res.* **2021**, *199*, 3553–3561. [CrossRef]
40. Puccinelli, M.; Pezzarossa, B.; Pintimalli, L.; Malorgio, F. Selenium biofortification of three wild species, *Rumex acetosa* L., *Plantago coronopus* L., and *Portulaca oleracea* L., grown as microgreens. *Agronomy* **2021**, *11*, 1155. [CrossRef]
41. Ceccanti, C.; Landi, M.; Incrocci, L.; Pardossi, A.; Venturi, F.; Taglieri, I.; Ferroni, G.; Guidi, L. Comparison of three domestications and wild-harvested plants for nutraceutical properties and sensory profiles in five wild edible herbs: Is domestication possible? *Foods* **2020**, *9*, 1065. [CrossRef]
42. Amir, R.; Munir, F.; Khan, M.; Iqbal, T. Use of plant hormones for the improvement of plant growth and production under salt stress. In *Salt Stress, Microbes, and Plant Interactions: Causes and Solution*, 2nd ed.; Akhtar, M.S., Ed.; Springer Nature: Singapore, 2019; Volume 1, pp. 59–90. [CrossRef]
43. Morsi, M.M.; Abdelmigid, H.M.; Aljoudi, N.G.S. Exogenous salicylic acid ameliorates the adverse effects of salt stress on antioxidant system in *Rosmarinus officinalis* L. *Egypt. J. Bot.* **2018**, *58*, 249–263. [CrossRef]
44. Islam, M.J.; Ryu, B.R.; Azad, M.O.K.; Rahman, M.H.; Rana, M.S.; Lim, J.D.; Lim, Y.S. Exogenous putrescine enhances salt tolerance and ginsenosides content in Korena ginseng (*Panax ginseng* Meyer) sprouts. *Plants* **2021**, *10*, 1313. [CrossRef]
45. Tounekti, T.; Hernández, I.; Müller, M.; Khemira, H.; Munné-Bosch, S. Kinetin applications alleviate salt stress and improve the antioxidant composition of leaf extracts in *Salvia officinalis*. *Plant Physiol. Biochem.* **2011**, *49*, 1165–1176. [CrossRef]
46. Li, S.; Jin, H.; Zhang, Q. The effect of exogenous spermidine concentration on polyamine metabolism and salt tolerance in zoysiagrass (*Zoysia japonica* Steud) subjected to short-term salinity stress. *Front. Plant Sci.* **2016**, *7*, 1221. [CrossRef] [PubMed]
47. Tiburcio, A.F.; Altabella, T.; Bitrián, M.; Alcázar, R. The roles of polyamines during the lifespan of plants: From development to stress. *Planta* **2014**, *240*, 1–18. [CrossRef] [PubMed]
48. Kamiab, F.; Tavassolian, I.; Hosseinifarahi, M. Biologia Futura: The role of polyamine in plant science. *Biol. Futur.* **2020**, *71*, 183–194. [CrossRef]
49. Filgueiras, C.C.; Martins, A.D.; Pereira, R.V.; Willett, D.S. The ecology of salicylic acid signaling: Primary, secondary and tertiary effects with application in agriculture. *Int. J. Mol. Sci.* **2019**, *20*, 5851. [CrossRef]
50. Li, S.M.; Zheng, H.X.; Zhang, X.S.; Sui, N. Cytokinins as central regulators during plant growth and stress response. *Plant Cell Rep.* **2021**, *40*, 271–282. [CrossRef]
51. Chen, D.; Shao, Q.; Yin, L.; Younis, A.; Zheng, B. Polyamine function in plants: Metabolism, regulation on development, and roles in abiotic stress responses. *Front. Plant Sci.* **2019**, *9*, 1945. [CrossRef]
52. Pottosin, I.; Shabala, S. Polyamines control of cation transport across plant membranes: Implications for ion homeostasis and abiotic stress signaling. *Front. Plant Sci.* **2014**, *5*, 154. [CrossRef] [PubMed]
53. Pál, M.; Szalai, G.; Janda, T. Speculation: Polyamines are important in abiotic stress signaling. *Plant Sci.* **2015**, *237*, 16–23. [CrossRef] [PubMed]
54. Ghabriche, R.; Ghnaya, T.; Mnasri, M.; Zaier, H.; Baioui, R.; Vromman, D.; Abdelly, C.; Lutts, S. Polyamine and tyramine involvement in NaCl-induced improvement of Cd resistance in the halophyte *Inula crithmoides* L. *J. Plant Physiol.* **2017**, *216*, 136–144. [CrossRef]
55. Ruiz-Carrasco, K.; Antognoni, F.; Coulibaly, A.K.; Lizardi, S.; Covarrubias, A.; Martínez, E.A.; Molina-Montenegro, M.A.; Biondi, S.; Zurita-Silva, A. Variation in salinity tolerance of four lowland genotypes of quinoa (*Chenopodium quinoa* Willd.) as assessed by growth, physiological traits, and sodium transporter gene expression. *Plant Physiol. Biochem.* **2011**, *49*, 1333–1341. [CrossRef]
56. Ruiz, K.B.; Rapparini, F.; Bertazza, G.; Silva, H.; Torrigiani, P.; Biondi, S. Comparing salt-induced responses at the transcript level in a salares and coastal-lowlands landrace of quinoa (*Chenopodium quinoa* Willd.). *Environ. Exp. Bot.* **2017**, *139*, 127–142. [CrossRef]
57. Zarranz Elso, M.; García-Jiménez, P.; Robaina, R.R. Endogenous polyamine content and photosynthetic performance under hypo-osmotic conditions reveal *Cymodocea nodosa* as an obligate halophyte. *Aquat. Biol.* **2012**, *17*, 7–17. [CrossRef]
58. Gharbi, E.; Martínez, J.P.; Benahmed, H.; Fauconnier, M.L.; Lutts, S.; Quinet, M. Salicylic acid differently impacts ethylene and polyamine synthesis in the glycophyte *Solanum lycopersicum* and the wild-related halophyte *Solanum chilense* exposed to mild salt stress. *Physiol. Plant* **2016**, *158*, 152–167. [CrossRef] [PubMed]
59. Zapata, P.J.; Serrano, M.; García-Legaz, M.F.; Pretel, M.T.; Botella, M.A. Short term effect of salt shock on ethylene and polyamines depends on plant salt sensitivity. *Front. Plant Sci.* **2017**, *8*, 855. [CrossRef] [PubMed]
60. Minocha, R.; Majumdar, R.; Minocha, S.C. Polyamines and abiotic stress in plants: A complex relationship. *Front. Plant Sci.* **2014**, *5*, 175. [CrossRef] [PubMed]
61. Pardo-Domènech, L.L.; Tifrea, A.; Grigore, M.N.; Boscaiu, M.; Vicente, O. Proline and glycine betaine accumulation in two succulent halophytes under natural and experimental conditions. *Plant Biosyst.* **2016**, *150*, 904–915. [CrossRef]

62. Pottosin, I.; Velarde-Buendía, A.M.; Bose, J.; Fuglsang, A.T.; Shabala, S. Polyamines cause plasma membrane depolarization, activate Ca^{2+}, and modulate H^+-ATPase pump activity in pea roots. *J. Exp. Bot.* **2014**, *65*, 2463–2472. [CrossRef] [PubMed]
63. Pottosin, I.; Olivas-Aguirre, M.; Dobrovinskaya, O.; Zepeda-Jazo, I.; Shabala, S. Modulation of ion transport across plant membranes by polyamines: Understanding specific modes of action under stress. *Front. Plant Sci.* **2021**, *11*, 616077. [CrossRef]
64. Kaur, G.; Asthir, B. Proline: A key player in plant abiotic stress tolerance. *Biol. Plant.* **2015**, *59*, 609–619. [CrossRef]
65. Ozgur, R.; Uzilday, B.; Sekmen, A.H.; Turkan, I. Reactive oxygen species regulation and antioxidant defence in halophytes. *Funct. Plant Biol.* **2013**, *40*, 832–847. [CrossRef]
66. Slama, I.; Abdelly, C.; Bouchereau, A.; Flowers, T.; Savouré, A. Diversity, distribution, and roles of osmoprotective compounds accumulated in halophytes under abiotic stress. *Ann. Bot.* **2015**, *115*, 433–447. [CrossRef]
67. Anwar, R.; Mattoo, A.K.; Avtar, K.H. Polyamine interactions with plant hormones: Crosstalk at several levels. In *Polyamines: A Universal Molecular Nexus for Growth, Survival, and Specialized Metabolism*, 1st ed.; Kusano, T., Suzuki, H., Eds.; Springer: Tokyo, Japan, 2015; pp. 267–302. [CrossRef]
68. Ksouri, R.; Megdiche, W.; Falleh, H.; Trabelsi, N.; Boulaaba, M.; Smaoui, A.; Abdelly, C. Influence of biological, environmental, and technical factors on phenolic content and antioxidant activities of Tunisian halophytes. *Comptes Rendus Biol.* **2008**, *331*, 865–873. [CrossRef]
69. Ksouri, R.; Ksouri, W.M.; Jallali, I.; Debez, A.; Magné, C.; Hiroko, I.; Abdelly, C. Medicinal halophytes: Potent source of health promoting biomolecules withy medical, nutraceutical and food applications. *Crit. Rev. Biotechnol.* **2012**, *32*, 289–326. [CrossRef] [PubMed]
70. Pereira, C.G.; Custódio, L.; Rodrigues, M.J.; Neng, N.R.; Nogueira, J.M.F.; Carlier, J.; Costa, M.C.; Varela, J.; Barreira, L. Profiling of antioxidant potential and phytoconstituents of *Plantago coronopus*. *Braz. J. Biol.* **2017**, *77*, 632–641. [CrossRef] [PubMed]
71. Barros, L.; Morales, P.; Carvalho, A.M.; Ferreira, I.C.F.R. Antioxidant potential of wild plant foods. In *Mediterranean Wild Edible Plants: Ethnobotany and Food Composition Tables*; Springer: New York, NY, USA, 2016; pp. 209–232. ISBN 978-1-49-393329-7.
72. Bueno, M.; Cordovilla, M.P. Polyamines in halophytes. *Front. Plant Sci.* **2019**, *10*, 439. [CrossRef] [PubMed]
73. Ioannidis, N.E.; Tsiavos, T.; Kotzabasis, K. Chemical bonding of chlorophylls and plant aminic axial ligands impact harvesting of visible light and quenching of fluorescence. *Photochem. Photobiol.* **2012**, *88*, 98–106. [CrossRef]
74. Sang, T.; Shan, X.; Li, B.; Shu, S.; Sun, J.; Guo, S. Comparative proteomic analysis reveals the positive effect of exogenous spermidine on photosynthesis and salinity tolerance in cucumber seedlings. *Plant Cell Rep.* **2016**, *35*, 1769–1782. [CrossRef]
75. Wu, Z.; Wang, J.; Yan, D.; Yuan, H.; Wang, Y.; He, Y.; Wang, X.; Li, Z.; Mei, J.; Hu, M.; et al. Exogenous spermidine improves salt tolerance of pecan-grafted seedlings via activating antioxidant system and inhibiting the enhancement of Na^+/K^+ ratio. *Acta Physiol. Plant.* **2020**, *42*, 83. [CrossRef]
76. Jiang, D.; Hou, J.; Gao, W.; Tong, X.; Li, M.; Chu, X.; Chen, G. Exogenous spermidine alleviates the adverse effects of aluminium toxicity on photosystem II through improved antioxidant system and endogenous polyamine contents. *Ecotoxicol. Environ. Saf.* **2021**, *207*, 111265. [CrossRef]
77. Zhou, C.; Sun, Y.; Ma, S.; Wang, J. Heterologous expression of *EsSPDS1* in tobacco plants improves drought tolerance with efficient reactive oxygen species scavenging systems. *South Afr. J. Bot.* **2015**, *96*, 19–28. [CrossRef]
78. Majumdar, R.; Barchi, B.; Turlapati, S.A.; Gagne, M.; Minocha, R.; Long, S.; Minocha, S.C. Glutamate, ornithine, arginine, proline, and polyamine metabolic interactions: The pathway is regulated at the post-transcriptional level. *Front. Plant Sci.* **2016**, *7*, 78. [CrossRef] [PubMed]
79. Lasanajak, Y.; Minocha, R.; Minocha, S.C.; Goyal, R.; Fatima, T.; Handa, A.K.; Mattoo, A.K. Enhanced flux of substrates into polyamine biosynthesis but not ethylene in tomato fruit engineered with yeast S-adenosylmethionine decarboxylase gene. *Amino Acids* **2014**, *46*, 729–742. [CrossRef]
80. Jia, T.; Hou, J.; Iqbal, M.Z.; Zhang, Y.; Cheng, B.; Feng, H.; Li, Z.; Liu, L.; Zhou, J.; Feng, G.; et al. Overexpression of the with clover *TrSAMDC1* gene enhanced salt and drought resistance in *Arabidopsis thaliana*. *Plant Physiol. Biochem.* **2021**, *165*, 147–160. [CrossRef]
81. Hoagland, D.R.; Arnon, D.I. The water-culture method for growing plants without soil. *Calif. Agric. Exp. Stn. Bull.* **1950**, *347*, 1–32.
82. Al Hassan, M.; López-Gresa, M.P.; Boscaiu, M.; Vicente, O. Stress tolerance mechanisms in *Juncus*: Responses to salinity and drought in three *Juncus* species adapted to different natural environments. *Funct. Plant Biol.* **2016**, *43*, 949–960. [CrossRef] [PubMed]
83. Dudonné, S.; Vitrac, X.; Coutière, P.; Woillez, M.; Mérillon, J.M. Comparative study of antioxidant properties and total phenolic content of 30 plant extracts of industrial interesting using DPPH, ABTS, FRAP, SOD, and ORAC assays. *J. Agric. Food Chem.* **2009**, *57*, 1768–1774. [CrossRef] [PubMed]
84. Dewanto, V.; Wu, X.; Adom, K.K.; Liu, R.H. Thermal processing enhances the nutritional value of tomatoes by increasing total antioxidant activity. *J. Agric. Food Chem.* **2002**, *50*, 3010–3014. [CrossRef] [PubMed]
85. Smith, M.A.; Davies, P.J. Separation and quantification of polyamines in plant tissue by high performance liquid chromatography of their dansyl derivatives. *Plant Physiol.* **1985**, *78*, 89–91. [CrossRef] [PubMed]
86. Bueno, M.; Lendínez, M.L.; Aparicio, C.; Cordovilla, M.P. Effect of salinity on polyamines and ethylene in *Atriplex prostrata* and *Plantago coronopus*. *Biol. Plant.* **2015**, *59*, 596–600. [CrossRef]

plants

MDPI

Article

Responses to Salinity in Four *Plantago* Species from Tunisia

Hela Belhaj Ltaeif [1,2,3], Anis Sakhraoui [3,4,5], Sara González-Orenga [1,6], Anbu Landa Faz [1,7], Monica Boscaiu [6], Oscar Vicente [1,*] and Slim Rouz [3]

1 Institute for the Conservation and Improvement of Valencian Agrodiversity (COMAV), Universitat Politècnica de València, Camino de Vera s/n, 46022 Valencia, Spain; belhajhela1@gmail.com (H.B.L.); sagonor@doctor.upv.es (S.G.-O.); anbu.landafaz@cinvestav.mx (A.L.F.)
2 National Agronomy Institute–Tunis, University of Carthage, Mahrajène, 1082 Tunis, Tunisia
3 Laboratory of Agricultural Production Systems and Sustainable Development (LR03AGR02), Department of Agricultural Production, Agricultural High School of Mograne, University of Carthage, 1121 Mograne-Zaghouan, Tunisia; anis.sakhraoui@esakef.u-jendouba.tn (A.S.); slim.rouz@esamg.u-carthage.tn (S.R.)
4 Agricultural High School of Kef, Jendouba University, 7119 Le Kef, Tunisia
5 Departamento de Biología Vegetal y Ecología, Universidad de Sevilla, Apartado 1095, 41080 Sevilla, Spain
6 Mediterranean Agroforestry Institute (IAM), Universitat Politècnica de València, Camino de Vera s/n, 46022 Valencia, Spain; mobosnea@eaf.upv.es
7 Center for Research and Advanced Studies of the National Polytechnic Institute, Av. Instituto Politécnico Nacional No. 2508, Colonia San Pedro Zacatenco, C.P. 07360 Ciudad de México D.F., Mexico
* Correspondence: ovicente@upvnet.upv.es

Citation: Ltaeif, H.B.; Sakhraoui, A.; González-Orenga, S.; Landa Faz, A.; Boscaiu, M.; Vicente, O.; Rouz, S. Responses to Salinity in Four *Plantago* Species from Tunisia. *Plants* **2021**, *10*, 1392. https://doi.org/10.3390/plants10071392

Academic Editor: Pedro Diaz-Vivancos

Received: 8 June 2021
Accepted: 2 July 2021
Published: 7 July 2021

Abstract: The genus *Plantago* is particularly interesting for studying the mechanisms of salt tolerance in plants, as it includes both halophytes and glycophytes, as well as species adapted to xeric environments. In this study, the salt stress responses of two halophytes, *P. crassifolia* and *P. coronopus*, were compared with those of two glycophytes, *P. ovata* and *P. afra*. Plants obtained by seed germination of the four species, collected in different regions of Tunisia, were subjected to increasing salinity treatments for one month under greenhouse conditions. Morphological traits and biochemical parameters, such as ion accumulation and the leaf contents of photosynthetic pigments, osmolytes, oxidative stress markers and antioxidant metabolites, were measured after the treatments. Salt-induced growth inhibition was more pronounced in *P. afra*, and only plants subjected to the lowest applied NaCl concentration (200 mM) survived until the end of the treatments. The biochemical responses were different in the two groups of plants; the halophytes accumulated higher Na^+ and proline concentrations, whereas MDA levels in their leaves decreased, indicating a lower level of oxidative stress. Overall, the results showed that *P. coronopus* and *P. crassifolia* are the most tolerant to salt stress, and *P. afra* is the most susceptible of the four species. *Plantago ovata* is also quite resistant, apparently by using specific mechanisms of tolerance that are more efficient than in the halophytes, such as a less pronounced inhibition of photosynthesis, the accumulation of higher levels of Cl^- ions in the leaves, or the activation of K^+ uptake and transport to the aerial part under high salinity conditions.

Keywords: salt stress; halophytes; growth responses; ion accumulation; osmolytes; oxidative stress biomarkers; antioxidants

1. Introduction

Global agricultural yields and food production are negatively affected by different environmental stress factors, especially drought and salinity [1,2]. These stressors inhibit plant growth and reproductive development, causing significant reductions in crop productivity and may even compromise yield entirely. Salinity is already affecting 25–30% of total cultivated land and 33% of irrigated land [3], although some estimates increase this percentage to more than 50% [4]. This situation is predicted to worsen shortly due to the consequences of climate change [5], as more cultivated areas will depend on irrigation and lower-quality water will be used, triggering an increase in the rate of secondary soil salinisation [6]. Salinity impairs plant growth and development due to its two components, osmotic stress and ion toxicity, inhibiting plant growth and cellular functions and, eventually, causing plant death [7–10]. Plants exposed to salt stress show morphological, physiological, metabolic, and molecular changes reflected, for example, in a delayed or completely inhibited seed germination, high seedling mortality [11] or a general inhibition of photosynthesis and growth [2,8,12]. Although most plants are glycophytes, susceptible to salinity, there is a small group of ca. 1500 species from different genera and families that are halophytes, which can survive and complete their life cycle on saline soils [13].

The genus *Plantago* L. (Plantaginaceae family) includes more than 250 annual and perennial herbs and subshrubs, distributed worldwide, except for tropical rainforest and the Antarctic. Some *Plantago* species are cosmopolitan, others have more limited geographical ranges, but the genus also includes local endemics [14,15]. There are many interesting aspects related to the taxonomy and evolutionary trends of this genus [16,17], but also concerning salt stress physiology and biochemistry, given that it includes several well-known halophytes [18–21].

Plantago coronopus L. is an annual or biennial species that ranges from North Africa and the Iberian Peninsula to SE Asia, reaching northern Europe through a narrow strip along the Atlantic coast [22,23]. It grows in different types of littoral and inland habitats, such as sand dunes, saline grasslands, salt marshes, scrublands, or human-disturbed areas [23], tolerating saline soils [24]. It is considered as a potential cash crop [24], an edible plant with nutraceutical [25] and antioxidant properties [26].

Plantago crassifolia Forsk. is a perennial species present only in the Mediterranean region. South African populations, previously ascribed to this species, are now considered as *P. carnosa* Lam, based on the analysis of the internal transcribed spacer (ITS) region of rRNA genes [27]. *Plantago crassifolia* is a typical halophyte, growing exclusively in saline habitats with moderate soil salinity and occupying interdune depressions and salt marsh edges [21,28]. It is reported as palatable fodder [29]. The two species, *P. coronopus* and *P. crassifolia*, are taxonomically related, belonging to the subgenus *Coronopus* (Lam. and DC.) Rahn, section *Coronopus* Lam [30].

Plantago ovata Forssk. is an annual or short-lived perennial species, ranging from the Canary Islands and SE Iberian Peninsula, across northern Africa, to India [31]. It was considered introduced in North America in the 18th century, but a molecular clock based on ITS and chloroplast DNA analysis dates a much earlier, non-anthropogenic introduction from the Old World, 200,000–650,000 years ago [32]. The species grows in dry areas on wasteland, annual pastures, almost always on somewhat nitrified soils, indifferent to soil pH, but has also been found, occasionally, in moderately saline soils [31]; in North America, it is present only in desert and Mediterranean habitats [33]. Due to the laxative properties of the seed mucilage, *P. ovata* is a well-known medicinal plant cultivated in many countries, with India as the leading producer [34].

Plantago afra L. (syn. *P. psyllium* L.) is an annual species with a wide geographic distribution, from the Canary Islands and the Iberian Peninsula, along the Mediterranean region, to Pakistan. It grows in annual grasslands, roadsides, and crop fields in semi-arid and arid areas [31]. Like *P. ovata*, it has medicinal applications and is cultivated in India, Pakistan, and Iran. The seed husks of *P. ovata* and *P. afra* are known by the name 'psyllium' and are a popular mild laxative used to relieve chronic constipation, bowel cancer and gastrointestinal irritation. Psyllium is also used as a dietary source of fibre to treat obesity and cholesterol reduction or as an antitussive and anti-inflammatory [35]. The two species belong to the subgenus *Psyllium* (Juss.) Harms and Reich; *P. ovata* is classified in the section *Albicans* Barnéoud, and *P. afra* in the section *Psyllium* (Juss.) Lam and DC [30].

This work aimed to compare the responses to salt stress in two typical halophytes of the genus *Plantago*, *P. crassifolia* and *P. coronopus*, and two other congeneric species, *P. ovata* and *P. afra*, more adapted to xeric environments. Although the resistance to salinity has been evaluated in many plants, comparative analyses in genetically related species adapted to different natural habitats are not so commonly performed and can provide insights into the most relevant mechanisms of salt tolerance in a particular genus. To address this question, inhibition of growth in the presence of increasing salt concentrations, up to 800 mM NaCl, was evaluated in plants of the four species as the most reliable parameter to establish their relative degree of tolerance. Then, these growth responses were correlated with the salt-induced changes in the levels of biochemical markers associated with specific stress response pathways, namely, monovalent ions (Na^+, Cl^-, K^+), specific osmolytes (proline and total soluble sugars), oxidative stress biomarkers (malondialdehyde, H_2O_2) and antioxidant compounds (phenolics and flavonoids).

2. Results

In this study, morphological, physiological, and biochemical traits were measured to analyse the impact of salinity stress treatments on four *Plantago* species (*P. coronopus*, *P. crassifolia*, *P. ovata* and *P. afra*).

A two-way ANOVA was performed to assess the effects of the factors 'species', 'treatment', and their interactions, on all measured parameters (Table 1). The 21 traits analysed displayed significant differences for the 'species' effect. Differences between treatments were also significant for all measured parameters, except K^+ levels in leaves, whereas leaf fresh weight was the only non-significant trait regarding the interaction of species and treatments.

Table 1. Results of two-way ANOVA (F ratios) for the independent factors 'Species' (S), 'Treatment' (T) and the interaction 'Species × Treatment' (S × T). Abbr.: ECs: electrical conductivity of the substrate; RL: root length; SL: shoot length; FW(%): leaf fresh weight (as a percentage of the control); RWC: root water content; LWC: leaf water content; Chl a: Chlorophyll a; Chl b: Chlorophyll b; Caro: carotenoids; Na^+ R: sodium in roots; Na^+ L: sodium in leaves; Cl^- R: chloride in roots; Cl^- L: chloride in leaves; K^+ R: potassium in roots; K^+ L: potassium in leaves; MDA: malondialdehyde; H_2O_2: hydrogen peroxide; TSS: total soluble sugars; Pro: proline; TPC: total phenolic compounds; TF: total flavonoids.

Variables	S	T	S × T
ECs	297.04 *	129.78 *	62.07 *
RL	205.23 *	34.43 *	70.59 *
SL	35.99 *	53.34 *	8.65 *
FW (%)	10.29 *	66.62 *	1.56
RWC	91.76 *	49.30 *	17.78 *
LWC	142.33 *	87.91 *	16.21 *
Chl a	17.68 *	45.54 *	8.04 *
Chl b	11.69 *	29.51 *	2.68 *
Caro	12.59 *	48.68 *	3.30 *
Na^+ R	239.91 *	26.67 *	22.49 *
Na^+ L	235.82 *	62.31 *	49.70 *
Cl^- R	23.76 *	10.69 *	7.04 *
Cl^- L	77.36 *	23.50 *	29.56 *
K^+ R	40.97 *	21.01 *	3.45 *
K^+ L	63.60 *	0.90	9.12 *
MDA	55.67 *	102.62 *	29.84 *
H_2O_2	102.17 *	11.32 *	22.67 *
TSS	7.75 *	58.23 *	20.62 *
Pro	192.94 *	63.45 *	33.07 *
TPC	40.68 *	10.35 *	24.94 *
TF	93.51 *	18.10 *	51.53 *

* Significant at the 95% confidence level.

2.1. Substrate Analysis

The electrical conductivity of the substrate in the pots ($EC_{1:5}$) at the end of the salt treatments increased in parallel with the NaCl concentration in the irrigation water, with significant differences between treatments (Table 2). EC reached the highest values in the pots watered with 800 mM NaCl, 9.57 dS m^{-1} for pots with *P. coronopus* and 9.45 dS m^{-1} for *P. crassifolia*. No significant differences were observed between different species for each salt concentration tested. EC was not determined (n.d.) for those treatments that resulted in the death of the plants, 800 mM NaCl for *P. ovata* or 400 mM and higher salt concentrations in the case of *P. afra* (Table 2).

Table 2. Electrical conductivity of the substrate in 1:5 soil:water suspensions ($EC_{1:5}$, dS m^{-1}), in pots watered for four weeks with the indicated NaCl concentrations. The values shown are means ± SE (n = 5). In each row, different letters indicate significant differences between treatments, according to the Tukey test ($p < 0.5$). n.d.: not determined.

	Control	200 mM	400 mM	600 mM	800 mM NaCl
P. crassifolia	0.48 ± 0.02 [a]	3.56 ± 0.18 [b]	5.46 ± 0.22 [c]	7.94 ± 0.50 [d]	9.45 ± 0.30 [e]
P. coronopus	0.45 ± 0.02 [a]	1.20 ± 0.09 [b]	5.66 ± 0.50 [c]	7.60 ± 0.10 [d]	9.57 ± 0.20 [e]
P. ovata	0.45 ± 0.02 [a]	1.51 ± 0.05 [b]	5.63 ± 0.14 [c]	8.51 ± 0.40 [d]	n.d.
P. afra	0.49 ± 0.05 [a]	2.15 ± 0.15 [b]	n.d.	n.d.	n.d.

2.2. Effects of Salt Stress on Plant Growth

Plantago ovata plants did not survive the four-week treatment with the highest salt concentration applied, 800 mM NaCl; therefore, samples from this treatment were not collected. In the case of the more salt-sensitive *P. afra*, only data from the control and the 200 mM NaCl treatments were obtained since the plants could not withstand 400 mM NaCl or higher salinities and died within the first two or three weeks of treatment. Although the halophytes, *P. crassifolia* and *P. coronopus*, survived all treatments without showing any apparent wilting symptoms, even in the presence of 800 mM NaCl, salt stress affected growth in all four species (Figure 1). For example, in all cases, root length increased in parallel with increasing external salinity (Figure 1a). Stimulation of root growth in response to salt stress seems to mimic the behaviour of the plants in nature, where high salt concentrations may induce roots to grow, searching for soil layers with lower salinity.

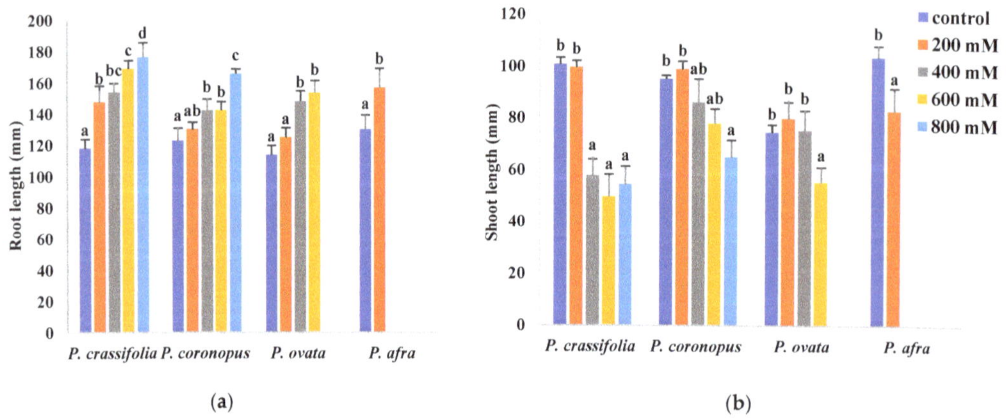

Figure 1. Root length (**a**) and shoot length (**b**) of the four selected *Plantago* species after four weeks of treatment with the indicated NaCl concentrations. The values shown are means ± SE (n = 5). For each species, different letters over the bars indicate significant differences between treatments, according to the Tukey test ($p < 0.5$).

All other measured parameters showed growth inhibition in response to the salt treatments in the four *Plantago* species. The experimental data also revealed that *P. ovata* is relatively resistant to salinity, even though it is not considered a typical halophytic species since it is not present in highly saline natural habitats.

Measurements of shoot length in plants treated with different salt concentrations allowed estimating the relative salt tolerance of the investigated species. According to this criterion, *P. coronopus* appears to be the most tolerant, with a statistically significant reduction of shoot length, compared to the non-stressed control, observed only in the presence of 800 mM NaCl. In *P. ovata*, no reduction was detected in plants subjected to the 200 and 400 mM NaCl treatments, whereas a significant decrease in shoot length was observed in *P. crassifolia* plants treated with 400 and higher NaCl concentrations. The most salt-sensitive species, *P. afra*, already showed inhibition of shoot growth at 200 mM NaCl, the only salt concentration that allowed survival of the plants under the conditions used in the experiments (Figure 1b).

Determination of the fresh weight of the aerial part of the plants confirmed the highest salt sensitivity of *P. afra*, for which a FW decrease of about 60% of the control was calculated in the presence of 200 mM NaCl. Growth inhibition between 200 and 600 mM NaCl followed a similar pattern for *P. coronopus* and *P. ovata*. The relative FW reduction at high salinities was comparatively lower in *P. crassifolia*, probably due to less water loss, given the succulent leaves of this species (Figure 2). Values in Figure 2 are shown as percentages of the corresponding non-stressed controls to better compare the four species, which have slightly different sizes.

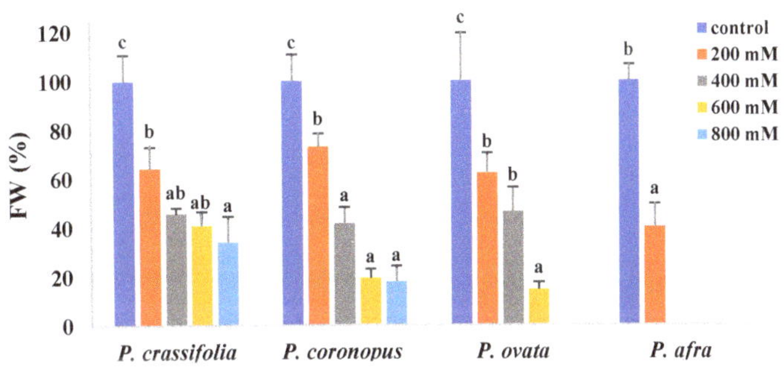

Figure 2. Fresh weight (FW) reduction of the aerial part of the plants in the four selected *Plantago* species after four weeks of treatment with the indicated NaCl concentrations. For each species, different letters over the bars indicate significant differences between treatments, according to the Tukey test ($p < 0.5$). Values are shown as percentages of the FW of the corresponding controls, taken as 100%.

The relative salt tolerance of the four *Plantago* species correlated with their resistance to salt-induced dehydration in roots and leaves (Figure 3). In roots, non-significant or slight reductions in water contents were observed for *P. crassifolia* and *P. coronopus*, which kept high, and similar, values even in the presence of 800 mM NaCl; root dehydration was relatively higher in *P. ovata* and, especially, in the least tolerant *P. afra* (Figure 3a). Salt-induced dehydration was more pronounced in the leaves than in the roots, although maintaining the same general pattern. Leaf water loss was slightly lower in the succulent *P. crassifolia* than in *P. coronopus* but substantially lower in the two halophytes than in the other two species (Figure 3b).

2.3. *Effects of Salt Stress on Photosynthetic Pigment Levels*

Inhibition of photosynthesis, partly due to degradation of photosynthetic pigments, is a general response of plants to abiotic stress. In the present study, a significant, concentration-dependent decrease in chlorophylls a and b and carotenoid contents, with respect to the non-stressed controls, has been observed in plants of the four selected *Plantago* species subjected to increasing salt treatments (Figure 4). In the presence of 200 mM NaCl, the lowest values of the three pigments were measured in *P. afra*, the most salt-sensitive species, whereas no significant reduction was observed in *P. ovata* or, for Chl b and carotenoids, in

P. coronopus. At higher salinities, *P. ovata* and the two halophytes showed similar qualitative patterns of variation, with the most substantial reduction in pigment contents generally detected in *P. crassifolia* and the least pronounced in *P. ovata* (Figure 4).

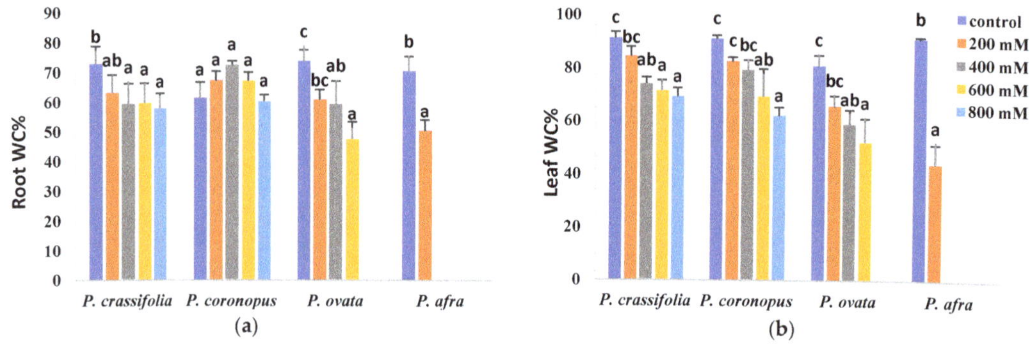

Figure 3. Water content percentage in roots (**a**) and leaves (**b**) in the four selected *Plantago* species, after four weeks of treatment with the indicated NaCl concentrations. The values shown are means ± SE (n = 5). For each species, different letters over the bars indicate significant differences between treatments, according to the Tukey test ($p < 0.5$).

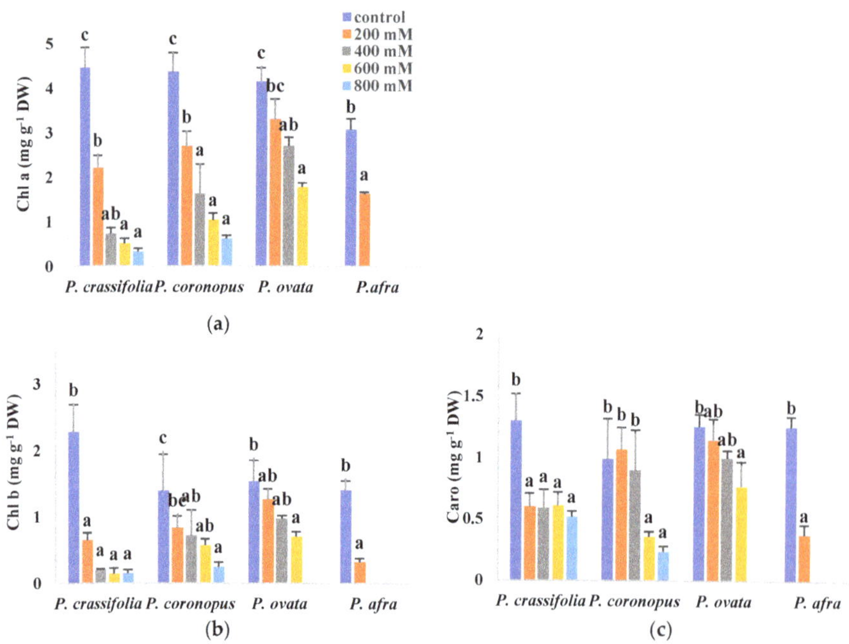

Figure 4. Chlorophyll a (Chl a) (**a**), chlorophyll b (Chl b) (**b**) and carotenoids (Caro) (**c**) contents in leaves of the four selected *Plantago* species, after four weeks of treatment with the indicated NaCl concentrations. The values shown are means ± SE (n = 5). For each species, different letters over the bars indicate significant differences between treatments, according to the Tukey test ($p < 0.5$).

2.4. Ion Accumulation

In the four investigated *Plantago* species, Na^+ and Cl^- concentrations increased in the roots and leaves of the plants, roughly in parallel with the increasing NaCl concentration in the irrigation water (Figure 5). For each species and specific salt treatment, the concentrations of Cl^- were consistently higher than those of Na^+ in both organs, and those of both ions were higher in the leaves than in the roots (Figure 5a–d), indicating the existence of mechanisms for their active transport to the aboveground organs. There were, however, differences in the accumulation patterns of the two ions. In roots, Na^+ contents were highest at all tested salinities (including the non-stressed controls), in *P. crassifolia*, and lowest in *P. afra*, whereas *P. coronopus* and *P. ovata* showed intermediate and similar values (Figure 5a). The same pattern of Na^+ accumulation was observed in the leaves, except that the absolute concentrations reached, at the same external salinity, were higher in *P. coronopus* than in *P. ovata*. Note that a 10-fold increase in Na^+ content was observed in the leaves of non-stressed *P. afra* plants with respect to the values measured in control roots (Figure 5b).

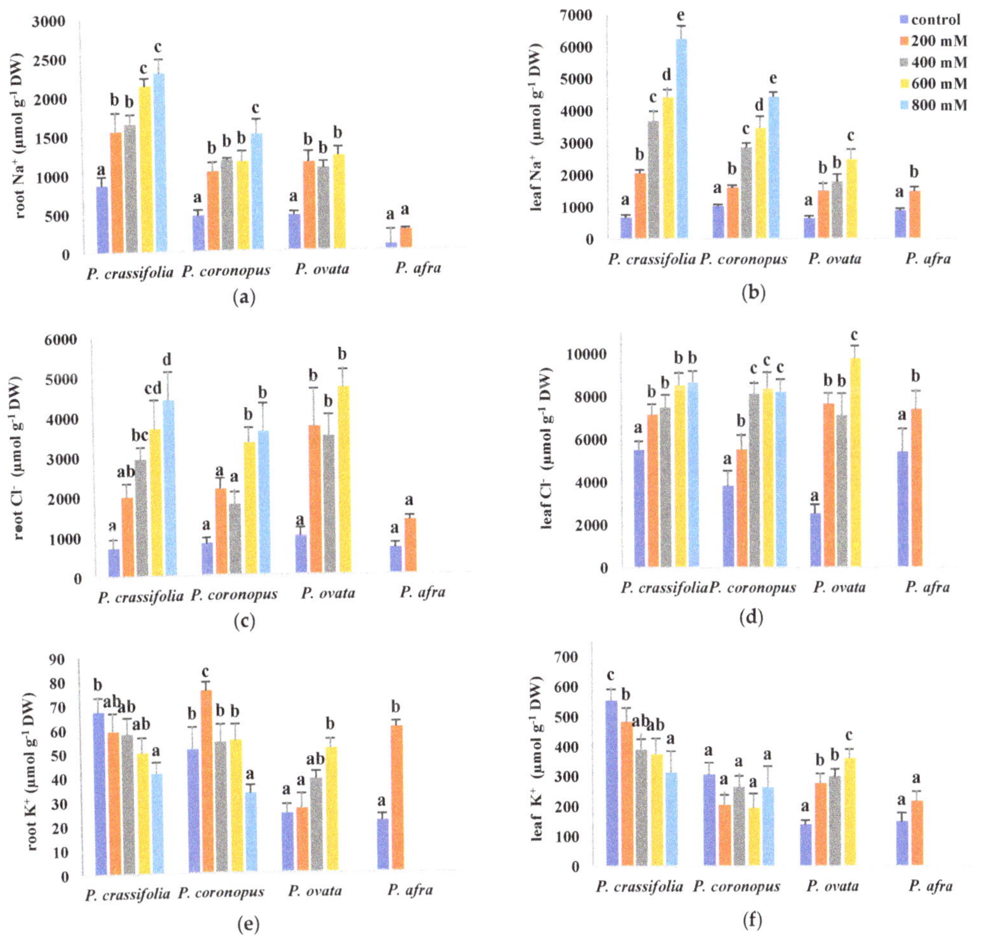

Figure 5. Root (**a**,**c**,**e**) and leaf (**b**,**d**,**f**) contents of sodium (Na^+), (**a**,**b**), chloride (Cl^-) (**c**,**d**) and potassium (K^+) (**e**,**f**) in plants of the four selected *Plantago* species, after four weeks of treatment with the indicated NaCl concentrations. The values shown are means ± SE (n = 5). For each species, different letters over the bars indicate significant differences between treatments, according to the Tukey test ($p < 0.5$).

Regarding Cl^- in roots, apart from the lowest levels found in *P. afra*, its accumulation patterns were somewhat different from those of Na^+, with *P. ovata* showing higher concentrations than *P. crassifolia* and *P. coronopus*, at the same salinity level (Figure 5c). Similar behaviour was observed in the leaves for the latter three species, whereas, contrary to roots, *P. afra* accumulated Cl^- to the same or even higher levels than the other species (Figure 5d). Nevertheless, the most striking feature was the huge Cl^- concentration determined in the leaves of control plants, especially in *P. crassifolia* and *P. afra*, not only in relation to the root contents (about eight-fold higher), but also in absolute terms (over 5 mmol g^{-1} DW) (Figure 5d).

Despite quantitative differences, Na^+ and Cl^- concentrations varied in the same way, qualitatively, in the four *Plantago* species in response to the salt treatments, always increasing with increasing salinity. However, the salt-induced changes in K^+ contents differed in the different taxa (Figure 5e, f). In *P. crassifolia*, K^+ levels decreased progressively, in roots and leaves, roughly in parallel with the increase of NaCl concentration in the irrigation water, whereas no significant variation was observed, in general, in *P. coronopus*, except for a significant reduction in roots in the presence of 800 mM NaCl. On the contrary, in *P. ovata* and *P. afra*, K^+ contents increased in response to the salt stress treatments. It should also be mentioned that, as for the other ions, K^+ levels were higher (five to ten-fold) in leaves than in roots in all four species (Figure 5e,f).

2.5. Salt Stress Effect on Osmolyte Contents

Proline (Pro), one of the most common plant osmolytes, accumulated in response to the salt treatments in the leaves of the two halophytes, *P. crassifolia* and *P. coronopus*. Pro reached maximum levels of about 50 μmol g^{-1} DW in the presence of 800 mM NaCl, representing an increase of five to six-fold over control values. Leaf Pro concentrations also increased in *P. ovata*, but only up to ~30 μmol g^{-1} DW at the highest concentration tested, 600 mM NaCl. In the most salt-sensitive species, *P. afra*, Pro remained extremely low, below one μmol g^{-1} DW (Figure 6a). However, total soluble sugars (TSS) contents showed different patterns of variation, increasing with increasing external salinity only in *P. ovata* and *P. afra* but decreasing in the halophytes (Figure 6b).

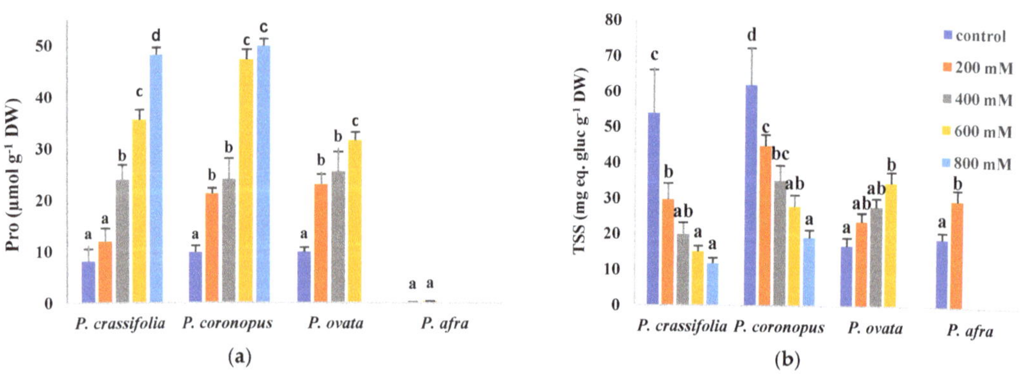

Figure 6. Leaf contents of proline (Pro) (**a**) and total soluble sugars (TSS) (**b**) in the four selected *Plantago* species, after four weeks of treatment with the indicated NaCl concentrations. The values shown are means ± SE (n = 5). For each species, different letters over the bars indicate significant differences between treatments, according to the Tukey test ($p < 0.5$). gluc: glucose.

2.6. Oxidative Stress Biochemical Markers

Salt-induced changes in the leaf levels of malondialdehyde (MDA) followed a similar pattern to those of TSS, increasing in parallel to the NaCl concentrations in the irrigation water in *P. ovata* and *P. afra* and progressively decreasing in *P. crassifolia* and *P. coronopus*

(Figure 7a). On the other hand, hydrogen peroxide leaf contents did not vary significantly in *P. afra* treated with 200 mM NaCl and increased significantly, in a concentration-dependent manner, in salt-treated plants of the other three species (Figure 7b).

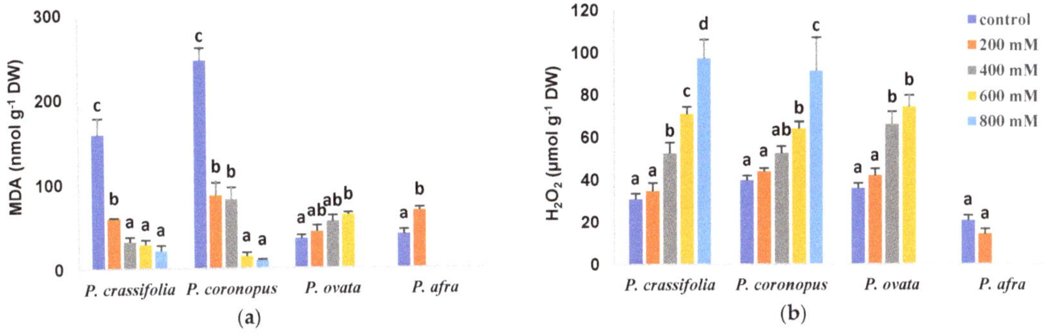

Figure 7. Leaf contents of malondialdehyde (MDA) (**a**) and hydrogen peroxide (H_2O_2) (**b**) in the four selected *Plantago* species, after four weeks of treatment with the indicated NaCl concentrations. The values shown are means ± SE (n = 5). For each species, different letters over the bars indicate significant differences between treatments, according to the Tukey test ($p < 0.5$).

2.7. Antioxidant Compounds

The leaf contents of total phenolic compounds (TPC) and total flavonoids (TF), as representative examples of non-enzymatic antioxidants, were measured in plants of the four investigated *Plantago* species (Figure 8). TPC levels increased in the four taxa in response to rising salinity, reaching the highest values in *P. coronopus* (9 mg equivalent of gallic acid per gram DW) and *P. crassifolia* (about 6.6 mg eq. GA g^{-1} DW), in the presence of 800 mM NaCl, which represent relative increases over the control, non-stressed plants of 2.7 and 2.4-fold, respectively. In *P. ovata* and *P. afra*, control TPC concentrations were lower than in their halophytic counterparts and, therefore, these species showed larger relative increases in response to salt stress (Figure 8a). TF contents also increased significantly with rising salinity, except for *P. afra*. For each NaCl concentration, both absolute TF levels and relative increases over control values were highest for *P. ovata* and lowest in *P. crassifolia* (Figure 8b).

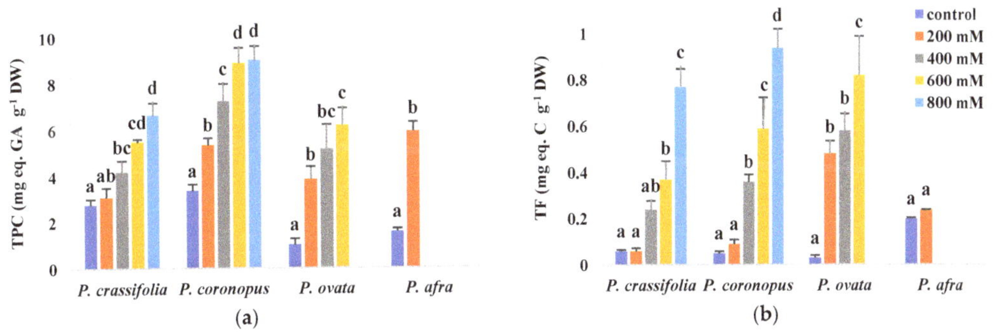

Figure 8. Leaf contents of total phenolic compounds (TPC) (**a**) and total flavonoids (TF) (**b**) in the four selected *Plantago* species, after four weeks of treatment with the indicated NaCl concentrations. The values shown are means ± SE (n = 5). For each species, different letters over the bars indicate significant differences between treatments, according to the Tukey test ($p < 0.5$). GA: gallic acid; C. catechin.

2.8. Principal Component Analysis

A Principal Component Analysis (PCA) was performed, including all variables measured in the four *Plantago* species, both growth parameters and the biochemical stress markers (Figure 9). Five components with an eigenvalue higher than one were detected. The first component (X-axis), explaining 60.6% of the total variance, was positively correlated with the electrical conductivity (EC) of the substrate; that is, with soil salinity (Figure 9a). Consequently, all variables that increased with increasing salinity, root length, Na^+ and Cl^- contents in roots and leaves, Pro, H_2O_2 and antioxidant compounds, were also positively correlated with the first component. On the other hand, a negative correlation was found with the rest of the growth parameters (in agreement with the observed salt-induced inhibition of growth) and with photosynthetic pigments, MDA and TSS, which generally decreased in response to the salt treatments. The second component, explaining an additional 12.9% of the total variability, was mostly correlated, negatively, with K^+ levels in roots and leaves (Figure 9a).

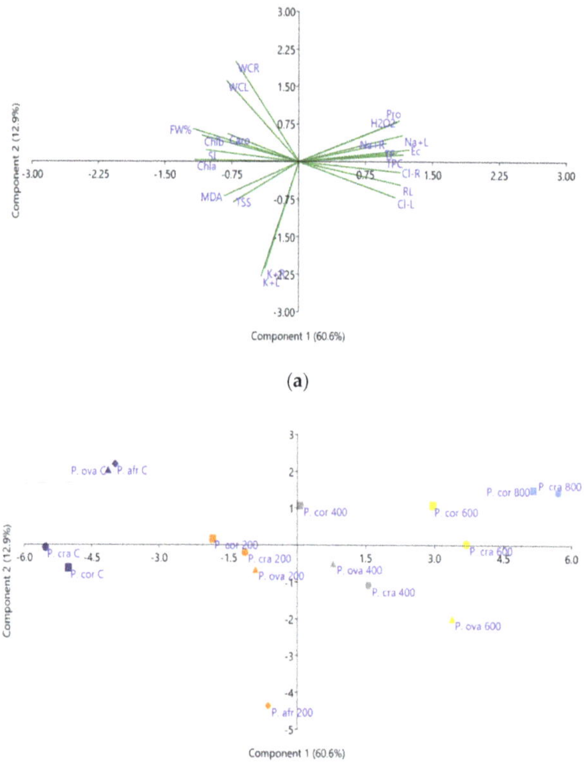

Figure 9. Loading plot of the principal component analysis (PCA) (**a**), and scatterplot of the PCA score (**b**), including all analysed traits in *P. coronopus* (P. cor), *P. crassifolia* (P. cra), *P. ovata* (P. ova) and *P. afra* (P. afr) subjected to treatments with 0 (control, C), 200, 400, 600 and 800 mM NaCl, for four weeks. The first and second principal components accounted for 60.6% and 12.9% of the total variation, respectively. Abbreviations: electrical conductivity of substrate (EC); root length (RL); stem length (SL); root water content (RWC); leaf water content (LWC); fresh weight of leaves (FW%); chlorophyll a (Chla) chlorophyll b (Chlb); total carotenoids (Caro); sodium content in leaves and roots (Na+L; Na+R); chloride content in leaves and roots (Cl-L; Cl-R); potassium content in leaves and roots (K+L; K+R), total soluble sugars (TSS), malondialdehyde (MDA), hydrogen peroxide (H2O2); proline (Pro); total phenolic compounds (TPC) and flavonoids (TF).

The scatterplot of the projected variables (Figure 9b) allowed a good separation of the salt treatments along the X-axis, from the non-stressed plants to those subjected to the highest salinity levels. A clear separation was also observed under control conditions between *P. crassifolia* and *P. coronopus* on the one side and *P. ovata* and *P. afra* on the other. Moreover, the response of *P. afra* to salt stress at 200 mM NaCl, the only treatment allowing survival of the plants of this less tolerant species, clearly differed from that of the other three taxa, in agreement with the relatively high salt tolerance of *P. ovata*, similar to that of the halophytes (Figure 9b).

3. Discussion

The four *Plantago* species analysed in the present work can be separated into two taxonomic groups; *P. crassifolia* and *P. coronopus* belong to the subgenus *Coronopus*, and *P. ovata* and *P. afra* are included in the subgenus *Psyllium*. Moreover, the first two species are defined as halophytes, whereas the other two are considered glycophytes. These differences are reflected in the positions of the four species in the scatterplot of the PCA score. Nevertheless, the present work results indicated that, although *P. afra* is indeed sensitive to salinity, *P. ovata* is quite salt-tolerant, apparently because it can use some specific tolerance mechanisms more efficiently than the halophytes.

As established for many different plant species, salt stress induces changes in root system morphology, growth rate and reproductive traits in *Plantago* [21,36,37]. However, the relative survival thresholds and the quantitative assessment of stress-induced growth inhibition are probably the most objective criteria to rank taxonomically related species according to their tolerance to different environmental stressors such as salinity or drought [21,38–40]. Of the four analysed *Plantago* species, *P. afra* was the most susceptible to salt stress as the plants survived the one-month treatment only in the presence of 200 mM NaCl, the lowest salt concentration tested. A previous study also reported that growth of *P. afra* was significantly inhibited at salinities higher than 100 mM NaCl and the plants did not survive the concentration of 300 mM [35]. Of the remaining species, *P. crassifolia* and *P. coronopus* were the most stress-tolerant, as reported in previous studies [21,24], which agrees with their ecology. Plants of *P. ovata* were relatively more affected by salinity than the two halophytes; still, under our experimental conditions, they survived all salt treatments except that at very high salinity, 800 mM NaCl. Indeed, this species has been considered moderately salt-tolerant [41], although marked differences between genotypes have been reported in the responses to salinity [42,43].

Reduced plant growth is one of the first and most general responses to stress. Accordingly, a general effect of growth inhibition in the presence of salt has been observed in all four *Plantago* species. However, the plant roots significantly increased in length in parallel to increasing salinity. According to Neumann [44], a rapid root elongation may occur in salt-stressed plants due to the massive production of young cells by stimulation of root meristem divisions. A more extensive root system penetrates deeper soil layers to obtain water and nutrients; this implies a higher water uptake capacity in tolerant plants, allowing ion dilution to help avoid reaching toxic levels in the cytosol [45]. Similar results have been reported, for example, in salt-stressed *P. major* plants, where primary roots were longer at all salinity levels compared to control plants [18,36]. The reduction in growth parameters of the plants' aerial parts, shoot length and fresh weight, and the level of leaf dehydration allowed us to establish the relative salt tolerance of the four species, as indicated above: *P. crassifoilia* ≅ *P. coronopus* > *P. ovata* >> *P. afra*.

Chlorophyll is a useful biochemical salt stress marker in plants, as high NaCl concentrations induce chlorophyll loss and necrosis of the leaves in many species [46,47]. Chlorophyll contents generally decrease in the presence of salt, often proportionally to the salt sensitivity of the plants, so that highly salt-tolerant halophytes may not show a reduction in chlorophyll levels under salinity conditions [12,48,49]. The decrease of photosynthetic pigments results from the inhibition of enzymes involved in chlorophyll biosynthesis and the fast breakdown of the pigments due to activation of chlorophyllase, responsible for chlorophyll degradation [50–52]. The selected *Plantago* species also showed this general pattern, a salt-induced, concentration-dependent decrease in photosynthetic pigment contents in response to the salt treatments. The reduction in the pigments' concentrations roughly corresponded to the relative salt tolerance of the plants, except that *P. ovata* appeared to be less affected than the halophytes, with chlorophylls a and b, and carotenoid contents significantly lower than control values only observed at high salinities.

One of the most significant differences in the mechanisms of salt stress response between glycophytes and halophytes regards managing the toxic ions present on saline soils. Glycophtes and monocotyledonous halophytes generally rely on reducing ion uptake through their roots or blocking their transport to the leaves. On the other hand, dicotyledonous halophytes activate the transport of toxic ions to the aboveground plant organs to be used for osmotic adjustment, but sequester them in the vacuoles to avoid their deleterious effects in the cytosol [7,53]. Ion compartmentalisation in the vacuoles is an extremely efficient mechanism, cheaper in energy consumption terms than the synthesis of organic osmolytes for ensuring an increased osmotic potential [54]. In the present work, we show that ion concentrations were consistently higher in leaves than in roots, at each salt concentration tested and in the four *Plantago* species, supporting the existence of these mechanisms of active ion transport to the leaves. Nevertheless, the patterns of accumulation of Na^+ and Cl^- differed quantitatively. Under salt stress, the glycophytes showed lower Na^+ content in roots and leaves than the halophytes, with the highest absolute values measured in *P. crassifolia* leaves. These findings indicate that Na^+ accumulation plays an essential role in the osmotic adjustment of halophytes of this genus subjected to high salinity conditions, as previously reported for these two species [21,24] and also for *P. maritima* [20,55–57].

Regarding Cl^- concentrations, the differences between species were not so pronounced as those of Na^+, neither in roots nor in leaves; the most relevant difference was that, under the same salinity conditions, the glycophyte *P. ovata* accumulated Cl^- to higher concentrations than the halophytes *P. crassifolia* and *P. coronopus*. The extremely high Cl^- concentration measured in leaves of the control, non-stressed plants is also remarkable. These data point to a constitutive defence mechanism against salt stress based on the accumulation of high leaf concentrations of this anion, even under low salinity conditions.

Concerning K^+, it is known that this 'physiological cation' plays an important role in plant growth and development, as well as in the maintenance of osmotic adjustment and cell turgor under stress [58]. A reduction of K^+ contents is generally observed under salt stress conditions, resulting from competition between Na^+ and K^+ for the same binding sites in proteins, including ion transporters [59]. Therefore, maintenance, or even increases in leaf K^+ levels in the presence of high Na^+ concentrations may contribute significantly to salt tolerance mechanisms. Indeed, activation of K^+ transport from roots to leaves at high salinities has been reported in some species, including glycophytes [60–62], and it is considered that salinity may enhance K^+ transport through the vascular bundles [9,63]. The analysed *Plantago* species differed in the patterns of K^+ transport and accumulation. The leaf K^+ contents did not vary with increasing salinity in *P. coronopus*, whereas they increased significantly in salt-treated *P. ovata* plants, probably contributing to the tolerance of this species.

Plants accumulate compatible solutes such as proline (Pro) and soluble sugars (TSS) to contribute to osmotic adjustment, and as osmoprotectants, under different stress conditions [53]. The accumulation of these metabolites is one of the best-known responses of plants to changes in the external osmotic potential [7,64]. Many reports showed that sorbitol is the primary physiological osmolyte in species of the genus *Plantago*, both salt-tolerant [24,55,65] and salt susceptible [56]. However, the differences in absolute sorbitol levels accumulated in response to salt treatments do not explain the different salt tolerance of the investigated species. In several halophytes of this genus, activation of Pro biosynthesis has been observed at high external salinity [21,28,57,66,67]. Pro can be considered, therefore, as a secondary functional osmolyte in salt-tolerant *Plantago* species. Pro is one of the most common compatible solutes involved in stress tolerance mechanisms in plants, accumulated in large quantities under high salinity stress (and other stressful conditions) in many plant species [68–70]. Apart from playing a major role in osmotic adjustment, Pro can act as an enzyme protectant, free radical scavenger, cytosolic pH stabiliser for subcellular structures and cell redox balancer [71]. In the present study, the NaCl treatments induced a significant, concentration-dependent increase in the leaf Pro contents, especially in the halophytes, *P. crassifolia* and *P. coronopus*, at the highest concentrations tested, 600–800 mM NaCl, but also, to a lesser extent, in *P. ovata*, in agreement with previous reports [72]. On the other hand, extremely low Pro concentrations were measured in the salt-sensitive *P. afra*.

Soluble sugars have also been shown to accumulate in plants in response to abiotic stresses, contributing to osmotic adjustment and playing additional regulatory functions [73,74]. However, TSS do not seem to have any relevant role in salt tolerance mechanisms in the investigated *Plantago* species, although they showed different accumulation patterns in the halophytes and the glycophytes. Thus, TSS increased significantly in *P. ovata*, but only at the highest salt concentration tested, 600 mM NaCl, and in *P. afra* in the presence of 200 mM NaCl; however, the differences with respect to the corresponding controls are too small to have any important osmotic effect. Soluble sugar contents, on the contrary, decreased with increasing salinity in *P. crassifolia* and *P. coronopus*. These differences are probably due to a more pronounced salt-induced inhibition of photosynthesis in the halophytes, as revealed by the stronger reduction in pigment levels as compared with *P. ovata*.

Salt stress increases the production of reactive oxygen species (ROS), which, when in excess, have deleterious effects by oxidation of nucleic acids, lipids, and proteins, inducing severe dysfunctions and even cell death [75]. Malondialdehyde (MDA) is a product of membrane lipid peroxidation widely used as a biomarker of oxidative stress [76]. Leaf MDA contents decreased with increasing salinity in *P. crassifolia* and *P. coronopus*; on the contrary, they increased slightly in *P. ovata* and *P. afra*, with statistically significant differences with the corresponding control in the presence of 600 and 200 mM NaCl, respectively. This finding indicates that the halophytes are better protected from salt-induced oxidative damage of cell membranes, probably because of more efficient defence mechanisms based on toxic ion compartmentalisation and osmolyte (Pro) accumulation, as discussed above. Similar results were reported from a comparative study between the salt-tolerant *P. maritima* and the glycophyte *P. media*, with a decrease of MDA in the former and a significant increase in the latter species [77].

H_2O_2 is a ubiquitous, moderately reactive ROS with an essential role as a signalling molecule in stress defence and adaptive responses [78]. Its variation patterns were strikingly similar in the salt-tolerant *Plantago* taxa, both the halophytes *P. crassifolia* and *P. coronopus* and *P. ovata*, increasing in parallel to the applied NaCl concentration. These data support the notion that H_2O_2 is indeed involved in the antioxidant mechanisms of tolerance in salt-tolerant *Plantago* species. On the contrary, in the salt-sensitive *P. afra*, no significant changes in H_2O_2 concentrations were detected in salt-treated plants with respect to the non-stressed controls.

Secondary metabolites with antioxidant properties play an important role in the tolerance of plants to salt stress [17]. Among these compounds, particular attention has been given to phenolic compounds and, especially, to the subgroup of flavonoids, because of their strong antioxidant activity [75,79]. It is known that salt stress triggers increased concentrations of phenolic compounds and flavonoids in *Plantago* [43,80] and that the level of antioxidant activity may be related to the degree of salt tolerance, being higher, for example, in the halophyte *P. maritima* in comparison to the salt-sensitive *P. media*, under waterlogging and salinity stresses. Moreover, differences between different species in their phenolic and flavonoid profiles have been proposed as chemotaxonomic markers in this genus [81]. In our experiments, total phenolic compounds and total flavonoids increased in response to the NaCl treatments, in a concentration-dependent manner, in all four *Plantago* species (except for flavonoids in *P. afra*); it is interesting to note that flavonoid levels were higher in *P. ovata* than in the halophytes at all salt concentrations tested in the former species. All these data agree with previous reports that propose *Plantago* species as a source of bioactive molecules, particularly useful for the prevention of oxidative stress-related diseases, or as functional foods [82,83].

4. Material and Methods

4.1. Plant Material

This study was conducted on four *Plantago* species, namely *P. coronopus, P. crassifolia, P. ovata* and *P. afra*. The seeds were collected from their natural habitats in grasslands and salt marshes from different contrasted geographical regions in Tunisia.

The corresponding collection sites are listed in Table 3. The geographic locations were recorded by a GPS Model Garmin 72. Seeds were collected, cleaned, dried and stored at 4 °C.

Table 3. Origin and bioclimatic zones of the studied *Plantago* species.

Species	Subgenus	Location	Latitude	Longitude	Bioclimatic Zones
P. coronopus L.	*Coronopus*	Hergla/Sousse	35°58′52.07″	10°31′38.14″	Semi-arid inferior
P. crassifolia Forsk.	*Coronopus*	Djerba/Mednine	33°49′52.96″	11°2′17.67″	Arid inferior
P. ovata Forsk.	*Psyllium*	Tataouine	32°55′8.63″	10°24′59.45″	Saharian superior
P. afra L.	*Psyllium*	Bouargoub	36°28′34.84″	10°36′46.35″	Semi-arid superior

After sterilisation with commercial bleach and several washes with distilled water, the seeds were sown on peat in 1 L pots placed in plastic trays (12 pots per tray). The trays were maintained in a germination chamber under long-day photoperiod (16 h of light), at 23 °C during the day and 17 °C at night, and 50–80% relative humidity. The pots were watered twice per week with deionised water.

4.2. Plant Growth, Salt Treatments and Plant Sampling

Salt treatments were started four weeks after sowing. Plants were watered twice a week with solutions of 0 (control), 200, 400, 600 or 800 mM NaCl in deionised water. Each treatment included five individual plants of each species as biological replicas. Plant material (the root and the aerial part of each plant) was harvested after four weeks, and several growth parameters were determined: root length (RL), stem length (SL), and fresh weight of roots (RFW) and leaves (LFW). Part of the fresh root and leaf material was weighed (FW), dried in an oven at 65 °C for ca. 72 h (until constant weight), and weighed again (dry weight, DW) to calculate the water content percentage of roots and leaves, as WC% = [(FW−DW) / FW] × 100.

4.3. Electrical Conductivity of the Substrate

The electrical conductivity of the substrate ($EC_{1:5}$) was measured at the end of the treatments. The samples were collected from five pots per species and treatment, and air-dried. Then, a substrate: deionised water (1:5) mix was prepared by stirring at 600 rpm at room temperature. The suspension was filtered through filter paper, and the EC was measured with a Crison 522 conductivity-meter (Crison Instruments, Barcelona, Spain) and expressed in dS m^{-1}.

4.4. Photosynthetic Pigments Determination

Chlorophyll a (Chl a), chlorophyll b (Chl b) and total carotenoid (Caro) contents were determined as previously described [84]. Fresh leaf material (0.1 g) was ground with liquid nitrogen, one ml of ice-cold 80% acetone was added, and the sample was shaken overnight at 4 °C in the dark. The extract was centrifuged at 13,300× *g*, at 4 °C, the supernatant was collected, and the absorbance was measured at 470, 645 and 663 nm. The following equations were used for the calculation of pigment concentrations, which were finally expressed in mg g^{-1} DW:

$$\text{Chl a } (\mu g/mL) = 12.21 \times (A_{663}) - 2.81 \times (A_{646})$$

$$\text{Chl b } (\mu g/mL) = 20.13 \times (A_{646}) - 5.03 \times (A_{663})$$

$$\text{Caro } (\mu g/mL) = (1000 \times A_{470} - 3.27 \times [\text{Chl a}] - 104 \times [\text{Chl b}])/227$$

These and all other UV/visible spectrophotometric assays described below were carried out using a UV-1600PC spectrophotometer (VWR, Llinars del Vallès, Barcelona, Spain).

4.5. Ion Content Measurements

Concentrations of sodium (Na^+), potassium (K^+), and chloride (Cl^-) were measured in the roots and leaves of plants sampled after the salt treatments, and in the corresponding non-stressed controls, according to Weimberg [85]. Dried material (ca. 0.1 g) was ground to a fine powder and extracted in 15 mL of MilliQ water, incubating the samples for one hour in a water bath, at 95 °C, followed by cooling to room temperature and filtration through a 0.45 µm Gelman nylon filter (Pall Corporation, Port Washington, NY, USA). The cations Na^+ and K^+ were quantified with a PFP7 flame photometer (Jenway Inc., Burlington, VT, USA) and the anion using a chlorimeter (Sherwood, model 926, Cambridge, UK).

4.6. Proline and Total Soluble Sugars Quantification

Proline (Pro) content was determined in fresh tissue by the ninhydrin-acetic acid method [86]. Free Pro was extracted in 3% aqueous sulphosalicylic acid, and the extract was mixed with acid ninhydrin solution, incubated at 95 °C for 1 h, cooled on ice and then extracted with two volumes of toluene. The absorbance of the organic phase was determined at 520 nm using toluene as a blank. Samples containing known amounts of Pro were assayed in parallel to obtain a standard curve. Pro concentration was expressed as µmol g^{-1} DW.

Total soluble sugars (TSS) were quantified according to Dubois et al. [87]. Fresh leaf material (ca. 0.1 g) was extracted in 3 mL of 80% (*v*/*v*) methanol on a rocker shaker for 24 h. The samples were vortexed and centrifuged at 13,300× *g* for 10 min, and the supernatants were collected and diluted 10-fold with water. The diluted samples were supplemented with concentrated sulphuric acid and 5% phenol, and the absorbance was measured at 490 nm. TSS contents were expressed as 'mg equivalent of glucose', used as the standard (mg eq. gluc g^{-1} DW).

4.7. Oxidative Stress Markers

Malondialdehyde (MDA) contents were determined following a previously reported procedure [88] with some modifications [89], using the same 80% methanol extracts prepared for TSS quantification. The samples were mixed with 0.5% thiobarbituric acid (TBA) dissolved in 20% trichloroacetic acid (TCA) (or with 20% TCA without TBA for the controls) and then incubated at 95 °C for 20 min. The reactions were stopped on ice, and the samples were centrifuged at 13,300× *g* for 10 min at 4 °C. Finally, the absorbance of the supernatants was determined at 440, 532 and 600 nm. MDA concentration was calculated using the equations previously described [89], based on the molar extinction coefficient at 532 nm of the MDA-TBA adduct (ε_{532} = 155 mM^{-1} cm^{-1}).

Measurement of the hydrogen peroxide (H_2O_2) content was carried out according to a previously published method [90]. H_2O_2 was extracted in a 0.1% (*w*/*v*) TCA solution from 0.1 g fresh leaf material. The extract was centrifuged at 13,300× *g* for 15 min, and the supernatant was collected and mixed with one volume of 10 mM potassium phosphate buffer (pH 7) and two volumes of 1 M KI. Finally, the absorbance of the sample was measured at 390 nm. H_2O_2 contents were expressed as µmol g^{-1} DW.

4.8. Non-Enzymatic Antioxidants

Total phenolic compounds (TPC) and total flavonoid (TF) contents were measured in the same 80% methanol extracts used for TSS and MDA quantification. TPC were determined as previously described [91] by reaction of the extracts with $NaHCO_3$ and the Folin-Ciocalteu reagent. The reaction mixtures were kept in the dark, at room temperature, for 90 min, and the absorbance was then measured at 765 nm. TPC concentration was expressed as equivalents of the gallic acid standard (mg eq. GA g^{-1} DW).

TF were determined by reaction with $AlCl_3$ under alkaline conditions after nitration of catechol groups with $NaNO_2$ [92]. The absorbance of the samples was read at 510 nm. Catechin was used as a standard to plot a calibration curve, and the results were expressed as catechin equivalents (mg eq. C g^{-1} DW).

4.9. Statistical Analysis

Each assay was conducted in a completely randomised design (CRD) with four genotypes and two treatments. Variance analysis was performed to determine the interaction between the different applied treatments and the different species. The measured parameters were subjected to a two-way analysis of variance (ANOVA test). The confidence interval was calculated at the threshold of 95% with mean comparison according to the Tukey test using 'PLAnt Breeding STATistical software' (PLABSTAT) [93], version 3A of 2011-06-14. Throughout the text, all values shown are means of five biological replicas (five individual plants) $\pm$ standard error (SE).

A Principal Components Analysis (PCA) was carried out on the correlation matrix using PAST software, version 4.03 [94]. The PCA was applied to the data matrix (21 morphological, physiological and biochemical traits $\times$ 4 *Plantago* species). The input data contained the mean values of all parameters measured under the different salt stress conditions. The cumulative variability of each parameter was calculated, as well as eigenvalues and principal component scores.

5. Conclusions

The four *Plantago* species analysed here can be clearly divided, by several criteria, into two groups: the halophytes *P. crassifolia* and *P. coronopus* and the glycophytes *P. ovata* and *P. afra*. The halophytes, as expected, are highly salt-tolerant, surviving one-month treatment at salinities as high as 800 mM NaCl. Despite not being considered a typical halophyte, *P. ovata* plants are nonetheless relatively resistant to salt, withstanding one month in the presence of 600 mM NaCl. *Plantago afra*, on the other hand, is the most salt-sensitive of the four species, surviving only the 200 mM NaCl treatment.

The most relevant tolerance mechanisms of *P. crassifolia* and *P. coronopus* are based on: (i) the active transport of Na^+ and Cl^- ions to the leaves, where they contribute to cellular osmotic balance under high salinity conditions, as 'inorganic osmolytes'; (ii) the accumulation of high leaf levels of the organic osmolyte proline; (iii) their relative resistance to the generation of oxidative stress causing membrane lipid peroxidation; and (iv) the salt-induced increase of the levels of antioxidant metabolites, such as phenolic compounds and flavonoids. In *P. ovata*, the efficiency of the above mechanisms is generally lower than in the halophytes, but this limitation is partly compensated by: (i) a more efficient transport to the aerial part and accumulation in the leaves of Cl^- ions; (ii) the activation of K^+ uptake and transport to the leaves under high salinity conditions; (iii) a less pronounced inhibition of photosynthesis, as indicated by the smaller reduction of photosynthetic pigments contents; and (iv) the accumulation of flavonoids in the leaves to relatively higher concentrations than in the halophytes, at salt concentrations of 200 to 600 mM NaCl. Apart from these induced mechanisms, constitutive responses contribute to salt tolerance in the three species, namely the accumulation in leaves of inorganic ions at high concentrations in control, non-stressed plants. Summarising, *P. ovata*, not considered a halophytic species, is nevertheless quite resistant to salt stress but using tolerance mechanisms somewhat different from those of the typical congeneric halophytes, *P. crassifolia* and *P. coronopus*.

This work confirms the usefulness of performing comparative studies on the responses to stress of taxonomically related species with different degrees of resistance to the particular stressful condition, to identify the most relevant tolerance mechanisms.

Author Contributions: Conceptualization, H.B.L., M.B., O.V. and S.R.; methodology, H.B.L., S.G.-O. and A.L.F.; software, H.B.L., A.S; validation, M.B., O.V. and S.R.; formal analysis, investigation, data curation, H.B.L. and A.S.; writing—original draft preparation, H.B.L.; writing—review and editing, M.B. and O.V.; funding acquisition, O.V. and M.B.; visualization, supervision, M.B., O.V. and S.R. All authors have read and agreed to the published version of the manuscript.

Funding: This research was supported by internal funds of Universitat Politècnica de València to O.V. and M.B.

Institutional Review Board Statement: Not applicable

Informed Consent Statement: Not applicable

Data Availability Statement: Data are contained within the article.

Acknowledgments: Hela Belhaj Ltaeif's and Anbu Landa-Faz's stays at the Universitat Politècnica de Valencia were supported by Scholarships from the Ministry of Higher Education and Scientific Research and University of Carthago, Tunisia, and from CONACYT, Mexico, respectively.

Conflicts of Interest: The authors declare no conflict of interest.

References

1. Parvaiz, A.; Satyawati, S. Salt stress and phyto-biochemical responses of plants—A review. *Plant Soil Environ.* **2008**, *54*, 88–99. [CrossRef]
2. Arzani, A.; Ashraf, M. Smart engineering of genetic resources for enhanced salinity tolerance in crop plants. *Crit. Rev. Plant Sci.* **2016**, *35*, 146–189. [CrossRef]
3. Shahid, S.A.; Zaman, M.; Heng, L. Soil salinity: Historical perspectives and a world overview of the problem. In *Guideline for Salinity Assessment, Mitigation and Adaptation Using Nuclear and Related Techniques*; Zaman, M., Shahid, S.A., Heng, L., Eds.; Springer: Cham, Germany, 2018. [CrossRef]
4. Pitman, M.G.; Láuchli, A. Global impact of salinity and agricultural ecosystems. In *Salinity: Environment-Plants–Molecules*; Läuchli, A., Lüttge, U., Eds.; Kluver Academic Publishers: Dordrecht, The Netherlands, 2002. [CrossRef]
5. IPCC. Intergovernmental panel on climate change. In Proceedings of the 5th Assessment Report, WGII, Climate Change: Impacts, Adaptation, and Vulnerability. Available online: http://www.ipcc.ch/report/ar5/wg2/ (accessed on 4 March 2021).
6. Fita, A.; Rodríguez-Burruezo, A.; Boscaiu, M.; Prohens, J.; Vicente, O. Breeding and domesticating crops adapted to drought and salinity: A new paradigm for increasing food production. *Front. Plant Sci.* **2015**, *6*, 978. [CrossRef] [PubMed]
7. Munns, R.; Tester, M. Mechanisms of salinity tolerance. *Annu. Rev. Plant Biol.* **2008**, *59*, 651–681. [CrossRef]
8. Gupta, B.; Huang, B. Mechanism of salinity tolerance in plants: Physiological, biochemical, and molecular characterization. *Int. J. Genomics* **2014**, *701596*. [CrossRef] [PubMed]
9. Isayenkov, S.V.; Maathuis, F.J.M. Plant salinity stress: Many unanswered questions remain. *Front. Plant Sci.* **2019**, *1*, 80. [CrossRef]
10. Van Zelm, E.; Zhang, Y.; Testerink, Y. Salt tolerance mechanisms of plants. *Annu. Rev. Plant Biol.* **2020**, *71*, 403–433. [CrossRef]
11. Muhammad, Z.; Hussain, F. Effect of NaCl salinity on the germination and seedling growth of some medicinal plants. *Pak. J. Bot.* **2010**, *42*, 889–897.
12. Ashraf, M.; Harris, P.J.C. Photosynthesis under stressful environments: An overview. *Photosynthetica* **2013**, *51*, 163–190. [CrossRef]
13. Santos, J.; Al-Azzawi, M.; Aronson, J.; Flowers, T.J. eHALOPH a database of salt-tolerant plants: Helping put halophytes to work. *Plant Cell Physiol.* **2016**, *57*, e10. [CrossRef]
14. Hassemer, G.; Shipunov, A.; Rønsted, N.A.H.; Meudt, H.M. Taxonomic and geographic novelties in the genus *Plantago* (Plantaginaceae) in Chile, including the description of a new species. *Phytotaxa* **2018**, *340*, 137–156. [CrossRef]
15. Hassemer, G.; Bruun-Lund, S.; Shipunov, A.; Briggs, B.G.; Meudt, H.M.; Rønsted, N.A.H. The application of high-throughput sequencing for taxonomy: The case of *Plantago* subg. *Plantago* (Plantaginaceae). *Mol. Phylogenet. Evol.* **2019**, *138*, 156–173. [CrossRef] [PubMed]
16. Chater, A.O.; Cartier, D. Plantago L. In *Flora Europaea 4, Plantaginaceae to Compositae*; Tutin, T.G., Heywood, V.H., Burges, N.A., Valentine., D.H., Eds.; Cambridge University Press: Cambridge, UK, 1976; Volume 4, pp. 38–44.
17. Ksouri, R.; Ksouri, W.M.; Jallali, I.; Debez, A.; Magné, C.; Hiroko, I.; Abdelly, C. Medicinal halophytes: Potent source of health promoting biomolecules with medical, nutraceutical and food applications. *Crit. Rev. Biotechnol.* **2012**, *32*, 289–326. [CrossRef]
18. Flanagan, L.B.; Jefferies, R.L. Effect of increased salinity on CO_2 assimilation, O_2 evolution and the δ13C values of leaves of *Plantago maritima* L. developed at low and high NaC1 levels. *Planta* **1989**, *78*, 377–384. [CrossRef] [PubMed]
19. Maathuis, F.J.; Prins, H.B. Patch clamp studies on root cell vacuoles of a salt-tolerant and a salt-sensitive *Plantago* species: Regulation of channel activity by salt stress. *Plant Physiol.* **1990**, *92*, 23–28. [CrossRef] [PubMed]

20. Staal, M.; Maathuis, F.J.M.; Elzenga, J.T.M.; Overbeek, J.H.M.; Prins, H.B.A. Na^+/H^+ antiport activity in tonoplast vesicles from roots of the salt-tolerant *Plantago maritima* and the salt-sensitive *Plantago media*. *Physiol. Plant.* **1991**, *82*, 179–184. [CrossRef]
21. Al Hassan, M.; Pacurar, A.; López-Gresa, M.P.; Donat-Torres, M.P.; Llinares, J.V.; Boscaiu, M.; Vicente, O. Effects of salt stress on three ecologically distinct *Plantago* Species. *PLoS ONE* **2016**, *11*, e0160236. [CrossRef] [PubMed]
22. Hultén, E.; Fries, M. *Atlas of North European Vascular Plants North of the Tropic of Cancer*; Maps 1–996; Koeltz Scientific Books: Königstein, Germany, 1986.
23. Villellas, J.; Ehrlén, J.; Olesen, J.M.; Braza, R.; García, M.B. Plant performance in central and northern peripheral populations of the widespread *Plantago coronopus*. *Ecography* **2013**, *36*, 36–145. [CrossRef]
24. Koyro, H.W. Effect on salinity on growth, photosynthesis, water relations and solute composition of the potential cash crop halophyte *Plantago coronopus* (L.). *Environ. Exp. Bot.* **2006**, *56*, 136–146. [CrossRef]
25. Ceccanti, C.; Landi, M.; Incrocci, L.; Pardossi, A.; Venturi, F.; Taglieri, I.; Ferroni, G.; Guidi, L. Comparison of three domestications and wild-harvested plants for nutraceutical properties and sensory profiles in five wild edible herbs: Is domestication possible? *Foods* **2020**, *9*, 1065. [CrossRef]
26. Boestfleisch, C.; Wagenseil, N.B.; Buhmann, A.K.; Seal, C.E.; Wade, E.M.; Muscolo, A.; Papenbrock, J. Manipulating the antioxidant capacity of halophytes to increase their cultural and economic value through saline cultivation. *AoB Plants* **2014**, *6*, plu046. [CrossRef]
27. Höpke, J.; Mucina, L.; Albach, D.C. Phylogenetic and morphometric analysis of *Plantago* section *Coronopus* (Plantaginaceae). *Taxon* **2019**, *68*, 315–339. [CrossRef]
28. Vicente, O.; Boscaiu, M.; Naranjo, M.A.; Estrelles, E.; Bellés, J.M.; Soriano, P. Responses to salt stress in the halophyte *Plantago crassifolia* (Plantaginaceae). *J. Arid Environ.* **2004**, *58*, 463–481. [CrossRef]
29. El Shaer, H.M.; Attia-Ismail, S.A. Halophytic and salt-tolerant feedstuffs in the Mediterranean Basin and Arab Region: An overview. In *Halophytic and Salt-Tolerant Feedstuffs: Impacts on Nutrition, Physiology and Reproduction of Livestock*; El Shaer, H.M., Squires, V.R., Eds.; CRC Press: Boca Raton, FL, USA, 2016.
30. Rønsted, N.; Chase, M.W.; Albaxh, D.C.; Bello, M.A. Phylogenetic relationships within *Plantago* (Plantaginaceae): Evidence from nuclear ribosomal ITS and plastid trnL-F sequence data. *Bot. J. Linn. Soc.* **2002**, *139*, 323–338. [CrossRef]
31. Pedról, J.; Plantago, L. *Flora ibérica 13 Plantaginaceae-Scrophulariaceae*; Benedí, C., Rico, E., Eds.; Real Jardín Botánico, CSIC: Madrid, Spain, 2009; Volume 13, pp. 4–38.
32. Meyers, S.C.; Liston, A. The biogeography of *Plantago ovata* Forssk. (Plantaginaceae). *Int. J. Plant Sci.* **2008**, *169*, 954–962. [CrossRef]
33. Shryock, D.F.; Washburn, L.K.; De Falco, L.A.; Esque, T.C. Genetic and habitat data for *Plantago ovata* in the Mojave Desert: U. S. *Geological Survey data release* **2020**. [CrossRef]
34. Talukder, P.; Talapatra, S.; Ghoshal, N.; Raychaudhuri, S. Antioxidant activity and high-performance liquid chromatographic analysis of phenolic compounds during in vitro callus culture of *Plantago ovata* Forsk. and effect of exogenous additives on accumulation of phenolic compounds. *J. Sci. Food Agric* **2016**, *96*, 232–244. [CrossRef] [PubMed]
35. Izadi-Darbandi, E.; Mehdikhani, H. Salinity effect on some of the morphophysiological traits of three *Plantago* species (*Plantago* spp.). *Sci. Hortic.* **2018**, *236*, 43–51. [CrossRef]
36. Rubinigg, M.; Wenisch, J.; Elzenga, J.T.M.; Stulen, I. NaCl salinity affects lateral root development in *Plantago martitima*. *Funct. Plant Biol.* **2004**, *31*, 775–780. [CrossRef]
37. Al Hassan, M.; Pacurar, A.; Gaspar, A.; Vicente, O.; Boscaiu, M. Growth and reproductive success under saline conditions of three *Plantago* species with different levels of stress tolerance. *Not. Bot. Horti Agrobot. Cluj Napoca* **2014**, *42*, 180–186. [CrossRef]
38. Orsini, F.; D'Urzo, M.P.; Inan, G.; Serra, S.; Oh, D.H.; Mickelbart, M.V.; Consiglio, F.; Li, X.; Jeong, J.C.; Yun, D.J.; et al. A comparative study of salt tolerance parameters in 11 wild relatives of *Arabidopsis thaliana*. *J. Exp. Bot.* **2010**, *61*, 3787–3798. [CrossRef]
39. Cicevan, R.; Al Hassan, M.; Sestras, A.F.; Prohens, J.; Vicente, O.; Sestras, R.E.; Boscaiu, M. Screening for drought tolerance in cultivars of the ornamental genus Tagetes (Asteraceae). *PeerJ* **2016**, *4*, e2133. [CrossRef] [PubMed]
40. Todea, I.M.; González-Orenga, S.; Plazas, M.; Sestras, A.F.; Prohens, T.J.; Vicente, O.; Sestras, R.E. Screening for salt and water stress tolerance in fir (*Abies alba*) populations. *Not. Bot. Horti Agrobot. Cluj Napoca* **2019**, *47*, 1063–1072. [CrossRef]
41. Kazal, V.; Bhushan, B.; Kumari Narval, R.; Kumar, S.; Pal, A. Evaluation of salt tolerance in different isabgol (*Plantago ovata* Forsk.) genotypes under differential salt stress imposition. *Indian J. Agric. Biochem.* **2017**, *30*, 41–49. [CrossRef]
42. Khaliq, R.; Zahoor, M.; Zafar, Z.U.; Athar, H.R. Growth responses of *Plantago ovata* L. to varying levels of NaCl. *Iran. J. Plant Physiol.* **2011**, *1*, 157–167.
43. Golkar, P.; Amooshahi, F.; Arzani, A. The effects of salt stress on physio-biochemical traits, total phenolic and mucilage content of *Plantago ovata* Forsk under in vitro conditions. *J. Appl. Bot. Food Qual.* **2017**, *90*, 224–231.
44. Neumann, P.M. Inhibition of root growth by salinity stress: Toxicity or an adaptive biophysical response. In *Structure and Function of Roots*; Baluska, F., Ciamporova, M., Gasparikova, O., Eds.; Kluwer Academic Publishers: Dordrecht, The Netherlands, 1995; pp. 299–304.
45. Chelli-Chaabouni, A.; Hammami, M.; Gouia, M.; Gargouri, K.; Gargouri, R.; Drira, N. Effect of Salt Stress on Pistacia atlantica Rootstock Seedlings in Nursery Conditions. In *XIV GREMPA Meeting on Pistachios and Almonds*; Zakynthinos, G., Ed.; CIHEAM/FAO/AUA/TEI Kalamatas/NAGREF: Zaragoza, Spain, 2010; pp. 135–140.

46. Dubey, S.; Bhargava, A.; Fuentes, F.; Shukla, S.; Srivastava, S. Effect of salinity stress on yield and quality parameters in flax (*Linum usitatissimum* L.). *Not. Bot. Horti Agrobot. Cluj Napoca* **2020**, *48*, 954–966. [CrossRef]
47. Rahneshan, Z.; Nasibi, F.; Moghadam, A.A. Effects of salinity stress on some growth, physiological, biochemical parameters and nutrients in two pistachio (*Pistacia vera* L.) rootstocks. *J. Plant Interact.* **2018**, *13*, 73–82. [CrossRef]
48. Stepien, P.; Johnson, G.N. Contrasting responses of photosynthesis to salt-stress in glycophyte *Arabidopsis* and halophyte *Thellungiella*: Role of the plastid terminal oxidase as an alternative electron sink. *Plant Physiol.* **2009**, *149*, 1154–1165. [CrossRef] [PubMed]
49. Kozminska, A.; Al Hassan, M.; Hanus-Fajerska, E.; Naranjo, M.A.; Bosciau, M.; Vicente, O. Comparative analysis of water deficit and salt tolerance mechanisms in Silene. *South Afr. J. Bot.* **2018**, *117*, 193–206. [CrossRef]
50. Ashraf, M. Some important physiological selection criteria for salt tolerance in plants. *Flora* **2004**, *199*, 361–376. [CrossRef]
51. Flexas, J.; Medrano, H. Drought inhibition of photosynthesis in C3 plants: Stomatal and non-stomatal limitations revised. *Ann. Bot.* **2002**, *89*, 183–189. [CrossRef] [PubMed]
52. Munné Bosch, S.; Allegre, I. Die and let live: Leaf senescence contributes to plant survival under drought stress. *Funct. Plant Biol.* **2004**, *31*, 203–216. [CrossRef] [PubMed]
53. Flowers, T.J.; Colmer, T.D. Salinity tolerance in halophytes. *New Phytol.* **2008**, *179*, 945–963. [CrossRef] [PubMed]
54. Raven, J.A. Regulation of pH and generation of osmolarity in vascular plants: A cost-benefit analysis in relation to efficiency of use of energy, nitrogen and water. *New Phytol.* **1985**, *101*, 25–77. [CrossRef]
55. Erdei, L.; Kuiper, P.J.C. The effect of salinity on growth, cation content, Na^+-Uptake and translocation in salt-sensitive and salt-tolerant *Plantago* species. *Physiol. Plant.* **1979**, *47*, 95–99. [CrossRef]
56. Königshofer, H. Changes in ion composition and hexitol content of different *Plantago* species under the influence of salt stress. *Plant Soil* **1983**, *72*, 289–296. [CrossRef]
57. Sleimi, N.; Guerfali, S.; Bankaji, I. Biochemical indicators of salt stress in *Plantago maritima*: Implications for environmental stress assessment. *Ecol. Indic.* **2015**, *48*, 570–577. [CrossRef]
58. Marschner, H. Ion uptake mechanisms of individual cells and roots: Short-distance transport. In *Mineral Nutrition of Higher Plants*, 2nd ed.; Academic Press: London, UK, 1995; pp. 6–78. [CrossRef]
59. Greenway, H.; Munns, R. Mechanisms of salt tolerance in nonhalophytes. *Ann. Rev. Plant Physiol.* **1980**, *31*, 149–190. [CrossRef]
60. Schiop, S.T.; Al Hassan, M.; Sestras, A.F.; Boscaiu, M.; Sestras, R.E.; Vicente, O. Identification of salt stress biomarkers in Romanian Carpathian populations of *Picea abies* (L.) Karst. *PLoS ONE* **2015**, *10*, e0135419. [CrossRef] [PubMed]
61. Arteaga, S.; Al Hassan, M.; Chaminda Bandara, W.M.; Yabor, L.; Llinares, J.V.; Boscaiu, M.; Vicente, O. Screening for salt tolerance in four local varieties of *Phaseolus lunatus* from Spain. *Agriculture* **2018**, *8*, 201. [CrossRef]
62. Borsai, O.; Al Hassan, M.; Negrușier, C.; Raigón, M.D.; Boscaiu, M.; Sestraș, R.E.; Vicente, O. Responses to salt stress in *Portulaca*: Insight into its tolerance mechanisms. *Plants* **2020**, *9*, 1660. [CrossRef] [PubMed]
63. Shabala, S.; Cuin, T.A. Potassium transport and plant salt tolerance. *Physiol. Plant.* **2008**, *133*, 651–669. [CrossRef] [PubMed]
64. Hasegawa, P.M.; Bressan, R.A.; Zhu, J.K.; Bohnert, H.J. Plant cellular and molecular responses to high salinity. *Annu. Rev. Plant Physiol. Plant Mol. Biol.* **2000**, *51*, 463–499. [CrossRef] [PubMed]
65. Ahmad, I.; Larher, F.; Stewart, G.R. Sorbitol, a compatible osmotic solute in *Plantago maritima*. *New Phytol.* **1979**, *82*, 671–678. [CrossRef]
66. Tipirdamaz, R.; Gagneul, D.; Duhazé, C.; Aïnouche, A.; Monnier, C.; Özkum, D.; Larher, F. Clustering of halophytes from an inland salt marsh in Turkey according to their ability to accumulate sodium and nitrogenous osmolytes. *Environ. Exp. Bot.* **2006**, *57*, 139–153. [CrossRef]
67. Pardo-Domènech, L.; Tifrea, A.; Grigore, M.N.; Boscaiu, M.; Vicente, O. Proline and glycine betaine accumulation in two succulent halophytes under natural and experimental conditions. *Plant Biosyst* **2016**, *150*, 904–915. [CrossRef]
68. Kishor, P.B.K.; Sangam, S.; Amrutha, R.N.; Laxmi, P.S.; Naidu, K.R.; Rao, K.R.S.S.; Rao, S.; Reddy, K.J.; Theriappan, P.; Sreenivasulu, N. Regulation of proline biosynthesis, degradation, uptake and transport in higher plants: Its implications in plant growth and abiotic stress tolerance. *Curr. Sci.* **2005**, *88*, 424–438.
69. Szabados, L.; Savouré, A. Proline: A multifunctional amino acid. *Trends Plant Sci.* **2010**, *15*, 89–97. [CrossRef]
70. Iqbal, N.; Umar, S.; Khan, N.A.; Khan, M.I.R. A new perspective of phytohormones in salinity tolerance regulation of proline metabolism. *Environ. Exp. Bot.* **2014**, *100*, 34–42. [CrossRef]
71. Liang, X.; Zhang, L.; Natarajan, S.K.; Becker, D.F. Proline mechanisms of stress survival. *Antioxid. Redox Signal.* **2013**, *19*, 998–1011. [CrossRef] [PubMed]
72. Karimi, N.; Modarresi, M.; Mollaei, S.E. Enhancement of compatible solute and secondary metabolites production in *Plantago ovata* Forsk. by salinity stress. *J. Med. Plant Res.* **2012**, *6*, 3495–3500.
73. Parida, A.K.; Das, A.B. Salt tolerance and salinity effects on plant: A review. *Ecotoxicol. Environ. Saf.* **2005**, *60*, 324–349. [CrossRef] [PubMed]
74. Gil, R.; Boscaiu, M.; Lull, C.; Bautista, I.; Lid, N.A.; Vicente, O. Are soluble carbohydrates ecologically relevant for salt tolerance in halophytes? *Funct. Plant. Biol.* **2013**, *40*, 805–818. [CrossRef] [PubMed]
75. Apel, K.; Hirt, H. Reactive oxygen species: Metabolism, oxidative stress, and signal transduction. *Annu. Rev. Plant Biol.* **2004**, *55*, 373–379. [CrossRef]

76. Del Rio, D.; Stewart, A.J.; Pellegrini, N. A review of recent studies on malondialdehyde as toxic molecule and biological marker of oxidative stress. *Nutr. Metab. Cardiovasc. Dis.* **2005**, *15*, 316–328. [CrossRef]
77. Sekmen, A.H.; Türkan, I.; Takio, S. Differential responses of antioxidative enzymes and lipid peroxidation to salt stress in salt-tolerant *Plantago maritima* and salt-sensitive *Plantago media*. *Physiol. Plant.* **2007**, *131*, 399–411. [CrossRef]
78. Das, K.; Roychoudhury, A. Reactive oxygen species (ROS) and response of antioxidants as ROS-scavengers during environmental stress in plants. *Front. Environ. Sci.* **2014**, *2*, 53. [CrossRef]
79. Vicente, O.; Boscaiu, M. Flavonoids: Antioxidant compounds for plant defence... and for a healthy human diet. *Not. Bot. Horti Agrobot. Cluj Napoca* **2018**, *46*, 14–21. [CrossRef]
80. Türkan, I.; Demiral, T.J.; Sekmen, A.H. The regulation of antioxidant enzymes in two *Plantago* species differing in salinity tolerance under combination of waterlogging and salinity. *Funct. Plant Biol.* **2012**, 40484–40493. [CrossRef] [PubMed]
81. Jurišić Grubešić, R.; Srečnik, G.; Kremer, D.; Vuković Rodríguez, J.; Nikolić, T.; Vladimir-Knežević, S. Simultaneous RP-HPLC-DAD separation and determination of flavonoids and phenolic acids in *Plantago*, L. species. *Chem. Biodiver.* **2013**, *10*, 1305–1316. [CrossRef] [PubMed]
82. Rodrigues, M.J.; Gangadhar, K.N.; Vizetto-Duarte, C.; Wubshet, S.G.; Nyberg, N.T.; Barreira, L.; Varela, J.; Custódio, L. Maritime halophyte species from southern Portugal as sources of bioactive molecules. *Mar. Drugs* **2014**, *12*, 2228. [CrossRef]
83. Pereira, C.G.; Custódio, L.; Rodrigues, M.J.; Neng, N.R.; Nogueira, J.M.F.; Carlier, J.; Costa, M.C.; Varela, J.; Barreira, L. Profiling of antioxidant potential and phytoconstituents of *Plantago coronopus*. *Braz. J. Biol.* **2017**, *77*, 3. [CrossRef] [PubMed]
84. Lichtenthaler, H.K.; Wellburn, A.R. Determination of total carotenoids and chlorophyll a and b of leaf extracts in different solvents. *Biochem. Soc. Trans.* **1983**, *11*, 591–592. [CrossRef]
85. Weimberg, R. Solute adjustments in leaves of two species of wheat at two different stages of growth in response to salinity. *Physiol. Plant.* **1987**, *70*, 381–388. [CrossRef]
86. Bates, L.S.; Waldren, R.P.; Teare, I.D. Rapid determination of free proline for water stress studies. *Plant Soil* **1973**, *39*, 205–207. [CrossRef]
87. Dubois, M.; Gilles, K.A.; Hamilton, J.K.; Rebers, P.A.; Smith, F. Colorimetric method for determination of sugars and related substances. *Anal. Chem.* **1956**, *28*, 350–356. [CrossRef]
88. Hodges, D.M.; Delong, J.M.; Forney, C.F.; Prange, R.K. Improving the thiobarbituric acid-reactive-substances assay for estimating lipid peroxidation in plant tissues containing anthocyanin and other interfering compounds. *Planta* **1999**, *207*, 604–611. [CrossRef]
89. Taulavuori, E.; Hellström, E.K.; Taulavuori, K.; Laine, K. Comparison of two methods used to analyse lipid peroxidation from *Vaccinium myrtillus* L. during snow removal, re-acclimation and cold acclimation. *J. Exp. Bot.* **2021**, *52*, 2375–2380. [CrossRef]
90. Loreto, F.; Velikova, V. Isoprene produced by leaves protects the photosynthetic apparatus against ozone damage, quenches ozone products and reduces lipid peroxidation of cellular membranes. *Plant Physiol.* **2001**, *127*, 1781–1787. [CrossRef]
91. Blainski, A.; Lopes, G.C.; Palazzodemello, J.C. Application and analysis of the Folin-Ciocalteu method for the determination of the total phenolic content from *Limonium brasiliense* L. *Molecules* **2013**, *18*, 6852–6865. [CrossRef] [PubMed]
92. Zhishen, J.; Mengcheng, T.; Jianming, W. The determination of flavonoid contents in mulberry and their scavenging effects on superoxide radicals. *Food Chem.* **1999**, *64*, 555–559. [CrossRef]
93. Utz, H.F. *Plabstat: A Computer Program for Statistical Analysis of Plant Breeding Experiments*; Version 3A; Universitat Hohenheim: Stuttgart, Germany, 2001.
94. Hammer, O.; Harper, D.A.T.; Rayan, P.D. Past: Palaeontogical statistics software package for education and data analysis. *Palaeontol. Electron.* **2011**, *4*, 9.

plants

MDPI

Article

Differential Salt Tolerance Strategies in Three Halophytes from the Same Ecological Habitat: Augmentation of Antioxidant Enzymes and Compounds

AbdEl-Mageed F. M. Ghanem [1], Elsayed Mohamed [1,*], Ahmed M. M. A. Kasem [1] and Abbas A. El-Ghamery [2]

1 Botany and Microbiology Department, Faculty of Science, Al-Azhar University, Assuit 71524, Egypt; AbdEl-MageedEldeeb.42@azhar.edu.eg (A.F.M.G.); amkasem@azhar.edu.eg (A.M.M.A.K.)
2 Botany & Microbiology Department, Faculty of Science (Cairo), Al-Azhar University, Madinat Nasr, Cairo 11751, Egypt; a.el_ghamery@azhar.edu.eg
* Correspondence: sayedmohamed@azhar.edu.eg

Abstract: Understanding the salt tolerance mechanism in obligate halophytes provides valuable information for conservation and re-habitation of saline areas. Here, we investigated the responses of three obligate halophytes namely *Arthrocnemum macrostachyum*, *Sarcocornia fruticosa* and *Salicornia europaea* to salt stress (0, 100, 200, 400 and 600 mM NaCl) during their vegetative growth with regard to biomass, ions contents (Na^+, K^+ and Ca^{2+}), chlorophyll contents, carotenoids, phenolic compounds, flavonoids, and superoxide dismutase, peroxidase and esterase activities. *S. europaea* showed the lowest biomass, root K^+ content, Chl a/b ratio, and carotenoids under salinity. This reduction of biomass is concomitant with the increase in proline contents and peroxidase activity. On the other hand, the promotion of growth under low salinity and maintenance under high salinity (200 and 400 Mm NaCl) in *A. Macrostachyum* and *S. fruticosa* are accompanied by an increase in Chl a/b ratio, carotenoids, phenolics contents, and esterase activity. Proline content was decreased under high salinity (400 and 600 mM NaCl) in both species compared to *S. europaea*, while peroxidase showed the lowest activity in both plants under all salt levels except under 600 mM NaCl in *Arthrocnemum macrostachyum* compared to *S. europaea*. These results suggest two differential strategies; (1) the salt tolerance is due to activation of antioxidant enzymes and biosynthesis of proline in *S. europaea*, (2) the salt tolerance in *A. macrostachyum*, *S. fruticosa* are due to rearrangement of chlorophyll ratio and biosynthesis of antioxidant compounds (carotenoids, phenolics and flavonoids) which their cost seem to need less energy than activation of antioxidant enzymes. The differential behavior in halophytes of the same habitat confirms that the tolerance mechanism in halophytes is species-specific which provides new insight about the restoration strategy of saline areas.

Keywords: halophytes; Amaranthaceae; salinity; antioxidant enzymes; phenolic compounds

Citation: Ghanem, A.F.M.; Mohamed, E.; Kasem, A.M.M.A.; El-Ghamery, A.A. Differential Salt Tolerance Strategies in Three Halophytes from the Same Ecological Habitat: Augmentation of Antioxidant Enzymes and Compounds. *Plants* **2021**, *10*, 1100. https://doi.org/10.3390/plants10061100

Academic Editors: Oscar Vicente and Marius-Nicusor Grigore

Received: 18 April 2021
Accepted: 28 May 2021
Published: 30 May 2021

1. Introduction

Soil salinization is a critical problem which influences agricultural activities and inhibits crop productivity. The food and agriculture organization (FAO) [1] reported approximately 831 million hectares (6% of total world land) were affected by salinity. Also, A high percentage of cultivated land around the world (more than 20%) is affected by salinity, and this percentage is daily increasing [2,3]. On the same side, population density, unfavorable environmental conditions and climate changes lead to reduce in cultivated lands [4]. Crop production decreasing with increasing population density could lead to famine around the world. Molecular biology and genetic engineering are powerful tools in the breeding of salt-tolerant crops. However, both approaches are slow, costly and they sometime fail to achieve the goal. Therefore, the cultivation of natural salt-tolerant plants as saline crops represents an easy and cheap solution for salt-affected areas [5,6].

Halophytes are plants that can maintain their biological activities and grow in salinity-affected soils [7]. One of the most popular effects of salinity is oxidative damage through the over generation of reactive oxygen species such as hydroxyl, superoxide and hydrogen peroxide [8]. Several morphological, physiological, biochemical and molecular changes have been observed to help halophytic plants to adapt to salinity [9–12]. These strategies depend on; maintaining the photosynthetic system through chlorophyll synthesis [13–15]; carotenoids enhancement or inhibition [16,17]; reactive oxygen species (ROS) production [8,18–21]; enzymatic antioxidant activation, such as superoxide dismutase (SOD) [20,22–24]; peroxidase [25] and Catalases [26,27]; non-enzymatic antioxidant synthesis, such as phenolic compound [28–32] and Flavonoids and [33]; osmoregulatory and compatible solutes synthesis [34], such as proline [35].

Amaranthaceae is a family of angiosperms which comprises about 165 genera and 2040 species [36] with a high number of xerohalophytes and halophytes around the world [37–39], 34 halophytic taxa are belonging to the family Chenopodiaceae/Amaranthaceae with a percentage of 22.08% of all halophytic angiosperms [40]. Among these taxa, *Arthrocnemum macrostachyum, Sarcocornia fruticosa* and *Salicornia europaea* are three halophytic plants distribute in the Mediterranean region [41]. *Arthrocnemum macrostachyum* is a perennial small shrub, erect to ascending stem, woody old stem and fleshy young stem, 30–40 cm in tall, like spike inflorescence, and papillose seeds. *Sarcocornia fruticosa* is a perennial sub shrub, erect to ascending stem, 20–80 cm in tall, and grey seeds covered with conical protuberances. *Salicornia europaea* is an erect annual herb with a cup-shaped branched stem, seed with conical protuberances [42]. These plants are considered cash crops due to their nutritional value and ecological importance in the phytoremediation of metals [43–45].

Fully understanding of salt tolerance mechanisms represents principle means in the management of the saline area and breeding of salt-tolerant cash crops [46,47]. Salt tolerance level is species-specific and the plant habitat contributes to the degree of salt tolerance and strategy among populations of the same species [24,48]. Mohamed et al. [22,23] reported that the Egyptian population of *Suaeda maritima* (Chenopodiaceae) has more salt tolerance than the Japanese population. Therefore, Egyptian Chenopodiaceae represents a unique genetic resource for saline agriculture application. To obtain more in-depth knowledge about the salt tolerance strategies of Chenopodiaceae, we hypothesized that Egyptian populations have unique salt tolerance levels and habitat of Mediterranean Sea influences on salt tolerance strategies of different species in this family. Our work aims to explore the salt tolerance strategies of three Egyptian Chenopods (Currently belong to Amaranthaceae) namely: *Arthrocnemum macrostachyum, Sarcocornia fruticosa* and *Salicornia europaea* from Damietta coast, through studying the effect of salt stress (0–600 Mm NaCl) on the growth parameters, chlorophyll contents, phenolic compounds, flavonoids, proline, malondialdehyde (MDA), esterase and antioxidant enzymes (superoxide dismutase, catalase and peroxidase activities).

2. Results

2.1. Effect of Salinity on Na^+, K^+ and Ca^{2+} Contents

While, K^+ content in the shoot system was slightly decreased at all saline concentrations except at 100 mM NaCl and the root system K^+ content was increased at all salt concentrations. The Na^+ and Ca^{2+} shoot and root contents were gradually increased by increasing salt concentrations in *A. macrostachyum*. In the case of *S. europaea*; shoot Na^+ and Ca^{2+} were increased at all concentrations, while K^+ content increased at 100 and 200 mM NaCl only. In the root system, Na^+ content increased in all concentrations, Ca^{2+} increased at 600 mM NaCl, but K^+ decreased at all concentrations. In *S. fruticosa* Na^+, Ca^{2+} and K^+ content decreased at all concentrations in the shoot system except at 600 mM NaCl, both Na^+ and Ca^{2+} were increased with respect to control. On the other hand, root Na^+ and Ca^{2+} contents increased with salinity and K^+ increased only at 200 and 400 mM NaCl (Table 1).

Table 1. Analysis of Na^+, K^+ and Ca^{2+} content (μmol/g DW) in root and shoot of namely *Salicornia europaea*, *Sarcocornia fruticosa* and *Arthrocnemum macrostachyum* under different salt concentrations. Different letters indicate significant differences.

Species	NaCl (mM)	Shoot			Root		
		Na^+ (μmol g^{-1} DW)	K^+ (μmol g^{-1} DW)	Ca^{2+} (μmol g^{-1} DW)	Na^+ (μmol g^{-1} DW)	K^+ (μmol g^{-1} DW)	Ca^{2+} (μmol g^{-1} DW)
S. europaea	0	2217 ± 5 m	387 ± 4 ef	293 ± 8 j	406 ± 19 k	493 ± 10 c	58.75 ± 6 bc
	100	8239 ± 12 cd	607 ± 11 a	2166 ± 22 b	985± 9 h	301 ± 13 g	39 ± 3 ef
	200	8637 ± 10 c	543 ± 11 b	2300± 14 a	1239 ± 12 g	338 ± 11 f	51.5 ± 4 cd
	400	8680 ± 8 c	362± 12 fg	2150 ± 16 b	1474 ± 25 f	370 ± 7 e	58 ± 5 bc
	600	7969 ± 10 d	372 ± 8 f	1925± 28 d	1670± 20 d	375 ± 8 e	66.5 ± 4 b
S. fruticosa	0	6420 ± 10 f	438± 11 d	1578 ± 4 e	330± 16 l	337 ± 11 f	13.5 ± 0.75 i
	100	5760 ± 11 g	428± 11 d	1291 ± 35 f	811 ± 38 i	294 ± 11 g	31 ± 3 fg
	200	5028 ± 15 j	264 ± 9 h	1127 ± 28 h	1006 ± 19 h	468 ± 9 cd	40 ± 5 ef
	400	4217 ± 15 k	211 ± 8 i	866± 35 i	1560 ± 22 e	397 ± 7 e	55 ± 3 bcd
	600	8913 ± 12 b	262 ± 9 h	2075 ± 28 c	1782 ± 22 c	296 ± 8 g	90 ± 2.5 a
A. macrostachyum	0	3130 ±22 l	414 ± 13 de	85 ± 7 k	537± 22 j	454 ± 11 d	16.5 ± 1 hi
	100	5057 ± 8 i	494 ± 9 c	1162 ± 7 g	1202 ± 15 g	671 ± 11 a	25.75 ± 0.75 gh
	200	5736 ± 17 h	336 ± 8 g	1239 ± 7 g	1570± 7 e	496 ± 7 c	46.5 ± 2.75 de
	400	7760 ± 12 de	367 ± 8 f	1664± 10 d	2338 ± 17 b	607 ± 11 b	56 ± 1.25 bcd
	600	10652 ± 25 a	276 ± 11 h	2187 ± 7 b	2693 ± 13 a	620 ± 9 b	64 ± 3 b

2.2. *Effect of Salinity on Growth Parameters*

2.2.1. Effect of Salinity on Biomass Production

Two-way ANOVA analysis for studied plants showed significant effects for the plant and species, and their interactions ($p < 0.001$) for all parameters (Table 2). These interactions support the different responses of the species to salinity.

Table 2. Two-way ANOVA of salinity, species, and their interaction on all tested parameters.

Parameters	Species	Species × Salinity	Salinity
shoot Fresh weight	***	***	***
shoot dry weight	***	***	***
Root fresh weight	***	***	***
Root Dry weight	***	***	***
Chl a	***	***	***
Chl b	***	***	***
Carotenoids	***	***	***
Chl a/b	***	***	**
MDA	***	***	***
Proline	***	***	***
phenol	***	***	***
Flavonoids	***	***	***

**: $p < 0.01$ and ***: $p < 0.001$.

One-way ANOVA showed that each species has its response for different parameters at applied saline concentrations. *A. macrostachyum* and *S. fruticosa* showed highest shoot and root fresh and dry weights with significant increasing at 100 mM NaCl and slightly increasing at 200 and 400 mM NaCl. At 600 mM NaCl, both species showed significant decreases in these parameters. In contrast, *S. europaea* showed non-significant difference in shoot fresh and dry weights and root dry weight at low and moderate salt treatments, but root fresh weight showed a significant decrease at all treatments, and all parameters were highly decreased at 600 mM NaCl (Figures 1–4).

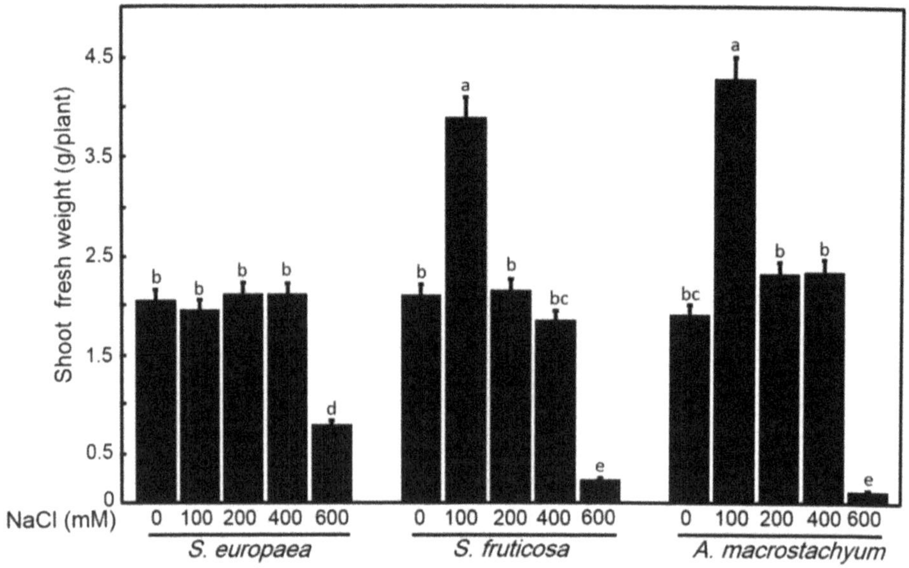

Figure 1. Shoot fresh weights of *S. europaea, S. fruticosa* and *A. macrostachyum* under different NaCl concentrations. Mean ± SE of three replicates. Different letters indicate significant differences ($p < 0.05$).

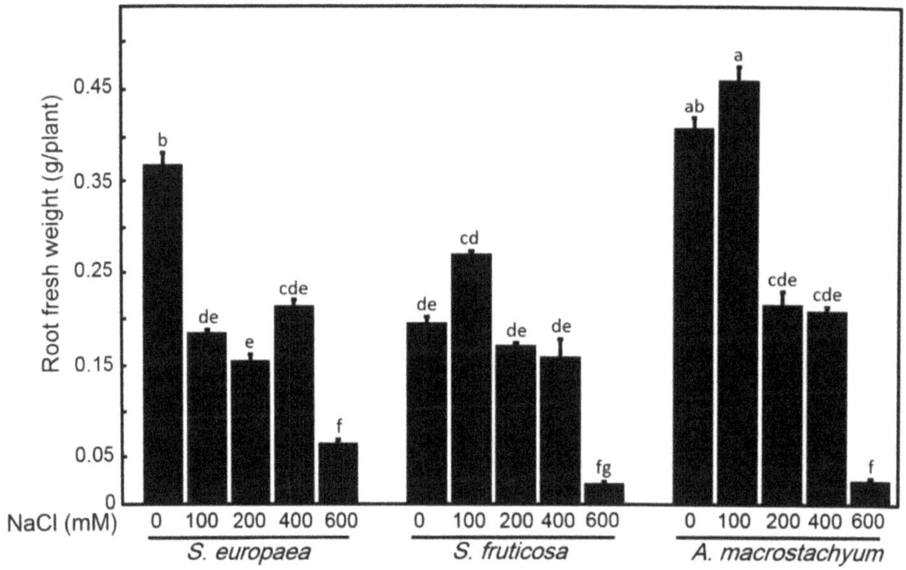

Figure 2. Root fresh weights of *S. europaea, S. fruticosa* and *A. macrostachyum* under different NaCl concentrations. Mean ± SE of three replicates. Different letters indicate significant differences ($p < 0.05$).

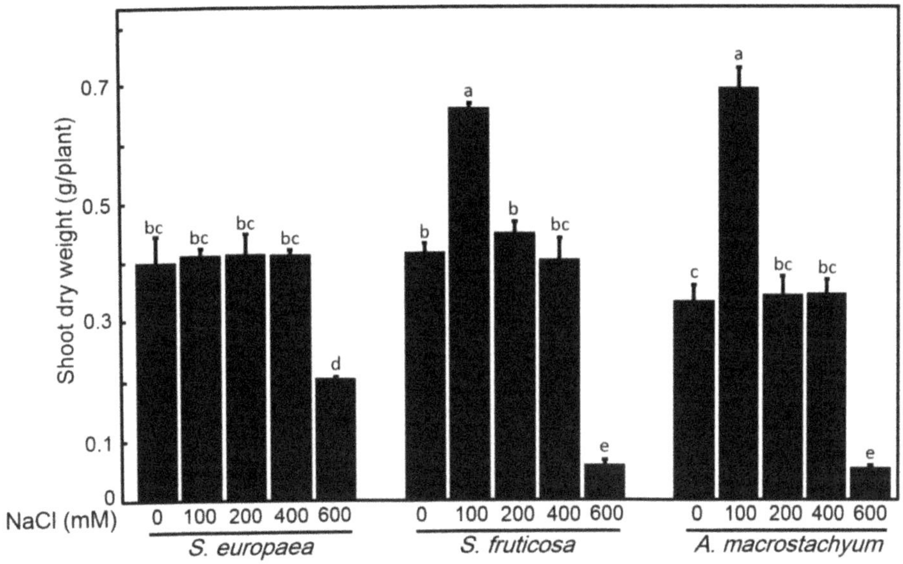

Figure 3. Shoot dry weights of *S. europaea, S. fruticosa* and *A. macrostachyum* under different NaCl concentrations. Mean ± SE of three replicates. Different letters indicate significant differences ($p < 0.05$).

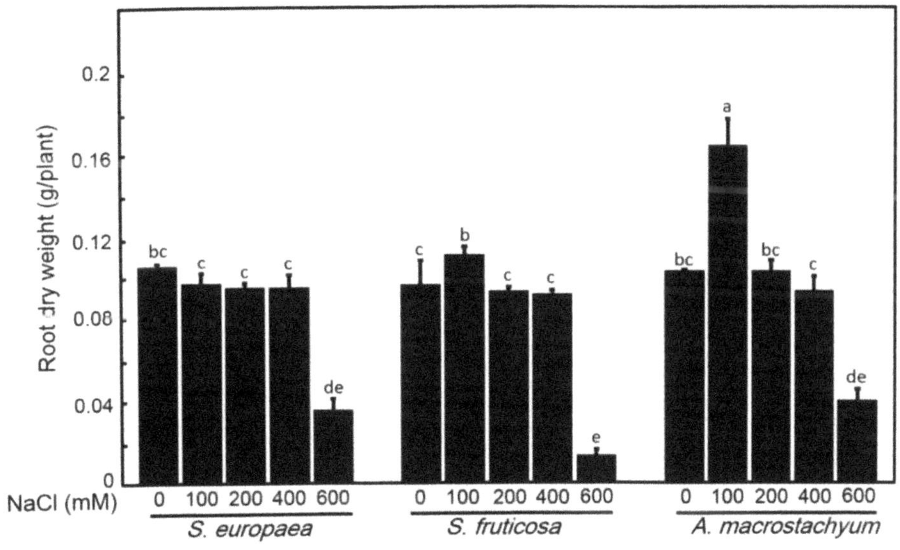

Figure 4. Root dry weights of *S. europaea, S. fruticosa* and *A. macrostachyum* under different NaCl concentrations. Mean ± SE of three replicates. Different letters indicate significant differences ($p < 0.05$).

2.2.2. Effect of Salinity on Chlorophyll Contents

Chlorophyll contents showed different responses to salt treatments in all studied species (Figures 5 and 6). Chlorophyll a contents showed non-significant differences in *A. macrostachyum* and *S. fruticosa* with slightly increasing at 200 mM NaCl in *A. macrostachyum* and at 100 mM in *S. fruticosa*, and it was significantly decreased at 600 mM NaCl in both species. In contrast, *S. europaea* showed slightly non-significant decreases at low and moderate treatments and significantly decreasing at higher concentrations. For chlorophyll b, while *A. macrostachyum* showed non-significant differences at low and moderate NaCl concentrations and significant decreases at 400 and 600 mM NaCl, *S. fruticosa* showed significant decreases at all salt treatments. In the case of *S. europaea*, chlorophyll b contents were significantly increased at low and moderate concentrations NaCl and significantly decreased at high NaCl concentrations. For chlorophyll a/b ratio, it was significantly increased at high salt concentrations in *A. macrostachyum*; and at all salt levels in *S. fruticosa*, and significantly decreased at all salt concentrations in *S. europaea* (Figure 7).

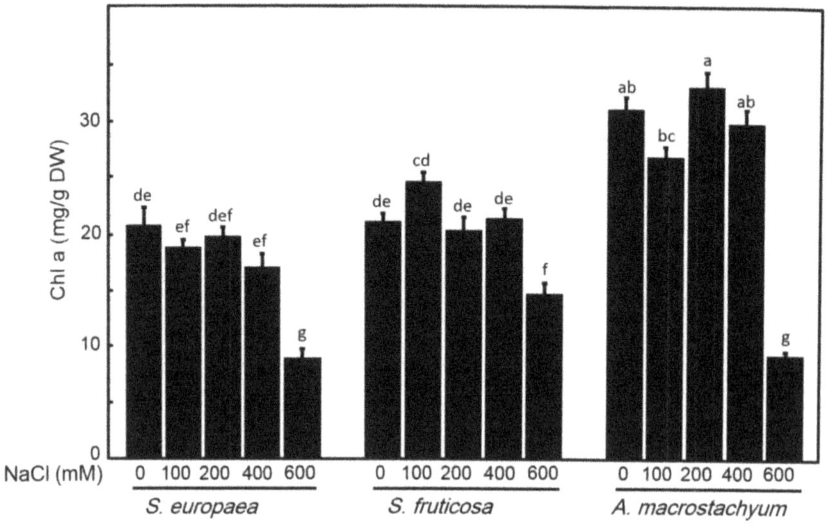

Figure 5. Chlorophyll a contents of *S. europaea*, *S. fruticosa* and *A. macrostachyum* under different NaCl concentrations. Mean ± SE of three replicates. Different letters indicate significant differences ($p < 0.05$).

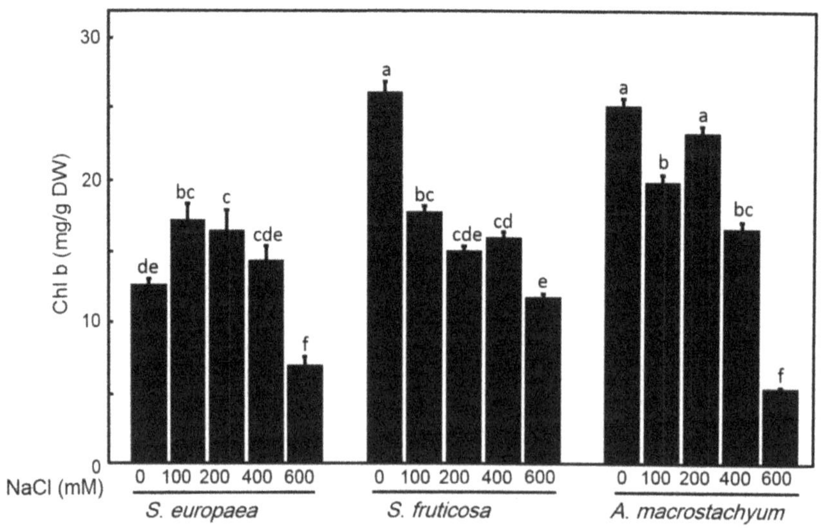

Figure 6. Chlorophyll b contents of *S. europaea*, *S. fruticosa* and *A. macrostachyum* under different NaCl concentrations. Mean ± SE of three replicates. Different letters indicate significant differences ($p < 0.05$).

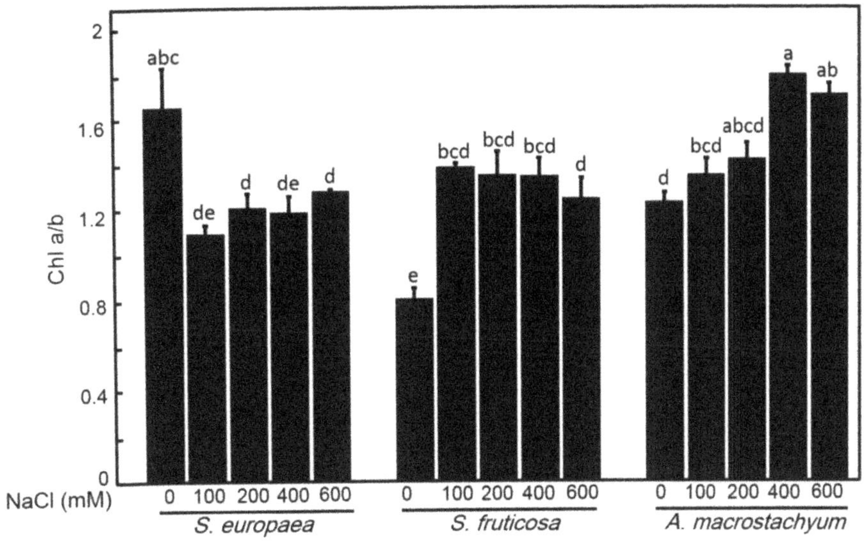

Figure 7. Chlorophyll a/b ratios of *S. europaea*, *S. fruticosa* and *A. macrostachyum* under different NaCl concentrations. Mean ± SE of three replicates. Different letters indicate significant differences ($p < 0.05$).

2.2.3. Effect of Salinity on Carotenoids

Carotenoids concentration showed significant increases with saline concentrations except at 100 and 600 mM NaCl which show non-significant differences in respect to control in *A. macrostachyum* while *S. fruticosa* showed non-significant increases under low and moderate salt concentrations, and a significant increase under high salt level. In contrast, *S. europaea*, showed significant decreases with increasing NaCl concentration (Figure 8).

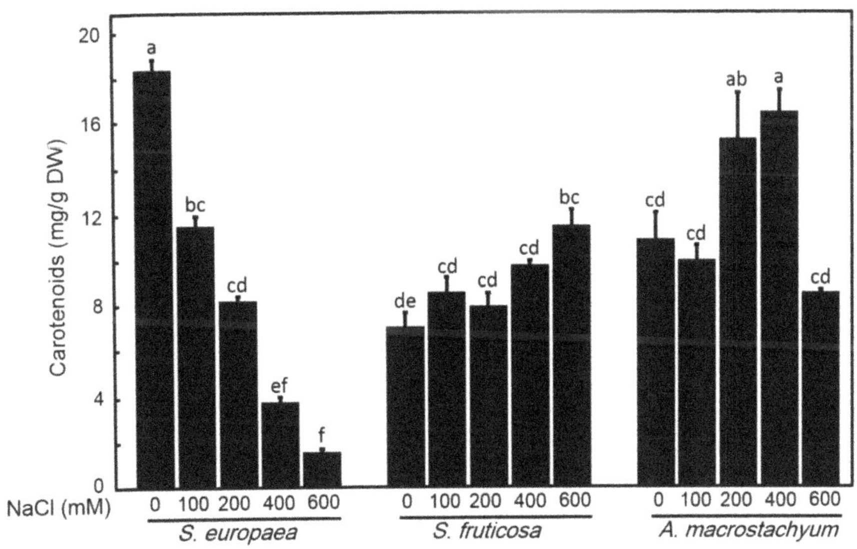

Figure 8. Carotenoids contents of *S. europaea*, *S. fruticosa* and *A. macrostachyum* under different NaCl concentrations. Mean ± SE of three replicates. Different letters indicate significant differences ($p < 0.05$).

2.2.4. Effect of Salinity on Total Phenolic Contents

Phenolic compound contents in *A. macrostachyum* significantly increased at moderate and high salinity levels (200 and 400 mM NaCl) with a slightly non-significant difference at 100 mM NaCl. In *S. fruticosa* and *S. europaea* slightly non-significant increases in phenolic contents were recorded at all saline concentrations. Interestingly, the three species showed significant decreases at 600 mM NaCl (Figure 9).

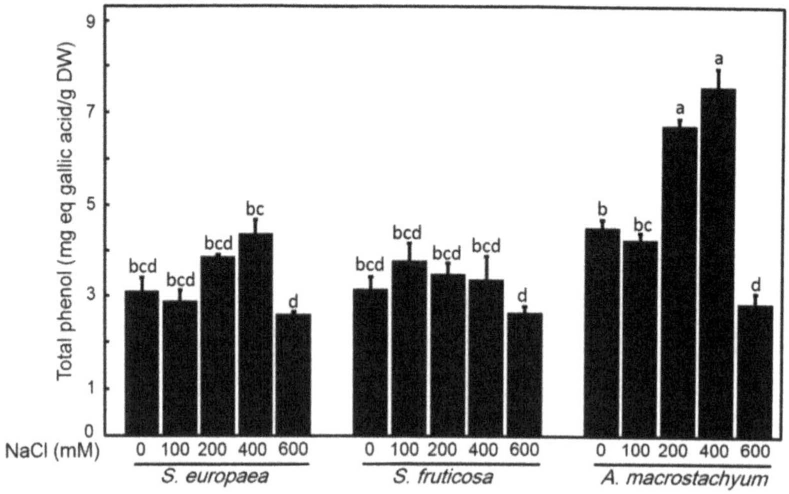

Figure 9. Total phenol contents of *S. europaea, S. fruticosa* and *A. macrostachyum* under different NaCl concentrations. Mean ± SE of three replicates. Different letters indicate significant differences ($p < 0.05$).

2.2.5. Effect of Salinity on Flavonoid Contents

A. macrostachyum showed significant increases in flavonoid contents at all treatments but significantly decreased at 100 and 600 mM NaCl. In contrast, *S. fruticosa* and *S. europaea* showed significant decreases in flavonoid contents with a slightly non-significant difference at 600 mM NaCl in *S. fruticosa* (Figure 10).

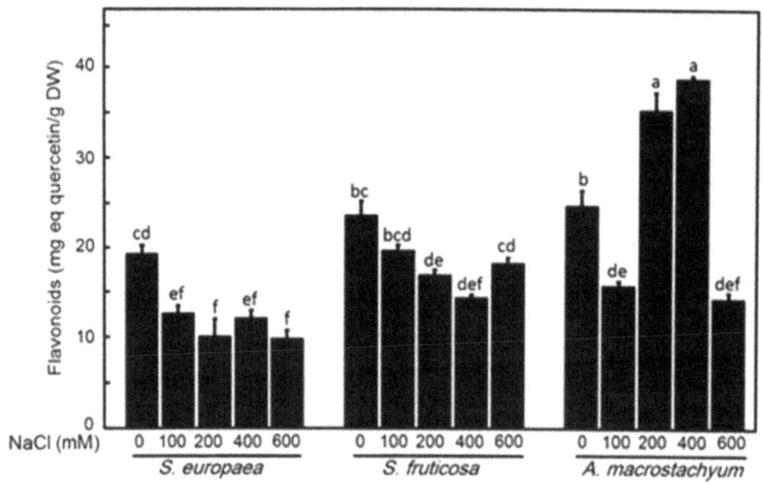

Figure 10. Total flavonoids contents of *S. europaea, S. fruticosa* and *A. macrostachyum* under different NaCl concentrations. Mean ± SE of three replicates. Different letters indicate significant differences ($p < 0.05$).

2.2.6. Effect of Salinity on Total Malondialdehyde (MDA) Content

A. macrostachyum and *S. fruticosa* showed significant increases in MDA concentrations in all treatments except at 600 mM NaCl, which observed a non-significant difference compared to control. *S. europaea* showed significant decreases at all salt concentrations except at 100 mM NaCl, which showed a non-significant difference in respect to control (Figure 11).

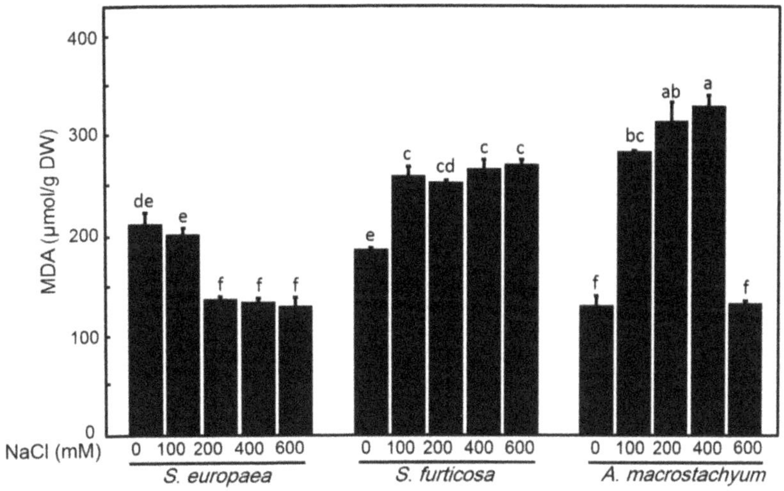

Figure 11. Malondialdehyde contents of *S. europaea*, *S. fruticosa* and *A. macrostachyum* under different NaCl concentrations. Mean ± SE of three replicates. Different letters indicate significant differences ($p < 0.05$).

2.2.7. Effect of Salinity on Proline Content

While *S. fruticosa* showed significant increases in proline content with increasing salt concentrations, Proline content in *A. macrostachyum* and *S. europaea* were significantly increased at 200, 400 and 600 mM NaCl only, and higher values of proline in *S. europaea* were recorded at 600 mM NaCl in respect to the other two species (Figure 12).

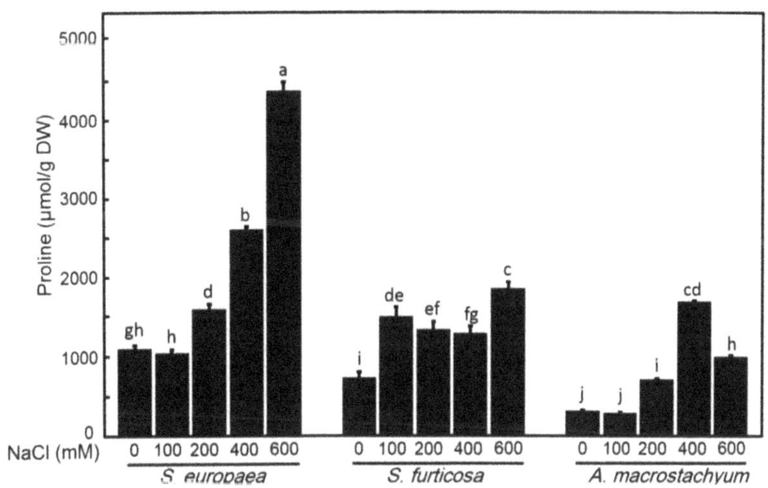

Figure 12. Proline contents of *S. europaea*, *S. fruticosa* and *A. macrostachyum under* different NaCl concentrations. Mean ± SE of three replicates. Different letters indicate significant differences ($p < 0.05$).

2.3. Isozymes Analysis

2.3.1. Esterases

The electrophoretic analysis using native PAGE showed two esterase loci in all studied species and under all treatments with different amounts and intensities. The highest intensities were observed in *A. macrostachyum* and *S. fruticosa* at 200 and 400 mM NaCl and 100 and 200 mM NaCl in *S. europaea* (Figure 13).

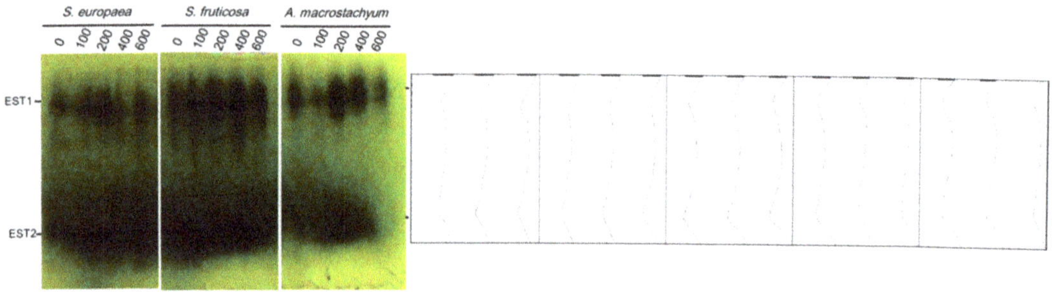

Figure 13. Esterase isozymes of *S. europaea*, *S. fruticosa* and *A. macrostachyum* under different NaCl concentrations.

2.3.2. SOD Isozymes

SOD activity was increased at high salt levels in *A. macrostachyum* and *S. europaea*, and under low and moderate salinity in *S. fruticosa* (Supplementary S1).

2.3.3. POD Isozymes

POD enzyme showed a unique locus in all studied plants and at all treatments. The highest intensities were recorded at 100 and 600 mM NaCl in *A. macrostachyum*, and at 200 mM NaCl in *S. europaea* which showed the highest POD activity in respect to the other two species. On the other hand, weak activity was observed at all treatments in *S. fruticosa* (Figure 14).

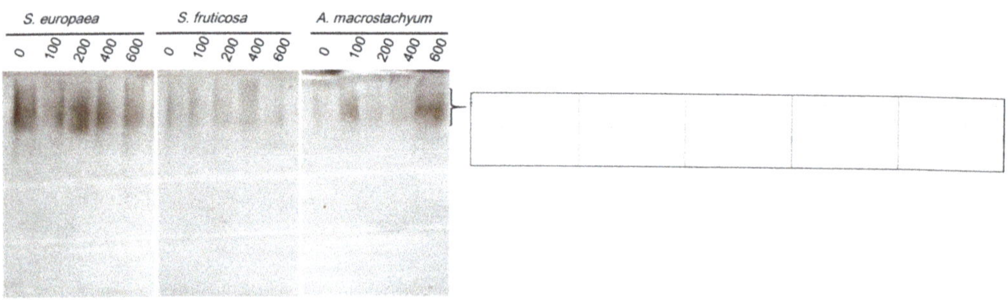

Figure 14. Peroxidase isozymes of *S. europaea*, *S. fruticosa* and *A. macrostachyum* under different NaCl concentrations.

Pearson correlation and principal component analysis.

For *Salicornia europaea*, growth parameters have positive correlations under salinity with Chl a, Chl b, and carotenoids, and negatively correlated with proline content (Table S1). In the same context, under 100 mM saline treatment, principal component analysis showed PC1 and PC2 described 50.9%, and 25.9% of the variance, respectively (Figure 15). Three groups were observed from this analysis; Growth parameters (shoot fresh weight, shoot dry weight. root fresh weight and root dry weight), Chl a, Chl b and carotenoids constructed the first group, Flavonoids, Chl a/b, MDA formed the second group, and both proline and phenolic compound represented the third group. For *Sarcocornia fruticosa*, growth parameters were positively correlated with Chl a, Chl b and Chl a/b but negatively correlated with carotenoids, MDA, phenolic compounds and flavonoids (Table S1). In contrast, PC1 and PC2 explained 64%, and 17.5% of the variance, respectively. Three groups were also visualized from this analysis; Growth parameters, Chl a, Chl a/b, MDA and proline formed the first group, Chl b and flavonoids represented the second group, and phenolics represented the third group. For *Arthrocnemum macrostachyum*, Our results showed significant negative correlations between most growth parameters with carotenoids, proline, flavonoids and phenolic compounds, Chl a/b and MDA, but positively correlated with Chl b (Table S1). On the other hand, Principal component analysis showed PC1 and PC2 described 58%, and 19.1% of the variance, respectively. Three groups were also noticed from PCA analysis; Growth parameters, carotenoids, phenolic compounds, Chl a/b and MDA formed the first group, Chl a, Chl b and flavonoids constructed the second group, and proline formed the third group.

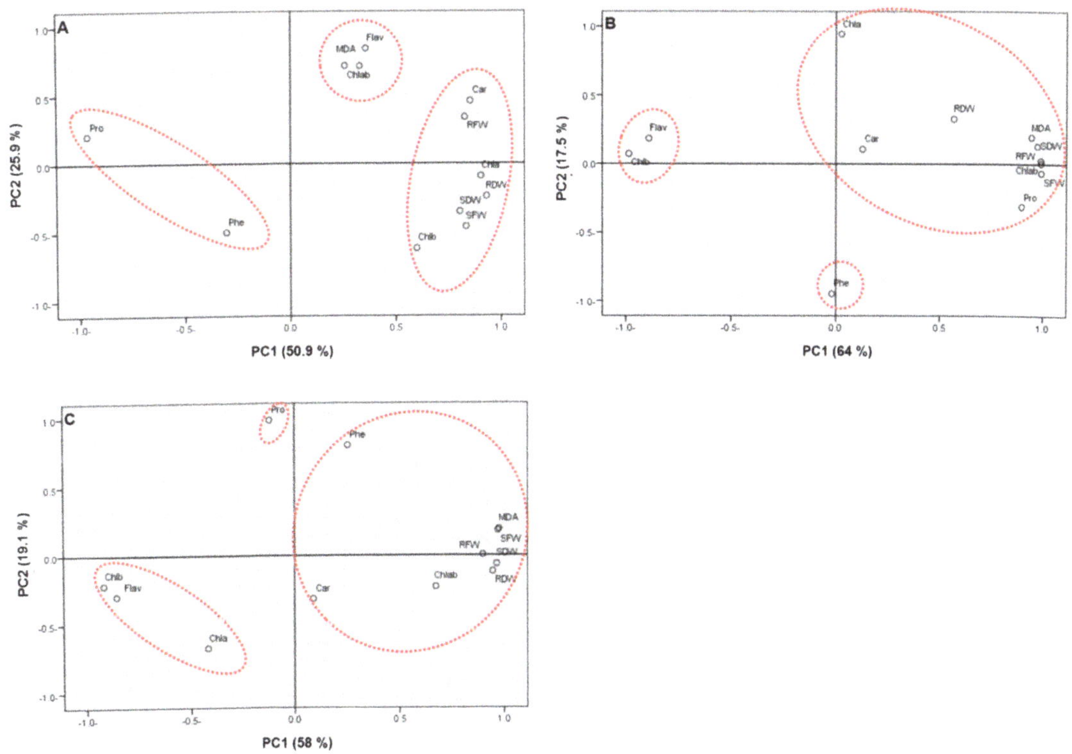

Figure 15. Principal components analysis bi-plot. Values of the studied parameters were analyzed under salinity (100 mM NaCl) with respect to control. Growth parameters (shoot fresh weight: SFW, root fresh weight: RFW, shoot dry weight: SDW, root dry weight: RDW), chlorophyll a: Chl a, chlorophyll b: Chl b, Chl a/b, carotenoids: Car, Proline: pro, malondialdehyde: MDA, phenolics: Phe, Flavonoids: Flav. In (**A**) *Salicornia europaea,* (**B**) *Sarcocornia fruticosa*, and (**C**) *Arthrocnemum macrostachyum*.

3. Discussion

Exploration of salt tolerance mechanisms of many halophytes species is of considerable value for the selection of suitable crops for saline agriculture. In this study, three halophytic species *Arthrocnemum macrostachyum*, *Sarcocornia fruticosa* and *Salicornia europaea* (Amaranthaceae/Chenopodiaceae) were collected from the same saline habitat and tested for their tolerances to salinity.

For growth criteria, *A. macrostachyum* and *S. fruticosa* improved their fresh and dry weight when grown under low and moderate salt concentrations, but their fresh weights were reduced at high salinity. *A. macrostachyum* had the optimum growth at 400 mM NaCl and its growth decreased at 600–1000 mM NaCl [49]. The same induction trend of growth under moderate salinity (170–510 mM NaCl) was observed in *A. macrostachyum* and *S. fruticosa* from Spain, with a decline trend under high salt conditions [14,50]. Also, García-Caparrós et al. [51] reported that total dry weight and relative growth rate of *S. fruticosa* decreased significantly under low and moderate salinity (100 and 200 mM NaCl) for 60 days. Therefore, our results suggest that Egyptian *A. Macrostachyum* and *S. fruticosa* need low salt levels for optimal growth, and they could maintain their growth under moderate and high salinity (200 and 400 Mm NaCl). The variation in salt tolerances of both plants in the previous studies might be because of the maternal habitats of these populations. Mohamed et al. [24] reported that maternal salinity plays important role in salt tolerance during the growth of *Zygophyllum ccocenium.*

On the other hand, *S. europaea* showed significant decreases in shoot fresh and dry weights at high salinity levels but slightly non-significant variation under moderate salinity. Ungar et al. [52] reported that *S. europaea* growth was increased under moderate salinity (170–510 mM NaCl). In contrast, *S. rubra* had the optimal growth in the absence of salt to 200 mM NaCl while its growth was inhibited with further increase of salt level. The decline of root biomass under moderate salinity suggests the severe effect of salinity on the root system than shoot and the adaptive strategy to avoid more uptakes of toxic ions [53].

Inorganic ions play role in maintain osmotic and turgor pressure in halophytes more than glycophytes, which predominantly depend on the increased synthesis of de novo compatible solutes [54]. Flowers et al. [55] reported that the Na^+ is one of the most important ions which play important role in adjusting cellular osmotic potential. Our results showed that *A. macrostachyum* and *S. europaea* Na^+ contents increased with increasing external NaCl concentrations while Na^+ content in *S. fruticosa* was only increased under high salinity. This increase in Na^+ has a role in maintain shoot osmotic and turgor. Redondo-Gomez et al. [14] and Khan et al. [56] reported increasing Na^+ content with increasing external NaCl concentrations in *A. macrostachyum* because halophytes have a unique ability for osmotic adjustment due to accumulation of Na^+ in vacuoles, and K^+ and organic solutes in the cytosol [57,58]. The stimulation of K^+ in halophytes root under saline conditions is well documented in many plants, such as *Suaeda monoica* and *Triglochin maritima* [59]. In the present study, while K^+ ions in roots were increased with increasing salinity in *A. macrostachyum* and *S. fruticosa* and declined in *S. europaea*, Shoot K^+ ions increased at low concentration and decreased at high and moderate concentrations. These results suggest that K^+ content can be used as a marker for discrimination between salt tolerance strategies in halophytes [58]. In the same context, Ca^{2+} increased with salinity, This increase is due to its vital role in salt adaptation through binding of Ca^{2+} with SOS_3 and subsequently activate SOS_2, this complex stimulates Na^+/H^+ antiporter which plays a crucial role in the regulation of Na^+ ions in the cytosol [60].

In saline habits, soil salinity and arid climate greatly affect the synthesis of pigments in plants [61] and salinity reduces the net photosynthetic rate [62]. Redondo-Gómez et al. [14] reported that *A. macrostachyum* can improve or adjust the rate of photosynthesis under saline conditions. Aghaleh et al. [17] and Akcin and Yalcin [63] reported that photosynthetic pigments of *S. europaea* from Iran and Turkey were affected by increasing soil salinity. Our data showed non-significant variations in chl a under all salinity levels except in *S. europaea* under very high salinity (600 Mm NaCl), and significant decreases of chl b were only observed in *S. fruticosa* under all saline concentration and in *A. macrostachyum* and *S. europaea* under high saline concentration (600 Mm NaCl). This result suggests that *S. fruticosa* has a differential response to salt stress compared to *A. macrostachyum* and *S. europaea*. The increase in Chl a/b ratio in *A. macrostachyum* and *S. fruticosa* suggests that both species had more adaptation to saline conditions than *S. europaea* [24,64].

Carotenoids play a vital role as non-enzymatic antioxidants in protecting photosynthetic system. Our results showed significant increases in carotenoids with the elevation of NaCl concentration in *A. macrostachyum* and *S. fruticosa*, except at low salinity level for both species and under 600 in *A. macrostachyum*. This increase in carotenoid concentration may be one strategy to maintain chlorophyll amount and not decreasing it with different salinity concentrations. A similar study confirmed the increase in carotenoids in *Nitraria retusa* was associated with increasing salt tolerance [65]. In contrast, Carotenoids in *S. europaea* decreased significantly at all treatments. Such decreases in carotenoid contents under salinity stress were reported in different plant species [17,61,63,66]. These results suggest that carotenoids play an important role in the salt tolerance of *A. macrostachyum* and *S. fruticosa* than in *S. europaea*.

Phenolic compounds are secondary metabolites that play an important role in protecting plants against oxidative stress [67]. Increasing phenolic compounds synthesis is considered one of the most important methods in water deficiency resistance [68]. The synchronous significant increase of phenolic compounds and flavonoids in *A. macrostachyum* at moderate and high salinity levels indicates the importance of these compounds in stress tolerance in a synergistic relationship with carotenoids that also showed significant increases with salinity. Król et al. [69] and Caliskan et al. [70] reported that the metabolism of phenylpropanoid and phenolic compounds accumulation were enhanced in different plant species in response to different environmental stress conditions. Along the same line, the non-significant decrease in chlorophyll content in *A. macrostachyum* at high salinity level is due to increase in phenolic compounds contents at the same salinity level. This was supported by the finding of Bhattacharya et al. [71] who reported that phenolic compounds play a vital role in the biosynthesis of lignin and pigments in plants. Also, *S. fruticosa* showed a slight increase in phenolic contents at moderate NaCl concentrations and non-significant differences at other concentrations. This indicates that moderate salinity stimulates the production of phenolic compounds in *S. fruticosa*. On other hand, a constant or slight increase in total phenolic contents in *S. europaea* at different salinity levels was associated with the decrease in carotenoids. These results may indicate the importance of phenolic compounds in the alleviation of deleterious effects of salt stress in *A. macrostachyum* and *S. fruticosa* [72–75].

A. macrostachyum showed significant increases in flavonoid contents at all treatments except at low salinity which had a significant decrease. In contrast, *S. fruticosa* and *S. europaea* showed significant decreases in flavonoid contents but not at high salinity level in *S. fruticosa* which showed a slightly non-significant decrease. Brown et al. [76] reported that flavonoids act as auxin transport inhibitors, therefore, high promotion of shoot growth under low salinity (100 Mm NaCl) in *A. macrostachyum* and *S. fruticosa* may be due to low flavonoids content. The positive performance of shoot growth, despite its low root biomass, may be due to the same previous reason.

Malondialdehyde (MDA) concentration expresses the extent of destruction in the membrane because it acts as a common end product of lipid peroxidation [19]. Jithesh et al. [77] and Mohamed et al. [23] reported the presence of a positive correlation between salinity stress and MDA content in halophytic plants. *A. macrostachyum* and *S. fruticosa* showed significant increases in MDA concentration in all treatments except at 600 mM NaCl in *A. macrostachyum*. This result is in agreement with Abd El-Maboud [75] who reported increasing in MDA concentration in *A. macrostachyum* in the summer season. On other hand, *S. europaea* showed a significant decrease in MDA content with no effect at low salinity concentrations. This result contradicts the reported increase in MDA in *S. europaea* collected from Iran with the increase in salinity level [17]. This decrease in MDA concentration in *S. europaea* may be due to an increase in peroxidase activity, which was often stored at the cytosol, peroxisome and vacuole [78,79]. The increasing of peroxidase activity plays an active role in free radical oxidative stress inhibition, which leads to protect the membrane and decrease lipid peroxidation. Also, decreasing MDA may be due to the increasing accumulation of proline content in *S. europaea* than the other two species, which act as non-enzymatic antioxidant enzymes and this suggestion is in agreement with proline having a role in ROS scavenger [80].

Proline accumulates in the cell as an osmoregulatory solution which plays an important role in the adaptation of halophytes to high salinity levels [81]. Increasing in the accumulation of proline in response to salinity stress was reported in different plant species by different researchers [82,83]. Increasing proline synthesis helps in decreasing water loss and ions' toxicity [84]. Our results showed a significant increase in proline contents at a high salinity level in all studied species. This increase may indicate upregulation of proline synthesis [85]. Increasing proline content in *S. europaea* at high salinity than the other two species may be important to compensate for the decrease in the carotenoids and flavonoids contents and to help as free radical scavengers.

Pectin is one of the basic components of a plant cell wall. It can be both methyl-esterified and acetyl-esterified. De-esterification occurs by specific esterases [86]. Esterase plays a vital role in avoiding the salt-induced imbalance in cell wall formation. Our results showed two esterase loci in all studied species and under all treatments with the higher intensities in *A. macrostachyum* and *S. fruticosa* at moderate salinity level, and at low and moderate salinity levels in *S. europaea*. These results are in agreement with Dasgupta et al. [87] who reported that esterase isoforms intensities were increased with elevating salt concentration. Mohamed et al. [25] found esterase has two isoforms in *Pancratium maritimum* and their intensities were increased under moderate saline concentration.

For *S. europaea* under salt stress (100 mM NaCl), Principal component analysis observed the arrangement of growth parameters, chlorophyll parameters, MDA and flavonoids on the positive X-axis, while proline and phenolic compounds grouped on the negative X-axis. This result suggests the salt tolerance of this species due to the accumulation of proline and phenolic compounds. In contrast, all parameters grouped on the positive X-axis, except Chl b and flavonoids were observed on the negative X-axis for *S. fruticosa*, and Ch b, flavonoids, Chl a and proline grouped on the negative axis for *A. macrostachyum*. These results confirm the growth promotion of both species due to increasing of Chl a/b ratio and the decline of flavonoids contents.

The promotion of growth parameters in *S. europaea* under 600 Mm NaCl compared to *A. macrostachyum* and *S. fruticosa* may be due to the decline of flavonoids accumulation in *Salicornia* under this salt level compared to the other two species. The decline in most parameters under 600 mM NaCl in *A. macrostachyum* suggests the deleterious effects of this concentration on this species.

Superoxide dismutase is considered the most important enzyme during the growth of plants under biotic and abiotic stress through catalyzing the dismutation of superoxide radicals into H_2O and Oxygen [88–90]. Nisar et al. [91] reported constitutive and decline of SOD activity in germinating black and brown *A. macrostachyum* seeds respectively under salinity. On the other hand, salinity induced promotion in SOD activity in *S. europaea* seedlings [92]. The induction of SOD under salinity was a prominent feature in halophytes such as *Suaeda maritima, Pancratium maritimum and Zygophyllum coccenium* [22–25]. In this study, SOD activity increased under high salinity in *A. macrostachyum* and *S. europaea*, and under moderate salinity in *S. fruticosa*. These results suggest a differential mechanism for SOD under salinity in these species.

POD enzyme has a major protective role for the cell against hydrogen peroxide which is produced under stress conditions [93,94]. Our study showed that POX enzyme has a stable faint locus at all salinity levels in *S. fruticosa*, and at high and moderate salinity in *A. macrostachyum* and *S. europaea*. The highest POD activity was recorded in *S. europaea* in respect to other species. This increase in both peroxidase and SOD activities under higher salinity may decrease free radical concentrations and protects membranes from lipid peroxidation, and hence the low MDA concentration in *S. europaea* than the other two species. Also, this indicates that POD is one of the most important strategies in salt tolerance in *S. europaea* more than the other two species.

From the foregoing discussion, the three halophytic species that are belonging to the same family and collected from the same saline ecological habitat showed differential mechanism to salt tolerance. The salt tolerance of *S. europaea* is derived from the promotion of proline level and peroxidase activity. The stable shoot and decline in root biomass suggest investment of energy in the promotion of antioxidant enzymes and compounds than use it in the growth process. This was supported statistically by the presence of a significant negative correlation between growth parameters and proline contents. In contrast, salt tolerance of *A. macrostachyum and S. fruticosa* is concomitant with rearrangement of chlorophyll contents, high level of carotenoids and phenolic compounds, and activation of esterase enzyme. This conclusion seems to be confirmed by the negative correlation between most of these compounds and the growth parameters of both species. The positive performance of both species' biomass, compared to *S. europaea,* suggests little energy was used in the salt tolerance mechanism in these plants. Also, a trade-off strategy between the growth process and defense system was noticed in the case of *S. europaea*. These results confirm differential salt tolerance strategies of different halophytes in the same habitat which provide valuable information in the selection of the best strategy in re-habitation of saline coastal areas.

4. Materials and Methods

4.1. Plant Seeds Collection

Inflorescences containing mature dry seeds of three species belong to Amaranthaceae family were collected from a halophytic region Damietta–Alexandria road during June 2018 and transported to the laboratory. Seeds were manually separated from the inflorescence and stored in paper bags until use. Studied species soil analysis was conducted according to Jackson [95]. The soil electrical conductivity was 15.325 ds/m and pH values were 9.36, Ca, Mg, Cl and HCO_3 concentrations were 0.035%, 0.01%, 0.4686% and 0.03355% respectively.

4.2. Growth Conditions

Seeds of studied plant species were surfaced sterilized using 70% ethyl alcohol for 30 s followed by 3.5% (*v*/*v*) Sodium hypochlorite for 5 min, then washed thoroughly with distilled water [22]. Sterilized seeds of each species were sown in 25 replicates plastic pots with 20 cm height and 10 cm in diameter containing sandy soil and irrigated with 150 mL of 20% MS medium. The germination was carried out under natural greenhouse conditions (temperature range 14–28 °C, humidity about 40%, and photoperiod 14: 10 light: dark) for 30 days. After this period, 15 plastic pots of each species with uniform seedlings size were chosen and divided to five groups; each group contains three replicates, and each replicate containing five plants. Five treatments were used in this experiment (0, 100, 200, 400, and 600 mM NaCl) and plants were irrigated with 1 L of 20% MS medium prepared in distilled water, 100, 200, 400 and 600 mM NaCl (150 mL weekly) for two months.

4.3. Determination of Na^+, K^+ and Ca^{2+}

Air-dried shoot and root were grounded to fine powders and 0.2 g of each sample were treated with 7:3 sulfuric: perchloric acid mixture. Cations' concentrations were determined according to Jackson [95].

4.4. Growth Parameters

4.4.1. Shoot and Root Fresh and Dry Weight Determination

For each treatment, five plants were used for the shoot and root fresh and dry weights determination. Plants were removed from the pots and washed under tap water to remove any dust then plants were dried using paper tissues. Aerial parts and root system were separated and weighed using sensitive balance, after that plants were dried using a hot air oven at 70 °C for 72 h until the weights become constant and reweighed to record dry weight.

4.4.2. Determination of Photosynthetic Pigments

For the determination of chlorophyll a, b and carotenoids, 0.1 g of plant tissue was homogenized in 10 mL of 80% acetone then centrifuged at 5000 rpm for 10 min. Supernatant absorbance was read at 663, 645, and 470 nm and photosynthetic pigment contents were calculated from the equations as described by Lichtenthaler and Wellburn [96].

4.4.3. Determination of Malondialdehyde (MDA) Content

Malondialdehyde (MDA) was determined according to Carmak and Horst [97] methods, 0.2 g of fresh plant aerial system were homogenized in 2 mL of 0.1% (*w*/*v*) trichloroacetic acid (TCA) at 4 °C. The homogenate was centrifuged for 10 min at 1000 rpm and to 0.5 mL of the supernatant, 3 mL of 0.5% (*v*/*v*) Thiobarbaturic acid (prepared in 20% TCA) was added. The mixture was incubated in 95 °C water bath with continuous shaking for 50 min, and then samples were placed in an ice bath until the temperature decreased to 25°C. The samples were re-centrifuged for 10 min at 10,000 rpm and the absorbance of the mixture was read at 532 nm. The non-specific absorption read at 600 nm was subtracted from all the readings and the MDA contents were calculated using the absorption coefficient as follows:

$$\text{MDA level (nmol)} = \Delta\ (\text{A } 532\text{ nm} - \text{A } 600\text{ nm})/1.56 \times 10^5 \quad (1)$$

4.4.4. Determination of Proline Content

Proline was determined using Bates et al. [98] method as follows; 0.5 g of fresh plant shoot were homogenized in 4 mL of 3.0% Sulphosalicylic acid. Then the homogenate was centrifuged for 10 min at 1000 rpm. To 1 mL of the supernatant 2 mL of acid Ninhydrin reagent and 2.0 mL of glacial acetic acid were added in a test tube, Then the mixture was incubated in a water bath at 100 °C for 60 min. then the mixture was cooled suddenly in an ice bath. After cooling, 4 mL of toluene were added to the solution mixture and vortex. The chromophore containing toluene (upper layer) was transferred to a new test tube. Finally, the absorbance was read at 520 nm using a spectrophotometer and Toluene as a blank. The concentration of proline was determined using the standard curve and expressed as mg g^{-1}.

4.4.5. Determination of Total Phenolic Compounds and Flavonoids

For the determination of phenolic compounds, 0.1 g of the shoot was homogenized in 10 mL of 70% acetone, then centrifuge at 5000 rpm for 10 min. To 1 mL of supernatant, 2 mL of sodium carbonate (15%) and 1 mL Folin–Ciocâlteu reagent (FCR) was added and the absorbance was recorded at 650 nm. Gallic acid was used as a standard for the determination of phenolic contents [99]. For total flavonoid, the aluminum trichloride method was used, to 1 mL of extract 2.5 mL of $AlCl_3$ reagent in ethanol 90% (20.0 mg/mL), then incubated at room temperature for 40 min. and the absorbance was recorded at 415 nm. Quercetine was used as a standard for flavonoids determination [100]. All absorbances were determined using Jenway 7315 spectrophotometer, Jenway Scientific Instrumental Company, UK.

4.5. Isozymes Analysis

4.5.1. Enzymes Extraction and Detection

For protein extraction; 0.2 g of plant aerial part tissue were macerated in 1 mL of 50 mM Tris HCl buffer (pH 6.8) containing 1 mM EDTA, 1 mM DDT, and 20 mg polyvinyl polypyrrolidone (PVPP) using chilled ceramic mortar and pestles. The homogenate was centrifuged at 10,000 rpm for 10 min at 4 °C. The supernatant was stored in 4 °C until used. The protein concentration was determined by spectrophotometry according to Lowry's method [101] using bovine serum albumin as a standard.

Native discontinuous system was prepared according to Laemmli [102] without adding Sodium dodecyl sulfate (SDS) and 50 μg from each sample were loaded directly without denaturation. The running voltage was started at 80 V for 30 min then increased to 120 V until the loading dye migration reached the bottom of the resolving gel.

For visualization of esterases, the gel was incubated in 100 mL of 100 mM Sodium phosphate buffer (pH 7) containing 40 mg α-Naphthyl acetate and 0.2 g fast blue RR for 30 min. in dark at 37 °C, and then the gel was fixed in 7% acetic acid solution [103].

For visualization Peroxidase activity, Seevers et al. [104] method was used, after electrophoresis gel was incubating for 30 min at 25 °C in 200 mM Sodium acetate buffer (pH 5) containing 3% H_2O_2 and 1.3 mM Benzidine, and the gel was fixed in 30% fixing solution.

For visualization Superoxide Dismutase (SOD), the gel was incubated in 200 mM K-phosphate buffer (pH 7.8) containing 0.1 mM riboflavin and 0.24 mM Nitroblue tetrazolium for 30 min, and then the gel was stained by exposure to fluorescence light.

All gels were photographed using Cannon kiss4 digital camera then transferred to a computer and converted into density profile using Image J program [105].

4.5.2. Statistical Analysis

All data were expressed as means with standard error, and Levene's test was used to investigate the homogeneity of variances of all data, then the data were subjected to one-way ANOVA and Tukey test. Two-way ANOVA was applied to determine the effect of salinity, species, and their interaction with all parameters. Principal component analysis was used to explore the correlation between growth parameters and studied organic compounds under salinity (100 mM NaCl). Also, Pearson's correlation coefficient was applied to investigate the correlation between all studied parameters under salinity treatments. All statistical analyses were carried out using SPSS 16.0 software. The means comparison was set at $p < 0.05$ and values denoted by the same letter are not significantly different.

Supplementary Materials: The following are available online at https://www.mdpi.com/article/10.3390/plants10061100/s1, Figure S1: Superoxide dismutase isozymes of *S. europaea*, *S. fruticosa* and *A. macrostachyum* under different NaCl concentrations, Table S1: The correlation coefficient of growth parameters (shoot fresh weight: SFW, root fresh weight: RFW, shoot dry weight: SDW, root dry weight: RDW), Chl a, Chl b, Chl a/b, carotenoids: Car, Proline: pro, malondialdehyde: MDA, phenolics: Phe, Flavonoids: Flav. in *Arthrocnemum macrostachyum*, *Sarcocornia fruticosa* and *Salicornia europaea* under salinity treatments.

Author Contributions: Conceptualization, A.F.M.G., E.M., A.M.M.A.K. and A.A.E.-G.; methodology, A.F.M.G., E.M., and A.M.M.A.K.; software, E.M..; validation, A.F.M.G., E.M., A.M.M.A.K. and A.A.E.-G.; formal analysis, A.F.M.G., E.M. and A.M.M.A.K.; investigation, A.A.E.-G., E.M. and A.M.M.A.K.; resources, A.F.M.G., E.M. and A.M.M.A.K.; data curation, A.F.M.G., E.M. and A.M.M.A.K.; writing—original draft preparation, A.F.M.G. and E.M.; writing—review and editing, E.M. and A.M.M.A.K.; visualization, E.M., A.M.M.A.K. and A.A.E.-G.; supervision, E.M., A.M.M.A.K. and A.A.E.-G.; project administration, A.M.M.A.K. and A.A.E.-G.; funding acquisition, A.F.M.G., E.M. and A.M.M.A.K. All authors have read and agreed to the published version of the manuscript.

Funding: This research received no external funding.

Institutional Review Board Statement: Not applicable.

Informed Consent Statement: Not applicable.

Data Availability Statement: Not applicable.

Conflicts of Interest: The authors declare no conflict of interest.

References

1. FAO. *Global Network on Integrated Soil Management for Sustainable Use of Salt-Affected Soils*; FAO Land and Plant Nutrition Management Service: Rome, Italy, 2005. Available online: http://www.fao.org/ag/agl/agll/spush (accessed on 29 May 2020).
2. Gupta, B.; Huang, B. Mechanism of salinity tolerance in plants: Physiological, biochemical, and molecular characterization. *Int. J. Genom.* **2014**, 701596. [CrossRef]
3. Shrivastava, P.; Kumar, R. Soil salinity: A serious environmental issue and plant growth promoting bacteria as one of the tools for its alleviation. *Saudi J. Biol. Sci.* **2015**, *22*, 123–131. [CrossRef] [PubMed]

4. Hasanuzzaman, M.; Nahar, K.; Fujita, M. Plant response to salt stress and role of exogenous protectants to mitigate salt-induced damages. In *Ecophysiology and Responses of Plants under Salt Stress*; Ahmad, P., Azooz, M.M., Prasad, M.N.V., Eds.; Springer: New York, NY, USA, 2013; pp. 25–87.
5. Yamaguchi, T.; Blumwald, E. Developing salt-tolerant crop plants: Challenges and opportunities. *Trends Plant Sci.* **2005**, *10*, 615–620. [CrossRef] [PubMed]
6. Herppich, W.B.; Huyskens-keil, S.; Kadau, R. Effects of short-term low-temperature storage on mechanical and chemical properties of white Asparagus cell walls. *J. Appl. Bot. Food Qual.* **2005**, *79*, 63–71.
7. Khan, Z.I.; Hussain, A.; Ashraf, M.; McDowell, L.R. Mineral status of soils and forages in southwestern Punjab-Pakistan: Micro-minerals. *Asian Aust. J. Anim. Sci.* **2006**, *19*, 1139–1147.
8. Zhang, X.; Yin, H.; Chen, S.; He, J.; Guo, S. Changes in antioxidant enzyme activity and transcript levels of related genes in *Limonium sinense* Kuntze seedlings under NaCl stress. *J. Chem.* **2014**, 749047. [CrossRef]
9. Poljakoff-Mayber, A. Morphological and anatomical changes in plants as a response to salinity stress. In *Plants in Saline Environment*; Poljakoff-Mayber, A., Gale, J., Eds.; Springer: New York, NY, USA, 1975; pp. 97–117.
10. Mansour, M.M.F. Nitrogen containing compounds and adaptation of plants to salinity stress. *Biol. Plant.* **2000**, *43*, 491–500. [CrossRef]
11. Munns, R.; James, R.A. Screening methods for salinity tolerance: A case study with tetraploid wheat. *Plant Soil* **2003**, *253*, 201–218. [CrossRef]
12. Hameed, M.; Ashraf, M.; Naz, N. Anatomical adaptations to salinity in cogon grass [*Imperata cylindrica* (L.) Raeuschel] from the Salt Range, Pakistan. *Plant Soil* **2009**, *322*, 229–238. [CrossRef]
13. Rivelli, A.R.; James, R.A.; Munns, R.; Condon, A.T. Effect of salinity on water relations and growth of wheat genotypes with contrasting sodium uptake. *Sciences* **2002**, *22*, 123–131. [CrossRef]
14. Redondo-Gómez, S.; Mateos-Naranjo, E.; Figueroa, M.E.; Davy, A.J. Salt stimulation of growth and photosynthesis in an extreme halophyte, *Arthrocnemum macrostachyum*. *Plant Biol.* **2010**, *12*, 79–87. [CrossRef]
15. Koyro, H.W.; Hussain, T.; Huchzermeyer, B.; Khan, M.A. Photosynthetic and growth responses of a perennial halophytic grass *Panicum turgidum* to increasing NaCl concentrations. *Environ. Exp. Bot.* **2013**, *91*, 22–29. [CrossRef]
16. Koyro, H.W. Effect of salinity on growth, photosynthesis, water relations and solute composition of the potential cash crop halophyte *Plantago coronopus* (L.). *Environ. Exp. Bot.* **2006**, *56*, 136–146. [CrossRef]
17. Aghaleh, M.; Niknam, V.; Ebrahimzadeh, H.; Razavi, K. Salt stress effects on growth, pigments, proteins and lipid peroxidation in *Salicornia persica* and *S. europaea*. *Biol. Plant.* **2009**, *53*, 243–248. [CrossRef]
18. Foreman, J.; Demidchik, V.; Bothwell, J.H.; Mylona, P.; Miedema, H.; Torres, M.A.; Linstead, P.; Costa, S.; Brownlee, C.; Jones, J.D.; et al. Reactive oxygen species produced by NADPH oxidase regulate plant cell growth. *Nature* **2003**, *27*, 442–446.
19. Esfandiari, E.; Shekari, F.; Shekari, F.; Esfandiari, M. The effect of salt stress on antioxidant enzymes activity and lipid peroxidation on the wheat seedling. *Not. Bot. Horti Agrobot. Cluj-Napoca* **2007**, *35*, 48–56.
20. Ozgur, R.; Uzilday, B.; Sekmen, A.H.; Turkan, I. Reactive oxygen species regulation and antioxidant defence in halophytes. *Funct. Plant Biol.* **2013**, *40*, 832–847. [CrossRef]
21. Zhang, Z.; Guo, J.; Zhao, Y. Identification and characterization of maize ACD6-like gene reveal ZmACD6 as the maize orthologue conferring resistance to *Ustilago maydis*. *Plant Signal Behav.* **2019**, *14*, e1651604. [CrossRef] [PubMed]
22. Mohamed, E.; Matsuda, R.; El-khatib, A.A.; Takechi, K.; Takano, H.; Takio, S. Characterization of the superoxide dismutase genes of the halophyte *Suaeda maritima* in Japan and Egypt. *Plant Cell Rep.* **2015**, *34*, 2099–2110. [CrossRef]
23. Mohamed, E.; Matsuda, R.; El-khatib, A.A.; Takechi, K.; Takano, H.; Takio, S. Differential tolerance to high salt with regard to cell growth and superoxide dismutase (SOD) activity in calluses of the halophyte *Suaeda maritima* from Japan and Egypt. *Plant Omics* **2016**, *9*, 81–89.
24. Mohamed, E.; Kasem, A.M.M.A.; Gobouri, A.A.; Elkelish, A.; Azab, E. Influence of Maternal Habitat on Salinity Tolerance of *Zygophyllum coccineum* with Regard to Seed Germination and Growth Parameters. *Plants* **2020**, *9*, 1504. [CrossRef] [PubMed]
25. Mohamed, E.; Kasem, A.M.M.A.; Farghali, K.A. Seed germination of Egyptian *Pancratium maritimum* under salinity with regard to cytology, antioxidant and reserve mobilization enzymes, and seed anatomy. *Flora* **2018**, *242*, 120–127. [CrossRef]
26. Kim, Y.H.; Kim, C.Y.; Song, W.K.; Park, D.S.; Kwon, S.Y.; Lee, H.S.; Bang, J.W.; Kwak, S.S. Overexpression of sweet potato swpa4 peroxidase results in increased hydrogen peroxide production and enhances stress tolerance in tobacco. *Planta* **2008**, *227*, 867–881. [CrossRef]
27. Ahmad, P.; Jaleel, C.A.; Salem, M.A.; Nabi, G.; Sharma, S. Roles of enzymatic and non-enzymatic antioxidants in plants during abiotic stress. *Crit. Rev. Biotechnol.* **2010**, *30*, 161–175. [CrossRef]
28. Caillet, S.; Salmiéri, S.; Lacroix, M. Evaluation of free radical-scavenging properties of commercial grape phenol extracts by a fast-colorimetric method. *Food Chem.* **2006**, *95*, 1–8. [CrossRef]
29. Amarowicz, R.; Weidner, S. Biological activity of grapevine phenolic compounds. In *Grapevine Molecular Physiology and Biotechnology*, 2nd ed.; Roubelakis-Angelakis, K.A., Ed.; Springer: New York, NY, USA, 2009; pp. 389–405.
30. Amarowicz, R.; Weidner, S.; Wojtowicz, I.; Karmac, M.; Kosinska, A.; Rybarczyk, A. Influence of low-temperature stress on changes in the composition of grapevine leaf phenolic compounds and their antioxidant properties. *Funct. Plant Sci. Biotechnol.* **2010**, *4*, 90–96.

31. Lattanzio, V. Phenolic compounds: Introduction. In *Natural Products*; Ramawat, K.G., Merillon, J.M., Eds.; Springer: Berlin/Heidelberg, Germany, 2013.
32. Pereira, A. Plant abiotic stress challenges from the changing environment. *Front. Plant Sci.* **2016**, *7*, 1123. [CrossRef]
33. Di Ferdinando, M.; Brunetti, C.; Fini, A.; Tattini, M. Flavonoids as Antioxidants in plants under Abiotic Stresses. In *Abiotic Stress Responses in Plants: Metabolism, Productivity and Sustainability*; Ahmad, P., Prasad, M.N.V., Eds.; Springer: New York, NY, USA, 2012; pp. 159–179.
34. Slama, I.; Abdelly, C.; Bouchereau, A.; Flowers, T.; Savouré, A. Diversity, distribution and roles of osmoprotective compounds accumulated in halophytes under abiotic stress. *Ann. Bot.* **2015**, *115*, 433–447. [CrossRef]
35. Kaur, G.; Asthir, B. Proline: A key player in plant abiotic stress tolerance. *Biol. Plant.* **2015**, *59*, 609–619. [CrossRef]
36. Christenhusz, M.J.M.; Byng, J.W. The number of known plant species in the world and its annual increase. *Phylotaxa* **2016**, *261*, 201–217. [CrossRef]
37. Breckle, S.W. Studies on halophytes from Iran and Afganistan. II Ecology of halophytes along salt gradients. *Proceeding R. Soc. Edinb.* **1986**, *89*, 203–215.
38. Aronson, J. Halophytes. In *A Database of Salt Tolerant Plant of the World*; Office of Arid Lands Studies, University Arizona Tucson: Tucson, AZ, USA, 1989.
39. Akhani, P.; Ghorbanli, M. A contribution to the halophytic vegetation and flora of Iran. In *Towards the Rational Use of High Salinity Tolerant Plants*; Lieth, H., Al-Masoom, A., Eds.; Kluwer Academic Publishers: Dodrecht, The Netherlands, 1993; Volume 1, pp. 35–44.
40. Öztürk, M.; Altay, V.; Altundağ, E.; Gücel, S. Halophytic plant diversity of unique habitats in Turkey: Salt mine caves of Çankırı and Iğdır. In *Halophytes for Food Security in Dry Lands*; Khan, M.A., Ozturk, M., Gul, B., Ahmed, M.Z., Eds.; Elsevier Academic Publisher: Oxford, GB, USA, 2016; pp. 291–315.
41. Winter, K.; Troughton, J.H.; Evenari, M.; Läuchli, A.; Lüttge, U. Mineral ion composition and occurrence of CAM-like diurnal malate fluctuations in plants of coastal and desert habitats of Israel and the Sinai. *Oecologia* **1976**, *25*, 125–143. [CrossRef]
42. Boulos, L. *Flora of Egypt*; Al Hadara Publication: Cairo, Egypt, 1999; Volume 1, p. 417.
43. Singh, D.; Buhmann, A.K.; Flowers, T.J.; Seal, C.E.; Papenbrock, J. Salicornia as a crop plant in temperate regions: Selection of genetically characterized ecotypes and optimization of their cultivation conditions. *AoB Plants* **2014**, *6*. [CrossRef]
44. Said, O.B.; DaSilva, M.M.; Hannier, F.; Beyrem, H.; Chícharo, L. Using *Sarcocornia fruticosa* and *Saccharomyces cerevisiae* to remediate metal contaminated sediments of the Ria Formosa lagoon (SE Portugal). *Ecohydrol. Hydrobiol.* **2019**, *19*, 588–597. [CrossRef]
45. El Naker, N.A.; Yousef, A.F.; Yousef, L.F. A review of *Arthrocnemum* (*Arthrocaulon*) *macrostachyum* chemical content and bioactivity. *Phytochem. Rev.* **2020**, *19*, 1427–1448. [CrossRef]
46. Xiong, L.; Zhu, J.K. Salt tolerance. *Arab. Book* **2002**, *1*, e0048. [CrossRef] [PubMed]
47. Cuevas, J.; Daliakopoulos, I.N.; del Moral, F.; Hueso, J.J.; Tsanis, I.K. A Review of Soil-Improving Cropping Systems for Soil Salinization. *Agronomy* **2019**, *9*, 295. [CrossRef]
48. Hanin, M.; Ebel, C.; Ngom, M.; Laplaze, L.; Masmoudi, K. New Insights on Plant Salt Tolerance Mechanisms and Their Potential Use for Breeding. *Front. Plant Sci.* **2016**, *7*, 1787. [CrossRef]
49. Khan, M.A.; Gul, B. Arthrocnemum macrostachyum: A potential case for agriculture using above seawater salinity. In *Prospects for Saline Agriculture*; Springer: Dordrecht, The Netherlands, 2002; pp. 353–364.
50. Redondo-Gómez, S.; Wharmby, C.; Castillo, J.M.; Mateos-Naranjo, E.; Luque, C.J.; DeCires, A.; Luque, T.; Davy, A.J.; Figueroa, E.M. Growth and photosynthetic responses to salinity in an extreme halophyte, *Sarcocornia fruticosa*. *Physiol. Plant.* **2006**, *128*, 116–124. [CrossRef]
51. García-Caparrós, P.; Llanderal, A.; Pestana, M.; Correia, P.J.; Lao, M.T. Nutritional and physiological responses of the dicotyledonous halophyte *Sarcocornia fruticosa* to salinity. *Aust. J. Bot.* **2017**, *65*, 573–581. [CrossRef]
52. Ungar, I.A.; Benner, D.K.; McGraw, D.C. Distribution and growth of *Salicornia europaea* on an inland salt pan. *Ecology* **1979**, *60*, 329–336. [CrossRef]
53. Cai, Z.Q.; Gao, Q. Comparative physiological and biochemical mechanisms of salt tolerance in five contrasting highland quinoa cultivars. *BMC Plant Biol.* **2020**, *20*, 1–15. [CrossRef] [PubMed]
54. Shabala, S. Learning from halophytes: Physiological basis and strategies to improve stress tolerance in crops. *Ann. Bot.* **2013**, *112*, 1209–1221. [CrossRef]
55. Flowers, T.J.; Munns, R.; Colmer, T.D. Sodium chloride toxicity and the cellular basis of salt tolerance in halophytes. *Ann. Bot.* **2015**, *115*, 419–431. [CrossRef]
56. Khan, M.A.; Ungar, I.A.; Showalter, A.M. Salt stimulation and tolerance in an intertidal stem succulent halophyte. *J. Plant Nutr.* **2005**, *28*, 1365–1374. [CrossRef]
57. Glenn, E.P.; Brown, J.J.; Blumwald, E. Salt tolerance and crop potential. *Crit. Rev. Plant Sci.* **1999**, *18*, 227–255. [CrossRef]
58. Shabala, S.; Mackay, A. Ion transport in halophytes. *Adv. Bot. Res.* **2011**, *57*, 151–199.
59. Marcum, K.B.; Murdoch, C.L. Salt tolerance of the coastal salt marsh grass, *Sporobolus virginicus* (L.) Kunth. *New Phytol.* **1992**, *120*, 281–288. [CrossRef]
60. Hadi, M.R.; Karimi, N. The role of calcium in plants' salt tolerance. *J. Plant Nutr.* **2012**, *35*, 2037–2054. [CrossRef]
61. Nunes-Nesi, A.; Sulpice, R.; Gibon, Y.; Fernie, A.R. The enigmatic contribution of mitochondrial function in photosynthesis. *J. Exp. Bot.* **2008**, *59*, 1675–1684. [CrossRef] [PubMed]

62. Burman, U.; Garg, B.; Kathju, S. Interactive Effects of Thiourea and Phosphorus on Cluster bean Under Water Stress. *Biol. Plant.* **2004**, *48*, 61–65. [CrossRef]
63. Akcin, A.; Yalcin, E. Effect of salinity stress on chlorophyll, carotenoid content, and proline in *Salicornia prostrata* Pall. and *Suaeda prostrata* Pall. subsp. prostrata (Amaranthaceae). *Braz. J. Bot.* **2016**, *39*, 101–106. [CrossRef]
64. Rabhi, M.; Castagna, A.; Remorini, D.; Scattino, C.; Smaoui, A.; Ranieri, A.; Abdelly, C. Photosynthetic responses to salinity in two obligate halophytes: *Sesuvium portulacastrum* and *Tecticornia indica*. *S. Afr. J. Bot.* **2012**, *79*, 39–47. [CrossRef]
65. Boughalleb, F.; Denden, M. Physiological and biochemical changes of two halophytes, *Nitraria retusa* (Forssk.) and *Atriplex halimus* (L.) under increasing salinity. *Agric. J.* **2011**, *6*, 327–339. [CrossRef]
66. Morsy, A.A.; Youssef, A.M.; Mosallam, H.A.; Hashem, A.M. Assessment of selected species along Alamein-Wadi El-Natrun Desert Road. *Egypt J. Appl. Sci. Res.* **2008**, *4*, 1276–1284.
67. Shao, H.B.; Chu, L.Y.; Jaleel, C.A.; Zhao, C.X. Water-deficit stress-induced anatomical changes in higher plants. *Comptes Rendus Biol.* **2008**, *331*, 215–225. [CrossRef] [PubMed]
68. Parida, A.K.; Das, A.B.; Sanada, Y.; Mohanty, P. Effects of salinity on biochemical components of the mangrove, *Aegiceras corniculatum*. *Aquat. Bot.* **2004**, *80*, 77–87. [CrossRef]
69. Król, A.; Amarowicz, R.; Weidner, S. Changes in the composition of phenolic compounds and antioxidant properties of grapevine roots and leaves (*Vitis vinifera* L.) under continuous of long-term drought stress. *Acta Physiol. Plant.* **2014**, *6*, 1491–1499.
70. Caliskan, O.; Bayazıt, S.; Oktem, M.; Ergul, A. Evaluation of the genetic diversity of pomegranate accessions from Turkey using new microsatellite markers. *Turk. J. Agric. For.* **2017**, *41*, 142–153. [CrossRef]
71. Bhattacharya, A.; Sood, P.; Citovsky, V. The roles of plant phenolics in defence and communication during agrobacterium and rhizobium infection. *Mol. Plant Pathol.* **2010**, *11*, 705–719. [CrossRef]
72. Weidner, S.; Karolak, M.; Karamać, M.; Kosińska, A.; Amarowicz, R. Phenolic compounds and properties of antioxidants in grapevine roots (*Vitis vinifera* L.) under drought stress followed by recovery. *Acta Soc. Bot. Pol.* **2009**, *78*, 97–103. [CrossRef]
73. Bettaieb, I.; Hamrouni-Sellami, I.; Bourgou, S.; Limam, F.; Marzouk, B. Drought effects on polyphenol composition and antioxidant activities in aerial parts of *Salvia officinalis* L. *Acta Physiol. Plant.* **2011**, *33*, 1103–1111. [CrossRef]
74. Naikoo, M.I.; Dar, M.I.; Raghib, F.; Jaleel, H.; Ahmad, B.; Raina, A.; Khan, F.A.; Naushin, F. Role and regulation of plants phenolics in abiotic stress tolerance: An overview. In *Plant Signaling Molecules: Role and Regulation Under Stressful Environments*; Iqbal, M., Ferrante, A., Redd, P.S., Khan, N.A., Eds.; Elsevier Publications: Amsterdam, The Netherlands, 2019; pp. 157–168.
75. Abd El-Maboud, M.M. Seasonal variations effect on antioxidant compounds and their role in the adaptation of some halophytes at Wadi Gharandal, Southwest Sinai. *Ann. Agric. Sci.* **2019**, *64*, 161–166. [CrossRef]
76. Brown, D.E.; Rashotte, A.M.; Murphy, A.S.; Normanly, J.; Tague, B.W.; Peer, W.A.; Taiz, L.; Muday, G.K. Flavonoids act as negative regulators of auxin transport in vivo in Arabidopsis. *Plant Physiol.* **2001**, *126*, 524–535. [CrossRef]
77. Jithesh, M.N.; Prashanth, S.R.; Sivaprakash, K.R.; Parida, A.K. Antioxidative response mechanisms in halophytes: Their role in stress defense. *J. Genet.* **2006**, *85*, 237–254. [CrossRef]
78. Prasad, T.K.; Anderson, M.D.; Stewart, C.R. Localization and characterization of peroxidases in the mitochondria of chilling acclimated maize seedlings. *Plant Physiol.* **1995**, *108*, 1597–1605. [CrossRef] [PubMed]
79. Mhamdi, A.; Hager, J.; Chaouch, S.; Queval, G.; Han, Y.; Taconnat, L.; Saindrenan, P.; Goua, H.; Issakidis Bourguet, E.; Renou, J.P.; et al. Arabidopsis Glutathione Reductase 1 plays a crucial role in leaf responses to intracellular hydrogen peroxide and in ensuring appropriate gene expression through both salicylic acid and jasmonic acid signaling pathways. *Plant Physiol.* **2010**, *153*, 1144–1160. [CrossRef]
80. Szabados, L.; Savoure, A. Proline: A multifunctional amino acid. *Trends Plant Sci.* **2010**, *15*, 89–97. [CrossRef]
81. Bohnert, H.J.; Nelson, D.E.; Jensen, R.G. Adaptations to environmental stresses. *Plant Cell* **1995**, *7*, 1099–1111. [CrossRef]
82. Ahmad, P.; John, R.; Sarwat, M.; Umar, S. Responses of proline, lipid peroxidation and antioxidative enzymes in two varieties of *Pisum sativum* L. under salt stress. *Int. J. Plant Prod.* **2012**, *2*, 353–366.
83. Mohamed, E.; Ansari, N.; Yadav, D.S.; Agrawal, M.; Agrawal, S.B. Salinity alleviates the toxicity level of ozone in a halophyte *Mesembryanthemum crystallinum* L. *Ecotoxicology* **2021**. [CrossRef] [PubMed]
84. Ashraf, M.; Foolad, M.R. Roles of glycine betaine and proline in improving plant abiotic stress resistance. *Environ. Exp. Bot.* **2007**, *59*, 206–216. [CrossRef]
85. Mann, A.; Kumar, A.; Saha, M.; Lata, C.; Kumar, A. Stress induced changes in osmoprotectants, ionic relations, antioxidants activities and protein profiling characterize *Sporobolus marginatus* Hochst. ex A. Rich. salt tolerance mechanism. *Indian J. Exp. Biol.* **2019**, *57*, 672–679.
86. Ce'cile, T.; Francoise, L.; Pierre, V.C. Polymorphism and modulation of cell wall esterase enzyme activities in the chicory root during the growing season. *J. Exp. Bot.* **2006**, *57*, 81–89.
87. Dasgupta, N.; Nandy, P.; Tiwari, C.; Das, S. Salinity-imposed changes of some isozymes and total leaf protein expression in five mangroves from two different habitats. *J. Plant Interact.* **2010**, *5*, 211–221. [CrossRef]
88. Bowler, C.; Van Montagu, M.; Inze, D. Superoxide dismutase and stress tolerance. *Annu. Rev. Plant Physiol. Plant Mol. Biol.* **1992**, *43*, 83–116. [CrossRef]
89. Mohamed, E.; Kasem, A.M.M.A.; El-khatib, A. Allelopathic potential of Egyptian halophytes *Arthrocnemum macrostachyum* and *Halocnemum strobilaceum* from two coastal areas. *Allelopath. J.* **2020**, *50*, 225–241. [CrossRef]

90. Sekmen, A.H.; Turkan, I.; Tanyolac, Z.O.; Ozfidan, C.; Dinc, A. Different antioxidant defense responses to salt stress during germination and vegetative stages of endemic halophyte *Gypsophila oblanceolata* Bark. *Environ. Exp. Bot.* **2012**, *77*, 63–76. [CrossRef]
91. Nisar, F.; Gul, B.; Khan, M.A.; Hameed, A. Heteromorphic seeds of coastal halophytes *Arthrocnemum macrostachyum* and *A. indicum* display differential patterns of hydrogen peroxide accumulation, lipid peroxidation and antioxidant activities under increasing salinity. *Plant Physiol. Biochem.* **2019**, *144*, 58–63. [CrossRef]
92. Aghaleh, M.; Niknam, V.; Ebrahimzadeh, H.; Razavi, K. Effect of salt stress on physiological and antioxidative responses in two species of *Salicornia* (*S. persica* and *S. europaea*). *Acta Physiol. Plant.* **2011**, *33*, 1261–1270. [CrossRef]
93. Cevahir, G.; Yentur, S.; Yazgan, M.; Unal, M.; Yilmazer, N. Peroxidase activity in relation to anthocyanin and chlorophyll content in juvenile and adult leaves of "mini-star" *Gazanla splendens*. *Pak. J. Bot.* **2004**, *36*, 603–609.
94. Graskova, I.A.; Zhivetyev, M.A.; Putalina, T.E.; Krasnobaev, V.A.; Voinikov, V.N. Activity and izoenzyme spectrum of peroxidase of some herbaceous plants from the bank of lake Baikal, growing under abiotic stress. *Electron. Sci. J. Investig. Russ. (Issledovano v Rosii)* **2010**, *2010*, 293–303.
95. Jackson, M.L. *Soil Chemical Analysis*; New Delhi Prentice Hall of India: New Delhi, India, 1973.
96. Lichtenthaler, H.K.; Wellburn, W.R. Determination of total carotenoids and chlorophylls a and b of leaf extracts in different solvents. *Biochem. Soc. Trans.* **1983**, *11*, 591–592. [CrossRef]
97. Carmak, I.; Horst, J.H. Effects of aluminum on lipid peroxidation, superoxide dismutase, catalase, and peroxidase activities in root tips of soybean (*Glycine max*). *Physiolgia Plant.* **1991**, *83*, 463–468. [CrossRef]
98. Bates, L.S.; Waldren, R.P.; Teare, I.D. Rapid determination of free proline for wate-stress studies. *Plant Soil* **1973**, *39*, 205–207. [CrossRef]
99. Bray, H.G.; Thorpe, W.V. Analysis of phenolic compounds of interest in metabolism. *Methods Biochem. Anal.* **1954**, *1*, 27–52.
100. Miliauskas, G.; Venskutonis, P.R.; Van Beek, T.A. Screening of radical scavenging activity of some medicinal and aromatic plant extracts. *Food Chem.* **2004**, *85*, 231–237. [CrossRef]
101. Lowry, O.H.; Rosenbrough, N.J.; Farr, A.L.; Randall, R.L. Protein Measurements with the Folin Phenol Reagent. *J. Biol. Chem.* **1951**, *193*, 265–275. [CrossRef]
102. Laemmli, U.K. Cleavage of structural proteins during the assembly of the head of bacteriophage T4. *Nature* **1970**, *227*, 680–685. [CrossRef] [PubMed]
103. Smith, H.H.; Hamill, D.E.; Weaver, E.A.; Thompson, K.H. Multiple molecular forms of peroxidases and esterases among Nicotiana species and amphidiploids. *J. Hered.* **1970**, *61*, 203–212. [CrossRef]
104. Seevers, P.M.; Daly, J.M.; Catedral, F.F. The role of peroxidase isozymes in resistance to wheat stem rust disease. *Plant Physiol.* **1971**, *48*, 353–360. [CrossRef] [PubMed]
105. Schneider, C.A.; Rasband, W.S.; Eliceiri, K.W. NIH image to imageJ: 25 years of image analysis. *Nature Methods* **2012**, *9*, 671–675. [CrossRef] [PubMed]

MDPI
St. Alban-Anlage 66
4052 Basel
Switzerland
Tel. +41 61 683 77 34
Fax +41 61 302 89 18
www.mdpi.com

Plants Editorial Office
E-mail: plants@mdpi.com
www.mdpi.com/journal/plants

www.ingramcontent.com/pod-product-compliance
Lightning Source LLC
Chambersburg PA
CBHW061840210926
51850CB00006B/935
9783036565736